# Paleoclimatology: Understanding Past Climate

# Paleoclimatology: Understanding Past Climate

Edited by **Loren Gilbert**

**SYRAWOOD**
PUBLISHING HOUSE

New York

Published by Syrawood Publishing House,
750 Third Avenue, 9th Floor,
New York, NY 10017, USA
www.syrawoodpublishinghouse.com

**Paleoclimatology: Understanding Past Climate**
Edited by Loren Gilbert

International Standard Book Number: 978-1-68286-126-4 (Hardback)

Printed in the United States of America.

# Contents

# Preface

Paleoclimatology is the field of study that aims to analyze the climate of the geologic past and climatic variability on earth. Some of the significant topics covered in this book are geological time scale, biogeochemistry, oceanography, remote sensing and measurement techniques to understand the climate of past, etc. The researches and case studies encompassed in this book are aimed at understanding past climate and climate changes in context of present and future climatic variability. This book will help new researchers by foregrounding their knowledge in this field.

Various studies have approached the subject by analyzing it with a single perspective, but the present book provides diverse methodologies and techniques to address this field. This book contains theories and applications needed for understanding the subject from different perspectives. The aim is to keep the readers informed about the progress in the field; therefore, the contributions were carefully examined to compile novel researches by specialists from across the globe.

Indeed, the job of the editor is the most crucial and challenging in compiling all chapters into a single book. In the end, I would extend my sincere thanks to the chapter authors for their profound work. I am also thankful for the support provided by my family and colleagues during the compilation of this book.

**Editor**

# Evaluating the dominant components of warming in Pliocene climate simulations

D. J. Hill[1,2], A. M. Haywood[1], D. J. Lunt[3], S. J. Hunter[1], F. J. Bragg[3], C. Contoux[4,5], C. Stepanek[6], L. Sohl[7], N. A. Rosenbloom[8], W.-L. Chan[9], Y. Kamae[10], Z. Zhang[11,12], A. Abe-Ouchi[9,13], M. A. Chandler[7], A. Jost[5], G. Lohmann[6], B. L. Otto-Bliesner[8], G. Ramstein[4], and H. Ueda[10]

[1] School of Earth and Environment, University of Leeds, Leeds, UK
[2] British Geological Survey, Keyworth, Nottingham, UK
[3] School of Geographical Sciences, University of Bristol, Bristol, UK
[4] Laboratoire des Sciences du Climat et de l'Environnement, Saclay, France
[5] Sisyphe, CNRS/UPMC Univ. Paris 06, Paris, France
[6] Alfred Wegener Institute Helmholtz Centre for Polar and Marine Research, Bremerhaven, Germany
[7] Columbia University – NASA/GISS, New York, NY, USA
[8] National Center for Atmospheric Research, Boulder, Colorado, USA
[9] Atmosphere and Ocean Research Institute, University of Tokyo, Kashiwa, Japan
[10] Graduate School of Life and Environmental Sciences, University of Tsukuba, Tsukuba, Japan
[11] UniResearch and Bjerknes Centre for Climate Research, Bergen, Norway
[12] Nansen-zhu International Research Centre, Institute of Atmospheric Physics, Chinese Academy of Sciences, Beijing, China
[13] Japan Agency for Marine-Earth Science and Technology, Yokohama, Japan

*Correspondence to:* D. J. Hill (eardjh@leeds.ac.uk)

**Abstract.** The Pliocene Model Intercomparison Project (PlioMIP) is the first coordinated climate model comparison for a warmer palaeoclimate with atmospheric $CO_2$ significantly higher than pre-industrial concentrations. The simulations of the mid-Pliocene warm period show global warming of between 1.8 and 3.6 °C above pre-industrial surface air temperatures, with significant polar amplification. Here we perform energy balance calculations on all eight of the coupled ocean–atmosphere simulations within PlioMIP Experiment 2 to evaluate the causes of the increased temperatures and differences between the models. In the tropics simulated warming is dominated by greenhouse gas increases, with the cloud component of planetary albedo enhancing the warming in most of the models, but by widely varying amounts. The responses to mid-Pliocene climate forcing in the Northern Hemisphere midlatitudes are substantially different between the climate models, with the only consistent response being a warming due to increased greenhouse gases. In the high latitudes all the energy balance components become important, but the dominant warming influence comes from the clear sky albedo, only partially offset by the increases in the cooling impact of cloud albedo. This demonstrates the importance of specified ice sheet and high latitude vegetation boundary conditions and simulated sea ice and snow albedo feedbacks. The largest components in the overall uncertainty are associated with clouds in the tropics and polar clear sky albedo, particularly in sea ice regions. These simulations show that albedo feedbacks, particularly those of sea ice and ice sheets, provide the most significant enhancements to high latitude warming in the Pliocene.

## 1 Introduction

Atmospheric carbon dioxide concentrations continue to rise due to anthropogenic emissions. The latest measurements show that annual mean concentrations have risen beyond 390 parts per million (Conway et al., 2012). The Pliocene

**Table 1.** Key model and experimental design parameters for each of the eight PlioMIP Experiment 2 simulations.

| GCM | Atmospheric resolution (° lat × ° long × levels) | Ocean resolution (° lat × ° long × levels) | Boundary conditions employed | Ocean initialization | Reference |
|---|---|---|---|---|---|
| CCSM4 | 0.9 × 1.25 × 26 | 1 × 1 × 60 | Alternate | PRISM3 (anomaly) | Rosenbloom et al. (2013) |
| COSMOS | 3.75 × 3.75 × 19 | 3 × 1.8 × 40 | Preferred | PRISM3 (anomaly) | Stepanek and Lohmann (2012) |
| GISS-E2-R | 2 × 2.5 × 40 | 1 × 1.25 × 32 | Preferred | PRISM3 | Chandler et al. (2013) |
| HadCM3 | 2.5 × 3.75 × 19 | 1.25 × 1.25 × 20 | Alternate | PRISM2 mPWP control | Bragg et al. (2012) |
| IPSLCM5A | 3.75 × 1.9 × 39 | 0.5–2 × 2 × 31 | Alternate | Pre-industrial control | Contoux et al. (2012) |
| MIROC4m | 2.8 × 2.8 × 20 | 0.5–1.4 × 1.4 × 43 | Preferred | PRISM3 | Chan et al. (2011) |
| MRI-CGCM 2.3 | 2.8 × 2.8 × 30 | 0.5–2 × 2.5 × 23 | Alternate | PRISM3 (anomaly) | Kamae and Ueda (2012) |
| NorESM-L | 3.75 × 3.75 × 26 | 3 × 3 × 30 | Alternate | Levitus | Zhang et al. (2012) |

was the last period of Earth history with similar to modern atmospheric $CO_2$ concentrations (Kürschner et al., 1996; Seki et al., 2010; Pagani et al., 2010; Bartoli et al., 2011). These were associated with elevated global temperatures in both the ocean (Dowsett et al., 2012) and on land (Salzmann et al., 2013). As the last period of global warmth before the climate transition into the bipolar ice age cycles of the Pleistocene, the mid-Pliocene warm period (mPWP) has been a target for both palaeoenvironmental data acquisition and palaeoclimate modelling over a number of years (Dowsett et al., 1992; Chandler et al., 1994; Haywood et al., 2009; Dowsett et al., 2010). Although a number of different general circulation models (GCMs) have been used to simulate Pliocene climates (Chandler et al., 1994; Sloan et al., 1996; Haywood et al., 2000, 2009; Dowsett et al., 2011), it is only recently that a coordinated multi-model experiment has been initiated, with standardized design for mid-Pliocene simulations (Haywood et al., 2010, 2011).

The Pliocene Model Intercomparison Project (PlioMIP) represents the first coordinated multi-model experiment to simulate a warmer than modern palaeoclimate, with high atmospheric $CO_2$ concentrations (405 ppmv). It has recently been added to the Paleoclimate Model Intercomparison Project (PMIP; Hill et al., 2012) and the first phase, incorporating two simulations, completed. This paper focuses on PlioMIP Experiment 2, designed for coupled ocean–atmosphere GCMs (Haywood et al., 2011). Although, many of the large-scale features of the simulated Pliocene climate have been well documented (Dowsett et al., 2012; Haywood et al., 2013; Zhang et al., 2013, 2013a,b; Salzmann et al., 2013), the causes of the simulated changes and differences

between the simulations have not been extensively explored prior to this study. In this paper the energy balance of the PlioMIP Experiment 2 simulations are analysed in order to understand the causes of Pliocene atmospheric warming and the latitudinal distribution of increased surface air temperatures. This analysis allows us to analyse the causes of the warming, both directly through the simulated energy balance components and through examination of Earth system component changes that are driving these. It is also important when discrepancies with available proxy reconstructions of Pliocene warming are considered (Salzmann et al., 2013).

## 2 Participating models

Eight different modelling groups have submitted simulations to PlioMIP Experiment 2. All of these models are coupled ocean–atmosphere GCMs, but range in complexity and spatial resolution. Table 1 contains the details of each of the models' simulation, including the resolution at which it was run, the boundary conditions employed and the model initialization. Each of the simulations is documented in much more detail in a separate paper within a special issue of *Geoscientific Model Development*, referenced in Table 1. The general climate sensitivity of the model and the annual mean global warming produced in its PlioMIP Experiment 2 simulation is detailed in Table 2. Further details about the models can also be found in Haywood et al. (2013) and the references therein.

**Table 2.** The climate sensitivity and global mean annual surface air temperature warming of each of the models with simulations in the PlioMIP Experiment 2 ensemble. Climate sensitivity is the equilibrium global warming for a doubling of atmospheric $CO_2$ – the values quoted here are a general value for each model and do not refer to the particular set-up and initialization procedures used for PlioMIP.

| GCM | Climate sensitivity (°C) | Mean annual mPWP SAT warming (°C) |
|---|---|---|
| CCSM4 | 3.2 | 1.86 |
| COSMOS | 4.1 | 3.60 |
| GISS-E2-R | 2.7 | 2.24 |
| HadCM3 | 3.1 | 3.27 |
| IPSLCM5A | 3.4 | 2.03 |
| MIROC4m | 4.1 | 3.46 |
| MRI-CGCM 2.3 | 3.2 | 1.84 |
| NorESM-L | 3.1 | 3.27 |

## 3 PlioMIP Experiment 2

PlioMIP uses the latest iteration of the PRISM (Pliocene Research, Interpretation and Synoptic Mapping) mid-Pliocene palaeoenvironmental reconstruction, PRISM3 (Dowsett et al., 2010), as the basis for the imposed model boundary conditions. This reconstruction represents the peak averaged (Dowsett and Poore, 1991) warm climate of the mid-Pliocene warm period (mPWP; 3.246–3.025 Ma; Dowsett et al., 2010) in the middle of the Piacenzian Stage. It incorporates sea surface temperatures, bottom water temperatures (Dowsett et al., 2009), vegetation (Salzmann et al., 2008), ice sheets (Hill et al., 2007, 2010), orography (Sohl et al., 2009) and a global land-sea mask equivalent to 25 m of sea level rise. The vegetation, ice sheets and orographic reconstructions are all required as boundary conditions within the models, although they must be translated onto the resolution of each individual model. Vegetation was reconstructed using the BIOME4 classification scheme (Kaplan, 2001) and must therefore be translated onto the vegetation scheme used by each model.

Although as part of PlioMIP a standard experimental design was implemented, it was appreciated that not all of the modelling groups would be able to perform the ideal mPWP experiment. As such, alternate boundary conditions were specified for those models that could not effectively change the land-sea mask from the present-day configuration. This meant that the ocean advance specified in low-lying coastal regions and West Antarctica as well as the filling of Hudson Bay were not included in some of the simulations (Table 1). Furthermore a choice was given concerning the initial state of the ocean between a specification of the PRISM3 three-dimensional ocean temperatures (Dowsett et al., 2009) and initialization with the same ocean temperatures as the pre-industrial control simulation (Haywood et al., 2011).

## 4 PlioMIP Experiment 2 global warming

Overall the PlioMIP models simulate mPWP annual mean global surface air temperature (SAT) increases of 1.8–3.6 °C (Table 2). Tropical temperatures increased by only 1.0–3.1 °C, while in the Arctic surface air temperatures increased by 3.5–13.2 °C (Fig. 1). Sea surface temperatures (SSTs) follow a similar pattern, but with a reduced magnitude of global warming and significantly greater warming in the North Pacific (Fig. 1d). The patterns of warming in the northern mid-latitudes and southern high latitudes are much more variable between the different models. Relative variation between the models is largest in the North Atlantic, midlatitude mountain regions and central Antarctica for SATs (Fig. 1c) and in the North Atlantic, North Pacific and sea ice areas of the Arctic and Southern oceans for SSTs (Fig. 1f).

The warming of the PlioMIP simulations is accompanied by increased precipitation (Haywood et al., 2013) and monsoonal activity (Zhang et al., 2013) and reductions in sea ice (Clark et al., 2013), although the Atlantic Meridional Overturning Circulation shows little response (Zhang et al., 2013b). Global mean temperature response (Table 2), as well as polar amplification (Salzmann et al., 2013), do not show a strong correlation to either the use of preferred or alternate boundary conditions or to the initial conditions of the ocean. Although land-sea masks vary between the different models in the Hudson Bay and West Antarctic region, they do not show the largest relative variance, suggesting that the alternatives used in PlioMIP Experiment 2 do not introduce significant biases.

## 5 Energy balance approach

Energy balance analyses have been used in many palaeoclimate simulations and ensembles to understand the simulated temperature changes (e.g. Donnadieu et al., 2006; Murakami et al., 2008). The results from each of the GCMs can be broken down in to the various components in the energy balance of each individual simulation. The approach taken builds on the energy balance modelling of Heinemann et al. (2009) and Lunt et al. (2012), where globally averaged temperatures are approximated using planetary albedo $\alpha$ and the effective longwave emissivity $\varepsilon$.

$$\frac{S_0}{4}(1 - \alpha) = \varepsilon \sigma T^4, \tag{1}$$

where $S_0$ is the total solar irradiance (1367 Wm$^{-2}$) and $\sigma$ is the Stefan–Boltzmann constant ($5.67 \times 10^{-8}$ Wm$^{-2}$ K$^{-4}$). Planetary albedo is the ratio of outgoing ($\uparrow$) to incoming ($\downarrow$) shortwave radiation at the top of the atmosphere (TOA) and effective longwave emissivity the ratio of TOA to surface (SURF) upward longwave radiation,

$$\alpha = \frac{SW_{TOA}^{\uparrow}}{SW_{TOA}^{\downarrow}}, \quad \varepsilon = \frac{LW_{TOA}^{\uparrow}}{LW_{SURF}^{\uparrow}}. \tag{2}$$

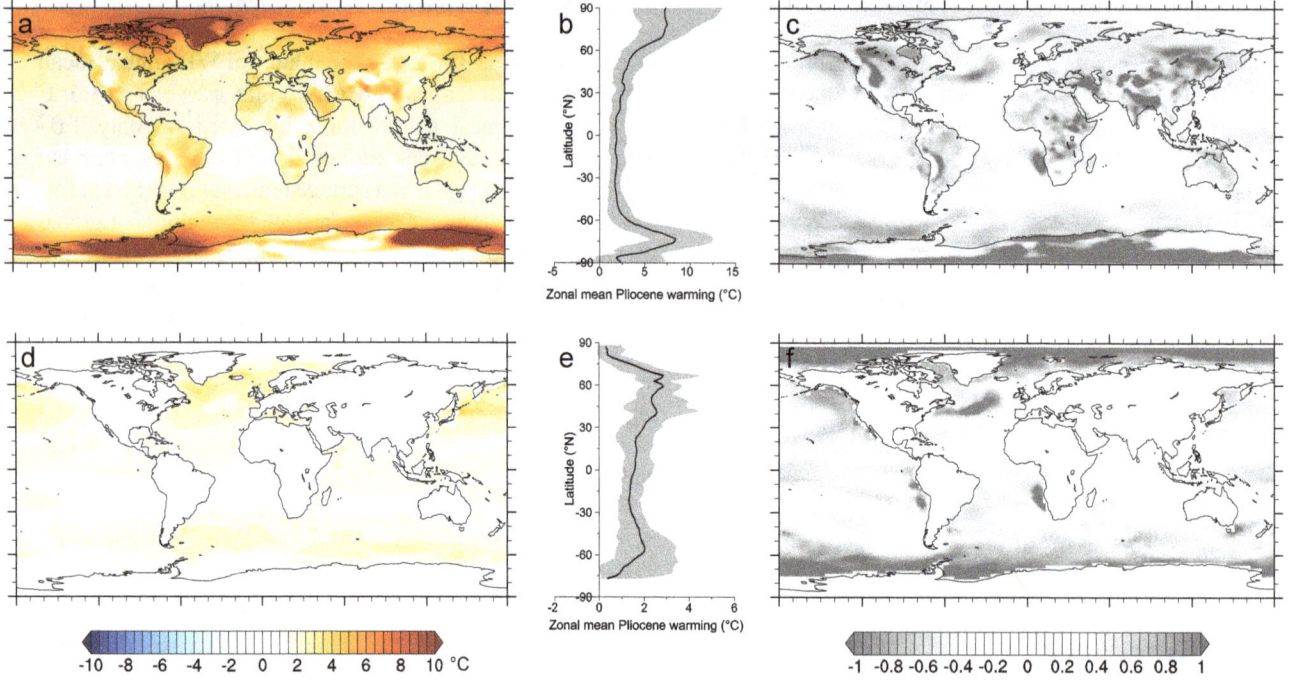

**Fig. 1.** Multi-model mean PlioMIP Experiment 2 warming between mid-Pliocene and pre-industrial simulations. (**a**) Annual mean surface air temperature (SAT) warming, (**b**) zonal mean SAT warming (solid line), with shading showing the range of model simulations, and (**c**) relative variance between the PlioMIP Experiment 2 simulations ($\sigma/\Delta$SAT). (**d**) Annual mean sea surface temperature (SST) warming, (**e**) zonal mean SST warming and (**f**) relative variance of SSTs.

This can be expanded to approximate the one dimensional, zonally averaged temperatures at each latitude of the model grid by including a component for the implied net meridional heat transport divergence ($H$).

$$SW^{\downarrow}_{TOA} (1 - \alpha) + H = \varepsilon \sigma T^4, \tag{3}$$

where

$$H = -\left(\left(SW^{\downarrow}_{TOA} - SW^{\uparrow}_{TOA}\right) - LW^{\uparrow}_{TOA}\right). \tag{4}$$

Thus the temperature at each latitude in a GCM experiment is given by

$$T = \left(\frac{SW^{\downarrow}_{TOA} (1 - \alpha) - H}{\varepsilon \sigma}\right)^{1/4} \equiv T(\varepsilon, \alpha, H). \tag{5}$$

By applying the notation of Lunt et al. (2012) to denote the pre-industrial control experiment as a second experiment represented by an apostrophe, the Pliocene surface air temperature warming ($\Delta T$) can be calculated by

$$\Delta T = T(\varepsilon, \alpha, H) - T(\varepsilon', \alpha', H'). \tag{6}$$

Due to their small changes relative to their absolute values, Pliocene warming can be approximated by a linear combination of changes in emissivity ($\Delta T_{\varepsilon}$), albedo ($\Delta T_{\alpha}$) and heat transport ($\Delta T_H$). However, these components can be

further broken down into the changes in the impact of atmospheric greenhouse gases ($\Delta T_{gg\varepsilon}$), clouds (via impacts on both emissivity, $\Delta T_{c\varepsilon}$, and albedo, $\Delta T_{c\alpha}$; see Sect. 6 for discussion) and clear sky albedo ($\Delta T_{cs\alpha}$; generally dominated by changes in surface albedo, but including atmospheric absorption and scattering components). In experiments and latitudes where changes in topography occur between the Pliocene and pre-industrial times, the impact of these changes in surface altitude ($\Delta T_{topo}$) must also be accounted for.

$$\Delta T \approx \Delta T_{gg\varepsilon} + \Delta T_{c\varepsilon} + \Delta T_{c\alpha} + \Delta T_{cs\alpha} + \Delta T_H + \Delta T_{topo} \tag{7}$$

Each of these components can be calculated from various combinations of Pliocene and pre-industrial albedos, emissivities and implied heat transports, although to differentiate between cloud and clear sky components some must be calculated in the clear sky case (denoted with a subscript cs).

$$\Delta T_{gg\varepsilon} = T(\varepsilon_{cs}, \alpha, H) - T(\varepsilon'_{cs}, \alpha, H) - \Delta T_{topo} \tag{8}$$

$$\Delta T_{c\varepsilon} = (T(\varepsilon, \alpha, H) - T(\varepsilon_{cs}, \alpha, H))$$
$$\quad - \left(T(\varepsilon', \alpha, H) - T(\varepsilon'_{cs}, \alpha, H)\right) \tag{9}$$

$$\Delta T_{c\alpha} = (T(\varepsilon, \alpha, H) - T(\varepsilon, \alpha_{cs}, H))$$
$$\quad - \left(T(\varepsilon, \alpha', H) - T(\varepsilon, \alpha'_{cs}, H)\right) \tag{10}$$

$$\Delta T_{cs\alpha} = T(\varepsilon, \alpha_{cs}, H) - T(\varepsilon, \alpha'_{cs}, H) \tag{11}$$

$$\Delta T_H = T(\varepsilon, \alpha, H) - T(\varepsilon, \alpha, H') \tag{12}$$

Changing the topography within a climate model can have many effects on the energy balance of the model, including changing circulation patterns, heat transport, surface conditions, cloud formation, storm generation, etc. Most of these features cannot be properly quantified without performing further simulations. However, a simple lapse rate correction will remove the direct impact on surface temperatures of changing the height of the surface. Although lapse rates vary over time and space, the impact of changing the topography in the Pliocene simulations ($\Delta T_{topo}$) can be approximated by multiplying the change in topography ($\Delta h$) by a constant atmospheric lapse rate ($\gamma \approx 5.5 \, \mathrm{K \, km^{-1}}$; Yang and Smith, 1985).

$$\Delta T_{topo} = \Delta h \cdot \gamma \qquad (13)$$

## 6 Treatment of clouds within the energy balance calculations

The energy balance calculations presented here split the planetary albedo and emissivity impacts into a component due to clouds and a clear sky one. In order to do this the clear sky radiation fluxes, which are the radiation fluxes that would have occurred had there not been any clouds, are used within the calculations. The cloud components are then the global temperature change due to the impact of clouds on planetary albedo and emissivity. If all else in the climate system remained the same except for the surface albedo then the clear sky albedo would show a change in temperature, but so also would the cloud albedo. This is because the impact of identical clouds on the total planetary albedo have changed due to the surface which they are covering being different. Similarly an increase in highly reflective cloud types would have a greater impact on temperature over the dark oceans than over a high albedo land surface.

A more thorough methodology for calculating the cloud and clear sky components would be to incorporate cloudiness into the calculations and thereby remove the reliance of cloud impact on the surface conditions. This requires an understanding of the impact of different modelled cloud types on the energy balance and, as we are considering a multimodel ensemble, the consistent application of this across all the simulations within the ensemble. Total cloud fractions are produced by each model, but these use different overlap functions, all of which are simplifications of the observed dependencies of cloud overlap on atmospheric dynamics (Wang and Dessler, 2006; Naud et al., 2008). Therefore, the total cloud fractions of the models are not only incomparable, but may also not be the best representation of cloud impacts on energy balance.

The impact of using the pragmatic methodology outlined above can be approximated using the surface downward shortwave radiation to approximate the impact of clouds. Figure 2 shows the albedo components of Pliocene warming in the HadCM3 model, using both this approximation and the

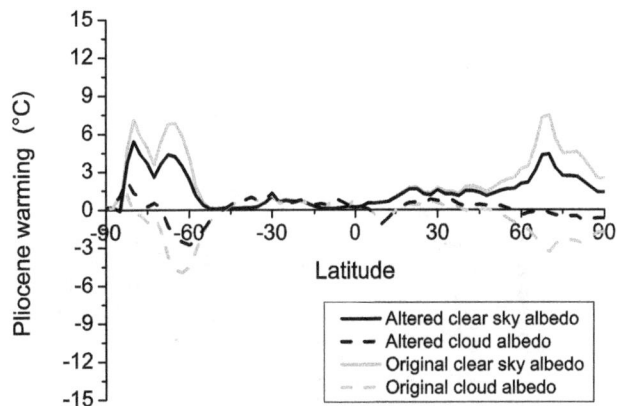

**Fig. 2.** Comparison of the impact on Pliocene warming of the components of planetary albedo under different formulations for the HadCM3 model. The grey lines are the formulations presented in this Sect. 5, while the black lines represent an altered calculation of the albedo components approximating the impact of clouds using the difference between the incoming radiation at the top of the atmosphere and the modelled downward shortwave radiation at the surface.

methodology outlined in Sect. 5. Although using the outlined energy balance approach produces greater magnitudes of impacts from clear sky and cloud albedo, as would be expected, the overall structure and the conclusions that would be drawn from these calculations remain.

## 7 Energy balance results for individual simulations

The energy balance calculations for each of the individual simulations within the PlioMIP Experiment 2 ensemble are shown in Fig. 3. The overall structure of the energy balance components is largely the same between all the simulations, although there are differences in the responses in each of the models and their magnitudes. In the tropics the warming is dominated by the greenhouse gas emissivity, with the other components having a small impact on warming. Those simulations showing greater tropical warming tend to also have a significant warming component from cloud impact on albedo. The midlatitudes are the region where the models show the least consistency, especially in the Northern Hemisphere. The only consistent component of the warming comes from greenhouse gas emissivity, although the cloud and clear sky albedos also tend to have a warming impact. In the high latitudes much of the warming comes from changes in the clear sky albedo, which are only partially offset by changes in the overall cloudy sky impact on planetary albedo. There are also slightly enhanced greenhouse gas warming and a warming impact of cloud emissivity. In the Southern Hemisphere the clear sky albedo warming often has a double peak, the first around 60° S, representing changes in the simulated Southern Ocean sea ice and the second over the

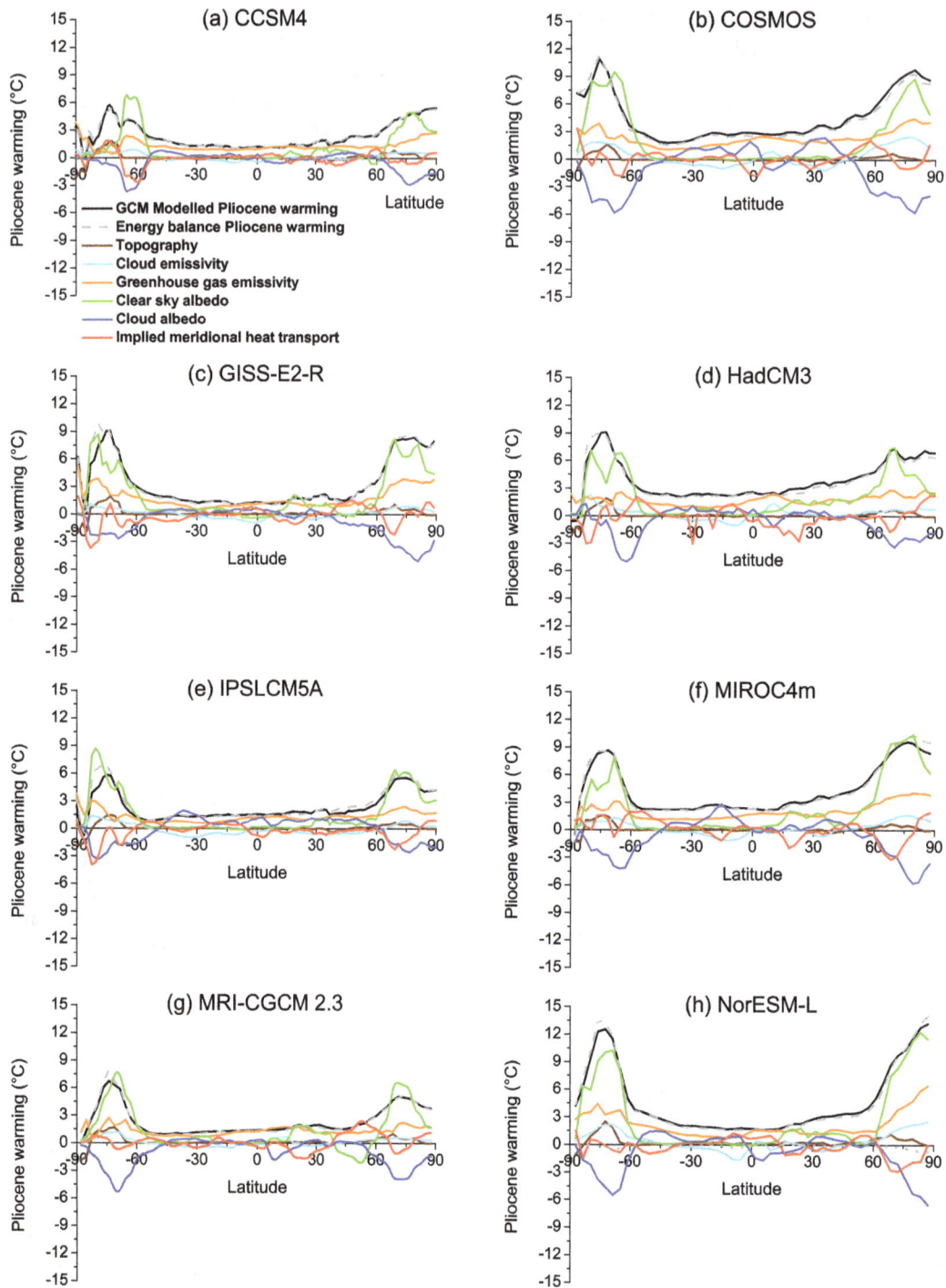

**Fig. 3.** Energy balance analysis for each of the eight PlioMIP Experiment 2 simulations, from (**a**) CCSM4, (**b**) COSMOS, (**c**) GISS-E2-R, (**d**) HadCM3, (**e**) IPSLCM5A, (**f**) MIROC4m, (**g**) MRI-CGCM 2.3 and (**h**) NorESM-L. Plots show the zonal mean warming, at each latitude in the model, from each of the energy balance components. The solid black line is the zonal mean surface air temperature increase from the GCM simulation, while the dashed grey line is the Pliocene warming approximated by the energy balance calculations.

Antarctic continents, from prescribed changes to the Antarctic ice sheets.

## 8   PlioMIP Experiment 2 energy balance

In order to evaluate the simulation of warm climates of the Pliocene in general, a simple mean of the energy balance components from each of the individual simulations within the PlioMIP Experiment ensemble has been performed. When combined with the range of values within the ensemble this allows an assessment of the general cause of warming within the PlioMIP simulations and the robustness of any conclusion that can be drawn. Figure 4 shows the ensemble mean of the various energy balance components along with the range from the eight simulations, while Fig. 5 shows the individual energy balance components for each of the PlioMIP simulations.

Clear sky albedo includes contributions from surface albedo changes and atmospheric absorption and scattering. The latter could become important, even in models with no mechanisms for changing atmospheric transparency, as atmospheric thickness can increase due to changes in surface altitude. In the PlioMIP simulations clear sky albedo shows little contribution to warming in the tropics and Southern Hemisphere midlatitudes. In the Northern Hemisphere midlatitudes most models show a warming due to clear sky albedo, apart from the MRI-CGCM 2.3 simulation that shows a cooling (Fig. 5a). In the polar regions, all the simulations show a strong warming signal from clear sky albedo, although the range in the magnitude of this warming is large. Changes in clear sky albedo mostly reflect changes on Earth's surface. Vegetation, snow and ice (both terrestrial ice masses and sea ice) are generally the main contributors to these changes. The warming found in the Northern Hemisphere, from 15–60° N is largely being driven by changes in the vegetation boundary conditions, particularly over the Sahara, Arabia and central Asia (Fig. 6). In the Arctic, warming due to clear sky albedo is primary driven by changes in ice sheet boundary conditions (reduced Greenland Ice Sheet) and changes in the predicted sea ice, but also by the poleward shift of the Arctic tree line (Salzmann et al., 2008). In the Southern Ocean and Antarctica the warming due to clear sky albedo has a double peak in most models, reflecting a reduction in the simulated Southern Ocean sea ice and a reduction in the prescribed Antarctic Ice Sheet (Fig. 7). Most models show reductions in sea ice coverage across the whole range of latitudes in which sea ice is modelled, except for the highest Northern Hemisphere latitudes and in CCSM close to Antarctica (Fig. 7c). The simulated changes in total cloud are relatively small and inconsistent between the models (Fig. 7e), so do not appear to introduce systematic biases.

All the simulations show a warming due to greenhouse gas emissivity of around 1–2 °C. These impacts are largely constant across latitudes, but with a slight polar amplification,

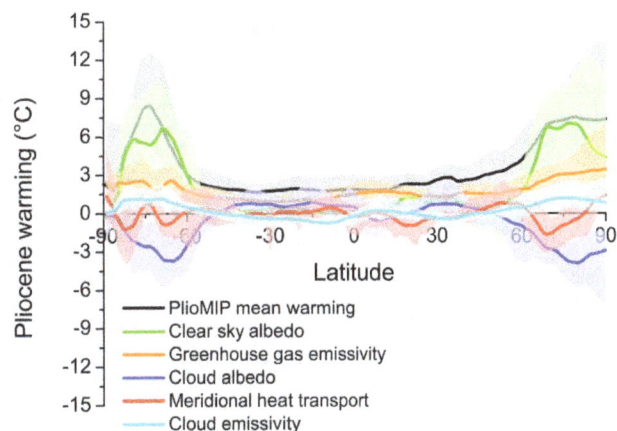

**Fig. 4.** Summary of the PlioMIP Experiment 2 energy balance analysis. Solid line shows the multi-model mean warming for each component, with the associated shading representing the range. These values have been interpolated onto a 1° latitude grid for comparison purposes.

especially in the Arctic (Fig. 5b). This is consistent with the prescribed increases in $CO_2$ (at 405 ppm for the mid-Pliocene, as opposed to 280 ppm in the pre-industrial simulations). The amplified high-latitude response is due to increases in the atmospheric water vapour predicted by the models. Differences in the simulation of this water vapour increase between different models explain why the range of temperature increases due to greenhouse gas warming is much higher in the polar regions.

The changes in the impact of clouds on planetary albedo are small in the tropics and midlatitudes. Different models seem to produce significantly different responses making the signal particularly noisy (Fig. 5c). However, the multi-model mean cloud albedo impact on warming appears to reflect some of the large-scale features of the PlioMIP simulations (Haywood et al., 2013). Between the Equator and ∼ 45° there is a general warming due to a reduction in the impact of cloud albedo, interrupted by a cooling in the Northern Hemisphere tropics. This cooling is due to an increase in cloud cover resulting from a northward shift of the Inter-Tropical Convergence Zone (Kamae et al., 2011). In the high latitudes the increased impact of clouds on planetary albedo, partly due to changes in the underlying surface albedo, leads to a significant cooling, peaking at between 3 and 6 °C in both hemispheres. Cloud emissivity shows a similar pattern of impacts, but in the opposite direction. However, the response is generally of a smaller magnitude (Fig. 5d).

Reconstruction of mid-Pliocene sea surface temperatures has led to increased heat transport in the North Atlantic being suggested as a primary driver of warming in the mid-Pliocene (Dowsett et al., 1992; Raymo et al., 1996). However, the implied overall meridional heat transport in the PlioMIP simulations, which integrates both oceanic and atmospheric transports, shows little coherent signal. The fact that there is only

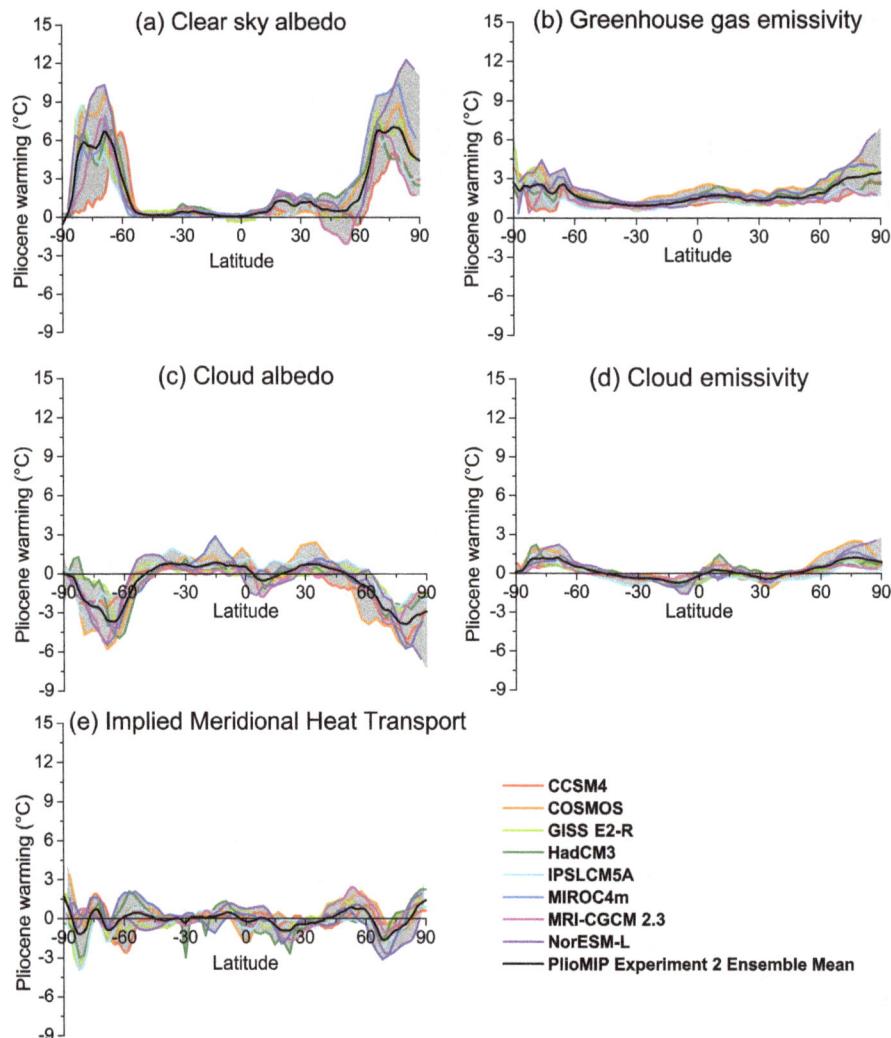

**Fig. 5.** Breakdown of the energy balance components, (**a**) clear sky albedo, (**b**) greenhouse gas emissivity, (**c**) cloud albedo, (**d**) cloud emissivity and (**e**) implied meridional heat transport. Solid black line shows the multi-model mean, range is shown by the grey shading and each of the individual model results is shown by the coloured solid lines.

one region where all of the simulations show a temperature change of the same direction suggests that the only robust conclusion that can be drawn about heat transport is a reduction of overall transport into the Arctic (Fig. 5e). This would be an expected result of reduced thermal gradients due to polar amplification in the Arctic region under climate warming. These energy balance calculations support analysis of the Atlantic Meridional Overturning Circulation in the PlioMIP ensemble, which shows that there is little change in the northward heat transport in the North Atlantic (Zhang et al., 2013b). This calls into question the role of ocean heat transport in the general warming of the mid-Pliocene. However, it may be important in the Pliocene variability of sea surface temperatures, which is particularly high in the North Atlantic (Dowsett et al., 2012).

## 9    Conclusions

The mid-Pliocene was probably the last time in Earth's history when atmospheric carbon dioxide concentrations were similar to today (Kürschner et al., 1996; Seki et al., 2010; Pagani et al., 2010; Bartoli et al., 2011). It has been the focus of palaeoenvironmental reconstructions and palaeoclimate model experiments for many years. However, the recently begun Pliocene Model Intercomparison Project is the first time that coordinated multi-model experiments, with common boundary conditions and experimental protocols, have been undertaken. The warming seen in the Pliocene has been well documented from a wide variety of sites from across the globe and using a number of different proxy techniques (Dowsett et al., 2012; Salzmann et al., 2013). Previous simulations of Pliocene warmth have been performed with only a single model and multi-model analyses have been severely

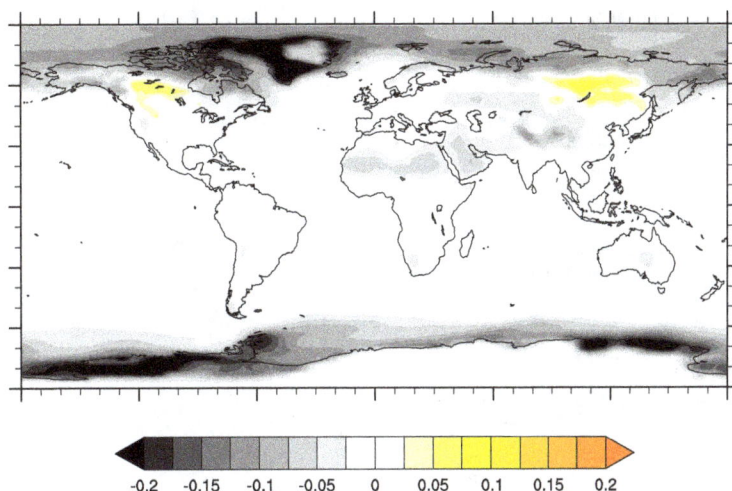

**Fig. 6.** Spatial distribution of multi-model mean changes in the clear sky albedo between mid-Pliocene and pre-industrial simulations. This primarily shows changes due to specified vegetation and global ice sheets and modelled sea ice and snow cover, although it includes an atmospheric component. Greyscale shows reductions in albedo, generally associated with warming and yellow shading indicates increases in albedo that would generally cause a reduction in surface air temperature. Increases in albedo closely follow the northward expansion of grassland imposed on the simulations (Haywood et al., 2010).

hampered by differing experimental designs (Haywood et al., 2009). For the first time a robust analysis of the causes of warming in Pliocene climate models is possible.

Energy balance calculations show that the tropical warming seen in all the models is primarily caused by greenhouse gas emissivity, with specified increases in atmospheric $CO_2$ concentration being the most important factor. Along with different sensitivity to the imposed $CO_2$ concentrations, changes in warming due to the cloud impact on planetary albedo drive differences between the models in the tropics. At polar latitudes all the energy balance components become important, but clear sky albedo is the dominant driver of the high levels of warming and polar amplification. This is largely due to reductions in the specified ice sheets and simulated sea ice, but in the Northern Hemisphere also reflects a northward shift in the treeline. The models show a very different response in the midlatitudes of the Northern Hemisphere, with large uncertainties in the relative contributions of the different energy balance components. This is particularly true for the North Atlantic and Kuroshio Current regions, where intermodel variability is highest and warming is simulated very differently (Haywood et al., 2013). A more complete picture of these currents, their strength and variability within the Pliocene, would enable a much better analysis of the skill of the models in these key regions.

Atmospheric $CO_2$ concentrations remain a significant uncertainty in Pliocene climate, with different proxy techniques producing values between pre-industrial levels and double pre-industrial levels (Kürschner et al., 1996; Seki et al., 2010; Pagani et al., 2010; Bartoli et al., 2011). As tropical warming is largely driven by this factor, simulations with accurate representation of low latitude clouds could provide some new

insight into the levels of $CO_2$ required to produce Pliocene climates (for example in a similar way to Lunt et al., 2012, for the Eocene). Alternatively, accurate reconstructions of surface temperatures and atmospheric $CO_2$ in combination with modelling studies could reveal the extent of changes to tropical cloud cover in the warmer Pliocene world.

Particularly strong warming in the high latitudes is driven by changes in albedo, especially from sea ice, ice sheets, snow cover and vegetation, which are only partially offset by extra cooling components from increased cloud albedo impacts. This is the region with the largest warming signal and also the largest uncertainties between the simulations. Therefore, improvements in the reconstruction of global ice cover and Arctic vegetation, along with improved data to evaluate the simulation of sea ice and high Arctic atmospheric and ocean temperatures, could significantly improve the simulations and allow much better constraints on total Pliocene warming. From the PlioMIP Experiment 2 simulations it appears that higher $CO_2$ concentrations warmed the planet during the Pliocene and drove large surface albedo feedbacks in the high latitudes through changes in sea ice, vegetation and ice sheets. The latter two of these factors are important components of long-term Earth system sensitivity, further supporting a long-term response of $CO_2$ greater than conventional climate sensitivity (Lunt et al., 2010; Haywood et al., 2013).

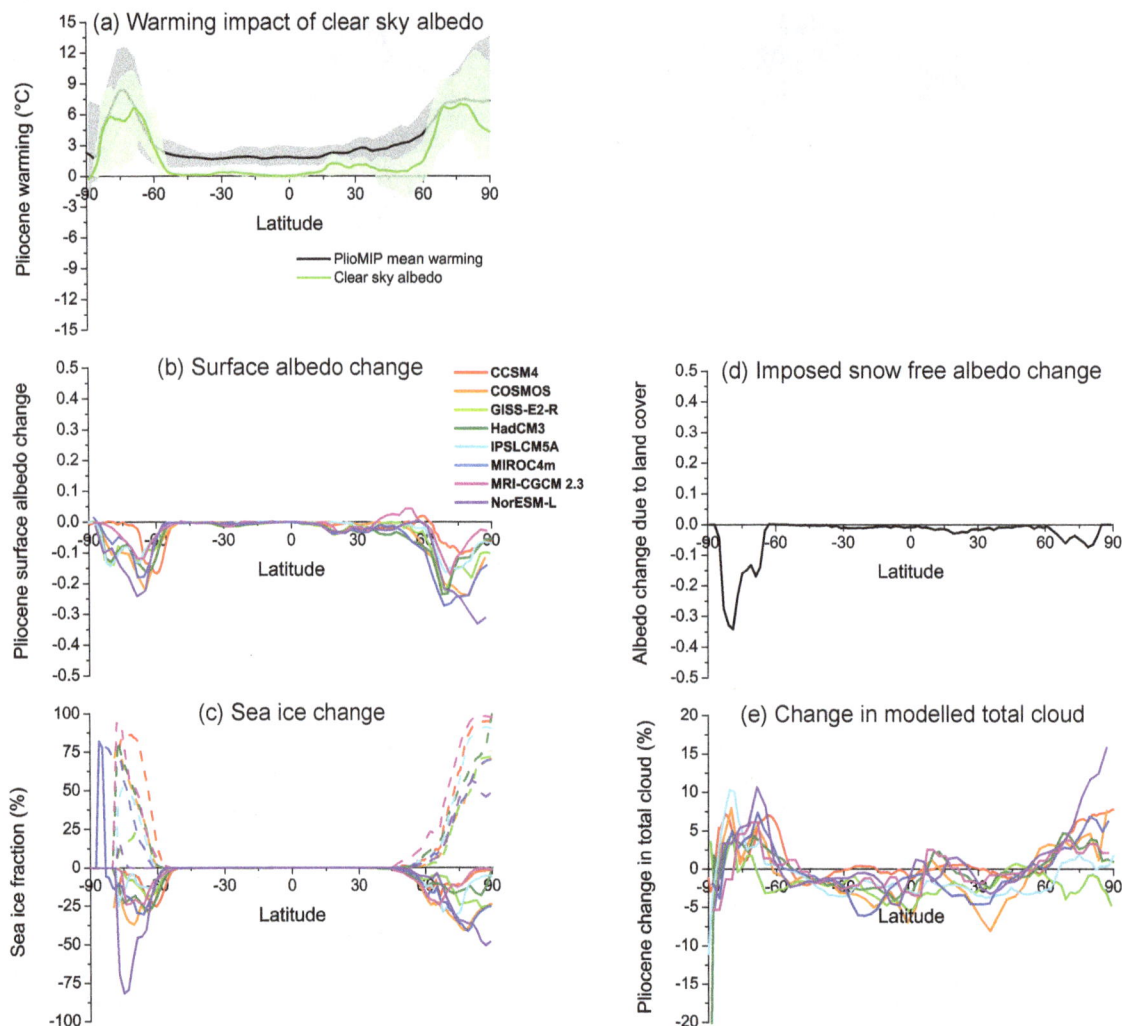

**Fig. 7.** Zonal mean of a series of factors that contribute to the clear sky albedo and its impact on Pliocene warming. (**a**) The total Pliocene warming and the clear sky albedo component of this (black and green respectively; see Fig. 4). (**b**) Total zonal surface albedo change in each of the 8 models. (**c**) Zonal mean sea ice percent coverage in the models, solid line is the Pliocene total and the dashed line is the change between Pliocene and pre-industrial experiments. (**d**) Imposed snow free albedo change, as calculated from the observed and reconstructed Pliocene mega-biomes, which include vegetation and ice sheet changes (Salzmann et al., 2008; Haywood et al., 2010). (**e**) Change in modelled total percent cloud cover for each model. These are calculated by each of the models using their own algorithm and cloud parameterizations, so are indicative rather than directly comparable.

*Acknowledgements*. D. J. Hill acknowledges the Leverhulme Trust for the award of an Early Career Fellowship and the National Centre for Atmospheric Science and the British Geological Survey for financial support. A. M. Haywood and S. J. Hunter acknowledge that the research leading to these results has received funding from the European Research Council under the European Union's Seventh Framework Programme (FP7/2007-2013)/ERC grant agreement no. 278636. A. M. Haywood acknowledges funding received from the Natural Environment Research Council (NERC Grant NE/I016287/1, and NE/G009112/1 along with D. J. Lunt). D. J. Lunt and F. J. Bragg acknowledge NERC grant NE/H006273/1. D. J. Lunt acknowledges Research Councils UK for the award of an RCUK fellowship and the Leverhulme Trust for the award of a Phillip Leverhulme Prize. The HadCM3

simulations were carried out using the computational facilities of the Advanced Computing Research Centre, University of Bristol – http://www.bris.ac.uk/acrc/. G. Lohmann received funding through the Helmholtz research programme PACES and the Helmholtz Climate Initiative REKLIM. C. Stepanek acknowledges financial support from the Helmholtz Graduate School for Polar and Marine Research and from REKLIM. Funding for L. Sohl and M. A. Chandler provided by NSF (National Science Foundation) Grant ATM0323516 and NASA Grant NNX10AU63A. B. L. Otto-Bliesner and N. A. Rosenbloom recognize that NCAR is sponsored by the US NSF and computing resources were provided by the Climate Simulation Laboratory at NCAR's Computational and Information Systems Laboratory (CISL), sponsored by the NSF and other agencies. W.-L. Chan and A. Abe-Ouchi would like to thank the Japan Society for the Promotion of Science for financial support and R. Ohgaito for advice on setting up the MIROC4m experiments on the Earth Simulator, JAMSTEC. The source code of the MRI model is provided by S. Yukimoto, O. Arakawa, and A. Kitoh of the Meteorological Research Institute, Japan. Z. Zhang acknowledges that the development of NorESM-L was supported by the Earth System Modelling (ESM) project funded by Statoil, Norway. Two anonymous reviewers are thanked for the improvements to the manuscript they inspired. Aisling Dolan is acknowledged for a beautiful title for this paper.

Edited by: C. Brierley

# References

Bartoli, G., Hönisch, B., and Zeebe, R. E.: Atmospheric $CO_2$ decline during the Pliocene intensification of Northern Hemisphere glaciations. Paleoceanography, 26, PA4213, doi:10.1029/2010PA002055, 2011.

Bragg, F. J., Lunt, D. J., and Haywood, A. M.: Mid-Pliocene climate modelled using the UK Hadley Centre Model: PlioMIP Experiments 1 and 2, Geosci. Model Dev., 5, 1109–1125, doi:10.5194/gmd-5-1109-2012, 2012.

Chan, W.-L., Abe-Ouchi, A., and Ohgaito, R.: Simulating the mid-Pliocene climate with the MIROC general circulation model: experimental design and initial results, Geosci. Model Dev., 4, 1035–1049, doi:10.5194/gmd-4-1035-2011, 2011.

Chandler, M., Rind, D., and Thompson, R.: Joint investigations of the middle Pliocene climate II: GISS GCM Northern Hemisphere results, Global Planet. Change, 9, 197–219, 1994.

Chandler, M. A., Sohl, L. E., Jonas, J. A., Dowsett, H. J., and Kelley, M.: Simulations of the mid-Pliocene Warm Period using two versions of the NASA/GISS ModelE2-R Coupled Model, Geosci. Model Dev., 6, 517–531, doi:10.5194/gmd-6-517-2013, 2013.

Clark, N. A., Williams, M., Hill, D. J., Quilty, P., Smellie, J., Zalasiewicz, J., Leng, M., and Ellis, M.: Fossil proxies of near-shore sea surface temperature and seasonality from the late Neogene Antarctic shelf, Naturwissenschaften, 100, 699–722, 2013.

Contoux, C., Ramstein, G., and Jost, A.: Modelling the mid-Pliocene Warm Period climate with the IPSL coupled model and its atmospheric component LMDZ5A, Geosci. Model Dev., 5, 903–917, doi:10.5194/gmd-5-903-2012, 2012.

Conway, T. J., Lang, P. M., and Masarie, K. A.: Atmospheric carbon dioxide dry air mole fractions from the NOAA ESRL Carbon Cycle Cooperative Global Air Sampling Network, 1968–2011, Version: 2012-08-15, available at: ftp://ftp.cmdl.noaa.gov/ccg/co2/flask/event/ (last access: 27 February 2013), 2012.

Donnadieu, Y., Pierrehumbert, R., Jacob, R., and Fluteau, F.: Modelling the primary control of paleogeography on Cretaceous climate, Earth Planet. Sc. Lett., 248, 426–437, 2006.

Dowsett, H. J. and Poore, R. Z.: Pliocene sea surface temperatures of the North Atlantic Ocean at 3.0 Ma, Quaternary Sci. Rev., 10, 189–204, 1991.

Dowsett, H. J., Cronin, T. M., Poore, R. Z., Thompson, R. S., Whatley, R. C., and Wood, A. M.: Micropaleontological evidence for increased meridional heat transport in the North Atlantic Ocean during the Pliocene, Science, 258, 1133–1135, 1992.

Dowsett, H. J., Robinson, M. M., and Foley, K. M.: Pliocene three-dimensional global ocean temperature reconstruction, Clim. Past, 5, 769–783, doi:10.5194/cp-5-769-2009, 2009.

Dowsett, H. J., Robinson, M., Haywood, A. M., Salzmann, U., Hill, D. J., Sohl, L., Chandler, M. A., Williams, M., Foley, K., and Stoll, D.: The PRISM3D Paleoenvironmental Reconstruction, Stratigraphy, 7, 123–139, 2010.

Dowsett, H. J., Haywood, A. M., Valdes, P. J., Robinson, M. M., Lunt, D. J., Hill, D. J., Stoll, D. K., and Foley, K. M.: Sea surface temperatures of the mid-Piacenzian Warm Period: A comparison of PRISM3 and HadCM3, Palaeogeogr. Palaeocl., 309, 83–91, 2011.

Dowsett, H. J., Robinson, M. M., Haywood, A. M., Hill, D. J., Dolan, A. M., Stoll, D. K., Chan, W. L., Abe-Ouchi, A., Chandler, M. A., Rosenbloom, N. A., Otto-Bleisner, B. L., Bragg, F. J., Lunt, D. J., Foley, K. M., and Riesselman, C. R.: Assessing confidence in Pliocene sea surface temperatures to evaluate predictive models, Nat. Clim. Change, 2, 365–371, 2012.

Haywood, A. M., Valdes, P. J., and Sellwood, B. W.: Global scale palaeoclimate reconstruction of the middle Pliocene climate using the UKMO GCM: initial results, Global Planet. Change, 25, 239–256, 2000.

Haywood, A. M., Chandler, M. A., Valdes, P. J., Salzmann, U., Lunt, D. J., and Dowsett, H. J.: Comparison of mid-Pliocene climate predictions produced by the HadAM3 and GCMAM3 General Circulation Models, Global Planet. Change, 66, 208–224, doi:10.1016/j.gloplacha.2008.12.014, 2009.

Haywood, A. M., Dowsett, H. J., Otto-Bleisner, B., Chandler, M. A., Dolan, A. M., Hill, D. J., Lunt, D. J., Robinson, M. M., Rosenbloom, N., Salzmann, U., and Sohl, L. E.: Pliocene Model Intercomparison Project (PlioMIP): experimental design and boundary conditions (Experiment 1), Geosci. Model Dev., 3, 227–242, doi:10.5194/gmd-3-227-2010, 2010.

Haywood, A. M., Dowsett, H. J., Robinson, M. M., Stoll, D. K., Dolan, A. M., Lunt, D. J., Otto-Bleisner, B., and Chandler, M. A.: Pliocene Model Intercomparison Project (PlioMIP): experimental design and boundary conditions (Experiment 2), Geosci. Model Dev., 4, 571–577, doi:10.5194/gmd-4-571-2011, 2011.

Haywood, A. M., Hill, D. J., Dolan, A. M., Otto-Bleisner, B. L., Bragg, F., Chan, W.-L., Chandler, M. A., Contoux, C., Dowsett, H. J., Jost, A., Kamae, Y., Lohmann, G., Lunt, D. J., Abe-Ouchi, A., Pickering, S. J., Ramstein, G., Rosenbloom, N. A., Salzmann, U., Sohl, L., Stepanek, C., Ueda, H., Yan, Q., and Zhang, Z.: Large-scale features of Pliocene climate: results from the Pliocene Model Intercomparison Project, Clim. Past, 9, 191–209, doi:10.5194/cp-9-191-2013, 2013.

Heinemann, M., Jungclaus, J. H., and Marotzke, J.: Warm Paleocene/Eocene climate as simulated in ECHAM5/MPI-OM, Clim. Past, 5, 785–802, doi:10.5194/cp-5-785-2009, 2009.

Hill, D. J., Haywood, A. M., Hindmarsh, R. C. M., and Valdes, P. J.: Characterizing ice sheets during the Pliocene: evidence from data and models, in: Deep-Time Perspectives on Climate Change: Marrying the signal from Computer Models and Biological Proxies, edited by: Williams, M., Haywood, A. M., Gregory, F. J., and Schmidt, D. N., The Micropalaeontological Society, Special Publications, The Geological Society, London, 517–538, 2007.

Hill, D. J., Dolan, A. M., Haywood, A. M., Hunter, S. J., and Stoll, D. K.: Sensitivity of the Greenland Ice Sheet to Pliocene sea surface temperatures, Stratigraphy, 7, 111–122, 2010.

Hill, D. J., Haywood, A. M., Lunt, D. J., Otto-Bliesner, B. L., Harrison, S. P., and Braconnot, P.: Paleoclimate modelling: an integrated component of climate change science, PAGES Newsletter, 20, 103, 2012.

Kamae, Y. and Ueda, H.: Mid-Pliocene global climate simulation with MRI-CGCM2.3: set-up and initial results of PlioMIP Experiments 1 and 2, Geosci. Model Dev., 5, 793–808, doi:10.5194/gmd-5-793-2012, 2012.

Kamae, Y., Ueda, H., and Kitoh, A.: Hadley and Walker circulations in the mid-Pliocene Warm Period simulated by an Atmospheric General Circulation Model, J. Meteorol. Soc. Jpn., 89, 475–493, 2011.

Kaplan, J. O.: Geophysical applications of vegetation modeling, PhD thesis, Lund University, Lund, 2001.

Kürschner, W. M., van der Burgh, J., Visscher, H., and Dilcher, D. L.: Oak leaves as biosensors of late Neogene and early Pleistocene paleoatmospheric $CO_2$ concentrations, Mar. Micropalaeontol., 27, 299–312, 1996.

Lunt, D. J., Haywood, A. M., Schmidt, G. A., Salzmann, U., Valdes, P. J., and Dowsett, H. J.: Earth system sensitivity inferred from Pliocene modelling and data, Nat. Geosci., 3, 60–64, 2010.

Lunt, D. J., Dunkley Jones, T., Heinemann, M., Huber, M., LeGrande, A., Winguth, A., Loptson, C., Marotzke, J., Roberts, C. D., Tindall, J., Valdes, P., and Winguth, C.: A model–data comparison for a multi-model ensemble of early Eocene atmosphere–ocean simulations: EoMIP, Clim. Past, 8, 1717–1736, doi:10.5194/cp-8-1717-2012, 2012.

Murakami, S., Ohgaito, R., Abe-Ouchi, A., Crucifix, M., and Otto-Bliesner, B. L.: Global-scale energy and freshwater balance in glacial climate: A comparison of three PMIP2 LGM simulations, J. Climate, 21, 5008–5033, 2008.

Naud, C. M., Del Genio, A., Mace, G. G., Benson, S., Clothiaux, E. E., and Kollias, P.: Impact of dynamics and atmospheric state on cloud vertical overlap, J. Climate, 21, 1758–1770, 2008.

Pagani, M., Liu, Z. H., LaRiviere, J., and Ravelo, A. C.: High Earth-system climate sensitivity determined from Pliocene carbon dioxide concentrations, Nat. Geosci., 3, 27–30, 2010.

Raymo, M. E., Grant, B., Horowitz, M., and Rau, G. H.: Mid-Pliocene warmth: Stronger greenhouse and stronger conveyor, Mar. Micropalaeontol., 27, 313–326, 1996.

Rosenbloom, N. A., Otto-Bliesner, B. L., Brady, E. C., and Lawrence, P. J.: Simulating the mid-Pliocene Warm Period with the CCSM4 model, Geosci. Model Dev., 6, 549–561, doi:10.5194/gmd-6-549-2013, 2013.

Salzmann, U., Haywood, A. M., Lunt, D. J., Valdes, P. J., and Hill, D. J.: A new global biome reconstruction and data-model comparison for the Middle Pliocene, Global Ecol. Biogeogr., 17, 432–447, 2008.

Salzmann, U., Dolan, A. M., Haywood, A. M., Chan, W.-L., Voss, J., Hill, D. J., Abe-Ouchi, A., Otto-Bliesner, B. L., Bragg, F. J., Chandler, M. A., Contoux, C., Dowsett, H. J., Jost, A., Kamae, Y., Lohmann, G., Lunt, D. J., Pickering, S. J., Pound, M. J., Ramstein, G., Rosenbloom, N. A., Sohl, L., Stepanek, C., Ueda, H., and Zhang, Z.: Challenges in quantifying Pliocene terrestrial warming revealed by data-model discord, Nat. Clim. Change, 3, 969–974, 2013.

Seki, O., Foster, G. L., Schmidt, D. N., Mackensen, A., Kawamura, K., and Pancost, R. D.: Alkenone and boron based Pliocene $pCO_2$ records, Earth Planet. Sc. Lett., 292, 201–211, doi:10.1016/j.epsl.2010.01.037, 2010.

Sloan, L. C., Crowley, T. J., and Pollard, D.: Modeling of middle Pliocene climate with the NCAR GENESIS general circulation model, Mar. Micropaleontol., 27, 51–61, 1996.

Sohl, L. E., Chandler, M. A., Schmunk, R. B., Mankoff, K., Jonas, J. A., Foley, K. M., and Dowsett, H. J.: PRISM3/GISS topographic reconstruction, US Geol. Surv. Data Series 419, US Geological Survey, Reston, VA, USA, 2009.

Stepanek, C. and Lohmann, G.: Modelling mid-Pliocene climate with COSMOS, Geosci. Model Dev., 5, 1221–1243, doi:10.5194/gmd-5-1221-2012, 2012.

Wang, L. and Dessler, A. E.: Instantaneous cloud overlap statistics in the tropical area revealed by ICESat/GLAS data, Geophys. Res. Lett., 33, L15804, doi:10.1029/2005GL024350, 2006.

Yang, S.-K. and Smith, G.L.: Further studies on atmospheric lapse rate regimes, J. Atmos. Sci., 42, 961–965, 1985.

Zhang, R., Yan, Q., Zhang, Z. S., Jiang, D., Otto-Bliesner, B. L., Haywood, A. M., Hill, D. J., Dolan, A. M., Stepanek, C., Lohmann, G., Contoux, C., Bragg, F., Chan, W.-L., Chandler, M. A., Jost, A., Kamae, Y., Abe-Ouchi, A., Ramstein, G., Rosenbloom, N. A., Sohl, L., and Ueda, H.: Mid-Pliocene East Asian monsoon climate simulated in the PlioMIP, Clim. Past, 9, 2085–2099, doi:10.5194/cp-9-2085-2013, 2013.

Zhang, Z. S., Nisancioglu, K., Bentsen, M., Tjiputra, J., Bethke, I., Yan, Q., Risebrobakken, B., Andersson, C., and Jansen, E.: Pre-industrial and mid-Pliocene simulations with NorESM-L, Geosci. Model Dev., 5, 523–533, doi:10.5194/gmd-5-523-2012, 2012.

Zhang, Z.-S., Nisancioglu, K. H., and Ninnemann, U. S.: Increased ventilation of Antarctic deep water during the warm mid-Pliocene, Nat. Commun., 4, 1499, doi:10.1038/ncomms2521, 2013a.

Zhang, Z.-S., Nisancioglu, K. H., Chandler, M. A., Haywood, A. M., Otto-Bliesner, B. L., Ramstein, G., Stepanek, C., Abe-Ouchi, A., Chan, W.-L., Bragg, F. J., Contoux, C., Dolan, A. M., Hill, D. J., Jost, A., Kamae, Y., Lohmann, G., Lunt, D. J., Rosenbloom, N. A., Sohl, L. E., and Ueda, H.: Mid-pliocene Atlantic Meridional Overturning Circulation not unlike modern, Clim. Past, 9, 1495–1504, doi:10.5194/cp-9-1495-2013, 2013b.

# Biogeochemical variability during the past 3.6 million years recorded by FTIR spectroscopy in the sediment record of Lake El'gygytgyn, Far East Russian Arctic

C. Meyer-Jacob[1], H. Vogel[2,3], A. C. Gebhardt[4], V. Wennrich[2], M. Melles[2], and P. Rosén[1,5]

[1]Department of Ecology and Environmental Science, Umeå University, 901 87 Umeå, Sweden
[2]Institute of Geology and Mineralogy, University of Cologne, Zuelpicher Str. 49a, 50674 Cologne, Germany
[3]Institute of Geological Sciences & Oeschger Centre for Climate Change Research, University of Bern, Baltzerstrasse 1 + 3, 3012 Bern, Switzerland
[4]Alfred Wegener Institute Helmholtz Centre for Polar and Marine Research, Columbusstraße, 27515 Bremerhaven, Germany
[5]Climate Impacts Research Centre (CIRC), Abisko Scientific Research Station, 981 07 Abisko, Sweden

*Correspondence to:* C. Meyer-Jacob (carsten.meyer-jacob@emg.umu.se)

**Abstract.** A number of studies have shown that Fourier transform infrared spectroscopy (FTIRS) can be applied to quantitatively assess lacustrine sediment constituents. In this study, we developed calibration models based on FTIRS for the quantitative determination of biogenic silica (BSi; $n = 420$; gradient: 0.9–56.5 %), total organic carbon (TOC; $n = 309$; gradient: 0–2.9 %), and total inorganic carbon (TIC; $n = 152$; gradient: 0–0.4 %) in a 318 m-long sediment record with a basal age of 3.6 million years from Lake El'gygytgyn, Far East Russian Arctic. The developed partial least squares (PLS) regression models yield high cross-validated (CV) $R^2_{CV} = 0.86$–0.91 and low root mean square error of cross-validation (RMSECV) (3.1–7.0 % of the gradient for the different properties). By applying these models to 6771 samples from the entire sediment record, we obtained detailed insight into bioproductivity variations in Lake El'gygytgyn throughout the middle to late Pliocene and Quaternary. High accumulation rates of BSi indicate a productivity maximum during the middle Pliocene (3.6–3.3 Ma), followed by gradually decreasing rates during the late Pliocene and Quaternary. The average BSi accumulation during the middle Pliocene was ∼ 3 times higher than maximum accumulation rates during the past 1.5 million years. The indicated progressive deterioration of environmental and climatic conditions in the Siberian Arctic starting at ca. 3.3 Ma is consistent with the first occurrence of glacial periods and the finally complete establishment of glacial–interglacial cycles during the Quaternary.

## 1 Introduction

The understanding of past environmental changes is of particular importance to facilitate the prediction of the magnitude and the regional implications of future environmental changes, especially in view of an anthropogenically forced global warming (IPCC, 2007). Lake sediment records are valuable archives preserving these changes. Records such as those from Lake Baikal (e.g. Williams et al., 1997), Lake Malawi (e.g. Scholz et al., 2006) and Lake Biwa (e.g. Fuji, 1988) extend several million years back in time and consist of several hundred metres of sediment. However, multiproxy analyses of such long records are very time-consuming and cost-intensive. Furthermore, high-resolution sampling of these records can be restricted by the amount of sample material available for the different analyses.

Fourier transform infrared spectroscopy (FTIRS) is a promising tool that copes with the above-mentioned problems due to its potential to analyse several components at once, simple sample pre-treatments, and the small sample size required for analysis (0.01 g). Because the excitation of polar bonds in molecules by IR radiation is wavenumber-specific depending on the structural and atomic composition of the molecules, it is possible to gather information about organic and minerogenic components from one single measurement. The technique has been applied to sediment to assess the concentration of silicate minerals (Sifeddine et al., 1994; Bertaux et al., 1996, 1998; Wirrmann et al., 2001) and carbonates (Mecozzi et al., 2001), as well as to characterise humic substances (Braguglia et al., 1995; Belzile et al., 1997; Calace et al., 1999, 2006; Mecozzi and Pietrantonio, 2006).

FTIRS has also successfully been applied in palaeolimnological studies to quantify biogeochemical properties such as biogenic silica, total organic carbon, total inorganic carbon, and total nitrogen (Vogel et al., 2008; Rosén et al., 2010). The developed FTIRS models of these approaches were based on site-specific and regional calibrations, respectively. Furthermore, Rosén et al. (2011) have shown that universally applicable models can be developed and applied to globally distributed lakes with considerably different settings. However, in these studies, reconstructions of selected properties by means of FTIRS have been conducted on sediment records of only a few metres' length ($< 17$ m) and restricted temporal range ($< 340$ ka). It remains uncertain whether calibration models based on IR spectral information can be developed for and applied on several-hundred-metre-long sediment sequences extending millions of years back in time. In particular, it is important to know whether or not variations in sediment composition arising from climatic, environmental or diagenetic changes bias the robustness of IR calibrations.

In this study, FTIRS is applied to a 318 m-long sediment record from Lake El'gygytgyn, Far East Russian Arctic ($67°30'$N, $172°5'$E) (Fig. 1), which was recovered during an ICDP (International Continental Scientific Drilling Program)-funded deep-drilling campaign in 2009 (ICDP site 5011-1; Melles et al., 2011). With its continuous formation and basal age of 3.6 million years (Melles et al., 2012; Nowaczyk et al., 2013), the record of Lake El'gygytgyn for the first time provides high-resolution insights into the climatic and environmental evolution of the Arctic during the Quaternary and late Pliocene.

Within the framework of this study, we test for the first time the applicability of IR-based calibration models for biogenic silica (BSi), total organic carbon (TOC), and total inorganic carbon (TIC) to a sediment record extending several million years back in time. Moreover, we provide new insights into the climatic and environmental evolution of the Arctic from the Pliocene to the present based upon variations in bioproductivity indicators in the sediment record of Lake El'gygytgyn.

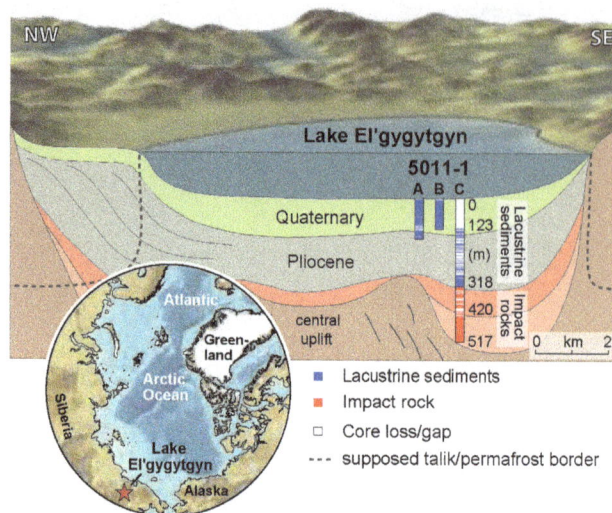

Fig. 1. Location of Lake El'gygytgyn in north-eastern Russia (inserted map) and schematic cross-section of the El'gygytgyn basin stratigraphy showing the recovery of holes A, B, and C at ICDP Site 5011-1 (modified after Melles et al., 2012).

## 2 Material and methods

### 2.1 Core recovery

Drilling operations were performed by Drilling, Observation and Sampling of the Earth's Continental Crust (DOSECC) Inc. using a GLAD-800 drilling system (Global Lake Drilling 800 m) from the lake ice cover in spring 2009 (Melles et al., 2011). Drill cores from three holes (A, B, and C) along with a percussion piston core (Lz1024) at ICDP site 5011-1, which is situated in the deepest part of the lake, form a core composite that penetrates down to the underlying impact breccia at 318 m below lake floor (mblf) (Fig. 1).

The composite profile of site 5011-1 was continuously subsampled in 2 cm intervals. FTIRS measurements were performed on 6771 of these samples. For the establishment of calibration models, an additional 255 samples were taken from core catchers ($\sim$ every 3 m) and core cuttings ($\sim$ every 1 m) of the drill cores ($n = 183$ for BSi, $n = 203$ for TOC and TIC), complemented by samples from 16.6 and 12.9 m-long percussion piston cores Lz1024 at site 5011-1 ($n = 204$ for BSi) (Juschus et al., 2005) and PG1351 about 2 km apart from site 5011-1 ($n = 83$ for BSi, $n = 156$ for TOC) (Melles et al., 2007).

### 2.2 Analytical methods

Conventionally measured concentrations of biogeochemical properties (BSi, TOC, and TIC) were used as reference data for the development of our calibration models. All sediment samples were freeze-dried and ground using either a swing mill (samples from drill cores of ICDP site 5011-1) or a planetary mill (samples from Lz1024 and PG1351). TC and

TIC in samples from site 5011-1 were determined by suspension method using a DIMATOC® 100 liquid analyser (Dimatec Corp.). TOC was then calculated by subtracting TIC from TC. The TOC content in samples from core PG1351 was analysed with a Metalyt-CS-1000-S (ELTRA Corp.) after sample pretreatment with HCL (10 %) at a temperature of 80 °C to remove carbonate. Concentrations of BSi in all sediment cores were obtained by applying the wet chemical leaching method according to Müller and Schneider (1993).

Accumulation rates ($AR_X$) of BSi ($AR_{BSi}$), TOC ($AR_{TOC}$), and TIC ($AR_{TIC}$) in $g\,m^{-2}\,yr^{-1}$ were calculated according to Eq. (1):

$$AR_X = SR \times DBD \times \%X \times 10^2, \qquad (1)$$

where SR is the sedimentation rate in $cm\,a^{-1}$, DBD is the dry bulk density in $g\,cm^{-3}$, and $\%X$ is the concentration of the property of interest. For each sample, the composite age was calculated using linear interpolation between the tie points of the age model (Nowaczyk et al., 2013) and sedimentation rates were calculated between the geomagnetic tie points. No accumulation rates could be calculated for the first 125 kyr of the sediment record, which originates from core Lz1024 (Melles et al., 2012), due to the lack of conventional density measurements, and for 84 samples of the remaining composite profile due to missing or erroneous density measurements.

Prior to the FTIR measurement, all samples were freeze-dried and ground; $0.011 \pm 0.0001$ g sample material was then mixed with $0.5 \pm 0.0005$ g of oven-dried spectroscopic grade potassium bromide (KBr) (Uvasol®, Merck Corp.), which does not influence the FTIR spectrum due to its transparency in the IR region. Afterwards the mixture was homogenised using a mortar and pestle. The low sample concentration (2.2 %) was chosen to avoid spectral distortions by very high absorbance and optical effects (Herbert et al., 1992; Griffiths and de Haseth, 2007). To gain constant measuring conditions and avoid variability caused by variations of temperature, the measurements were performed in a temperature-controlled laboratory with a constant temperature of $25 \pm 0.2$ °C. Samples were stored in the same room at least 5 h prior to the measurement. An IFS 66v/S FTIR spectrometer (Bruker Optics Inc.) equipped with a diffuse reflectance accessory (Harrick Inc.) and a Vertex 70 equipped with a HTS-XT accessory unit (Bruker Optics Inc.) were used for the analysis. Each sample was scanned 64 times at a resolution of $4\,cm^{-1}$ (reciprocal centimetres) for the wavenumber range from 3750 to $450\,cm^{-1}$.

## 2.3 Numerical analyses

Baseline correction and multiplicative scatter correction (MSC) were applied to normalise the recorded FTIR spectra and to remove spectral variations caused by noise (Geladi et al., 1985; Martens and Næs, 1989). We used partial least squares (PLS) regression to develop calibration models between FTIR spectral information and the corresponding conventionally measured BSi, TOC, and TIC concentrations. Conventionally measured primary sediment properties were square-root-transformed prior to analysis. All calibration models are based on the spectral range from 3750 to $450\,cm^{-1}$. Previous studies have shown that models using absorption bands specific for the property of interest can exhibit similar statistical performances as models based on the entire measured IR range (Rosén et al., 2010, 2011). However, by restricting the spectral range to component-specific absorption bands, the resulting models are more sensitive to overlapping absorption bands of other sediment compounds with similar spectral features. Calibration models based on the entire measured IR range, in contrast, consider spectral regions positively and negatively correlated with the property of interest and thus are taking changes in the overall sediment composition into account.

Internal and external validations were applied to quantitatively evaluate the performance of the developed PLS models. The internal validation was performed by sevenfold cross-validation (CV), which evaluates how much of the variation in the primary data is predicted by the model (predictive power) and determines the appropriate model complexity (number of significant PLS components). In this evaluation, a calibration model based on 6/7 of the available data was applied to the remaining 1/7 of the data to estimate its prediction ability. This procedure was then repeated 7 times until all samples of the data set were predicted solely based on their spectral information. The resulting cross-validated coefficient of determination $R^2_{CV}$ and root mean square error of cross-validation (RMSECV) were used to evaluate the internal model performance. $R^2_{CV}$ is a measure of the goodness of prediction of a model based on the predictive residual sum of squares (PRESS) (i.e. the squared difference between the predicted and observed values) and the residual sum of squares (SS) of the previous component ($R^2_{CV} = 1 - PRESS/SS$). A component was considered significant when $R^2_{CV} > 0$ (i.e. the component contributes to the predictive power of the model). The significance of the PLS models was tested by ANOVA (analysis of variance) of the cross-validated residuals (CV-ANOVA) (Eriksson et al., 2008) and response permutation, which estimates the degree of overfit and overprediction. Only models with a $p$ value lower than 0.001 resulting from the CV-ANOVA and valid according to the evaluation criteria of response permutations presented by Eriksson et al. (2006) were used in this study. The external validation was performed on 50 samples randomly chosen and equally distributed over the entire composite profile of ICDP site 5011-1. The coefficient of determination ($R^2$) between conventionally measured and FTIRS-inferred concentrations of biogeochemical properties as well as the root mean square error of prediction (RMSEP) were considered to estimate the prediction ability of the developed calibration applied to samples not included in the calibration set.

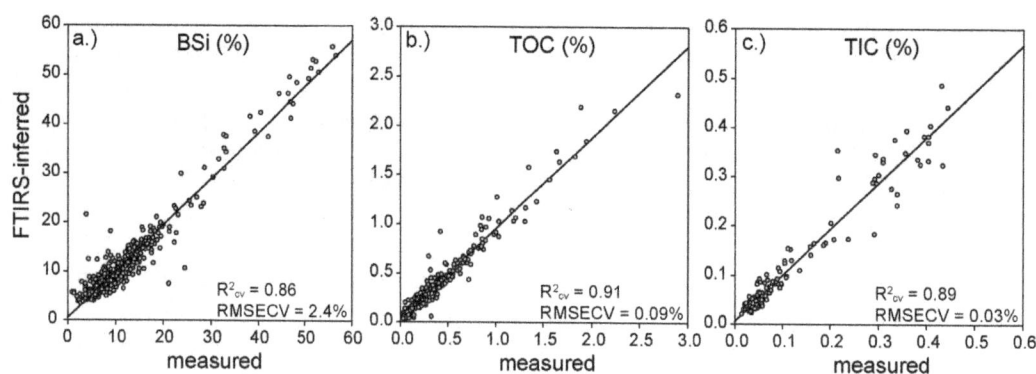

**Fig. 2.** Conventionally measured ($x$ axis) versus FTIRS-inferred concentrations ($y$ axis) of (**a**) biogenic silica (BSi), (**b**) total organic carbon (TOC), and (**c**) total inorganic carbon (TIC) with the cross-validated coefficient of determination ($R^2_{CV}$) and root mean square error of cross-validation (RMSECV) resulting from the internal validation of the developed calibration models. The calibration models are based on sediment samples from percussion piston cores PG1351 ($n = 83$ for BSi, $n = 156$ for TOC) and Lz1024 ($n = 204$ for BSi) and samples distributed throughout the entire drill core composite profile of ICDP site 5011-1 ($n = 133$ for BSi, $n = 153$ for TOC, $n = 152$ for TIC).

We performed multivariate data analyses using SIMCA-P 12.0 (Umetrics AB, Umeå, Sweden). OPUS 5.5 (Bruker Optics Inc.) was used for the visualisation of single FTIR spectra. For more information about the numerical analysis in this approach see Vogel et al. (2008) and Rosén et al. (2010).

## 3 Results and discussion

### 3.1 Statistical performance of FTIRS models

All three models – the 2-component FTIRS-BSi model, the 8-component FTIRS-TOC model, and the 8-component FTIRS-TIC model – show a strong correlation between FTIRS-inferred and conventionally measured concentrations ($R^2_{CV} = 0.86$ for BSi; $R^2_{CV} = 0.91$ for TOC; $R^2_{CV} = 0.89$ for TIC) (Table 1, Fig. 2). Corresponding RMSECV are low: 2.4 % for BSi (4.3 % of the gradient), 0.09 % for TOC (3.1 % of the gradient), and 0.03 % for TIC (2.1 % of the gradient). The external validation showed good statistical performance. The $R^2$ values are 0.94 for BSi and 0.83 for TOC, and the RMSEP values are 2.1 % for BSi (7.0 % of the gradient) and 0.13 % for TOC (7.9 % of the gradient), which indicate both a high prediction ability of the general trend and high prediction accuracies with respect to the absolute values of the proxy of interest. In comparison with the BSi and TOC models, the external validation of the TIC model yields a slightly poorer statistical performance ($R^2 = 0.86$, RMSECV = 0.08 % (21.1 % of the gradient)) (Table 1), indicating a high prediction ability of the general trend but a lower prediction accuracy with respect to absolute TIC values. For the TIC calibration model, one sample was excluded from the calibration set due to its high leverage on the model (TIC concentration > 3 times higher than the concentration of the next neighbouring observation). The use of spectral regions specific for the component of interest to improve

**Table 1.** Statistical performance of developed calibration models for biogenic silica (BSi), total organic carbon (TOC), and total inorganic carbon (TIC) based on sediments from Lake El'gygytgyn.

| Statistics | BSi (%) | TOC (%) | TIC (%) |
|---|---|---|---|
| Calibration set | | | |
| PLS components | 2 | 8 | 8 |
| Samples ($n$) | 420 | 309 | 152 |
| Min | 0.9 | 0.02 | 0.01 |
| Max | 56.5 | 2.89 | 0.44 |
| Gradient | 55.6 | 2.87 | 0.43 |
| Mean | 12.1 | 0.41 | 0.11 |
| Included wavenumbers (cm$^{-1}$) | 450–3750 | 450–3750 | 450–3750 |
| Internal validation | | | |
| $R^2_{CV}$ | 0.86 | 0.91 | 0.89 |
| RMSECV | 2.4 | 0.09 | 0.03 |
| RMSECV (% gradient) | 4.3 | 3.1 | 7.0 |
| External validation set | | | |
| Min | 1.2 | 0.04 | 0.02 |
| Max | 36.9 | 1.68 | 0.40 |
| Gradient | 35.7 | 1.64 | 0.38 |
| Mean | 10.1 | 0.37 | 0.09 |
| External validation | | | |
| $R^2$ | 0.94 | 0.83 | 0.86 |
| RMSEP | 2.1 | 0.13 | 0.08 |
| RMSEP (% gradient) | 7.0 | 7.9 | 21.1 |

the model performance as suggested by Rosén et al. (2010, 2011) improved neither the fit to the measured concentrations nor the model prediction accuracy for any of the proxies of interest.

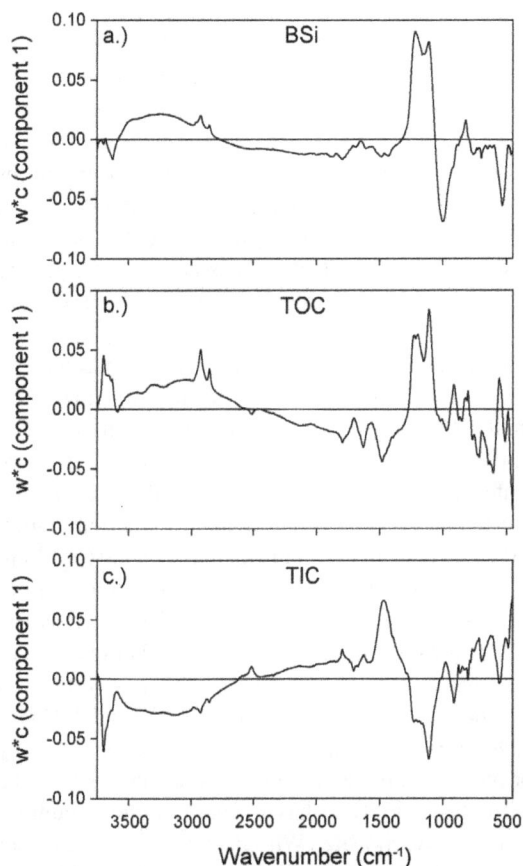

**Fig. 3.** Loadings of the developed FTIRS-calibration models for **(a)** biogenic silica (BSi), **(b)** total organic carbon (TOC), and **(c)** total inorganic carbon (TIC) showing the contribution of each wavenumber to the partial least squares (PLS) regression model. Positive values indicate wavenumbers positively correlated with the property of interest and negative values indicate wavenumbers negatively correlated with the property of interest. Loadings are expressed by weight vectors ($w \times c$) of the first PLS model component ($x$ axis) and the corresponding spectral range ($y$ axis). The calibration models are based on sediment samples from percussion piston cores PG1351 ($n = 83$ for BSi, $n = 156$ for TOC) and Lz1024 ($n = 204$ for BSi) and samples distributed throughout the entire drill core composite profile of ICDP site 5011-1 ($n = 133$ for BSi, $n = 153$ for TOC, $n = 152$ for TIC).

## 3.2   Spectral information

Pronounced absorption bands in the loadings of the FTIRS models can be related to known absorption bands of organic and minerogenic compounds. For the BSi model, the most important spectral regions positively correlated with the BSi concentration are situated within the ranges of 3580–2750, 1320–1060, and 860–790 cm$^{-1}$ (Fig. 3). Previous studies have shown that these regions can be attributed to absorptions caused by molecular vibrations of biogenic silica. Absorptions at around 1100 cm$^{-1}$ are assigned to the asymmetric stretching vibration mode of the [SiO$_4$] tetrahedron,

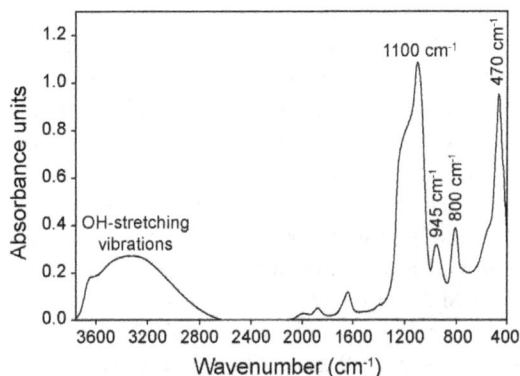

**Fig. 4.** FTIR spectrum of biogenic silica (BSi) from a sediment sample of Lake El'gygytgyn (core Lz1024, sample depth: 1.5 m, sample age: 22 ka) that was purified according to Chapligin et al. (2012). The spectrum shows distinct absorption bands at around 1100, 945, 800 and 471 cm$^{-1}$, which are associated with asymmetric Si-O-Si stretching, Si-OH molecular, symmetric Si-O-Si stretching, and Si-O-Si bending vibrations, respectively.

while the absorption band at around 800 cm$^{-1}$ is caused by symmetric Si-O-Si stretching vibrations (Moenke, 1974b; Patwardhan et al., 2006). A FTIR spectrum of purified BSi from Lake El'gygytgyn sediment shows two additional absorption bands at around 471 and 945 cm$^{-1}$ (Fig. 4), which are not pronounced in the loadings. The band at around 471 cm$^{-1}$ is associated with the bending vibration mode of the [SiO$_4$] tetrahedron and the band at around 945 cm$^{-1}$ is related to Si-OH molecular vibrations (Schmidt et al., 2001; Gendron-Badou et al., 2003). Instead, these spectral regions show a negative correlation to the BSi concentration due to the fact that other silicates (e.g. feldspars, clay minerals) absorb radiation in this part of the IR spectrum as well (Farmer, 1974; Moenke, 1974b). This can result in an anti-correlation to the BSi content, because measurements of concentrations of biogeochemical properties reflect the relative proportion of biogeochemical and minerogenic compounds. The positive loading values in the spectral regions between 3580 and 2750 cm$^{-1}$ can be linked to stretching vibrations of OH molecules (Moenke, 1974a), which are embedded in the molecular structure of BSi. The loadings of our BSi calibration model are consistent with loading plots for calibration models reported by Vogel et al. (2008) and Rosén et al. (2010, 2011).

The most important absorption band in the loadings of the TOC model is situated between 1275 and 1050 cm$^{-1}$ (Fig. 3). Absorption of IR radiation in this region can be linked to known absorption bands of organic compounds. For instance, absorptions between 1265 and 1230 cm$^{-1}$ are attributed to C-O stretching vibrations of ethers or carboxyl groups (Cocozza et al., 2003; Mecozzi and Pietrantonio, 2006). Absorptions at around 1125 cm$^{-1}$ are due to C-O-C stretching vibrations of complex carbohydrates, and absorptions between 1070 and 1040 cm$^{-1}$ are ascribed to C-O stretching vibrations of carbohydrates or polysaccharides (Calace et al., 1999; Chapman et al., 2001; Cocozza et al., 2003). However, the similarity of the loading plots of TOC and BSi indicates an interdependency between both proxies. Sediment samples of Lake El'gygytgyn used for the calibration models are in general characterised by low TOC concentrations (mean: 0.41 %) and high BSi concentrations (mean: 12.1 %). It is likely that the positive loading values from 1275 to 1050 $cm^{-1}$ are therefore attributed to Si-O molecular vibrations of BSi rather than to vibrations caused by organic matter. Positive loading values between 3000 and 2800 $cm^{-1}$ with distinct peaks at 2920 and 2850 $cm^{-1}$ can be attributed to stretching vibrations of C-H molecules in -CH, -$CH_2$ and -$CH_3$ groups of aliphatic and protein chains (Chapman et al., 2001; Cocozza et al., 2003; Mecozzi and Pietrantonio, 2006). Absorption in this part of the IR spectrum is characteristic of organic matter, because an overlap with specific absorption bands of BSi can be excluded. The positive values at around 3700 $cm^{-1}$ and between 3575 and 2560 $cm^{-1}$ are related to absorptions caused by hydroxide groups, which are common in organic matter. However, because an influence of BSi on the calibration model for TOC is likely, it is difficult to determine how much of the absorption is caused by hydroxide contained in organic matter (direct relationship) and how much of the absorption results from hydroxide groups associated with BSi (indirect relationship).

Distinctive positive absorption peaks in the loadings of the TIC calibration model at around 2515, 1795, 875, and 715 $cm^{-1}$ as well as the broad absorption band at around 1460 $cm^{-1}$ (Fig. 3) correspond well with known absorption bands caused by molecular vibrations of carbonates (Huang and Kerr, 1960; Mecozzi et al., 2001). These bands match absorption bands reported for calibration models developed by Vogel et al. (2008) and Rosén et al. (2010, 2011), as well as FTIR spectra of carbonate-rich sediments (Rosén et al., 2010) corroborating the robustness of FTIRS calibration models. The negative loading values between 3730 and 2630 $cm^{-1}$ and between 1270 and 1025 $cm^{-1}$ can partly be explained by an indirect relationship between the occurrence of TIC and BSi. The appearance of carbonates in the analysed samples is almost exclusively restricted to the lowermost part of the sediment record and accompanied by low BSi concentrations, which result in a negative correlation of spectral regions related to molecular vibrations of BSi to the TIC content.

## 3.3 Variability of the biogeochemical properties throughout the last 3.6 million years

The FTIRS-inferred concentrations for BSi and TOC strongly vary throughout the sediment record of Lake El'gygytgyn and range between 1 and 56 % for BSi and between 0 and 2.2 % for TOC (Fig. 5b and c). High BSi concentrations (> 25 %) occur during several periods of lighter values in the global marine $\delta^{18}O$ record (Lisiecki and Raymo,

2005), i.e. in sediments deposited during the marine isotope stages (MIS) KM5, KM3, G17, 101, 93, 87, 77, 57, 55, 49, 31, 11c, and 9c. Highest BSi concentrations (up to 56 %) occur in interglacial MIS 11c sediments, whereas lowest concentrations occur in the basal part of the record, which was deposited prior to 3.4 Ma. Elevated TOC concentrations (> 0.5 %), in contrast, are found in sediments associated with both warm and cold periods throughout the entire sediment sequence. Particularly low TOC values are most frequently inferred for sediments from the basal part of the sediment sequence, which was formed prior to 3.5 Ma.

FTIRS-inferred concentrations of TIC vary between 0 and 5.2 % (Fig. 5d). TIC predominately occurs in samples deposited in the basal part of the sediment record between 3.28 and 3.58 Ma, where it varies strongly. Due to the lack of carbonate-rich samples in the TIC calibration model, FTIRS-inferred concentrations outside of the calibration range (0.01–0.44 %) must be considered with care and are potentially over-/underestimated. Sediments formed within the first $\sim 40$ kyr after lake formation are characterised by TIC concentrations constantly above 0.2 %. Samples accumulated during the past 3.28 Myr generally show variations between 0 and 0.1 %. Due to the RMSECV and RMSEP of 0.03/0.08 % and the limit of detection of approximately 0.05 % for the conventional analysis of TIC, upon which the FTIRS calibration model is based, most of these samples should be free of carbonate. Within this period, only four sample levels deposited during MIS 48, 38, 7d, and 6 contain detectable TIC concentrations.

Because the sedimentation rate changes significantly during the past 3.6 Myr – from 45 cm $kyr^{-1}$ during the period 3.6–3.3 Ma to 4–5 cm $kyr^{-1}$ during the past 3.3 Myr (Nowaczyk et al., 2013) – accumulation rates were calculated to evaluate the palaeobioproductivity and carbonate accumulation in Lake El'gygytgyn. Accumulation rates of BSi ($AR_{BSi}$) and TOC ($AR_{TOC}$) show strong variations throughout the sediment record varying between 1 and 70 and between 0 and 5.2 g $m^{-2}$ $yr^{-1}$, respectively (Fig. 5b and c). The initial sedimentation during the first $\sim 10$ kyr is characterised by relatively high but rapidly decreasing $AR_{BSi}$ from 39 to 13 g $m^{-2}$ $yr^{-1}$ and $AR_{TOC}$ from 0.7 to less than 0.1 g $m^{-2}$ $yr^{-1}$. The high $AR_{BSi}$ at the base of the record does not necessarily reflect a high primary production by diatoms, because sediments deposited after the meteorite impact probably also contain other amorphous silicates formed by the impact (e.g. ashes, glasses). IR absorption caused by these components can have strong similarities to absorption bands of biogenic silica and therefore may lead to poor estimates by the BSi-FTIRS calibration model. The increased $AR_{TOC}$ is most likely an artefact of the higher uncertainty for very low FTIRS-inferred values (RMSECV = 0.09 %) and the high sedimentation rate during this period, in which FTIRS-inferred TOC concentrations do not exceed 0.1 %. Further, it can be assumed that biological activity was strongly impaired during the initial sedimentation in the El'gygytgyn impact

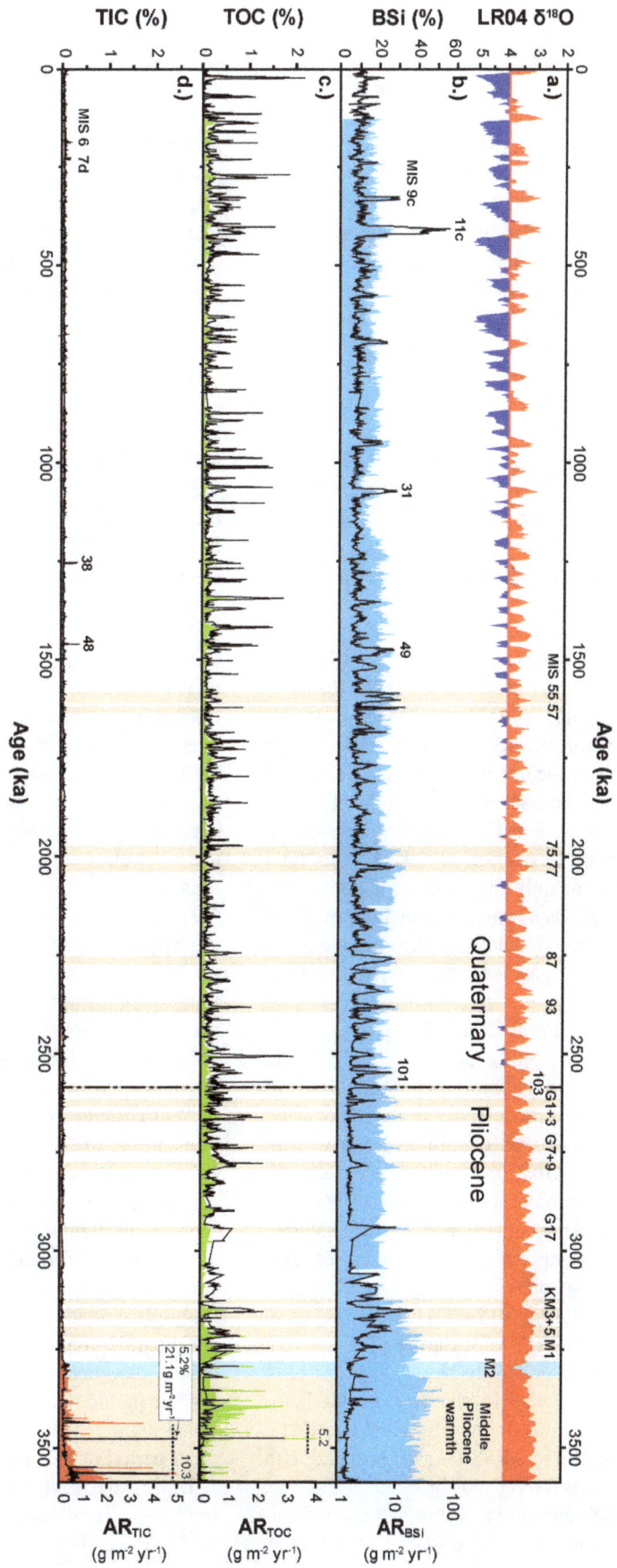

**Fig. 5.** (a) LR04 global marine isotope stack for the past 3.6 Myr (Lisiecki and Raymo, 2005) compared with (b) FTIRS-inferred biogenic silica (BSi) concentrations (line plot) and accumulation rates (ARBSi; area plot; note logarithmic scale), (c) FTIRS-inferred total organic carbon (TOC) concentrations (line plot) and accumulation rates (ARTOC; area plot), and (d) FTIRS-inferred total inorganic carbon (TIC) concentrations (line plot) and accumulation rates (ARTIC; area plot) in the Lake El'gygytgyn sediment record. Periods of high ARBSi are highlighted with orange bars (3.15 to 0.125 Ma: ARBSi > 10 g m$^{-2}$ yr$^{-1}$). Dashed lines indicate breaks in the scale with maximum values plotted above.

structure as a consequence of the meteorite impact. This is corroborated by the almost entire absence of microfossils in the first post-impact sediments (Brigham-Grette et al., 2013).

In Lake El'gygytgyn, aquatic bioproductivity is strongly connected to temperature variations (e.g. Melles et al., 2007). First, temperature controls the light availability via extent and duration of annual lake ice cover. Second, it controls the in-lake recycling of nutrients. Nutrients are transported from the sediment–water interface to the photic zone when the water column is fully mixed during the turnover after ice cover disintegration. And third, temperature also controls the allochthonous nutrient supply, which depends on the rate of chemical weathering, soil-forming processes, and the vegetation cover in the catchment.

The period from 3.58 to 3.32 Ma represents the longest period of constantly high $AR_{BSi}$ at Lake El'gygytgyn, with an average rate of $24 \, g \, m^{-2} \, yr^{-1}$, indicating favourable conditions for bioproductivity (Fig. 5b). This period of high BSi deposition partly coincides with the early Pliocene warm period ($\sim 5$–$3$ Ma), which was characterised by $\sim 2$–$3 \, °C$ higher global surface temperatures compared to today (Haywood and Valdes, 2004; Brierley et al., 2009), an atmospheric $pCO_2$ of $\sim 400$ ppmv (Pagani et al., 2010; Seki et al., 2010), reduced high-latitude ice sheet extents (Hill et al., 2010; McKay et al., 2012), and a global sea level 10–40 m above present (Raymo et al., 2011). Pollen-based temperature and precipitation reconstructions from the El'gygytgyn sediment record suggest regional temperatures of 7–8 °C and precipitation values of around $400 \, mm \, yr^{-1}$ higher than today for the period between 3.6 and 3.4 Ma (Brigham-Grette et al., 2013). Furthermore, this period at Lake El'gygytgyn is characterised by a significantly larger size of the dominant planktonic diatom (*Pliocaenicus* sp.) compared to any other interval in the sediment record, indicating increased nutrient availability and reduced seasonal ice cover (Brigham-Grette et al., 2013).

The productivity maximum at Lake El'gygytgyn is followed by a period of lower $AR_{BSi}$ from 3.32 to 3.28 Ma (MIS M2), in which the average accumulation rate drops to $12 \, g \, m^{-2} \, yr^{-1}$. A reduced bioproductivity during this period is consistent with cooler and dryer conditions inferred from the pollen record (Brigham-Grette et al., 2013). Marine sediment records suggest a coincident cooling of sea surface temperatures in the North Atlantic between 2–3 °C (De Schepper et al., 2009) and 6 °C (Lawrence et al., 2009) and a sea level lowstand of $\sim 65$ m below present (Dwyer and Chandler, 2009). After the cooling during MIS M2, $AR_{BSi}$ rise again, reaching rates of up to $32 \, g \, m^{-2} \, yr^{-1}$ during MIS M1, but subsequently decrease to rates mostly below $10 \, g \, m^{-2} \, yr^{-1}$ in the period between 3.28 and 3.15 Ma. This decline is interrupted by two $AR_{BSi}$ maxima during MIS KM5 and KM3. These isotope stages as well as MIS M1 are associated with global sea level highstands (Dwyer and Chandler, 2009). The stepwise decrease in BSi deposition and thus the deterioration of environmental conditions for in-lake bioproduction

from 3.32 Ma onwards is consistent with the onset of ice sheet expansion and cooling of coastal waters in Antarctica at about 3.3 Ma (McKay et al., 2012) and with decreasing $pCO_2$ concentrations from 3.2 Ma on (Seki et al., 2010). The fluctuations in bioproductivity between 3.6 and 3.15 Ma as deduced from $AR_{BSi}$ variations are well supported by variations in TOC accumulation, with peaks in $AR_{TOC}$ being largely correlated with periods of lighter values in the global marine $\delta^{18}O$ record and increased $AR_{BSi}$ (Fig. 5c).

Compared to the bioproductivity maximum during the middle Pliocene and the high but decreasing $AR_{BSi}$ from 3.3 Ma onwards, accumulation rates of BSi are considerably lower during the late Pliocene and Quaternary (Fig. 5b). From 3.15 to 1.8 Ma, the average BSi accumulation rate amounts to $5.2 \, g \, m^{-2} \, yr^{-1}$, but with peak values ($AR_{BSi} > 10 \, g \, m^{-2} \, yr^{-1}$) occurring during the warm phases MIS G17, G9, G7, G3, G1, 103, 93, 87, 77, and 75. According to pollen-based temperature and precipitation reconstruction from Lake El'gygytgyn, the period 3.26 to 2.2 Ma can be described as a protracted period of warm and moist conditions, approximately 3–6 °C warmer and between 100 and 200 mm wetter than today (Melles et al., 2012; Brigham-Grette et al., 2013). However, a gradual climatic deterioration is indicated by the increased occurrence of summers cooler than today from 2.5 Ma and the gradually increasing frequency of glacial periods between 2.3 and 1.8 Ma.

BSi fluxes decrease further during the period from 1.8 to 0.125 Ma, when glacial–interglacial cycles are well expressed at Lake El'gygytgyn (Melles et al., 2012), suggesting a further long-term deterioration of environmental and climatic conditions for bioproductivity in the Siberian Arctic. This period is characterised by average $AR_{BSi}$ of $3.9 \, g \, m^{-2} \, yr^{-1}$, and $AR_{BSi} > 10 \, g \, m^{-2} \, yr^{-1}$ only occur during MIS 57 and 55. From 1.5 Ma onwards, maximum $AR_{BSi}$ no longer exceed $8.5 \, g \, m^{-2} \, yr^{-1}$. Interestingly, accumulation rates during the so-called "super interglacials" MIS 31 and 11c do not exhibit outstanding high $AR_{BSi}$, despite pollen-based climate reconstructions suggesting conditions up to 4–5 °C warmer and $\sim 300$ mm wetter compared to present (Melles et al., 2012). Although the BSi concentrations during MIS 11c are by far the highest in the entire Lake El'gygytgyn sediment record, $AR_{BSi}$ are not comparatively elevated during this period. This may be best explained by an intensified catchment stabilisation, e.g. by a dense vegetation cover and soil formation, which has considerably reduced the supply of mineral matter from the catchment. A detailed multi-proxy study focusing on MIS 11c at Lake El'gygytgyn, provided by Vogel et al. (2013), shows similarities to the last deglaciation and interhemispheric climate connectivity during this period. However, the processes leading to the exceptional composition of MIS 11c sediments compared to any other period of the sediment record remain ambiguous.

During the past 3.15 Myr, elevated $AR_{TOC}$ are increasingly associated with heavier values in the global marine $\delta^{18}O$ record (Fig. 5c), reflecting the increase in frequency of glacial periods during the Quaternary. In the Lake El'gygytgyn sediments, the TOC amount is strongly connected to the duration of lake ice cover (Melles et al., 2007). During glacial times with perennial ice cover, which requires mean annual temperatures of at least 3.3 ($\pm$0.9) °C below today (Nolan, 2013), mixing of the water column is hampered, leading to oxygen-depleted bottom waters and enhanced organic matter preservation in the sediment. During warmer periods, in contrast, the annual ice cover disintegration allows for wind- and density-driven mixing that leads to an oxygenation of the bottom water and enhanced organic matter decomposition. However, elevated $AR_{TOC}$ also occur during particularly warm interglacials, when the bioproductivity is very high and the organic matter decomposition incomplete (Melles et al., 2007, 2012). This is particularly the case for MIS G9, G7, and G3.

The accumulation rates of TIC ($AR_{TIC}$) confirm the pattern already shown by the TIC concentrations with high carbonate values at the core base, a successive decrease to carbonate absence by 3.28 Ma, and four single occurrences during MIS 48, 38, 7d, and 6 (Fig. 5d). The significant concentration and accumulation of carbonate in the basal sediments might be connected to the origin of the lake depression by the meteorite impact 3.58 $\pm$ 0.04 Ma (Layer, 2000). The impact may have induced hydrothermal activity, which is common after impact events, where the melted/heated target material acting as a heat source interacts with near-surface $H_2O$ (Naumov, 2002). Amongst others, calcite can be precipitated in such a hydrothermal system (Osinski et al., 2005). This process is also indicated by the existence of secondary calcite in the underlaying impact breccia and bedrock (Raschke et al., 2013). Therefore, we assume that carbonate deposition during the early stage of Lake El'gygytgyn is related to erosion and dissolution of secondary/hydrothermal carbonates in the catchment and redeposition and reprecipitation in the lake.

The four glacials/stadials during the Quaternary with exceptional carbonate deposition (MIS 48, 38, 7d, and 6) show very low BSi (means: 5.1–6.4 %) and medium TOC concentrations (means: 0.39–0.78 %). According to the sediment classification after Melles et al. (2007, 2012), these periods represent peak cold glacial conditions with permanent lake ice cover and anoxic bottom waters. Sediments of this climate mode are further characterised by elevated but low total sulphur concentrations due to sulphate reduction and by enhanced bacterial methanogenesis (Melles et al., 2007). Under anoxic/reducing conditions, carbonate phases like rhodochrosite ($MnCO_3$) and siderite ($FeCO_3$) can form as a result of microbially mediated organic matter oxidation and Mn and Fe oxyhydroxide reduction. The occurrence of siderite is restricted to non-sulphidic, methanic environments, whereas rhodochrosite occurs in both sulphidic and non-sulphidic environments (Glasby and Schulz, 1999, and references therein). Due to the low sedimentation rate and the overall low organic content in the Lake

El'gygytgyn sediment record, formation of these carbonates types appears to be exceptional. Further studies are required to confirm the carbonate type and to better understand the processes leading to carbonate accumulation during these specific glacials/stadials.

## 4   Conclusions

The results of our study for the first time demonstrate that robust FTIRS calibration models can be developed for the quantitative assessment of biogeochemical properties in very long (> 300 m) sediment records extending several million years back in time. The developed models relating FTIR spectral information to conventional measurements for TOC, TIC and BSi show a good statistical performance for the 3.6 Myr-old sediment record of Lake El'gygytgyn.

By applying these models in higher resolution to the sediment sequence, we obtained a detailed reconstruction of variations in the aquatic bioproductivity in Lake El'gygytgyn, which is primarily triggered by climate changes. Our results show a productivity maximum during the middle Pliocene (3.6–3.3 Ma) indicated by very high accumulation rates of BSi. After a stepwise drop of in-lake bioproductivity between 3.3 and 3.15 Ma, $AR_{BSi}$ gradually decreased during the past 3.15 Myr. The indicated progressive deterioration of environmental and climatic conditions in the Siberian Arctic coincides with the increased occurrence of glacial periods culminating in the full establishment of high-amplitude glacial–interglacial cycles during the Quaternary. Compared to average BSi accumulation rates during the middle Pliocene, maximum BSi accumulation rates are 1.4 and 2.8 times lower during the period 3.15–1.5 Ma and 1.5–0.125 Ma, respectively. The agreement of the $AR_{BSi}$ with general climate patterns highlights the climate sensitivity of BSi at Lake El'gygytgyn. Furthermore, the occurrence of carbonates, predominately in basal sediments of the record (3.6–3.3 Ma), could be shown by means of FTIRS.

Our study demonstrates that FTIRS is a fast and cost-effective analytical alternative to conventional methods for the quantitative estimation of biogeochemical properties – such as BSi, TOC, or TIC – that allows the determination of several proxies from one single measurement. Its successful application to a record extending several million years back in time corroborates the potential of the technique for other deep-drilling projects dealing with long lacustrine or marine sediment successions.

*Acknowledgements.* The El'gygytgyn Drilling Project was funded by the International Continental Scientific Drilling Program (ICDP), the US National Science Foundation (NSF), the German Federal Ministry of Education and Research (BMBF), Alfred Wegener Institute (AWI) and GeoForschungsZentrum Potsdam (GFZ), the Russian Academy of Sciences Far East Branch (RAS FEB), the Russian Foundation of Basic Research (RFBR), and the

Austrian Federal Ministry of Science and Research (BMWF). The Russian GLAD 800 drilling system was developed and operated by DOSECC Inc., and LacCore at the University of Minnesota handled core curation. We would like to thank Per Persson (Umeå University) for providing access to FTIR equipment, Nicole Mantke (University of Cologne) for laboratory assistance, and Annika Holmgren, Carin Olofsson, and Cecilia Rydberg (Umeå University) for FTIRS assistance. Funding for the FTIRS research was provided by the Swedish Research Council (VR), FORMAS, and the Kempe Foundation.

Edited by: P. Minyuk

# References

Belzile, N., Joly, H. A., and Li, H.: Characterization of humic substances extracted from Canadian lake sediments, Can. J. Chem., 75, 14–27, 1997.

Bertaux, J., Ledru, M. P., Soubiès, F., and Sondag, F.: The use of quantitative mineralogy linked to palynological studies in paleoenvironmental reconstruction: the case study of the "Lagoa Campestre" lake, salitre, Minas Gerais, Brazil, C. R. Acad. Sci. Paris, 323, 65–71, 1996.

Bertaux, J., Fröhlich, F., and Ildefonse, P.: Multicomponent analysis of FTIR spectra: quantification of amorphous and crystallized mineral phases in synthetic and natural sediments, J. Sediment. Res., 68, 440–447, 1998.

Braguglia, C. M., Campanella, L., Petronio, B. M., and Scerbo, R.: Sedimentary humic acids in the continental margin of the Ross Sea (Antarctica), Int. J. Environ. Anal. Chem., 60, 61–70, 1995.

Brierley, C. M., Fedorov, A. V., Liu, Z., Herbert, T. D., Lawrence, K. T., and LaRiviere, J. P.: Greatly expanded tropical warm pool and weakened Hadley circulation in the early Pliocene, Science, 323, 1714–1718, 2009.

Brigham-Grette, J., Melles, M., Minyuk, P., Andreev, A., Tarasov, P., DeConto, R., Koenig, S., Nowaczyk, N., Wennrich, V., Rosén, P., Haltia-Hovi, E., Cook, T., Gebhardt, C., Meyer-Jacob, C., Snyder, J., and Herzschuh, U.: Pliocene Warmth, Polar Amplification, and Stepped Pleistocene Cooling recorded in NE Arctic Russia, Science, 340, 1421–1427, 2013.

Calace, N., Capolei, M., Lucchese, M., and Petronio, B. M.: The structural composition of humic compounds as indicator of organic carbon sources, Talanta, 49, 277–284, 1999.

Calace, N., Cardellicchio, N., Petronio, B. M., Pietrantonio, M., and Pietroletti, M.: Sedimentary humic substances in the Northern Adriatic Sea (Mediterranean Sea), Mar. Environ. Res., 61, 40–58, 2006.

Chapligin, B., Meyer, H., Bryan, A., Snyder, J., and Kemnitz, H.: Assessment of purification and contamination correction methods for analysing the oxygen isotope composition from biogenic silica, Chem. Geol., 300–301, 185–199, 2012.

Chapman, S. J., Campbell, C. D., Fraser, A. R., and Puri, G.: FTIR spectroscopy of peat in and bordering Scots pine woodland: relationship with chemical and biological properties, Soil Biol. Biochem., 33, 1193–1200, 2001.

Cocozza, C., D'Orazio, V., Miano, T. M., and Shotyk, W.: Characterization of solid and aqueous phases of a peat bog profile using molecular fluorescence spectroscopy, ESR and FT-IR, and comparison with physical properties, Org. Geochem., 34, 49–60, 2003.

De Schepper, S., Head, M. J., and Groeneveld, J.: North Atlantic Current variability through marine isotope stage M2 (circa 3.3 Ma) during the mid-Pliocene, Paleoceanography, 24, PA4206, doi:10.1029/2008PA001725, 2009.

Dwyer, G. S. and Chandler, M. A.: Mid-Pliocene sea level and continental ice volume based on coupled benthic Mg/Ca paleotemperatures and oxygen isotopes, Philos. T. Roy. Soc. A, 367, 157–168, 2009.

Eriksson, L., Johansson, E., Kettaneh-Wold, N., Trygg, J., Wikström, C., and Wold, S.: Multi- and megavariate data analysis – Part I: Basic principles and applications, 2nd Edn., Umetrics AB, Umeå, Sweden, 2006.

Eriksson, L., Trygg, J., and Wold, S.: CV-ANOVA for significance testing of PLS and OPLS$^{\circledR}$ models, J. Chemometr., 22, 594–600, 2008.

Farmer, V. C.: The layer silicates, in: The Infrared Spectra of Minerals, edited by: Farmer, V. C., Mineralogical Society Monograph 4, Adlard & Son, Dorking, Surrey, 331–363, 1974.

Fuji, N.: Palaeovegetation and palaeoclimate changes around Lake Biwa, Japan during the last ca. 3 million years, Quaternary Sci. Rev., 7, 21–28, 1988.

Geladi, P., MacDougall, D., and Martens, H.: Linearization and scatter-correction for near-infrared reflectance spectra of meat, Appl. Spectrosc., 39, 491–500, 1985.

Gendron-Badou, A., Coradin, T., Maquet, J., Fröhlich, F., and Livage, J.: Spectroscopic characterization of biogenic silica, J. Non-Cryst. Solids, 316, 331–337, 2003.

Glasby, G. P. and Schultz, H. D.: $E_H$, pH diagrams for Mn, Fe, Co, Ni, Cu and As under seawater conditions: applications of two new types of $E_H$, pH diagrams to the study of specific problems in marine geochemistry, Aquatic Geochem., 5, 227–248, 1999.

Griffiths, P. R. and de Haseth, J. A.: Fourier transform infrared spectroscopy, Wiley, New York, 2007.

Haywood, A. M. and Vales, P. J.: Modelling Pliocene warmth: contribution of atmosphere, oceans and cryosphere, Earth Planet. Sc. Lett., 218, 363–377, 2004.

Herbert, T. D., Brian, A. D., and Burnett, C.: Precise major component determination in deep-sea sediments using Fourier transform infrared spectroscopy, Geochim. Cosmochim. Acta, 56, 1759–1763, 1992.

Hill, D. J., Dolan, A. M., Haywood, A. M., Hunter, S. J., and Stoll, D. K.: Sensitivity of the Greenland Ice Sheet to Pliocene sea surface temperatures, Stratigraphy, 7, 111–121, 2010.

Huang, C. K. and Kerr, P. F.: Infrared study of the carbonate minerals, Am. Mineral., 45, 311–324, 1960.

IPCC: Climate change 2007: The physical science basis, in: Contribution of Working Group I to the Fourth Assessment Report of the Intergovernmental Panel on Climate Change, edited by: Solomon, S., Qin, D., Manning, M., Chen, Z., Marquis, M., Averyt, K. B., Tignor, M., and Miller, H. L., Cambridge University Press, Cambridge and New York, 2007.

Juschus, O., Wennrich, V., Quart, S., Minyuk, P., Melles, M., Gebhardt, C., and Niessen, F.: New long record Lz1024, in: The expedition El'gygytgyn lake 2003 (Siberian Arctic), edited by: Melles, M., Minyuk, P., Brigham-Grette, J., and Juschus, O., Reports Polar Mar. Res., 509, 110–113, 2005.

Lawrence, K. T., Herbert, T. D., Brown, C. M., Raymo, M. E., and Haywood, A. M.: High-amplitude variations in North Atlantic sea surface temperature during the early Pliocene warm period, Paleoceanography, 24, PA2218, doi:10.1029/2008PA001669, 2009.

Layer, P. W.: Argon-40/argon-39 age of the El'gygytgyn impact event, Chukotk, Russia, Meteorit. Planet. Sci., 35, 591–599, 2000.

Lisiecki, L. E. and Raymo, M. E.: A Pliocene-Pleistocene stack of 57 globally distributed benthic $\delta^{18}O$ records, Paleoceanography, 20, PA1003, doi:10.1029/2004PA001071, 2005.

Martens, H. and Næs, T.: Multivariate Calibration, John Wiley & Sons, Chichester, New York, Brisbane, Toronto, Singapore, 1989.

McKay, R., Naish, T., Carter, L., Riesselman, C., Dunbar, R., Sjunneskog, C., Winter, D., Sangiorgi, F., Warren, C., Pagani, M., Schouten, S., Willmott, V., Levy, R., DeConto, R., and Powell, R. D.: Antarctic and Southern Ocean influences on late Pliocene global cooling, P. Natl. Acad. Sci. USA, 109, 6423–6428, 2012.

Mecozzi, M. and Pietrantonio, E.: Carbohydrates proteins and lipids in fulvic and humic acids of sediments and its relationships with mucilanginous aggregates in the Italian seas, Mar. Chem., 101, 27–39, 2006.

Mecozzi, M., Pietrantonio, E., Amici, M., and Romanelli, G.: Determination of carbonate in marine solid samples by FTIR-ATR spectroscopy, Analyst, 126, 144–146, 2001.

Melles, M., Brigham-Grette, J., Glushkova, O. Y., Minyuk, P. S., Nowacyk, N. R., and Hubberten, H.-W.: Sedimentary geochemistry of core PG1351 from Lake El'gygytgyn – a sensitive record of climate variability in the East Siberian Arctic during the past three glacial-interglacial cycles, J. Paleolimnol., 37, 89–104, 2007.

Melles, M., Brigham-Grette, J., Minyuk, P., Koeberl, C., Andreev, A., Cook, T., Federov, G., Gebhardt, C., Haltia-Hovi, E., Kukkonen, M., Nowaczyk, N., Schwamborn, G., Wennrich, V., and the El'gygytgyn Scientific Party: the El'gygytgyn Scientific Drilling Project – conquering Arctic challenges through continental drilling, Sci. Drill., 11, 29–40, 2011.

Melles, M., Brigham-Grette, J., Minyuk, P. S., Nowaczyk, N. R., Wennrich, V., DeConto, R. M., Anderson, A., Andreev, A. A., Coletti, A., Cook, T. L., Haltia-Hovi, E., Kukkonen, M., Lozhkin, A. V., Rosén, P., Tarasov, P., Vogel, H., and Wagner, B.: 2.8 million years of Arctic climate change from Lake El'gygytgyn, NE Russia, Science, 337, 315–320, 2012.

Moenke, H. H. W.: Vibrational spectra and the crystal-chemical classification of minerals, in: The Infrared Spectra of Minerals, edited by: Farmer, V. C., Mineralogical Society Monograph 4, Adlard & Son, Dorking, Surrey, 111–118, 1974a.

Moenke, H. H. W.: Silica, the three-dimensional silicates, borosilicates and beryllium silicates, in: The Infrared Spectra of Minerals, edited by: Farmer, V. C., Mineralogical Society Monograph 4, Adlard & Son, Dorking, Surrey, 365–382, 1974b.

Müller, P. J. and Schneider, J.: An automated leaching method for the determination of opal in sediments and particulate matter, Deep-Sea Res., 40, 424–444, 1993.

Naumov, M. V.: Impact-generated hydrothermal systems: data from Popigai, Kara, and Puchezh–Katunki impact structures, in: Impacts in Precambrian Shields, edited by: Plado, J. and Pesonen, L. J., Springer-Verlag, Berlin, 117–171, 2002.

Nolan, M.: Quantitative and qualitative constraints on hind-casting the formation of multiyear lake-ice covers at Lake El'gygytgyn, Clim. Past, 9, 1253–1269, doi:10.5194/cp-9-1253-2013, 2013.

Nowaczyk, N. R., Haltia, E. M., Ulbricht, D., Wennrich, V., Sauerbrey, M. A., Rosén, P., Vogel, H., Francke, A., Meyer-Jacob, C., Andreev, A. A., and Lozhkin, A. V.: Chronology of Lake El'gygytgyn sediments – a combined magnetostratigraphic, palaeoclimatic and orbital tuning study based on multi-parameter analyses, Clim. Past, 9, 2413-2432, 2013, http://www.clim-past.net/9/2413/2013/.

Osinski, G. R., Lee, P., Parnell, J., Spray, J. G., and Baron, M.: A case study of impact-induced hydrothermal activity: the Haughton impact structure, Devon Island, Canadian High Arctic, Meteorit. Planet. Sci., 40, 1859–1877, 2005.

Pagani, M., Liu, Z., LaRiviere, J., and Ravelo, A. C.: High earth-system climate sensitivity determined from Pliocene carbon dioxide concentrations, Nat. Geosci., 3, 27–30, 2010.

Patwardhan, S. V., Maheshwari, R., Mukherjee, N., and Clarson, S. J.: Conformation and assembly of polypeptide scaffolds in templating the synthesis of silica: an example of a polylysine macromolecular "switch", Biomacromolecules, 7, 491–497, 2006.

Raschke, U., Schmitt, R. T., and Reimold, W. U.: Petrography and geochemistry of impactites and volcanic bedrock in the ICDP drill core D1c from Lake El'gygytgyn, NE Russia, Meteorit. Planet. Sci., 48, 1251–1286, 2013.

Raymo, M. E., Mitrovica, J. X., O'Leary, M. J., DeConto, R. M., and Hearty, P. J.: Departures from eustasy in Pliocene sea-level records, Nat. Geosci., 4, 328–332, 2011.

Rosén, P., Vogel, H., Cunningham, L., Reuss, N., Conley, D. J., and Persson, P.: Fourier transform infrared spectroscopy, a new method for rapid determination of total organic and inorganic and biogenic silica concentration in lake sediments, J. Paleolimnol., 43, 247–259, 2010.

Rosén, P., Vogel, H., Cunningham, L., Hahn, A., Hausmann, S., Pienitz, R., Zolitschka, B., Wagner, B., and Persson, P.: Universally applicable model for the quantitative determination of lake sediment composition using Fourier transform infrared spectroscopy, Environ. Sci. Technol., 45, 8858–8865, 2011.

Schmidt, M., Botz, R., Rickert, D., Bohrmann, G., Hall, S. R., and Mann, S.: Oxygen isotope of marine diatoms and relations to opal-A maturation, Geochim. Cosmochim. Acta, 65, 201–211, 2001.

Scholz, C. A., Cohen, A. S., Johnson, T. C., King, J. W., and Moran, K.: The 2005 Lake Malawi drilling project, Sci. Drill., 2, 17–19, 2006.

Seki, O., Foster, G. L., Schmidt, D. N., Mackensen, A., Kawamura, K., and Pancost, R. D.: Alkenone and boron-based Pliocene $p\text{CO}_2$ records, Earth Planet. Sc. Lett., 292, 201–211, 2010.

Sifeddine, A., Fröhlich, F., Fournier, M., Martin, L., Servant, M., Soubiès, F., Turcq, B., Suguio, K., and Volkmer-Ribeiro, C.: La sedimentation lacustre indicateur de changements des paléoenvironnements au cours des 30 000 dernières années (Carajasm Amazonie, Brésil), C. R. Acad. Sci. Paris, 318, 1645–1652, 1994.

Vogel, H., Rosén, P., Wagner, B., Melles, M., and Persson, P.: Fourier transform infrared spectroscopy, a new cost-effective tool for qualitative analysis of biogeochemical properties in long sediment records, J. Paleolimnol., 40, 689–702, 2008.

Vogel, H., Meyer-Jacob, C., Melles, M., Brigham-Grette, J., Andreev, A. A., Wennrich, V., Tarasov, P. E., and Rosén, P.: Detailed insight into Arctic climatic variability during MIS 11c at Lake El'gygytgyn, NE Russia, Clim. Past, 9, 1467-1479, doi:10.5194/cp-9-1467-2013, 2013.

Williams, D. F., Peck, J., Karabanov, E. B., Prokopenko, A. A., Kravchinsky, V., King, J., and Kuzmin, M. I.: Lake Baikal record of continental climate response to orbital insolation during the past 5 million years, Science, 278, 1114–1117, 1997.

Wirrmann, D., Bertaux, J., and Kossoni, A.: Late Holocene paleoclimatic changes in Western Central Africa inferred from mineral abundance in dated sediments from Lake Ossa (Southwest Cameroon), Quaternary Res., 56, 275–287, 2001.

# 3

# Salinity changes in the Agulhas leakage area recorded by stable hydrogen isotopes of $C_{37}$ alkenones during Termination I and II

**S. Kasper[1], M. T. J. van der Meer[1], A. Mets[1], R. Zahn[2,3], J. S. Sinninghe Damsté[1], and S. Schouten[1]**

[1]NIOZ Royal Netherlands Institute for Sea Research, Department of Marine Organic Biogeochemistry, P.O. Box 59, 1790 AB Den Burg (Texel), the Netherlands
[2]Institució Catalana de Recerca i Estudis Avançats, ICREA, Barcelona, Spain
[3]Universitat Autònoma de Barcelona, Institut de Ciència i Tecnologia Ambientals (ICTA) and Departament de Física, 08193 Bellaterra, Spain

*Correspondence to:* S. Kasper (sebastian.kasper@nioz.nl)

**Abstract.** At the southern tip of Africa, the Agulhas Current reflects back into the Indian Ocean causing so-called "Agulhas rings" to spin off and release relatively warm and saline water into the South Atlantic Ocean. Previous reconstructions of the dynamics of the Agulhas Current, based on paleo-sea surface temperature and sea surface salinity proxies, inferred that Agulhas leakage from the Indian Ocean to the South Atlantic was reduced during glacial stages as a consequence of shifted wind fields and a northwards migration of the subtropical front. Subsequently, this might have led to a buildup of warm saline water in the southern Indian Ocean. To investigate this latter hypothesis, we reconstructed sea surface salinity changes using alkenone $\delta D$, and paleo-sea surface temperature using $TEX_{86}^H$ and $U_{37}^{K'}$, from two sediment cores (MD02-2594, MD96-2080) located in the Agulhas leakage area during Termination I and II. Both $U_{37}^{K'}$ and $TEX_{86}^H$ temperature reconstructions indicate an abrupt warming during the glacial terminations, while a shift to more negative $\delta D_{alkenone}$ values of approximately 14‰ during glacial Termination I and II is also observed. Approximately half of the isotopic shift can be attributed to the change in global ice volume, while the residual isotopic shift is attributed to changes in salinity, suggesting relatively high salinities at the core sites during glacials, with subsequent freshening during glacial terminations. Approximate estimations suggest that $\delta D_{alkenone}$ represents a salinity change of ca. 1.7–1.9 during Termination I and Termination II. These estimations are in good agreement with the proposed changes in salinity derived from previously reported combined planktonic Foraminifera $\delta^{18}O$ values and Mg/Ca-based temperature reconstructions. Our results confirm that the $\delta D$ of alkenones is a potentially suitable tool to reconstruct salinity changes independent of planktonic Foraminifera $\delta^{18}O$.

# 1 Introduction

Approximately 2–15 Sv of warm and saline Indian Ocean water is annually released into the South Atlantic Ocean by the Agulhas Current, an ocean current system that is confined by the subtropical front (STF) and the southern African coast (Lutjeharms, 2006). The Agulhas Current is fed by warm, saline Indian Ocean water from two sources: the Mozambique Channel, between Madagascar and the East African coast, and the East Madagascar Current, which merges with the Mozambique Channel flow at approximately 28° S (Penven et al., 2006). When this warm, saline water reaches the Agulhas corridor at the tip of Africa, the vast majority is transported back into the Indian Ocean via the Agulhas return current (Fig. 1). However, between five and seven rings of warm, saline water are released into the Atlantic Ocean per year, termed Agulhas leakage (Lutjeharms, 2006). These Agulhas rings of Indian Ocean waters have been shown to play an important role in the heat and salt budget in the Atlantic Ocean, thereby impacting the Atlantic Meridional Overturning Circulation (AMOC) (Peeters et al., 2004; Biastoch et al., 2008; Bard and Rickaby, 2009; Haarsma et al., 2011; van Sebille et al., 2011).

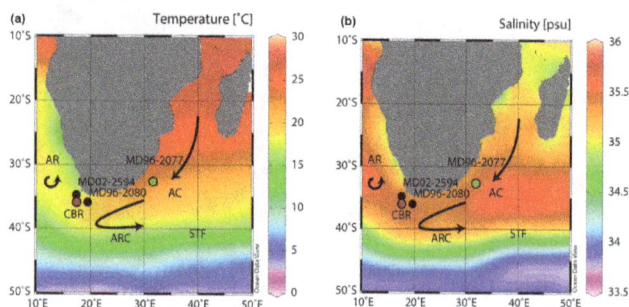

**Fig. 1.** Location of the cores MD02-2594 and MD96-2080 (black dots) and reference site Cape Basin record (CBR, red dot) (Peeters et al., 2004), MD96-2077 (green dot) (Bard and Rickaby, 2009) and oceanographic setting on (**a**) a map of modern sea surface temperatures and (**b**) a map of modern sea surface salinity. Agulhas Current (AC), Agulhas return current (ARC), Agulhas rings (AR) and subtropical front (STF). The underlying maps of modern sea surface temperatures and salinity were compiled with high-resolution CTD data from http://www.nodc.noaa.gov and the Ocean Data View software version 4.3.7 by Schlitzer, R., Ocean Data View (http://odv.awi.de), 2010.

The magnitude of Agulhas leakage into the Atlantic Ocean depends on the strength of the Agulhas Current as well as the position of the retroflection (Lutjeharms, 2006). However, the effect of Agulhas Current strength on the Agulhas leakage efficiency is still debated. For instance, Rouault et al. (2009) suggested that, based on recent temperature observations and modeling experiments, increased Agulhas leakage of warm and saline waters into the South Atlantic Ocean can be associated with increased Agulhas Current transport, while modeling experiments performed by van Sebille et al. (2009) suggested increased Agulhas leakage to be associated with a weakened Agulhas Current.

Previous studies have shown that during glacial stages a weakened and more variable Agulhas Current occurs together with reduced Agulhas leakage (Peeters et al., 2004; Franzese et al., 2006). Peeters et al. (2004) found relatively low contributions of "Agulhas leakage fauna" in the Cape Basin, suggesting a reduced Agulhas leakage during glacial stages (Rau et al., 2002; Peeters et al., 2004), coinciding with low sea surface temperatures (SSTs). This suggests a restriction in the Agulhas leakage during cold periods (Peeters et al., 2004). Furthermore, deep-ocean stable carbon isotope gradients have been applied in combination with sea surface temperature reconstructions as indicators of a connection between deep-water ventilation and Agulhas leakage strength (Bard and Rickaby, 2009). The results demonstrate that a reduced leakage typically correlates with reduced deep ventilation (Bard and Rickaby, 2009).

A northward shift of the STF and eastward forcing of the retroflection during glacial periods may have led to an increased back transport of warm, saline water into the Indian Ocean during glacial periods (Peeters et al., 2004).

Martinez-Mendez et al. (2010) showed increased sea water oxygen isotope ($\delta^{18}O_{sw}$) values, derived from paired planktonic Foraminifera $\delta^{18}O$ and Mg/Ca analysis of *Globigerina bulloides*, in the Agulhas leakage area throughout marine isotope stage 6 (MIS6) and marine isotope stage 3 (MIS3) and early marine isotope stage 2 (MIS2). These elevated $\delta^{18}O_{sw}$ values are likely indicative for increased salinity (Martinez-Mendez et al., 2010). However, it should be noted that salinity reconstructions, based on planktonic Foraminifera $\delta^{18}O_{sw}$ values, carry some uncertainties that are difficult to constrain, e.g., assumed constancy for the transfer functions of $\delta^{18}O_{sw}$ to salinity over space and time (Rohling and Bigg, 1998; Rohling, 2000).

Martinez-Mendez et al. (2010) further reported that reconstructed SST, derived from the planktonic Foraminifera Mg/Ca of *G. bulloides*, displayed a gradual warming trend starting in the early MIS6 and MIS2 (Martinez-Mendez et al., 2010). This is, however, in contradiction with temperature reconstructions based on $U_{37}^{K'}$ paleothermometry (Peeters et al., 2004; Martinez-Mendez et al., 2010), which showed cooler sea surface temperatures during glacial periods, followed by a rapid warming at the onset of the interglacial stages. These differences may be related to uncertainties associated with the different temperature proxies (Bard, 2001). Planktonic foraminiferal Mg/Ca ratios have been shown to reflect not only temperature, but also salinity (Ferguson et al., 2008; Arbuszewski et al., 2010, Hönisch et al., 2013). This has been demonstrated in high-salinity environments such as the Mediterranean Sea (Ferguson et al., 2008), and in open ocean settings such as the tropical Atlantic Ocean (Arbuszewski et al., 2010). Furthermore, $U_{37}^{K'}$–SST relationships are derived from photosynthetic haptophyte algae with different growth seasons and (depth) habitats than the planktonic Foraminifera *G. bulloides*, potentially recording different temperature ranges (Prahl and Wakeham, 1987; Bard, 2001).

We use the hydrogen isotope composition of the combined $C_{37:2-3}$ alkenones ($\delta D_{alkenone}$), produced by haptophyte algae, as a proxy for relative changes in sea surface salinity (SSS). Culture experiments for two common open ocean haptophyte species – *Emiliania huxleyi* and *Gephyrocapsa oceanica* – have shown that the hydrogen isotope composition of alkenones is mainly dependent on salinity, the hydrogen isotope composition of the growth media and, to a lesser extent, growth rate (Englebrecht and Sachs, 2005; Schouten et al., 2006). Furthermore, van der Meer et al. (2013) showed that measuring the combined $C_{37:2-3}$ $\delta D_{alkenone}$ rather than separated $C_{37:2}$ and $C_{37:3}$ alkenones yields a more robust water $\delta D$ and salinity signal, possibly by reducing biosynthetic effects related to the synthesis of the $C_{37:3}$ alkenones from the $C_{37:2}$ (Rontani et al., 2006). Application of the hydrogen isotope composition of alkenones has resulted in reasonable salinity reconstructions for the eastern Mediterranean and Black Sea (van der Meer et al., 2007, 2008) and hydrological reconstructions in the Panama Basin (Pahnke et al., 2007). Here, we apply the $\delta D_{alkenone}$ to estimate relative salinities of the Agulhas system focusing on Termination I and Termination II using the same cores used by Martinez-Mendez et al. (2010), situated in the Agulhas leakage area, off the coast of South Africa (Fig. 1). In order to assess the effect of growth rates on $\delta D_{alkenone}$, we measure the stable carbon isotope composition of the combined $C_{37:2-3}$ alkenones ($\delta^{13}C_{alkenones}$) on samples from glacials and interglacials (Rau et al., 1996; Bidigare et al., 1997; Schouten et al., 2006). Furthermore, we reconstruct SST using the $TEX_{86}^H$ proxy (Schouten et al., 2002; Kim et al., 2010) and compare this with the $U_{37}^{K'}$ and Mg/Ca record of the planktonic Foraminifera G. bulloides for the same sediment cores (Martinez-Mendez et al., 2010).

## 1.1 Material and methods

Sediment samples were taken from cores MD96-2080 (36°19.2' S, 19°28.2' E; 2488 m water depth) and MD02-2594 (34°42.6' S, 17°20.3' E; 2440 m water depth) from the Agulhas bank slope off the coast of southern South Africa (Fig. 1). Core MD02-2594 was taken during the RV Marion Dufresne cruise MD128 SWAF (Giraudeau et al., 2003). Core MD96-2080 was obtained during the IMAGES II Campaign NAUSICAA (Bertrand, 1997). Age models and records of Globigerina bulloides $\delta^{18}O$ and Mg/Ca have previously been established for both sediment cores (Martinez-Mendez et al., 2008, 2010). The sampled interval of core MD02-2594 covered the period 3 to 42 ka (MIS1 to mid-MIS3) and included Termination I. Core MD96-2080 covered the period between 117 and 182 ka (MIS5e to MIS6) and included Termination II (Tables 1, 2).

## 1.2 Sample preparation

Sediment samples were freeze-dried and homogenized with a mortar and pestle. The homogenized material was then extracted using the accelerated solvent extractor method (ASE) with dichloromethane (DCM) : methanol 9 : 1 (v/v) and a pressure of 1000 psi in three extraction cycles. The total lipid extract was separated over an $Al_2O_3$ column into a apolar, ketone and polar fraction using hexane : DCM 9 : 1, hexane : DCM 1 : 1 and DCM : methanol 1 : 1, respectively. The ketone fraction was analyzed for $U_{37}^{K'}$ using gas chromatography. Gas chromatography/high-temperature conversion/isotope ratio mass spectrometry (GC/TC/irMS) was used to measure the combined hydrogen isotope composition of the di- and tri-unsaturated $C_{37}$ alkenones. The polar fraction was analyzed for $TEX_{86}^H$ using high-performance liquid chromatography mass spectrometry (HPLC/MS). Stable carbon isotopes of the combined di- and tri-unsaturated $C_{37}$ alkenones were analyzed using GC/combustion/irMS.

**Table 1.** Results for combined $C_{37:2-3}$ alkenone stable hydrogen isotope ($\delta D_{alkenone}$, ‰), stable carbon isotope ($\delta^{13}C_{alkenone}$, ‰), $TEX_{86}^H$ SST and $U_{37}^{K'}$ SST analyses for core MD96-2080.

| Age (ka) | $U_{37}^{K'}$ | $TEX_{86}^H$ | $\delta D$ alkenone (‰)[a] | $\delta^{13}C$ alkenone (‰)[a] |
|---|---|---|---|---|
| 117.7 | 0.819 | −0.276 | −197 ± 2 | |
| 119.9 | 0.819 | −0.272 | −196 ± 2 | |
| 123.1 | 0.856 | −0.277 | −180 | |
| 125.4 | 0.864 | −0.278 | −193 ± 4 | −24.7 ± 0.5 |
| 126.5 | 0.833 | −0.273 | −194 ± 1 | −25.4 ± 0.2 |
| 127.0 | 0.829 | −0.274 | −198 ± 0 | |
| 127.9 | 0.845 | −0.275 | −188 ± 1 | −25.1 ± 0.8 |
| 128.7 | 0.851 | −0.266 | −194 ± 4 | −25.6 ± 0.4 |
| 129.8 | 0.828 | −0.248 | −194 ± 0 | |
| 131.3 | 0.821 | −0.253 | −192 ± 3 | |
| 132.8 | 0.798 | −0.250 | −189 ± 2 | −24.2 ± 0.4 |
| 133.6 | 0.819 | −0.238 | −190 ± 0 | −23.9 ± 0.1 |
| 135.1 | 0.776 | −0.235 | −182 ± 2 | |
| 137.3 | 0.750 | −0.251 | −182 ± 4 | −24.5 ± 0.1 |
| 137.7 | 0.749 | −0.256 | −183 ± 1 | −24.1 ± 0.4 |
| 141.2 | 0.709 | −0.321 | −179 ± 1 | |
| 142.2 | 0.710 | −0.320 | −177 ± 3 | |
| 145.0 | 0.717 | −0.325 | −180 ± 1 | −24.9 ± 0.1 |
| 147.5 | 0.726 | −0.309 | | |
| 151.5 | 0.707 | −0.314 | −184 ± 2 | −24.8 ± 0.6 |
| 152.0 | 0.734 | −0.314 | −182 ± 0 | |
| 154.7 | 0.724 | −0.330 | −186 ± 1 | −24.6 ± 0.2 |
| 158.0 | 0.738 | −0.290 | −177 ± 0 | |
| 162.3 | 0.745 | −0.332 | −179 ± 3 | |
| 168.8 | 0.745 | −0.314 | −191 ± 2 | |
| 177.9 | 0.714 | −0.292 | −190 ± 3 | |
| 182.7 | 0.707 | −0.303 | −187 ± 3 | |

[a] The error is defined as the range of duplicated measurements.

## 1.3 $U_{37}^{K'}$ analysis

Ketone fractions were analyzed by gas chromatography (GC) using an Agilent 6890 gas chromatograph with a flame ionization detector and a Agilent CP Sil-5 fused silica capillary column (50 m × 0.32 mm, film thickness = 0.12 $\mu$m) with helium as the carrier gas. The GC oven was programmed to increase the temperature subsequently from 70 to 130 °C at 20 °C min$^{-1}$, and then at 4 °C min$^{-1}$ to 320 °C, at which it was held isothermal for 10 min. $U_{37}^{K'}$ values were calculated according to Prahl and Wakeham (1987). Subsequently, SST was calculated using the core top calibration established by Müller et al. (1998).

## 1.4 $\delta D$ of alkenone analysis

Alkenone hydrogen isotope analyses were carried out on a Thermo Finnigan DELTA$^{Plus}$ XL GC/TC/irMS. The temperature conditions of the GC increased from 70 to 145 °C at 20 °C min$^{-1}$, then to 320 °C at 4 °C min$^{-1}$, at which it was held isothermal for 13 min using an Agilent CP Sil-5

**Table 2.** Results for combined $C_{37:2-3}$ alkenone stable hydrogen isotope ($\delta D_{\text{alkenone}}$, ‰), stable carbon isotope ($\delta^{13}C_{\text{alkenone}}$, ‰) $TEX_{86}^H$ SST and $U_{37}^{K'}$ SST analyses for core MD02-2594.

| Age (ka) | $U_{37}^{K'}$ | $TEX_{86}^H$ | $\delta D$ alkenone (‰)[a] | $\delta^{13}C$ alkenone (‰)[a] |
|---|---|---|---|---|
| 3.5 | 0.724 | −0.305 | −192 ± 2 | −24.1 ± 0.1 |
| 5.6 | 0.723 | −0.307 | −184 ± 2 | |
| 6.3 | 0.746 | −0.309 | −189 ± 4 | |
| 6.9 | 0.736 | −0.311 | −182 ± 1 | |
| 8.6 | 0.753 | −0.308 | −191 ± 3 | |
| 12.0 | 0.762 | −0.282 | −178 ± 2 | −24.1 ± 0.8 |
| 18.0 | 0.685 | −0.308 | −179 ± 1 | |
| 18.5 | 0.677 | −0.366 | −171 ± 2 | |
| 21.1 | 0.672 | −0.336 | −173 ± 3 | −24.2 ± 1.1 |
| 22.3 | 0.686 | −0.352 | −178 ± 2 | |
| 26.7 | 0.686 | −0.328 | −166 ± 2 | |
| 28.4 | 0.682 | −0.333 | −173 ± 0 | |
| 29.5 | 0.686 | −0.339 | −172 ± 1 | |
| 30.8 | 0.715 | −0.313 | −179 ± 2 | |
| 31.5 | 0.692 | −0.332 | −173 ± 2 | −23.6 ± 0.1 |
| 32.0 | 0.696 | −0.315 | −177 ± 1 | |
| 37.4 | 0.728 | −0.312 | −171 ± 3 | |
| 42.1 | 0.676 | −0.343 | −180 ± 2 | |

[a] The error is defined as the range of duplicated measurements.

column (25 m × 0.32 mm) with a film thickness of 0.4 µm and 1 mL min$^{-1}$ helium at constant flow. The thermal conversion temperature was set to 1425 °C. The $H_3^+$ correction factor was determined daily and ranged between 10 and 14. Isotopic values for alkenones were standardized against pulses of $H_2$ reference gas, which was injected three times at the beginning and two times at the end of each run. A set of standard $n$-alkanes with known isotopic composition (Mixture B prepared by Arndt Schimmelmann, University of Indiana) was analyzed daily prior to each sample batch in order to monitor the system performance. Samples were only analyzed when the alkanes in Mix B had an average deviation from their off-line determined value of < 5 ‰. Squalane was co-injected as an internal standard with each sample to monitor the precision of the alkenone isotope values. The squalane standard yielded an average $\delta D$ value of −167 ‰ ± 4.5, which compared favorably with its offline determined $\delta D$ value of −170 ‰. Alkenone $\delta D$ values were measured as the combined peak of the $C_{37:2}$ and $C_{37:3}$ alkenones (van der Meer et al., 2013), and fractions were analyzed in duplicates if a sufficient amount of sample material was available. Standard deviations of replicate analyses varied from ±0.1 ‰ to ±5.9 ‰.

## 1.5 $\delta^{13}C$ of alkenone analyses

Combined $C_{37:2-3}$ alkenones were analyzed using a Thermo Delta V isotope ratio monitoring mass spectrometer coupled to an Agilent 6890 GC. Samples were dissolved in hexane and analyzed using a GC temperature program starting at 70 °C, then increasing to 130 °C at 20 °C min$^{-1}$, and then to 320 °C at 4 °C min$^{-1}$, where it was held for 20 min. The stable carbon isotope compositions for $\delta^{13}C_{\text{alkenone}}$ are reported relative to Vienna Pee Dee Belemnite (VDPB). The $\delta^{13}C_{\text{alkenone}}$ values are averages of at least two runs. GC-irMS performance was checked daily by analyzing a standard mixture of $n$-alkanes and fatty acids, including two fully perdeuterated alkanes with a known isotopic composition. These perdeuterated alkanes were also co-injected with every sample analysis and yielded an average $\delta^{13}C$ value of −32.5 ± 0.5 ‰ and −27.0 ± 0.5 ‰ for $n$-$C_{20}$ and $n$-$C_{24}$, respectively. This compared favorably with their offline determined $\delta^{13}C$ values of −32.7 ‰ and −27.0 ‰ for $n$-$C_{20}$ and $n$-$C_{24}$, respectively.

## 1.6 $TEX_{86}^H$ analysis

Analyses for $TEX_{86}^H$ were performed as described by Schouten et al. (2007). In summary, an Agilent 1100 series HPLC/MS equipped with an auto-injector and Agilent ChemStation chromatography manager software was used. Separation was achieved on an Alltech Prevail Cyano column (2.1 × 150 mm, 3 µm), maintained at 30 °C. Glycerol dibiphytanyl glycerol tetraethers (GDGTs) were eluted with 99 % hexane and 1 % propanol for 5 min, followed by a linear gradient to 1.8 % propanol in 45 min, followed by back-flushing hexane/propanol (9 : 1, v / v) at 0.2 mL min$^{-1}$ for 10 min. Detection was achieved using atmospheric pressure positive ion chemical ionization mass spectrometry (APCI-MS) of the eluent. Conditions for the Agilent 1100 APCI-MS were as follows: nebulizer pressure of 60 psi, vaporizer temperature of 400 °C, drying gas ($N_2$) flow of 6 L min$^{-1}$ and temperature 200 °C, capillary voltage of −3 kV and a corona of 5 µA (∼ 3.2 kV). GDGTs were detected by single ion monitoring (SIM) of their $[M + H]^+$ ions (dwell time = 234 ms) (Schouten et al., 2007) and quantified by integration of the peak areas. The $TEX_{86}^H$ values and absolute temperatures were calculated according to Kim et al. (2010). This calibration is recommended for temperature reconstruction above 15 °C (Kim et al., 2010) and therefore appears to be the most suitable model for reconstructing subtropical temperatures, as found in the Agulhas leakage area.

## 2 Results

### 2.1 Sea surface temperature proxies

The $U_{37}^{K'}$ record of MD96-2080 indicated constant temperatures of approximately 20.7 ± 0.5 °C throughout MIS6 (138–182 ka) (Table 1, Fig. 2). With the onset of Termination II at approximately 138 ka, temperatures began to increase to a maximum of ∼ 25 °C during early MIS5e (∼ 125 ka), followed by a slight decrease in temperature towards ∼ 23 °C at about 120 ka. The $U_{37}^{K'}$ SST record for MD02-2594 indicated

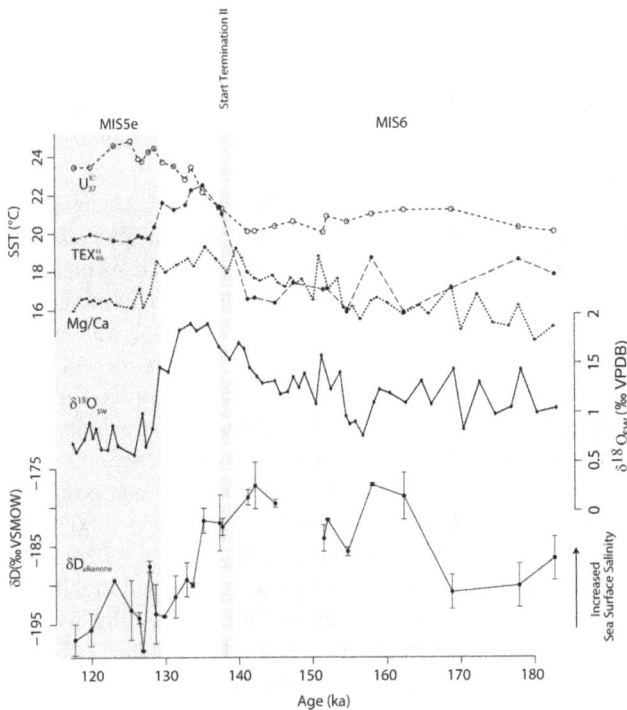

**Fig. 2.** Reconstructed SST of $U_{37}^{K'}$ (dashed line, open circles), $TEX_{86}^{H}$ (dashed line, closed circles), Mg/Ca of *G. bulloides* (dotted line, closed circles, Martinez-Mendez et al., 2010), reconstructed $\delta^{18}O_{sw}$ from *G. bulloides* (solid line, diamonds, Martinez-Mendez et al., 2010) and hydrogen isotope composition of $C_{37:2-3}$ alkenones (solid line, closed circles) for core MD96-2080.

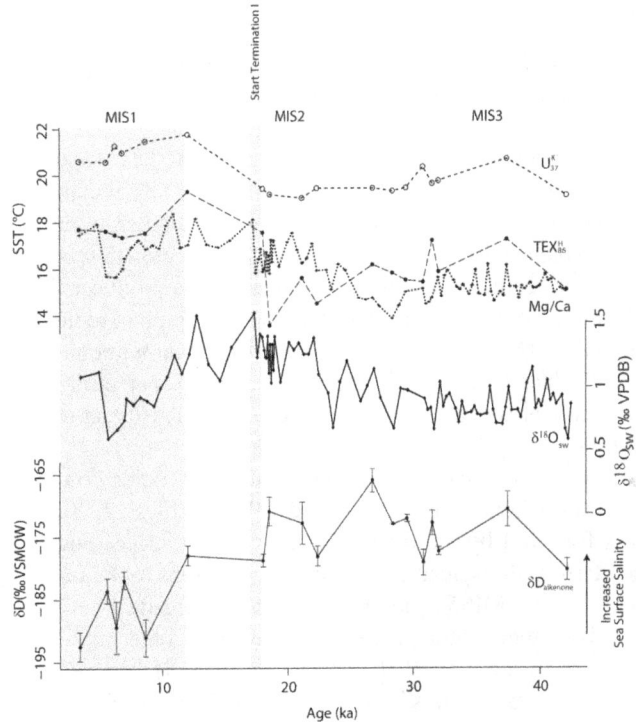

**Fig. 3.** Reconstructed SST of $U_{37}^{K'}$ (dashed line, open circles), $TEX_{86}^{H}$ (dashed line, closed circles), Mg/Ca of *G. bulloides* (dotted line, closed circles, Martinez-Mendez et al., 2010), reconstructed $\delta^{18}O_{sw}$ from *G. bulloides* (solid line, diamonds, Martinez-Mendez et al., 2010) and hydrogen isotope composition of $C_{37:2-3}$ alkenones (solid line, closed circles) for core MD02-2594.

relatively constant temperatures of $20 \pm 0.5\,°C$ throughout MIS3 and early MIS2 (Table 2, Fig. 3). With the onset of Termination I during mid-MIS2 ($\sim 18$ ka), temperatures showed a slight warming towards $\sim 22\,°C$ at the beginning of MIS1. Throughout MIS1, temperatures slightly decreased to approximately $21\,°C$.

The overall pattern in the $TEX_{86}^{H}$ records for the two cores indicated that absolute temperatures were cooler by up to $5\,°C$ compared to $U_{37}^{K'}$ temperatures during glacial and interglacial stages MIS5e and MIS6, as well as MIS3/MIS2 and MIS1. During MIS6, $TEX_{86}^{H}$ temperatures were relatively stable at $17 \pm 1\,°C$, although less stable than $U_{37}^{K'}$ SST (Table 1, Fig. 2). At the initial termination of the glacial MIS6 (ca. 135–138 ka), reconstructed $TEX_{86}^{H}$ temperatures increased rapidly to about $22\,°C$, which is similar in absolute terms compared to $U_{37}^{K'}$ SST (Fig. 2). Subsequently, $TEX_{86}^{H}$ SST decreased rapidly to $\sim 20\,°C$ and remained constant during MIS5e. During MIS3 and early MIS2, $TEX_{86}^{H}$ SST showed a trend towards cooler temperatures from approximately $17\,°C$ at $\sim 38$ ka to $14\,°C$ at the start of Termination I (18 ka) (Table 2, Fig. 3). At the onset of Termination I, $TEX_{86}^{H}$ SSTs abruptly increased. Temperatures reached a maximum of $19\,°C$ at the beginning of MIS1 (11 ka) and de-

creased again to relatively constant temperatures of ca. $18\,°C$ throughout MIS1. This trend is comparable to the $U_{37}^{K'}$ SST record, albeit with a negative offset of approximately $4\,°C$ throughout MIS3/MIS2 and MIS1, and approximately $2\,°C$ during the glacial termination phase (11–18 ka) (Fig. 3).

## 2.2  Stable hydrogen and carbon isotope composition of $C_{37:2-3}$ alkenones

The $\delta D_{alkenone}$ values in core MD96-2080 ranged between $-177$ and $-198\,‰$ (Table 1). During early MIS6 (169 to 181 ka), $\delta D_{alkenone}$ values were approximately $-189 \pm 2\,‰$ and shifted to more positive values of ca. $-177\,‰$ at 158 ka. In the time interval between 142 and 162 ka (MIS6), $\delta D_{alkenone}$ decreased to $-180 \pm 3\,‰$. During glacial Termination II (130 to 138 ka), the $\delta D_{alkenone}$ values decreased abruptly to approximately $-194 \pm 3\,‰$ for MIS5e (118–130 ka) (Fig. 2). The $\delta D_{alkenone}$ values in core MD02-2594 ranged between $-166$ and $-192\,‰$ (Table 2). In the time interval from 12 to 41 ka (MIS3 and MIS2), values for $\delta D_{alkenone}$ were relatively constant at approximately $-174 \pm 4\,‰$ (Fig. 3). At ca. 11 ka (onset of MIS1), $\delta D_{alkenone}$ values shifted to more negative values with an average of $-188 \pm 5\,‰$ for the time period of 3 to 9 ka (MIS1) (Fig. 3).

## 3  Discussion

### 3.1  Development of sea surface temperatures during Termination I and Termination II

Application of the $U_{37}^{K'}$ proxy resulted in higher reconstructed temperatures compared to the $TEX_{86}^H$ and Mg/Ca SST reconstructions (Figs. 2, 3). Difference in absolute temperatures may be explained by a variety of reasons such as growth seasons and/or depth habitats between the alkenone producers, the Thaumarchaeota (GDGT producers) and the planktonic Foraminifera (Müller et al., 1998; Bard, 2001; Karner et al., 2001; Wuchter et al., 2006; Lee et al., 2008; Saher et al., 2009; dos Santos et al., 2010; Fallett et al., 2011; Huguet et al., 2011).

Both $U_{37}^{K'}$ and $TEX_{86}^H$ temperatures suggest cooler conditions throughout stages MIS6 and MIS3/2 (Figs. 2, 3). This is followed by an abrupt warming during succeeding glacial terminations, leading to warmer conditions in the interglacial stages MIS5e and MIS1. The pattern fits well with $U_{37}^{K'}$-derived temperature reconstructions from other sediment cores in the area of this study site (Fig. 4) (Schneider et al., 1995; Peeters et al., 2004; Bard and Rickaby, 2009). Temperature reconstructions based on Mg/Ca and $TEX_{86}^H$ show that maximum temperatures occurred during the deglaciations. However, $U_{37}^{K'}$ temperature reconstructions showed that the SST maximum occurred approximately 5 ka later than in Mg/Ca- and $TEX_{86}^H$-based reconstructions. This temperature increase could possibly point towards an increased influence of warm Indian Ocean waters, and hence increased Agulhas leakage. This increased leakage in turn is likely related to a southward shift of the subtropical front due to shifting wind fields and a southward migration of the land ice shields (Peeters et al., 2004). Nevertheless, these observations do not necessarily imply a buildup of warm Indian Ocean waters prior to glacial terminations at the core site.

Strikingly, the timing of the beginning of the warming trend reflected in the Foraminifera Mg/Ca record of *G. bulloides* (Martinez-Mendez et al., 2010) is different from the $U_{37}^{K'}$ and $TEX_{86}^H$ records, as well as other $U_{37}^{K'}$ records in the region (Fig. 4) (Peeters et al., 2004).The Mg/Ca SST record identifies a warming trend starting in the early glacial periods and gradually extending over the glacial termination phases (Martinez-Mendez et al., 2010). Furthermore, a recent study by Marino et al. (2013) also showed a discrepancy between Mg/Ca of the planktonic Foraminifera *Globigerinoides ruber* and $U_{37}^{K'}$ SST. It has been reported that changes in salinity can also affect the Mg/Ca ratios in Foraminifera shells, specifically during glacial periods when salinity was likely elevated (Ferguson et al., 2008; Arbuszewski et al., 2010). Thus, the observed trends in foraminiferal Mg/Ca may result from a combined salinity and temperature signal (Hönisch et al., 2013).

### 3.2  Salinity changes during Termination I and Termination II

The $C_{37:2-3}$ alkenone hydrogen isotope records consistently show a substantial decrease toward more deuterium-depleted values during the glacial terminations and the interglacial stages MIS5e and MIS1 (Figs. 2, 3). We quantified changes from glacial to interglacial stages by averaging time intervals from before and after each termination. We observed average shifts in $\delta D_{alkenone}$ of approximately 14‰ for both Termination I and II (Table 3). These shifts in the $\delta D_{alkenone}$ values can be caused by a number of factors such as decreasing $\delta D$ of sea water ($\delta D_{sw}$) as an effect of decreasing global ice volume during the terminations (Rohling, 2000), ocean salinity, algal growth rate, haptophyte species composition (Schouten et al., 2006), differences in the hydrogen isotope composition of the $C_{37:2}$ and $C_{37:3}$ alkenones (D'Andrea et al., 2007; Schwab and Sachs, 2009). However, the latter factor is likely unimportant, as for the range of $U_{37}^{K'}$ values observed in this study we expect a maximum difference of 4‰ in the hydrogen isotope composition between $C_{37:2}$ and $C_{37:3}$, which falls within the accuracy of the GC/TC/irMS (van der Meer et al., 2013).

In order to assess changes in growth rate for the haptophytes, we measured the stable carbon isotope composition of the combined $C_{37:2-3}$ alkenones ($\delta^{13}C_{alkenone}$) (Tables 1, 2). Our results show relatively small changes of about −0.6‰ and −0.3‰ in $\delta^{13}C_{alkenone}$ during Termination I and II, respectively. The fractionation of stable carbon isotopes is mainly controlled by physiological factors like growth rate, cell size and geometry, as well as by the supply of dissolved $CO_2$ (Rau et al., 1996; Bidigare et al., 1997). The more depleted alkenone $\delta^{13}C$ values during interglacials suggest either slightly higher dissolved $CO_2$ concentrations or lower growth rates. We suggest that higher dissolved $CO_2$ concentrations likely explain the more depleted alkenone $\delta^{13}C$ values as $CO_2$ concentrations were higher during interglacials compared to glacials (Curry and Crowley, 1987). Since reduced growth rates would result in decreasing hydrogen isotopic fractionation, our observed increase in hydrogen isotopic fractionation during interglacials compared to glacials cannot be explained by growth rate changes (Schouten et al., 2006).

Species changes could also explain the observed hydrogen isotope shift. Assemblage studies in the Agulhas leakage have shown an increasing abundance of the predominant haptophyte *E. huxleyi* from the beginning of MIS7 towards MIS1, with a maximum relative abundance observed at the onset of MIS1 (Flores et al., 1999). *G. oceanica*, however, reaches maximum relative abundances during Termination II (Flores et al., 1999). Changes in the coccolithophore assemblage toward a larger fraction of *G. oceanica* could have resulted in more negative $\delta D_{alkenone}$ values during that time since *G. oceanica* fractionates more against D than *E. huxleyi* (Schouten et al., 2006). However, the abundance of

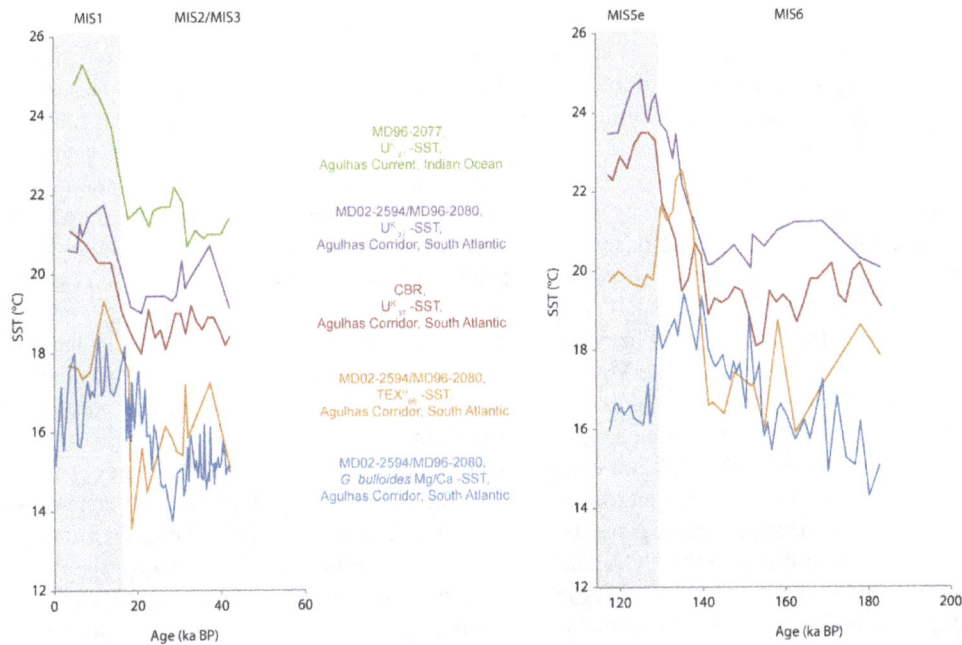

**Fig. 4.** Comparison of $U^{K'}_{37}$ SST, $TEX^{H}_{86}$ SST and *G. bulloides* Mg/Ca SST (Martinez-Mendez et al., 2010) of **(a)** MIS3-1 (MD02-2594) and **(b)** MIS6-5 (MD96-2080) (see Fig. 1 for core location) with $U^{K'}_{37}$ SST of MD96-2077 in the Agulhas Current, Indian Ocean (Bard and Rickaby, 2009) and $U^{K'}_{37}$ SST of Cape Basin record in the Agulhas corridor, South Atlantic (Peeters et al., 2004).

**Table 3.** Average values for the hydrogen isotope composition of the $C_{37:2-3}$ alkenones, the global $\delta^{18}O_{ice\ vol.}$ (Waelbroeck et al., 2002), $\delta D_{ice\ vol.}$ (Srivastava et al., 2010) and the local $\delta D_{sw}$ derived from $\delta^{18}O$ of *G. bulloides* (Martinez-Mendez et al., 2010) for the intervals of before and after Termination I and II.

|  | Time interval | $\delta D_{alkenone}$(‰) | $\delta^{18}O_{ice\ vol.}$(‰) | $\delta D_{ice\ vol.}$(‰) | $\delta D_{sw}$(‰) |
|---|---|---|---|---|---|
| Term. I | before (3.5–8.6 ka, $n = 5$) | $-188 \pm 5$ | $0.06 \pm 0.05$ | $0.7 \pm 0.4$ | $-0.4 \pm 0.9$ |
|  | after (18–37.4 ka, $n = 11$) | $-174 \pm 4$ | $0.83 \pm 016$ | $6.3 \pm 1.2$ | $6.5 \pm 2.6$ |
| Term. II | before (117.7–129.8 ka, $n = 9$) | $-194 \pm 3$ | $0.00 \pm 0.05$ | $0.3 \pm 0.4$ | $-0.5 \pm 2.7$ |
|  | after (141.2–162.3 ka, $n = 8$) | $-180 \pm 3$ | $0.88 \pm 013$ | $6.7 \pm 0.9$ | $9.6 \pm 2.4$ |

*G. oceanica* never exceeded the relative abundance of *E. huxleyi*, and it is therefore unlikely that species composition changes had a large impact on $\delta D_{alkenone}$ values during Termination II. During Termination I *E. huxleyi* reached its maximum relative abundances compared to *G. oceanica* (Flores et al., 1999) possibly resulting in more positive $\delta D_{alkenone}$ values rather than the observed trend toward more depleted values (Schouten et al., 2006). Therefore, the observed trends towards more depleted values in the $\delta D_{alkenone}$ during the glacial terminations are most likely not affected significantly by changes in the coccolithophore species composition.

Thus, our observed isotope shifts can likely only be explained by a shift in the $\delta D$ of water, through global ice volume changes, and/or salinity. We estimate the effect of changes in global ice volume on the $\delta D_{alkenone}$ by using the global mean ocean $\delta^{18}O_{sw}$ record based on benthic Foraminifera (Waelbroeck et al., 2002) and calculated an equivalent $\delta D_{sw}$ record by applying a local Indian Ocean meteoric waterline (Srivastava et al., 2010). Changes in global $\delta D_{sw}$ due to the ice volume effect are estimated to be approximately $-6$‰ during both terminations (Table 3). This shift is smaller than that observed in $\delta D_{alkenone}$, suggesting an increase in hydrogen isotopic fractionation during the two terminations.

The residual $\delta D_{alkenone}$ shift likely reflects changes in sea surface salinity. We find alkenones relatively enriched in D during the glacials MIS6 and MIS2/3 and relatively depleted in D during MIS5e and MIS1, suggesting lower salinities during interglacials compared to glacials (Figs. 2, 3). Indeed, reconstructed $\delta^{18}O_{sw}$ from the planktonic Foraminifera *G. bulloides* also indicates higher salinity throughout the glacials MIS6 and MIS3/MIS2 (Figs. 2, 3) (Martinez-Mendez et al., 2010). However, alkenone $\delta D$ values begin to shift toward more depleted values shortly before the start of Termination II, at approximately 135 ka, whereas the initial freshening recorded in the $\delta^{18}O_{sw}$ begins at about 133 ka (Fig. 2). Similar diverging trends are noted for reconstructed $\delta^{18}O_{sw}$ derived from the planktonic Foraminifera *G. ruber* (Marino et al., 2013). The most depleted values are reached in both reconstructed $\delta^{18}O_{sw}$ and $\delta D_{alkenone}$ during early MIS5e at about 128 ka (Fig. 2). Higher salinity conditions are also observed throughout MIS3/MIS2 in reconstructed $\delta^{18}O_{sw}$ and $\delta D_{alkenone}$ (Fig. 3) followed by freshening trends starting during early Termination I (∼ 18 ka). The offset in timing of the start of the freshening trends between the different proxies is similar to that observed in the Mg/Ca and $U_{37}^{K'}$ temperature records. This might be explained by differences in depth habitat between haptophyte algae and the Foraminifera and/or salinity effects on Mg/Ca and consequently the reconstructed $\delta^{18}O_{sw}$ record. Despite the discrepancy in the timing of the start of the freshening events, an overall increase in salinity is recorded in both $\delta^{18}O_{sw}$ and $\delta D_{alkenone}$ during glacial stages MIS6 and MIS3/5. This is followed by a rapid decrease in salinity during the terminations. The fresher conditions prevail during the subsequent interglacial stages.

Absolute salinity estimates are difficult to obtain from the $\delta D_{alkenone}$ due to the uncertainties in both the slope and intercept of the culture calibrations and other variables (Rohling, 2007). However, by estimating relative salinity changes only, using the slope of the $\delta D_{alkenone}$–salinity relationship, we avoid uncertainties related to the intercept. This provides an added advantage that the slopes for *E. huxleyi* or *G. oceanica* are nearly identical (i.e., 4.8‰ and 4.2‰ $\delta D_{alkenone}$ per salinity unit, respectively; Schouten et al., 2006). Estimations for relative salinity changes from $\delta D_{alkenone}$ result in a freshening trend of approximately 1.7–1.9 salinity units during the course of Termination I and II. These results are in fairly good agreement with the estimated salinity shift of 1.2–1.5 based on combined Mg/Ca SST estimates and $\delta^{18}O$ of *G. bulloides* for these time periods (Martinez-Mendez et al., 2010). However, they both seem to differ from the $\delta^{18}O_{sw}$ record based on *G. ruber*, which indicates no salinity difference between glacial–interglacial and only an intermittent shift during the terminations (Marino et al., 2013).

### 3.3  Paleoceanographic implications

Based on the reconstructed SST and relative SSS records, we suggest that increased salinity during glacial periods and subsequent freshening during the glacial terminations can be explained by the efficiency of the Agulhas leakage (i.e., the volume transport of water from the Indian Ocean to the South Atlantic). According to Peeters et al. (2004), $U_{37}^{K'}$ SST maxima correspond to maximum Agulhas leakage, as seen in the planktonic Foraminifera Agulhas leakage fauna, during glacial Termination I and II. We observe enriched $\delta D_{alkenone}$ values, suggesting increased salinity, coinciding with reduced Agulhas leakage during glacial stages

MIS6 and MIS2/3. We suggest that with reduced throughflow and increased residence time of Indian Ocean water, the surface waters become relatively more saline and cooler in the Agulhas region, including the Agulhas leakage area. In contrast, with higher transport rates the surface waters will retain more of the original temperature and salinity resulting in the reconstructed lower salinities and higher temperatures. Thus, temperature and salinity are likely decoupled in this setting. In this case, heat loss is enhanced when water masses flow polewards. Salinity, however, will be retained, and during low through-flow situations in relatively dry glacial periods, evaporation will increase sea surface salinity. At the same time, the limited precipitation and river runoff in this region will not counteract this increase sufficiently. However, the absolute amount of salt that is released into the Atlantic Ocean would still increase during terminations due to the increased flow, even though the surface waters become less saline. Consequently, this would lead to increasing the Atlantic Meridional Overturning Circulation (Bard and Rickaby, 2009; Haarsma et al., 2011).

### 4  Conclusions

In this study, we analyzed two sediment cores from the Agulhas leakage area covering Termination I and Termination II. We combined $TEX_{86}^H$ and $U_{37}^{K'}$ SST reconstructions with a previously reported SST record based on Mg/Ca of the planktonic Foraminifera *G. bulloides* (Martinez-Mendez et al., 2010). Sea surface temperatures reconstructed from three different proxies indicated relatively low temperature conditions throughout the late glacials MIS6 and MIS2/3 in the Agulhas leakage area, and at the onset of the deglaciation (Termination I and II) temperatures increase significantly. Relative salinity changes were reconstructed using $\delta D_{alkenone}$, which showed a shift from more positive values to more negative values during Termination I and Termination II suggesting elevated salinities during glacial periods, with subsequent freshening during glacial terminations. Similar trends in glacial to interglacial salinity changes were also observed based on planktonic Foraminifera $\delta^{18}O_{sw}$ reconstructions. Estimated salinity changes, based on $\delta D_{alkenone}$, range from 1.7 to 1.9 salinity units for Termination I and II. This is in fairly good agreement with salinity shifts based on the paired Mg/Ca and $\delta^{18}O$ approach of the planktonic Foraminifera *G. bulloides*. Our results therefore suggest an increased release of slightly less saline Indian Ocean water to the South Atlantic Ocean during the terminations than during the glacials, but with a net increase in salt transport during interglacials due to the higher throughflow.

*Acknowledgements.* We acknowledge financial support from the Seventh Framework Programme PEOPLE Work Programme, grant 238512 (Marie Curie Initial Training Network GATEWAYS). The Netherlands Organization for Scientific Research (NWO) is acknowledged for funding Marcel van der Meer (VIDI) and Stefan Schouten (VICI). S. Kasper would like to thank C. A. Grove (Royal NIOZ, the Netherlands) and D. Chivall (Royal NIOZ, the Netherlands) for their input on this manuscript.

Edited by: G. M. Ganssen

# References

Arbuszewski, J., deMenocal, P., Kaplan, A., and Farmer, E. C.: On the fidelity of shell-derived $\delta^{18}O_{sw}$ estimates, Earth Planet Sci. Lett., 300, 185–196, doi:10.1016/j.epsl.2010.10.035, 2010.

Bard, E.: Comparison of alkenone estimates with other paleotemperature proxies, Geochem. Geophys. Geosyst., 2, 1002, doi:10.1029/2000gc000050, 2001.

Bard, E. and Rickaby, R. E. M.: Migration of the subtropical front as a modulator of glacial climate, Nature, 460, 380–U393, doi:10.1038/nature08189, 2009.

Bertrand, P.: Les rapport de campagne a la mer a bord du Marion Dufresne – Campagne NAUSICAA – Images II – MD105 du 20/10/96 au 25/11/96, Inst. Fr. pour la Rech. et la Technol. Polaires, France, Plouzane, 1997.

Biastoch, A., Boning, C. W., and Lutjeharms, J. R. E.: Agulhas leakage dynamics affects decadal variability in Atlantic overturning circulation, Nature, 456, 489–492, 2008.

Bidigare, R. R., Fluegge, A., Freeman, K. H., Hanson, K. L., Hayes, J. M., Hollander, D., Jasper, J. P., King, L. L., Laws, E. A., Milder, J., Millero, F. J., Pancost, R., Popp, B. N., Steinberg, P. A., and Wakeham, S. G.: Consistent fractionation of $^{13}C$ in nature and in the laboratory: Growth-rate effects in some haptophyte algae, Global Biogeochem. Cy., 11, 279–292, 1997.

Curry, W. B. and Crowley, T. J.: The $\delta^{13}C$ of equatorial Atlantic surface waters: Implications for Ice Age pCO2 levels, Paleoceanography, 2, 489–517, doi:10.1029/PA002i005p00489, 1987.

D'Andrea, W. J., Liu, Z., Alexandre, M. D. R., Wattley, S., Herbert, T. D., and Huang, Y.: An Efficient Method for Isolating Individual Long-Chain Alkenones for Compound-Specific Hydrogen Isotope Analysis, Anal. Chem., 79, 3430–3435, 2007.

dos Santos, R. A. L., Prange, M., Castaneda, I. S., Schefuss, E., Mulitza, S., Schulz, M., Niedermeyer, E. M., Damste, J. S. S., and Schouten, S.: Glacial-interglacial variability in Atlantic meridional overturning circulation and thermocline adjustments in the tropical North Atlantic, Earth Planet Sci. Lett., 300, 407–414, doi:10.1016/j.epsl.2010.10.030, 2010.

Englebrecht, A. C. and Sachs, J. P.: Determination of sediment provenance at drift sites using hydrogen isotopes and unsaturation ratios in alkenones, Geochim. Cosmochim. Ac., 69, 4253–4265, 2005.

Fallet, U., Ullgren, J. E., Castañeda, I. S., van Aken, H. M., Schouten, S., Ridderinkhof, H., and Brummer, G.-J. A.: Contrasting variability in foraminiferal and organic paleotemperature proxies in sedimenting particles of the Mozambique Channel (SW Indian Ocean), Geochim. Cosmochim. Ac., 75, 5834–5848, doi:10.1016/j.gca.2011.08.009, 2011.

Ferguson, J. E., Henderson, G. M., Kucera, M., and Rickaby, R. E. M.: Systematic change of foraminiferal Mg/Ca ratios across a strong salinity gradient, Earth Planet Sci. Lett., 265, 153–166, doi:10.1016/j.epsl.2007.10.011, 2008.

Flores, J. A., Gersonde, R., and Sierro, F. J.: Pleistocene fluctuations in the Agulhas Current Retroflection based on the calcareous plankton record, Mar. Micropaleontol., 37, 1–22, 1999.

Franzese, A. M., Hemming, S. R., Goldstein, S. L., and Anderson, R. F.: Reduced Agulhas Leakage during the Last Glacial Maximum inferred from an integrated provenance and flux study, Earth Planet Sci. Lett., 250, 72–88, doi:10.1016/j.epsl.2006.07.002, 2006.

Giraudeau, J., Balut, Y., Hall, I. R., Mazaud, A., and Zahn, R.: SWAF-MDI128 Scientific Report, Inst. Polaire Fr., Plouzane, France, 108 pp., 2003.

Haarsma, R. J., Campos, E. J. D., Drijfhout, S., Hazeleger, W., and Severijns, C.: Impacts of interruption of the Agulhas leakage on the tropical Atlantic in coupled ocean-atmosphere simulations, Clim. Dynam., 36, 989–1003, doi:10.1007/s00382-009-0692-7, 2011.

Hönisch, B., Allen, K. A., Lea, D. W., Spero, H. J., Eggins, S. M., Arbuszewski, J., deMenocal, P., Rosenthal, Y., Russell, A. D., and Elderfield, H.: The influence of salinity on Mg/Ca in planktic foraminifers – Evidence from cultures, core-top sediments and complementary $\delta^{18}O$, Geochim. Cosmochim. Ac., 121, 196–213, doi:10.1016/j.gca.2013.07.028, 2013.

Huguet, C., Martrat, B., Grimalt, J. O., Damste, J. S. S., and Schouten, S.: Coherent millennial-scale patterns in $U^{K'}_{37}$ and $TEX^{H}_{86}$ temperature records during the penultimate interglacial-to-glacial cycle in the western Mediterranean, Paleoceanography, 26, Pa2218, doi:10.1029/2010pa002048, 2011.

Karner, M. B., DeLong, E. F., and Karl, D. M.: Archaeal dominance in the mesopelagic zone of the Pacific Ocean, Nature, 409, 507–510, 2001.

Kim, J. H., van der Meer, J., Schouten, S., Helmke, P., Willmott, V., Sangiorgi, F., Koc, N., Hopmans, E. C., and Damste, J. S. S.: New indices and calibrations derived from the distribution of crenarchaeal isoprenoid tetraether lipids: Implications for past sea surface temperature reconstructions, Geochim. Cosmochim. Ac., 74, 4639–4654, doi:10.1016/j.gca.2010.05.027, 2010.

Lee, K. E., Kim, J.-H., Wilke, I., Helmke, P., and Schouten, S.: A study of the alkenone, TEX86, and planktonic foraminifera in the Benguela Upwelling System: Implications for past sea surface temperature estimates, Geochem. Geophys. Geosyst., 9, Q10019, doi:10.1029/2008gc002056, 2008.

Lutjeharms, J. R. E.: The Agulhas Current, Springer, Berlin, 329 pp., 2006.

Marino, G., Zahn, R., Ziegler, M., Purcell, C., Knorr, G., Hall, I. R., Ziveri, P., and Elderfield, H.: Agulhas salt-leakage oscillations during abrupt climate changes of the Late Pleistocene, Paleoceanography, 28, 599–606, doi:10.1002/palo.20038, 2013.

Martinez-Mendez, G., Zahn, R., Hall, I. R., Pena, L. D., and Cacho, I.: 345,000-year-long multi-proxy records off South Africa document variable contributions of Northern versus Southern Component Water to the Deep South Atlantic, Earth Planet Sci. Lett., 267, 309–321, 2008.

Martinez-Mendez, G., Zahn, R., Hall, I. R., Peeters, F. J. C., Pena, L. D., Cacho, I., and Negre, C.: Contrasting multi-proxy reconstructions of surface ocean hydrography in the Agulhas Corridor and implications for the Agulhas Leakage during the last 345,000 years, Paleoceanography, 25, 12, Pa4227, doi:10.1029/2009pa001879, 2010.

Müller, P. J., Kirst, G., Ruhland, G., von Storch, I., and Rosell-
Melé, A.: Calibration of the alkenone paleotemperature index
$U_{37}^{K'}$ based on core-tops from the eastern South Atlantic and
the global ocean (60° N–60° S), Geochim. Cosmochim. Ac., 62,
1757–1772, 1998.

Pahnke, K., Sachs, J. P., Keigwin, L., Timmermann, A., and Xie,
S.-P.: Eastern tropical Pacific hydrologic changes during the past
27,000 years from D/H ratios in alkenones, Paleoceanography,
22, PA4214, doi:10.1029/2007pa001468, 2007.

Peeters, F. J. C., Acheson, R., Brummer, G. J. A., de Ruijter, W.
P. M., Schneider, R. R., Ganssen, G. M., Ufkes, E., and Kroon,
D.: Vigorous exchange between the Indian and Atlantic oceans
at the end of the past five glacial periods, Nature, 430, 661–665,
doi:10.1038/nature02785, 2004.

Penven, P., Lutjeharms, J. R. E., and Florenchie, P.: Madagascar:
A pacemaker for the Agulhas Current system?, Geophys. Res.
Lett., 33, L17609, doi:10.1029/2006gl026854, 2006.

Prahl, F. G. and Wakeham, S. G.: Calibration of Unsaturation Patterns in Long-Chain Ketone Compositions for Paleotemperature
Assessment, Nature, 330, 367–369, 1987.

Rau, A. J., Rogers, J., Lutjeharms, J. R. E., Giraudeau, J., Lee-
Thorp, J. A., Chen, M. T., and Waelbroeck, C.: A 450-kyr record
of hydrological conditions on the western Agulhas Bank Slope,
south of Africa, Mar. Geol., 180, 183–201, 2002.

Rau, G. H., Riebesell, U., and WolfGladrow, D.: A model of photosynthetic $^{13}$C fractionation by marine phytoplankton based on
diffusive molecular $CO_2$ uptake, Mar. Ecol. Progr. Series, 133,
275–285, doi:10.3354/meps133275, 1996.

Rohling, E. J.: Paleosalinity: confidence limits and future applications, Mar. Geol., 163, 1–11, 2000.

Rohling, E. J.: Progress in paleosalinity: Overview and pre-
sentation of a new approach, Paleoceanography, 22, PA3215,
doi:10.1029/2007pa001437, 2007.

Rohling, E. J. and Bigg, G. R.: Paleosalinity and $\delta^{18}$O: A critical
assessment, J. Geophys. Res.-Oceans, 103, 1307–1318, 1998.

Rontani, J.-F., Prahl, F. G., and Volkman, J. K.: Re-examination of
the double bond positions in alkenones and derivatives: biosynthetic implications, J. Phycol., 42, 800–813, doi:10.1111/j.1529-
8817.2006.00251.x, 2006.

Rouault, M., Penven, P., and Pohl, B.: Warming in the Agulhas Current system since the 1980's, Geophys. Res. Lett., 36, L12602,
doi:10.1029/2009gl037987, 2009.

Saher, M. H., Rostek, F., Jung, S. J. A., Bard, E., Schneider, R.
R., Greaves, M., Ganssen, G. M., Elderfield, H., and Kroon, D.:
Western Arabian Sea SST during the penultimate interglacial:
A comparison of $U^{K}{}'_{37}$ and Mg/Ca paleothermometry, Paleo-
ceanography, 24, PA2212, doi:10.1029/2007pa001557, 2009.

Schneider, R. R., Muller, P. J., and Ruhland, G.: Late Quaternary
surface circulation in the east equatorial South Atlantic: Evidence from Alkenone sea surface temperatures, Paleoceanography, 10, 197–219, 1995.

Schouten, S., Hopmans, E. C., Schefuß, E., and Sinninghe Damsté,
J. S.: Distributional variations in marine crenarchaeotal membrane lipids: a new tool for reconstructing ancient sea water temperatures?, Earth Planet Sci. Lett., 204, 265–274, 2002.

Schouten, S., Ossebaar, J., Schreiber, K., Kienhuis, M. V. M.,
Langer, G., Benthien, A., and Bijma, J.: The effect of temperature, salinity and growth rate on the stable hydrogen isotopic
composition of long chain alkenones produced by Emiliania huxleyi and Gephyrocapsa oceanica, Biogeosciences, 3, 113–119,
doi:10.5194/bg-3-113-2006, 2006.

Schouten, S., Huguet, C., Hopmans, E. C., Kienhuis, M. V. M.,
and Sinninghe Damsté, J. S.: Analytical Methodology for TEX$_{86}$
Paleothermometry by High-Performance Liquid Chromatog-
raphy/Atmospheric Pressure Chemical Ionization-Mass Spectrometry, Anal. Chem., 79, 2940–2944, doi:10.1021/ac062339v,
2007.

Schwab, V. F. and Sachs, J. P.: The measurement of D/H ratio in
alkenones and their isotopic heterogeneity, Org. Geochem., 40,
111–118, 2009.

Srivastava, R., Ramesh, R., Jani, R. A., Anilkumar, N., and Sud-
hakar, M.: Stable oxygen, hydrogen isotope ratios and salinity
variations of the surface Southern Indian Ocean waters, Curr. Sci.
India, 99, 1395–1399, 2010.

van der Meer, M. T. J., Baas, M., Rijpstra, W. I. C., Marino, G.,
Rohling, E. J., Sinninghe Damsté, J. S., and Schouten, S.: Hydrogen isotopic compositions of long-chain alkenones record freshwater flooding of the Eastern Mediterranean at the onset of sapro-
pel deposition, Earth Planet Sci. Lett., 262, 594–600, 2007.

van der Meer, M. T. J., Sangiorgi, F., Baas, M., Brinkhuis, H.,
Sinninghe Damsté, J. S., and Schouten, S.: Molecular isotopic
and dinoflagellate evidence for Late Holocene freshening of the
Black Sea, Earth Planet Sci. Lett., 267, 426–434, 2008.

van der Meer, M. T. J., Benthien, A., Bijma, J., Schouten, S., and
Sinninghe Damsté, J. S.: Alkenone distribution impacts the hydrogen isotopic composition of the $C_{37:2}$ and $C_{37:3}$ alkan-2-ones
in Emiliania huxleyi, Geochim. Cosmochim. Ac., 111, 162–166,
doi:10.1016/j.gca.2012.10.041, 2013.

van Sebille, E., Biastoch, A., van Leeuwen, P. J., and de Ruijter, W.
P. M.: A weaker Agulhas Current leads to more Agulhas leakage,
Geophys. Res. Lett., 36, L03601, doi:10.1029/2008gl036614,
2009.

van Sebille, E., Beal, L. M., and Johns, W. E.: Advective Time
Scales of Agulhas Leakage to the North Atlantic in Sur-
face Drifter Observations and the 3D OFES Model, J. Phys.
Oceanogr., 41, 1026–1034, doi:10.1175/2010jpo4602.1, 2011.

Waelbroeck, C., Labeyrie, L., Michel, E., Duplessy, J. C., Mc-
Manus, J. F., Lambeck, K., Balbon, E., and Labracherie, M.: Sea-
level and deep water temperature changes derived from benthic
foraminifera isotopic records, Quaternary Sci. Rev., 21, 295–305,
2002.

Wuchter, C., Schouten, S., Wakeham, S. G., and Sinninghe
Damsté, J. S.: Archaeal tetraether membrane lipid fluxes in
the northeastern Pacific and the Arabian Sea: Implications
for TEX$_{86}$ paleothermometry, Paleoceanography, 21, PA4208,
doi:10.1029/2006pa001279, 2006.

**4**

# Similarity estimators for irregular and age-uncertain time series

**K. Rehfeld**[1,2] **and J. Kurths**[1,2,3]

[1]Potsdam Institute for Climate Impact Research, P.O. Box 601203, 14412 Potsdam, Germany
[2]Department of Physics, Humboldt-Universität zu Berlin, Newtonstr. 15, 12489 Berlin, Germany
[3]Institute for Complex Systems and Mathematical Biology, University of Aberdeen, Aberdeen AB243UE, UK

*Correspondence to:* K. Rehfeld (krehfeld@awi.de)

**Abstract.** Paleoclimate time series are often irregularly sampled and age uncertain, which is an important technical challenge to overcome for successful reconstruction of past climate variability and dynamics. Visual comparison and interpolation-based linear correlation approaches have been used to infer dependencies from such proxy time series. While the first is subjective, not measurable and not suitable for the comparison of many data sets at a time, the latter introduces interpolation bias, and both face difficulties if the underlying dependencies are nonlinear.

In this paper we investigate similarity estimators that could be suitable for the quantitative investigation of dependencies in irregular and age-uncertain time series. We compare the Gaussian-kernel-based cross-correlation (gXCF, Rehfeld et al., 2011) and mutual information (gMI, Rehfeld et al., 2013) against their interpolation-based counterparts and the new event synchronization function (ESF). We test the efficiency of the methods in estimating coupling strength and coupling lag numerically, using ensembles of synthetic stalagmites with short, autocorrelated, linear and nonlinearly coupled proxy time series, and in the application to real stalagmite time series.

In the linear test case, coupling strength increases are identified consistently for all estimators, while in the nonlinear test case the correlation-based approaches fail. The lag at which the time series are coupled is identified correctly as the maximum of the similarity functions in around 60–55 % (in the linear case) to 53–42 % (for the nonlinear processes) of the cases when the dating of the synthetic stalagmite is perfectly precise. If the age uncertainty increases beyond 5 % of the time series length, however, the true coupling lag is not identified more often than the others for which the similarity function was estimated. Age uncertainty contributes up to

half of the uncertainty in the similarity estimation process. Time series irregularity contributes less, particularly for the adapted Gaussian-kernel-based estimators and the event synchronization function. The introduced link strength concept summarizes the hypothesis test results and balances the individual strengths of the estimators: while gXCF is particularly suitable for short and irregular time series, gMI and the ESF can identify nonlinear dependencies. ESF could, in particular, be suitable to study extreme event dynamics in paleoclimate records. Programs to analyze paleoclimatic time series for significant dependencies are included in a freely available software toolbox.

## 1 Introduction

Time series are often used to assess the properties of the processes that generated them, in climate science (Rehfeld et al., 2011) but also in many other scientific fields ranging from ecology (Lhermitte et al., 2011) to astrophysics (Scargle, 1989). Time series similarity measures quantify the degree of statistical association and are, particularly in the geoscientific context, often equated with Pearson correlation (Chatfield, 2004). They help to identify the strength of dependencies between climate processes and potential lead–lag relationships. For modern-day weather stations, both daily temperature and the time of observations are logged precisely. To identify relationships between distant weather evolution, time series of temperature anomalies can be compared. Paleoclimate data are crucial to investigate climate interrelationships beyond the instrumental record. Paleoclimate time series are, however, more challenging than the data sources in other disciplines: neither observation time nor the climatic variable are

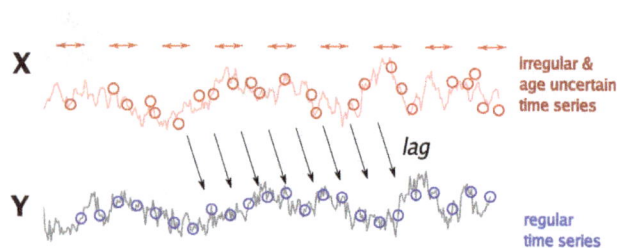

**Fig. 1.** Illustration: assume that the climatic process $Y$ is driven by process $X$ at a given lag. They are sampled by a paleoclimate proxy archive ($X$) and an automatized measurement device ($Y$), resulting in corresponding time series. A typical task in paleoclimate data analysis is to estimate the strength of statistical association between such time series; the delay time can hint at physical driving mechanisms.

known precisely. Both have to be reconstructed, resulting in irregular and age-uncertain time series, because variability in the growth of the archive impacts on the temporal resolution of the resulting proxy time series (Fig. 1). The dependency of reconstructed paleoclimate time series, and their relationship to global or external forcing, is often inferred from similarities, coinciding maxima/minima or trends, between graphical visualizations of the time series (for example in Zhang et al., 2008, 2011; Cheng et al., 2012; Sinha et al., 2011). Visual comparison is, however, inherently subjective, cannot be quantified and tested in a hypothesis test and will not suffice with the growing number of paleoclimatic data sets available.

Standard statistical techniques, such as estimating the Pearson correlation (XC), cannot readily be applied when the sampling of the time series is irregular. XC is, in principle, computed by taking the arithmetic mean over the products of coeval, centralized and standardized observations and reflects the goodness of a linear fit to the scatter plot of the data. If the two time series to be correlated are irregular, coeval observations are only given in the special case that both time series have the same timescale. In practice, this would arise only if, for example, two proxies were measured on the same samples. In the general case the irregularity precludes the direct computation.

Interpolating the time series to a regular coinciding timescale, however, results in a loss of high-frequency variability and a spectral bias towards low frequencies (Schulz and Stattegger, 1997). In a comparison of correlation analysis techniques the Gaussian-kernel-based Pearson correlation was identified as a reliable and robust estimator for irregular time series (Rehfeld et al., 2011). However, relationships in the climate system are not always linear, and therefore not necessarily identifiable by linear techniques such as Pearson correlation. This is not a problem in the geosciences alone, and similarity measures that can capture nonlinear interrelationships exist. Mutual information (MI), an entropy-based measure, has been used to investigate nonlinear dependencies of processes from observations (Donges et al., 2009; Runge et al., 2012; Hlinka et al., 2013). In this mea-

sure, the joint and marginal distributions of processes $X$ and $Y$ are evaluated. Its advantage is that it is model free and able to quantify nonlinear dependencies, but it is symmetric, $MI(X, Y) = MI(-X, Y)$, and more difficult to quantify as the quantification bias changes considerably for different sample sizes and estimator techniques (Khan et al., 2007; Kraskov et al., 2004). It has been adapted and tested for irregular and autocorrelated time series (Rehfeld et al., 2013) in a Gaussian-kernel-based variant. Both MI and XC depend on the notion of a scatter plot between the data.

An alternative, especially in the analysis of extreme events, could be found in the measure of event synchronization (ES, Quian Quiroga et al., 2002), which is not based on the available time series, but the relative timing of distinguished events in two time series. Originally conceived for neurophysiological signals, it has become a popular measure to investigate dependencies in precipitation time series (Malik et al., 2010, 2011; Rheinwalt et al., 2012), but it has not been tested for its suitability on short and autocorrelated time series. In its original form it provides a measure for the strength of synchronization and for the direction of a potential coupling between the processes generating the events, but not for the lag of the potential coupling. Although stated differently in the original paper, ES does not require regular observation intervals.

A number for an individual correlation coefficient can be interpreted, when its level of significance is determined as well. For the usually short and autocorrelated paleoclimatic time series, this can be done by bootstrapping the result (Mudelsee, 2002), or by testing the similarity for mutually uncorrelated surrogate time series with similar autocorrelation properties (Rehfeld et al., 2011, 2013). The values of the different estimators, however, cannot be compared directly, as they vary on different scales. In this paper we evaluate the impact of age uncertainty and time series irregularity on the accuracy of the estimators.

Furthermore we propose the concept of a *link strength*, to summarize the hypothesis test results of different estimators. If no outcome is significant, it is zero, if three out of five employed estimators yield a significant similarity, the link strength is 3/5 and if all tests for null correlation were rejected the link strength is equal to unity. The advantage of this approach lies in its robustness due to the different estimators, and in the easy consideration of uncertain data sets. If the uncertainty of the time series can be modeled, for example using the Monte Carlo techniques in age modeling software such as StalAge (Scholz and Hoffmann, 2011) or COPRA (Breitenbach et al., 2012), it can be incorporated in the link strength considerations in a straightforward manner.

In this paper we will investigate how well each of these estimators identify the strength and the delay time of actual coupling between paleoclimatic processes from irregular and age-uncertain time series. First we review the similarity measures (XC, MI), and develop a event synchronization function (ESF) based on the concept of ES. We simulate artificial stalagmites with linearly and nonlinearly coupled proxy time series based on autoregressive (AR) and threshold-autoregressive (TAR) models. Using these and the stalagmite time series from Dandak (Sinha et al., 2007; Berkelhammer et al., 2010) and Wanxiang (Zhang et al., 2008) caves, we investigate how the similarity estimators perform for irregular, age-uncertain and autocorrelated time series, and how they are impacted by age uncertainty.

## 2 Methods

In this section we first give necessary definitions for time series and similarity measures, and derive the ESF and the link strength concept.

### 2.1 Time series

Time series are a collection of measurements of specific properties of a dynamical process, together with the time when the observation (or measurement) took place. The individual data points of the series are often regarded as observations of processes, which may be deterministic, stochastic, or a combination of both. In classical time series analysis the observation times of the process $X_t$ are expected to be regular and certain, and the observation values to be measured exactly.

In contrast to this, for irregular time series no unique sampling rate can be defined, and the observation times cannot be directly related to an index anymore, but have to be given explicitly for each measurement.

**Definition 1 (Irregular time series)** *An irregular time series $x(t) = (t_i, x_i)$ is defined by its observation times $t_i$ and the respective observations $x_i$, where $i = 1, \ldots, N$. The two vectors have a common length $N_x$, with $t_1^x < t_2^x < \cdots < t_{N_x}^x$ as observation times.*

In the following we focus on the age-uncertain paleoclimate proxy time series for which a growth model of the archive has been combined with pointwise age information, for example from uranium/thorium measurements. Input data to this age modeling are (i) a dating table with its entries containing depths, associated age estimates and their uncertainties, usually given as standard deviations, and (ii) the proxy observations.

**Definition 2 (Dating table)** *A dating table $\mathbb{D} = (D_i, T_i, \sigma_{T_i})_{i=1,\ldots,N_{dat}}$ contains $N_{dat}$ pointwise age estimates $T_i$ taken at depths $D_i$ and their corresponding age standard deviations $\sigma_{T_i}$.*

**Definition 3 (Proxy observation series)** *Proxy observation series $\mathbf{X}^d = (d_j, x_j)$ are given for $j = 1, \ldots, N_{obs}$ measurement depths $d_j$ and proxy measurements $x_j$.*

For paleoclimate archives, the ages at few depths are estimated, with some uncertainty. Age models are then created to

interpolate from these few dates to a time axis for the proxy time series, which is sampled much more densely in depth than the dating table. Thus, an age model is defined here as one potential depth–age relationship $t_i(z_i)$ out of the possible ensemble of age models $\mathbb{T}$. For Monte Carlo (MC) age modeling, whole *ensembles* of age models, $\mathbb{T}$ are created, sampling the probability space inherent in the dating table (cf. def. 2). By convention, usually the *most likely* age model is selected as the time axis for proxy time series (Breitenbach et al., 2012; Scholz and Hoffmann, 2011). Finally the dating table is combined with the proxy observation series using a single-age model to form a time-uncertain time series.

### 2.2 Estimating similarity of irregular time series

Similarity measures reflect statistical properties of time series, which may not reflect the same climatic parameters. Different estimators focus on different characteristic properties related to the distributions of the observations. We summarize them in Table 1.

Assume that the processes $X$ and $Y$ generated time series $x(t)$ and $y(t)$. These processes, and the time series, are similar if, for example, coeval minima or maxima were observed. Comparison can then give information about functional relationships between processes underlying time series: given that two processes $X$ and $Y$ are not independent, there may either be a causal relationship or they are both driven by a global *common driver*, or there are unobservable intermediate processes, as illustrated in Fig. 2. A significant similarity estimate may therefore arise for such physical reasons – or as a false positive of the statistical test. If a transfer function between these two processes exists in a form $Y_t = \mathcal{F}(X_{t+\ell})$, this results in a repetition of a pattern, though maybe distorted, that occurs in $X_t$ at $t_0$ and in $Y_t$ at a time $t = t_0 + \ell$ later. A similarity estimator can help identify $\mathcal{F}$ and quantifies the similarities in the contemporary evolution of two time series:

**Definition 4 (Similarity estimator)** *A similarity estimator $S = \mathcal{F}((t^x, x)(t^y, y))$ reflects the similarity between $x(t)$ and $y(t)$ to a numeric value in an interval $[a, b]$, $S : x(t) \times y(t) \rightarrow [a, b]$.*

For most similarity measures $a = -1$, $b = 1$ is considered, but for different estimators different bounds exist. Here we only require that the relationship between true dependency

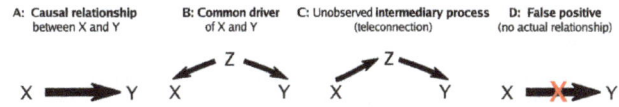

**Fig. 2.** Significant similarities between the time series at two locations, X and Y, can arise from **(a)** direct physical coupling, **(b)** a teleconnection, **(c)** a common driving mechanism or **(d)** by chance as false positives.

**Table 1.** Properties, parameters and references of the similarity estimator algorithms for irregularly sampled time series developed and tested in this paper.

| Estimator (abbr.) | Quantif. property | Parameter choice | References |
|---|---|---|---|
| 1 (gXCF) | Gaussian-kernel-based XCF (goodness of linear fit to scatter plot) | $h = 0.25$ | Rehfeld et al. (2011); Babu and Stoica (2010) |
| 2 (tiXCF) | interpolation + Pearson correlation (goodness of linear fit to scatter plot) | $\Delta t = \max(\Delta t^x, \Delta t^y)$ | Rehfeld et al. (2011); basics, for example, in Chatfield (2004) |
| 3 (gMI) | Gaussian-kernel-based MI (relative non-randomness in joint vs. marginal distribution) | $h = 0.5, \tau = 3$ | Rehfeld et al. (2013); basics, for example, in Cover and Thomas (2006) |
| 4 (iMI) | interpolation + MI (relative non-randomness in joint vs. marginal distribution) | $\Delta t = \max(\Delta t^x, \Delta t^y)$, $n_{\text{bins}} = 10$ | Rehfeld et al. (2013); basics, for example, in Cover and Thomas (2006) |
| 5 (ESF) | Relative timing of extreme events | $q = 0.8$ | based on Quian Quiroga et al. (2002); Malik et al. (2010) |

and estimated similarity is monotonically increasing, which is what we test for using artificially generated time series. If the delay time $\ell$ in the transfer function is nonzero, a similarity function gives the similarity between two time series for increasing delay:

**Definition 5 (Similarity function)** *A similarity function $S(\ell)$ gives the estimated similarity over different lag times $\ell$:*

$$S(\ell) = S(\ell \cdot \Delta t) = f\left((t^x, x), (t^y + \ell \cdot \Delta t, y)\right). \quad (1)$$

*The spacing of the lag vector is uniform and depends on the mean time resolution of the time series: $\Delta t = \max(\Delta t_x, \Delta t_y)$. To indicate that we are focusing on bivariate similarity we also use the alternative notation $S(X, Y)$ which does not explicitly refer to the possible lags.*

Similarity measures as required in this context should be symmetric, reflexive, translation and scale invariant (Batyrshin et al., 2012). The estimators presented here fulfill these requirements.

### 2.2.1 Kernel-based estimators for Pearson correlation

Pearson correlation is defined as the mean over coeval products of standardized observations (Chatfield, 2004). For irregular time series the inter-sampling time intervals vary and the classical definition cannot be applied. Rehfeld et al. (2011) tested different correlation estimators for irregular time series and found that a Gaussian-kernel-based estimator performed best. In the definition of the correlation function $\hat{\rho}(k\Delta t)$ at the lag $k\Delta t$:

$$\hat{\rho}(k\Delta t) = \frac{\sum_{i=1}^{N^x}\sum_{j=1}^{N^y} x_i y_j b_k(t^y_j - t^x_i)}{\sum_{i=1}^{N}\sum_{j=1}^{N} b_k(t^y_j - t^x_i)}, \quad (2)$$

the *kernel* $b_k(t^y_j - t^x_i)$ weights those products higher whose time lag lies closer to $k\Delta t$:

$$b_k(d) = \frac{1}{\sqrt{2\pi}h} e^{-|d|^2/2h^2}, \quad (3)$$

where $h = \Delta t/4$ or 0.25 for the rescaled time axis, $t^x_i = t^{\text{orig}}_i/\Delta^x_t$, and $d$ denotes the distance between the product inter-observation time and the desired lag, $d = t^y_j - t^x_i - k\Delta t$; $k$ denotes the lag index. The standard width parameter $h$ is chosen to result in a main lobe width of $\Delta t$, the mean sampling interval or common sampling period in the bivariate case. Note that the observations have to be standardized to zero mean and unit variance before the analysis.

### 2.2.2 Kernel-based estimators for mutual information

Mutual information $I(X, Y) = I_{xy}$ is a measure of the dependency (linear or nonlinear) between two random variables, $X$ and $Y$. This measure from information theory can be interpreted as the uncertainty reduction in variable $X$, given that $Y$ was observed. It is symmetric, that is, relationships of opposite sign but the same association strength, correlation and anti-correlation give the same MI. By definition, the measure yields a null result if, and only if, the two random variables, in this case time series of observations, are independent (Kraskov et al., 2004; Cover and Thomas, 2006).

While more complex estimators exist (e.g., Kraskov et al., 2004), the simplest estimator is

$$\hat{I}_{xy} = \sum_{x,y} p_{x,y} \log \frac{p_{x,y}}{p_x p_y}, \quad (4)$$

where $p_{x,y}$ is the two-dimensional joint probability density function of the variables $X$ and $Y$ and $p_x$ resp. $p_y$ are the one-dimensional probability distributions of $X$ resp. $Y$. The unit of measurement of MI depends on the *logarithm* chosen in the estimator: it is measured in *bits* if the logarithmic base 2 is chosen, and in *nats* for the natural logarithm.

In case of irregular sampling, however, the bivariate observation set $(X_t, Y_t)$ at regular observation points $t$ that are required for a scatter plot is not available. In standard interpolation procedures, both $(t_x, x)$ and $(t_y, y)$ would be resampled to obtain a bivariate set of observations with regular observation time intervals, $(t_r, x_r, y_r)$. This is undesirable for paleoclimate records (a) because every interpolation routine involves an assumption on the dynamics of the underlying process, and this is difficult to justify for climate data, and (b) it reduces the observable variability in the process (Schulz and Stattegger, 1997; Stoica and Sandgren, 2006; Babu and Stoica, 2010).

There are two main points where this problem can be addressed: either by reconstructing bivariate observations while avoiding variance reduction as much as possible or by a modification of the joint distribution, for example by introducing weights proportional to the sampling time distance similar to the Gaussian-kernel-based XC (Rehfeld et al., 2011). For MI the latter is difficult to achieve. But following the former solution, the probabilities required for Eq. (4) are straightforward to derive from relative frequencies.

Algorithmically, this can be described as follows:

1. A local reconstruction of the signal is performed by estimating for each point $i$ in the time series $X = (t^x, x)$ a corresponding observation from $Y = (t^y, y)$, by estimating a local, observation-time weighted mean $y_j^{lr}$ around a time point $t_i^x$ in $Y$,

$$y_j^{lr} = \sum_{i=1}^{N_y} b_k(d) y_i , \qquad (5)$$

with the Gaussian-kernel-based local weight $b_k(d)$ defined as in Eq. (3). For MI the standard deviation of the Gaussian weight function is set to $h = 0.5$. If there are less than five observations $y_i$ available in a time window $\pm 3\Delta t$ around $t_i^x$ this reconstruction is not performed. Repeating this for each time point $j = 1, \dots, N^x$ in $X$ one obtains a new, bivariate set of observations

$$Y^x = (t_i^x, x_i, y_i^{lr}) .$$

2. Afterwards the procedure is repeated by stepping through $t_j^y$, which yields

$$X^y = (t_j^y, x_j^{lr}, y_j) .$$

3. The local reconstruction $Y^x$ and the original observations $Y$ are then concatenated into one series $Y^r = \{Y \cup Y^x\}$ combining locally reconstructed and original observations. Similarly, a time series $X^r = (X \cup X^y)$ is obtained.

4. Based on this set of bivariate observations $(X^r, Y^r)$ the joint density of $X$ and $Y$ can be estimated using standard binning estimators for MI.

The reconstructed set of bivariate observations can also be used to construct Gaussian-weighted scatter plots, where the size of the marker reflects the amount of weight placed on the reconstructed observation (cf. Figs. 4b and 5b). MI is difficult to estimate in practice, first and foremost because of the large bias effects produced in the inference of the joint and marginal probabilities. Elaborate algorithms have been devised to improve this (described, for example, in Kraskov et al., 2004; Papana and Kugiumtzis, 2009; Roulston, 1999), but no straightforward solution to this has been found yet. We have tested several algorithms and finally resorted to the most simple equidistant *binning estimator* (Kraskov et al., 2004), due to its computational efficiency and simplicity. Bias effects are predominantly tied to the temporal sampling and length of the time series due to the occurrence of empty bins. Thus, if necessary, we can estimate and subtract the bias using uncorrelated processes with the same observation times as in $X$ and $Y$. However, for use as a similarity measure comparable to XCF and ES in the context of paleoclimate networks, we only require that the estimated MI be proportional to the actual association strength. For bivariate normally distributed and linearly correlated $X$ and $Y$, MI is by definition proportional to their estimated correlation coefficient $r_{xy}^2$:

$$I_{xy} = -\frac{1}{2} \log(1 - r_{xy}^2) , \qquad (6)$$

and can, by inversion of this equation, be scaled to the positive range of the correlation coefficient so that $\hat{I} \in [0, 1]$ (Nazareth et al., 2007). The expected value for mutual information of these processes at the lag of coupling is then given by $MI(X(t), Y(t+l)) = -0.5 \log(1 - r_{xy}^2)$. For the evaluation of the joint and marginal distributions, $n_{bins} = 10$ equidistant bins were employed. In principle, the number of bins should be adapted to the respective length of the time series involved, to reduce bias effects from empty bins.

### 2.2.3 Event synchronization function

The concept of event synchronization (ES) was introduced by Quian Quiroga et al. (2002). The motivation behind the development was to obtain a simple, fast method that quantifies the synchronization between time series where certain *events* can be distinguished. The primary application was focused on neurophysiological signals (Quian Quiroga et al., 2002; Kreuz et al., 2009), but it was also applied later for the investigation of rainfall patterns in the Asian monsoon domain (Malik et al., 2010, 2011) and Europe (Rheinwalt et al., 2012).

The main idea behind ES is that two time series are synchronized, if events in time series $x$ occur close in time to events in time series $y$. Considering the temporal order of the events (e.g., if an event in $y$ occurred *before* one in $x$), it is also possible to infer which process is *leading*. In the following we will define the event synchronization function, ESF, further developing the ES concept (Quian Quiroga et al., 2002; Malik et al., 2010).

Given two time series $(t^x, x)$ and $(t^y, y)$ that represent observations of autocorrelated stochastic processes, *events* are given by the set of observations that are considered *extreme*, in that their observation value lies above or below the $q/2$ resp. $(1 - q/2)$ percentiles of the distributions of $x$ and $y$. The actual *value* of the observation at the event points is not relevant for the further analysis. Once the events are defined, only the observation *times* are considered in the event time vectors $t_x^*$ and $t_x^*$. Next a temporal threshold $\tau$ is defined to evaluate the relationship between the events in $X$ and $Y$ with a maximum separation time:

$$\tau = \max\left(\Delta t^x, \min(\Delta t_x^*, \Delta t_y^*)/2\right). \tag{7}$$

Here, $\Delta t^x$ is the mean sampling rate of $X$, and $\Delta t_x^*$ and $\Delta t_y^*$ are the inter-event times in $X$ and $Y$, respectively.

Subsequently, the co-occurrence of events in $X$ and $Y$ is counted and summed for all events as

$$C(X|Y) = \sum_{l=1}^{N_x} \sum_{m=1}^{N_y} \mathbf{J}_{lm}^{xy}, \tag{8}$$

where $N_x$ and $N_y$, respectively, give the total numbers of events in $X$ and $Y$. The counter variable $\mathbf{J}_{lm}^{xy}$ is defined as

$$\mathbf{J}_{lm}^{xy} = \begin{cases} 1 & \text{if } 0 < t_l^x - t_m^y < +\tau \\ 1/2 & \text{if } t_l^x - t_m^y = 0 \\ 0 & \text{otherwise}. \end{cases} \tag{9}$$

$C(Y|X)$ is obtained by exchanging $X$ vs. $Y$ in the above expression, and combining both,

$$Q_{xy} = Q_{xy}(X, Y) = \frac{C(X|Y) + C(Y|X)}{\sqrt{N_x, N_y}}, \tag{10}$$

gives the *strength* of the event synchronization and

$$q_{xy} = \frac{C(X|Y) - C(Y|X)}{\sqrt{N_x, N_y}} \tag{11}$$

the *direction* of the association. Unless double counting of events occurs, these are normalized to $0 \le Q \le 1$ resp. $-1 \le q \le 1$. $Q = 1$ corresponds to completely synchronous occurrence of events in $X$ and $Y$, and $q = 1$ implies that all events in $Y$ precede those in $X$.

For the previous studies (Quian Quiroga et al., 2002; Malik et al., 2010, 2011) local definitions of the temporal threshold $\tau$ were used, preventing, in most cases, events from being double counted, and adapting it to the local inter-event rate. The chosen definition of $\tau$ is motivated by the fact that, to

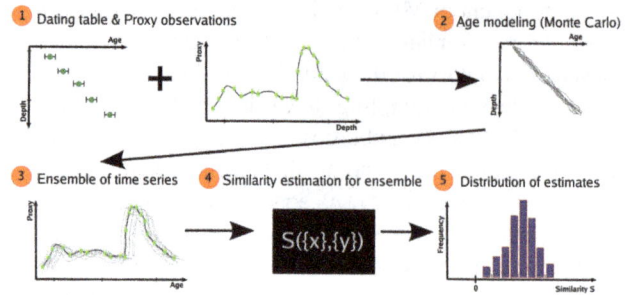

**Fig. 3.** How much age uncertainty is allowed to still enable reliable similarity estimation? Artificial stalagmites with increasing standard deviations of the ages are evaluated.

be able to compare the results for ES to those obtained from MI and XCF, a similarity function over the *delay* is needed. Thus, the delay $\tau$ cannot be allowed to be arbitrarily large or small, as in Malik et al. (2010) or Quian Quiroga et al. (2002).

The ESF is obtained by shifting the observation times of time series $X$ according to the desired lag:

$$\text{ES}(k\Delta t) = Q_{xy}((t_x - k\Delta t, x), (t_y, y)), \tag{12}$$

which, using the delay time $\tau$ from Eq. (7), makes it possible to use the ESF as a similarity function.

## 2.3  An approach to similarity assessment of time-uncertain time series

Age uncertainty is a key obstacle to be overcome for a comprehensive understanding of past earth system dynamics. To investigate the potential dependency structure of paleoclimate processes $X$ and $Y$ as they are reflected in natural archives, the contribution of age uncertainty to the uncertainty of the similarity $S(X, Y)$ is important.

Thus the aim is to estimate the distribution $p(S(X, Y))$ of similarity for given data sets $X$ and $Y$, where

$$X = \left[\mathbb{D}^x = \{D^x, T^x, \sigma_{T^x},\} Y^d = \{d^x, x\}\right] \quad \text{and} \tag{13}$$

$$Y = \left[\mathbb{D}^y = \{D^y, T^y, \sigma_{T^y}\}, X^d = \{d^y, y\}\right]. \tag{14}$$

Both input data sets consist of a dating table (Def. 2) $\mathbb{D}$ with dating depth vector $D$, the corresponding estimated ages $T$ and their uncertainties $\sigma_{Ty}$ and a set of proxy measurements $X^d$ resp. $Y^d$ (Def. 3), visualized as Step 1 in Fig. 3. The smoothing resulting from the size of the samples in depth direction, $\sigma_D$, is assumed to be negligible here. The input proxy measurements are mapped to observation times in the *age modeling* process. In general, algorithms to assess similarity between time series are not capable of processing *probability distributions* or *confidence intervals* instead of singleton values, neither for the observation times nor for the measurement values.

For Pearson correlation, an analytical approach to propagate the uncertainty around the input data into the correlation estimate is possible. However, Pearson correlation alone is insufficient to characterize similarity between paleoclimate time series in general and in the context of paleoclimate networks. Therefore, a Monte Carlo-based approach based on time series ensembles which are obtained via age modeling is used here, to keep the flexibility regarding similarity estimators:

1. In a first step the input data sets $X$ and $Y$ are processed. The monotonicity of the depth control variables, $d$ and $D$ is checked.

2. A Monte Carlo simulation for the uncertain age estimates in the dating table is performed: $N_{\text{ens}}$ ages are drawn from $\mathcal{N}(T_i^X, \sigma_{T_i^X})$ and $\mathcal{N}(T_j^Y, \sigma_{T_j^Y})$, respectively, for all $i = 1, \ldots, N_{dtg}^X$ pointwise age estimates corresponding to $j = 1, \ldots, N_{dtg}^Y$ entries in the dating table. This results in dating matrices $\hat{\mathbf{X}}$ and $\hat{\mathbf{Y}}$ with $N_{\text{ens}}$ columns containing the sampled ages. If no distribution of ages is otherwise given, the ages are expected to be Gaussian distributed with the given standard deviation.

3. The age estimates in each column and $\hat{\mathbf{X}}$ ($\hat{\mathbf{Y}}$) are interpolated to the depths of the proxy observations: $\mathbf{T} = \mathtt{interp}(D, \hat{\mathbf{X}}, d)$ which results in a matrix of reconstruction observation times $\mathbf{T}$. We used conventional linear interpolation of the ages in COPRA. Thus we obtain an ensemble of possible age–depth relationships $\{\mathbf{T}, d\}$ and an ensemble of proxy time series $\{\mathbf{T}, x\}$.

4. Each of the members of the ensemble of proxy time series is used as an input to the similarity statistic $S(X, Y)$. This results in a distribution of estimates $p(S(\hat{\mathbf{X}}, \hat{\mathbf{Y}}))$.

5. Analysis of distribution $S(\hat{\mathbf{X}}, \hat{\mathbf{Y}})$: apart from inspection of mean, variance and skewness of this distribution, a hypothesis test can be conducted, comparing $S(\hat{\mathbf{X}}, \hat{\mathbf{Y}})$ with a distribution obtained from suitable surrogate time series $S(\hat{\mathbf{X}}^*, \hat{\mathbf{Y}}^*)$.

This approach is general in the sense that it is independent of the specific function $\mathcal{F}([\hat{\mathbf{X}}, \hat{\mathbf{Y}}])$ that maps the uncertain input to some output estimate. Apart from $\mathcal{F} = S$, $\mathcal{F}$ may represent any bivariate statistic, and with minor modification is also applicable to calculate the influence of sampling uncertainty on univariate statistics, like the autocorrelation coefficients or persistence times (Rehfeld et al., 2011; Mudelsee, 2002). Bivariate similarity assessment is often concerned with estimation of a potential *coupling strength* $\alpha$ (hinting towards the same process of origin) and/or the *lag of coupling* $\ell$ for model-building. For Pearson correlation, the ratio of shared vs. total variance between two linearly correlated processes at a given lag $\ell$, $S(\ell)$, is given in the maximum of the cross-correlation function. While the relation to the overall variance of the processes does not necessarily hold by definition for other similarity measures, they, too, will observe the maximum of their similarity function $\max(\hat{S})$, at the lag of coupling $\ell$.

### 2.3.1 Synthetic data

"True" growth histories for two synthetic stalagmites $SS1$ and $SS2$ and according climate histories are obtained via simulation. These pseudo-archives are then "dated", climate histories are "sampled". Then the age modeling procedure is performed and its output is fed into similarity estimation. Finally, we assess how much of the similarity that was originally present in the climate history is still recognizable significantly, considering the uncertainties. The test strategy is illustrated in Fig. 3.

### 2.3.2 The synthetic stalagmite

A synthetic (or virtual) stalagmite is grown for the sensitivity analysis. The main parameters controlled are

- the growth rate $\lambda$ in mm yr$^{-1}$,

- the total length of the stalagmite (in mm),

- the type of accumulation (linear growth, or growth modeled via randomly distributed accumulation rates).

A growth rate of $\mu(\lambda(z)) = 1\,\text{mm yr}^{-1}$ is chosen. Linear growth may be a reasonable first order approximation (Telford et al., 2004), but microscopically, the growth rates of natural archives vary. Therefore, Gamma-distributed accumulation times are drawn for each depth $z_i = \{0, \ldots, Z\}$mm of the stalagmite, with the sampling time step mean $\mu(\lambda(z))$ determined by the desired growth rate and shape and scale parameters $\alpha$ and $\beta$ as $\Gamma(\alpha, \beta) = \Gamma(\alpha, \mu(\lambda(z))/\alpha)$. This way, the mean sampling rate can be kept constant, even when the irregularity of the sampling distribution is changed (Rehfeld et al., 2011). The cumulative sum of the accumulation times then gives the "true" ages of the archive at the depths $z_i$: $t_i^{\text{true}}(z_i) = \sum_{j=1}^{i} \lambda_i$.

### 2.3.3 The simulated climate history

We attach each synthetic stalagmite SS1 and SS2 to a climate history. The climate/pseudo-proxy simulation is based on the assumption that SS1 lies in an area whose climate is controlling that around SS2, through a teleconnection or, for example, by being situated downstream of the same monsoon branch (cf. Fig. 2). We simulate climate variability using two different coupling schemes, one linear, one nonlinear, to investigate how the proposed methods perform.

## Linearly coupled AR(1) processes

Assuming that the archive SS2 samples the same climate variability as SS1, in the same way though at a later time, we model such a causal sequence using coupled AR(1) processes. Then, the *true* proxy history of climate as recorded in SS1 is given by

$$X(t_i^{\text{true}}, z_i) = \phi X(t_{i-1}^{\text{true}}) + \sigma_\varepsilon \varepsilon_i, \tag{15}$$

and it determines part of the proxy history of $SS2$:

$$Y(t_i^{\text{true}}, z_i) = \alpha X(t_{i-\ell}^{\text{true}}) + \sigma_\xi \xi_i. \tag{16}$$

Here, $\varepsilon$ and $\xi$ are additional Gaussian white noise whose variances $\sigma_\varepsilon$ and $\sigma_\xi$ are scaled such that the variances of $X$ and $Y$ are equal to unity. $\alpha \in [-1, 1]$ is the coupling strength between SS1 and SS2 and $\phi$ the autocorrelation of SS1. Since there is no autocorrelative term in $Y_t$, the true similarity $S(X, Y)$ is equal to the cross-correlation: $S(X, Y) = \rho_{xy} = \alpha$ (Rehfeld et al., 2011).

## Nonlinear threshold-AR(1) processes

Let us assume that SS1 samples climate variability in a certain place, and that this can be modeled as in Eq. (15). Then the climate variability in another place, where SS2 is located, could be controlled in a nonlinear manner: the processes are negatively correlated, similar to Eq. (16) with $\alpha < 0$. If, however, a threshold in the climate system is exceeded, $X(t) > \tau$, the correlation changes and might even become positive. Such a multi-scale behavior can be modeled using threshold-AR processes (TAR, Tsay, 1989), which are similar to the regime-dependent AR models Zwiers and Storch (1990) used to model the behavior of the Southern Oscillation. Assume that the negative coupling $\alpha$ below the threshold $\tau$, here $\tau = 0$, for $X(t-1) \leqslant \tau$ turns into a positive correlation, with the same magnitude, for $X(t-1) > \tau$. Then the proxy history of SS2 can be modeled as

$$Y(t_i^{\text{true}}, z_i) = \alpha \kappa X(t_{i-\ell}^{\text{true}}) + \sigma(t^{\text{true}}) \xi_i, \tag{17}$$

where the $\kappa = -1$ if $X(t-1) \leqslant \tau$ and $\kappa = 1$ when $X(t-1) > \tau$. For convenience, the variance of the innovation term $\xi$ is scaled such that the overall variance of $Y$ is equal to unity in both cases.

### 2.3.4 "Dating" of the synthetic stalagmite

Mimicking the real-life situation, the *true* growth history of the synthetic stalagmite $z(t_{\text{true}})$ is, in the following, inaccessible. The stalagmite is subjected to *dating* along its depth. The dating table contains the dating depths $D$, the estimated age at these depths $T_j$, the proxy measurement sample width $\sigma_D$ and the age uncertainty $\sigma_T$.

In real life, the stalagmite would be dated using radiometric dating techniques based on uranium-thorium (Sinha et al.,

2007; Dykoski et al., 2005; Breitenbach et al., 2012) or radiocarbon (Yadava et al., 2004; Webster et al., 2007), yielding an estimate of $T(z_j)$ at a few points. The corresponding dating uncertainty, in reality dependent on many factors from initial isotope concentrations, overall age of the core, dating technique, lab and contamination (Fairchild and Baker, 2012), often lies between 0.1 to 0.5 % of the age for stalagmites, but may be considerably higher.

For the synthetic stalagmites, dating "samples" are taken at equidistant depths $D_j$ and the center points of the assumed age distribution are taken directly from the *true* age–depth relationship. The age uncertainty, however, is modeled as increasing proportionally with age, as $p \cdot T_j$. $p$ here denotes the (im-)precision of the dating and is varied in the following numerical experiments.

### 2.3.5 Age modeling for SS1 and SS2

Age modeling aims to reconstruct the "true" depth–age relationship that is inaccessible in real paleoclimate archives.

Based on the synthetic stalagmite dating tables $\mathbf{D}^x$ and $\mathbf{D}^y$ for SS1 and SS2, the "observation times" for the proxy observations $X^d$ and $Y^d$, $t^x$ and $t^y$, are constructed by interpolation from the known ages (see Eq. 13). In Monte Carlo-based numerical frameworks such as StalAge (Scholz and Hoffmann, 2011) or COPRA (Breitenbach et al., 2012), an ensemble of age models $\mathbf{T} = \{t_k, z_k\}^{k=1,...,N_{\text{ens}}}$ is created, which, in their entirety, reflect the age uncertainty of the estimated depth–age relationship. Based on this ensemble of age models, the uncertainty in the similarity estimates can be inferred, as is visible in Fig. 3.

In summary, the test plan is thus as follows:

1. Simulate a growth history $z(t)$ of a synthetic stalagmite of length $Z$ mm, corresponding to a "true" age–depth relationship $t_i^{\text{true}}(z_i)$, resp. $z_i(t^{\text{true}})$. For this, assume gamma-distributed growth and an accumulation rate $\lambda = 1$ mm yr$^{-1}$. $Z$ can be varied to study the influence of changing time series length.

2. Simulate proxy histories $\{T, x\}^{\text{SS1}}$ and $\{T, y\}^{\text{SS2}}$ according to the *true* growth history using coupled autoregressive processes (cf. Eqs. 16 and 17). Forget the true growth history.

3. Sample the true growth history at the dating depths and infer corresponding uncertainties.

4. Create $N_{\text{ens}}$ surrogate dating tables for SS1 and SS2 with increasing uncertainty of the ages according to the (im)precision $p$ (i.e., an ensemble of dating tables).

5. Assess if the estimates $S(\hat{\mathbf{X}}, \hat{\mathbf{Y}})$ are statistically significant for the given uncertainty, and how they are influenced by sampling heterogeneity and time uncertainty.

The core of the COPRA algorithm is used for MC simulations. $N_{\text{ens}} = 2000$ MC iterations are used to sample the

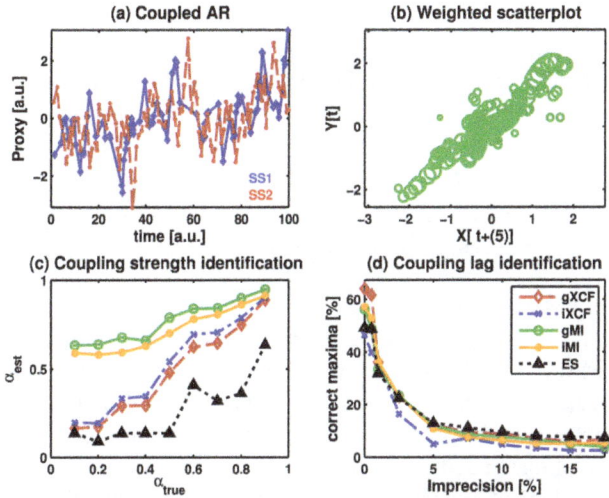

**Fig. 4.** Testing the similarity measures: for linearly coupled AR time series (cf. Eq. 16) from two synthetic stalagmites, SS1 and SS2, we give the sample time series (**a**) and the Gaussian weighted scatter plot (**b**). We check the monotonicity of the estimators with increasing coupling strength (**c**) and how often the maximum of the similarity function correctly coincides with the lag of coupling (**d**).

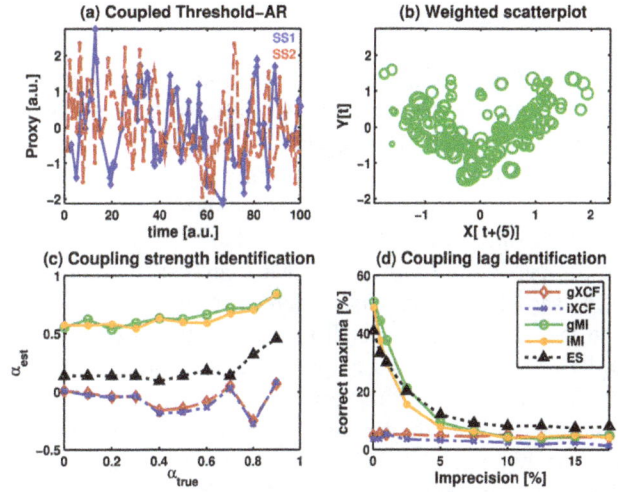

**Fig. 5.** Testing the similarity measures for nonlinear threshold-AR time series (cf. Eq. 17). For caption please refer to Fig. 4.

probability space and linear interpolation is employed to infer ages between point estimates of the age at depth.

## 3   Tests on synthetic stalagmites

We evaluate the performance of the different estimators described in Sect. 2, for which parameter choices and references are given in Table 1.

### 3.1   Characterization of linear proxy dependency

We first consider the linear dependency case, where the proxy history of SS1 is linearly correlated with that of SS2 a lag time $\ell$ later. We chose a length for the stalagmite of $L = 100$ mm for which we expect the time series to be roughly 100 yr long (cf. Sect. 2.3.2) and linearly correlated, as in Fig. 4a. For each test 100 time series were generated from AR1 processes (cf. Sect> 2.3.3), where process $Y$ is coupled to process $X$ at an intrinsic lag $\ell$ and with a coupling strength $\alpha$. The autocorrelation parameter was set to $\phi = 0.8$, the coupling lag to $\ell = 5$ and the coupling parameter to $\alpha = 0.6$. For such stochastic processes, the true similarity function is single peaked, with its peak height determined by $\alpha$, and its location on the lag-axis by the coupling lag $\ell$. The time series are irregular, therefore a direct scatter plot of the data is not possible. Figure 4b shows a weighted scatter plot where the time series have been reconstructed using Gaussian weights, as for the MI estimation in Sect. 2.2.2.

The tests were guided by two questions: do the similarity estimators reflect the actual similarity (here, the coupling strength at lag $\ell$, $\alpha$) truthfully and monotonically? and, how

well do they identify the lag of coupling $\ell$ as the maximum of the similarity function?

To answer the first question, we fix the imprecision at zero (at the dating points) and vary the coupling strength by setting the parameter $\alpha$ in Eq. (16) to values from 0.1 to 1. The results are given in Fig. 4c. The expected value of the similarity, $\alpha_{est}$, and the variance of the estimate are computed from the mean and standard deviations of the estimated, $\alpha_{est,i}$, for 100 realizations for each value of the coupling parameter. Each of the similarity measures returns estimates whose expectation values increase monotonically with the actual similarity, $\alpha_{true}$ in Eq. (16), except for the ESF, which has a single reversal which may be due to the low number of MC realizations (100) for each point in this diagram.

In practical data analysis, the potential lag and strength of (primary) coupling, identified as the maximum of the similarity function is of interest (e.g., for model-building). If no age uncertainty exists at the dating points, the maximum of the similarity function is correctly identified in 50–60% of the ensemble cases. When timescale uncertainty exists in the time series, this becomes difficult quickly (Fig. 4d). When the fraction of correct identifications has dropped to $\frac{1}{n_\ell} \approx 0.05$, where $n_\ell$ is the number of lags for which $S(\ell)$ has been estimated, the maxima of the similarity functions are perfectly uncorrelated. This limit is approached as an imprecision of more than 10 % is reached. Increasing imprecision contained in the time series also results in increasing estimation error (i.e., root mean square error(RMSE)) for the similarity at the lag of coupling, $S(\ell)$ (results not shown). When the stalagmite length is increased, the time series length increases and both the RMSE and the false identification rate decreases for all estimators.

## 3.2   Nonlinear dependencies

For the nonlinear TAR model, the time series in Fig. 5a are not as straightforward to compare visually as the linearly coupled ones in Fig. 4a. The weighted scatter plot for these time series in Fig. 5b shows the two different slopes of the positive and negative correlation regimes above and below the threshold value of zero.

The comparison of true vs. estimated coupling strength $\alpha$ in Fig. 5c shows no monotonous behavior for the linear correlation measures and no overall increase of their expected similarity estimates with the coupling strength. The MI estimators retain a monotonic increase, starting from a considerable bias value, while the ESF increases monotonically, but does not show consistent similarity estimate increases until the coupling strength is rather large. The monotonicity and linearity of the response for gMI, iMI and ESF improve considerably when the time series are chosen longer, that is, with a length of 200 or more (results not shown).

In the identification of the maximum lag the Gaussian MI succeeds most often for imprecisions up to 2.5 %. For more imprecise data sets the ESF remains stable, while the other measures perform worse and worse. The linear estimators, gXCF and iXCF do not identify the maxima correctly, neither the coupling strength, nor the lag of coupling.

## 3.3   Error source attribution

Age uncertainty has a considerable impact on the accuracy of similarity estimates, as we have shown in the previous section. But to what extent can this impact be attributed to the short length of the time series, or the time series irregularity that results from the increasing age uncertainty? The uncertainty around the ages in the dating table is, in Monte Carlo-based age–depth modeling, reflected by drawing different "dates" from distributions around these ages for each MC realization. These realizations will therefore have different partial slopes between any date $D_i$ and $D_{i+1}$. This corresponds to different estimated growth rates for the individual segments of the synthetic core. At a proxy sampling rate over depth that is constant, this will lead to uneven observation times for the time series which correspond to the MC realizations, and this irregularity increases with the age uncertainty. The RMSE of $S(\ell)$ is, however, also dependent on the irregularity of the time series, as it was shown for both XCF and MI previously (Rehfeld et al., 2011, 2013).

To separate these sources of uncertainty, $M = 2000$ realizations of coupled climate histories, as defined in 2.3.2, were generated in three different ways: *age uncertain*, *irregularly* and *regularly sampled*. The age-uncertain ensembles were the direct product of the age modeling efforts, as in the previous sections and with same parameter settings ($\phi = 0.8$, $\alpha = 0.9$, $\ell = 5$) For the irregular data set the proxy histories were re-generated with the true coupling strength on the irregular timescales of the age modeling output. To assess the

impact of regular sampling, regular time series of the same length, average temporal spacing and coupling scheme were also simulated. We evaluated the performance of the different estimators for the different sampling schemes at increasing dating imprecision using the *root mean square error* (RMSE) of the estimators for the target coupling parameter $\alpha$:

$$\mathrm{RMSE}(\alpha_{\mathrm{est}}) = \sqrt{\mathrm{var}(\alpha_{\mathrm{est}}) + \mathrm{bias}(\alpha_{\mathrm{est}})^2} , \qquad (18)$$

where $\mathrm{bias}(\alpha_{\mathrm{est}}) = \alpha_{\mathrm{true}} - \alpha_{\mathrm{est}}$.

We did this separately for each sampling scheme to obtain the $\mathrm{RMSE}_{\mathrm{reg}}$, the "baseline" RMSE for each estimator under regular sampling, $\mathrm{RMSE}_{\mathrm{irreg}}$ for the irregularly sampled ensembles and the $\mathrm{RMSE}_{\mathrm{au}}$ for the age-uncertain ensemble. Coupling strength, autocorrelation and time series length were fixed to the same values for the three different sampling schemes. To improve the comparability for the MI estimators, the bias offset was estimated from mutually uncorrelated time series with the same autocorrelation and length and subtracted prior to the conversion to the XCF scale.

Based on the assumption that the RMSE should increase from regular to irregular to age-uncertain time series, $\mathrm{RMSE}_{\mathrm{reg}} < \mathrm{RMSE}_{\mathrm{irreg}} < \mathrm{RMSE}_{\mathrm{au}}$, the "baseline" contribution is estimated from regular time series as $\mathrm{RMSE}_{\mathrm{reg}}$, the additional contribution from timescale irregularity as $\mathrm{RMSE}_{\mathrm{irreg}} - \mathrm{RMSE}_{\mathrm{reg}}$ and the additional RMSE of the age-uncertain time series' similarity as $\mathrm{RMSE}_{\mathrm{au}} - \mathrm{RMSE}_{\mathrm{irreg}}$.

The results, averaged over the realistic imprecision values (the 2nd–5th points in Figs. 4d and 5d), are given in Fig. 6.

Ideally the RMSE should of course be as small as possible. For the linear (CAR) case in Fig. 3.3, the smallest RMSE is observed for the ESF and the gXCF, the largest – by far – for the interpolation-based iXCF. While the regular (estimator) bias is low for the correlation estimators, the contribution of increasing irregularity of the time series sampling (due to the uncertain inputs) is non-negligible particularly for the interpolation-based cases. The age uncertainty alone accounts for additional, but generally smaller, error. While a large amount of the uncertainty of the interpolation-based estimators, iMI and iXCF, is due to sampling irregularity, ES has a large RMSE for regular time series, which is even higher than that for regular to slightly irregular time series. Therefore the contribution of irregular sampling to the cumulative uncertainty, as depicted in Fig. 3.3, is negative, thus improving the estimation efficiency!

In the nonlinear (TAR) case the picture is quite different. The correlation-based estimators are not able to tell the coupling strength, regardless of the sampling scheme. The gMI estimator ranks lowest, with a lower uncertainty contribution from irregular sampling compared to the iMI estimator. The ESF, again, improves its accuracy when the time series are irregular. The overall error level is higher than for the linear case.

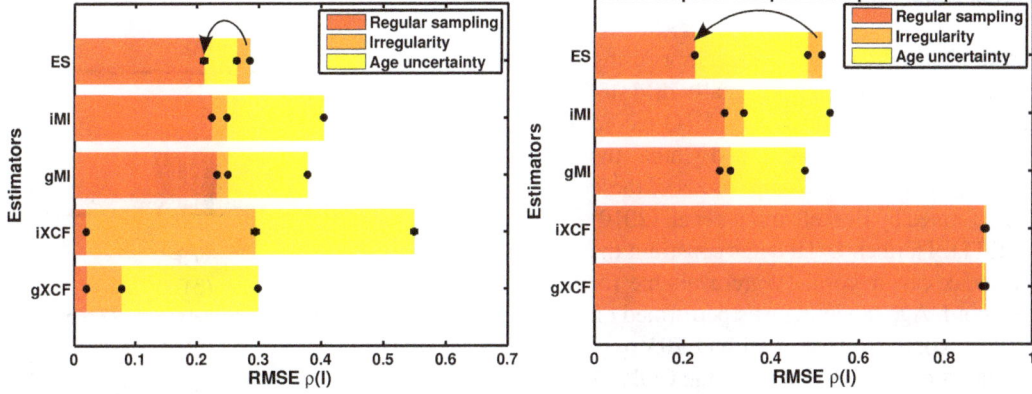

**Fig. 6.** Attribution of the uncertainty to its sources for **(a)** the linear CAR model and **(b)** the nonlinear TAR model: general (estimator) error in red, error introduced via irregular sampling (orange) and additional error due to the age uncertainty (yellow). The source-dependent RMSE was averaged over the second through to fifth imprecision levels given in Figs. 4d and 5d, as these correspond to the error levels most likely found in real-world studies. Errorbars indicate the associated standard deviation. For event synchronization the RMSE is lower for irregular than regular sampling, folding the irregular part of the bar backwards.

**Fig. 7.** The link strength concept: for each similarity estimator, significant results result in a link between the time series. The sum of these links determine the strength, or weight, of the link.

### 3.4 The link strength concept

Each of the tested similarity estimators comes with different underlying assumptions, estimator bias and variance, and they refer to different properties of the time series: the goodness of a linear fit to the joint distribution (XCF), the sharpness of the joint vs. the marginal distributions (MI) or the relative positions of extreme points, or events, in the time series (ES).

Therefore direct results obtained from the different estimators are difficult to compare, and they respond to coupling strength increases differently (Figs. 4c and 5c). The MI estimates, to this end, have to be converted to the XCF scale and thus are bound to the interval $[0, 1]$, not $[-1, 1]$ as for XC. This, together with the substantial and non-negative bias, induces a different proportionality between the actual coupling and the inferred association strength. Inferred ES, on the other hand, increases nonlinearly, but monotonically, with the coupling.

The main use of similarity measures is to assess the association strength between dynamics of processes. This can only be interpreted properly, if the significance of this estimate is known. To unify the results obtained from different similarity estimators, we propose to use a *link strength* $p(X, Y)$, to ho-

mogenize and summarize the results obtained for individual similarity measures.

The *link strength* $p(X, Y)$ for two observed time series $X$ and $Y$ is defined as the relative frequency of significant estimates from the $N_{\text{sim}}$ employed estimators $S_i$:

$$p_{\text{sim}}^q(X, Y) = \frac{\sum_{i=1}^{N_{\text{sim}}} P_i(X, Y)}{N_{\text{sim}}}, \tag{19}$$

as illustrated in Fig. 7. The link strength of the individual estimators, $P_i^q(X, Y)$ is recorded on a binary scale:

$$P_i^q(X, Y) = \begin{cases} 1 & \text{if } S_i \text{ symmetric and } S_i^{xy} > S_i^{\text{hi},xy} \\ 1 & \text{if } S_i \text{ asymmetric and} \\ & \left(S_i^{xy} > S_i^{\text{hi},xy}\right) | \left(S_i^{xy} < S_i^{\text{lo},xy}\right), \\ 0 & \text{otherwise}, \end{cases} \tag{20}$$

where $S^{\text{hi/lo}}$ refer to the critical values of a hypothesis test, the null hypothesis being that both $X$ and $Y$ are autocorrelated, but mutually uncorrelated, Gaussian distributed stochastic processes. The significance $q$ determines the critical values $S_i^{\text{hi},xy}$ and $S_i^{\text{lo},xy}$ which are obtained from the $q_{\text{hi}} = 1 - 0.5q$ and $q_{\text{lo}} = 0.5q$ quantiles of surrogate similarity estimates $S_i(X^*, Y^*)$.

Independent AR(1) surrogate time series $X^*$ and $Y^*$ are generated on the same time axes as $X$ and $Y$ according to Eq. (15). The individual AR(1) persistence time for actual paleoclimate data can be obtained using an efficient least-squares fitting algorithm (Rehfeld et al., 2011; Mudelsee, 2002). The link strength can be extended to incorporate age uncertainties by computing the similarities for $N_{\text{mc}}$ realizations of an age model and adding a second summation over these in Eq. (19).

## 4  Application to real stalagmite data

Now after having ensured the efficacy of the estimators using synthetic data sets, we apply the estimators to real-world stalagmite data sets from India, (the Dandak cave $\delta^{18}O$ record originally published in Sinha et al., 2007), and China (the Wanxiang record, Zhang et al., 2008). Comparisons of these data sets have been performed by Berkelhammer et al. (2010) and Rehfeld et al. (2011). Thirteen U/Th dates constrain the age model of the Dandak cave record, 19 are available for the Wanxiang cave record. Age modeling was performed on the full proxy data sets, comprising of 1875 and 703 oxygen isotope measurements over depth and using the COPRA algorithm with 1000 realizations (Breitenbach et al., 2012). The time series were cut to the overlapping time period from 600 to 1550 AD and detrended by subtracting the long-term mean, estimated using a Gaussian kernel smoother with a width $W$ of 1000 yr.

Berkelhammer et al. (2010) determined an averaged correlation of 0.27 for 50 yr overlapping time windows, while Rehfeld et al. (2011) found a lag zero correlation coefficient of 0.290 and 0.295 for iXCF and gXCF, respectively. This correlation was found to be significant at the 95 % level in the two-sided test for zero correlation, the null hypothesis being that the time series are autocorrelated but mutually uncorrelated.

Does this correlation persist, when the age uncertainties are considered in the analysis? We estimated the similarities for the two records considering all five estimators of Table 1 and for the original records as well as the results from age modeling, and give the results in Fig. 8. The histograms of similarity estimates for 100 realizations of the age models show a considerable spread. The mean similarity for the correlation estimators (indicated by the solid red line in Figs. 8a and 8b) is higher than that of the 95 % quantile of the surrogate distribution. The mean gMI estimate (8c) is close to the critical value, while the iMI (8d) and ES (8e) results lie well below. The median link strength (red line in Fig. 8f) is equal to 0.4. In contrast, the original age models published by Berkelhammer et al. (2010) and Zhang et al. (2008) yield significant results for all estimators except the ESF, resulting in an overall link strength of 0.8.

When we compute the similarities using the COPRA ensembles for the more sparse Dandak $\delta^{18}O$ time series published earlier (Sinha et al., 2007) the outcome is quite different – the link strength is only 0.2.

## 5  Discussion

Age uncertainty clearly affects all estimators of similarity for time series, and it is an illusion that it would be possible to mitigate the effects of uncertainty on the time axis for any type of analysis depending on observation times. Even if the observation – or accumulation – time of a grown archive is

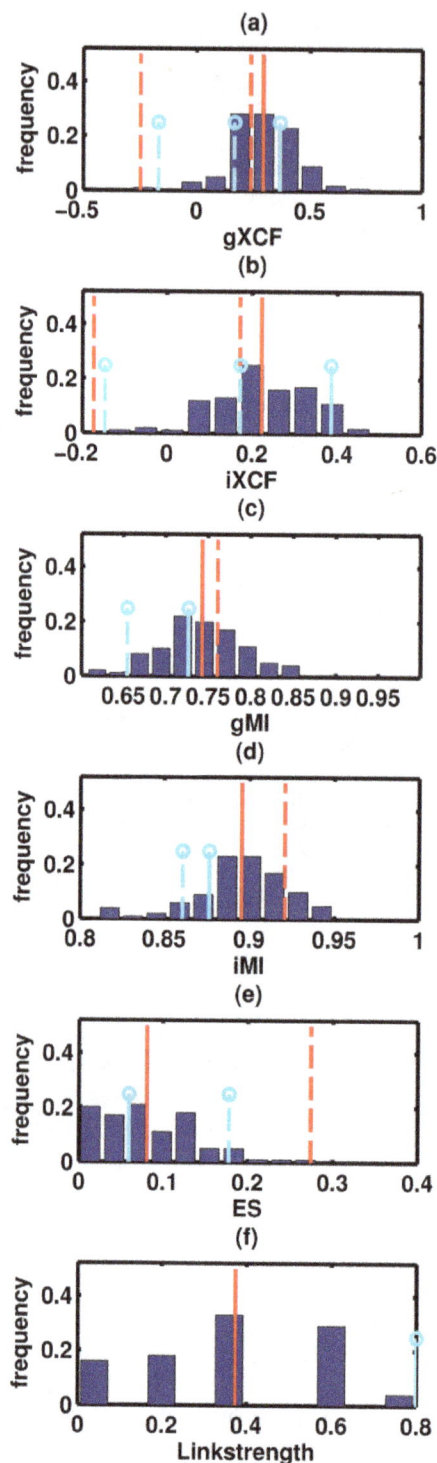

**Fig. 8.** Estimated lag zero similarities and link strength between the Dandak and Wanxiang cave records for the overlapping time period. The results for the age-uncertain ensembles are given in the dark blue histograms. The red solid line refers to the mean of these estimates, the light blue stem to the results for the mean timescale. The dashed lines refer to the respective confidence intervals.

known precisely at some depths, an observation time reconstruction from age modeling requires an assumption on the accumulation behavior which, necessarily, will be wrong to some extent, as stochasticity and irregularity in the growth will always be present. This is a fact not challenged by the choice of a different interpolation routine (e.g., to a continuous cubic spline), which is often preferred by geoscientists (Breitenbach et al., 2012; Scholz and Hoffmann, 2011). On the positive side, and although counterintuitive, incorporating (small) age uncertainty in the analysis might even improve the estimate when a deterministic (thus necessarily wrong) assumption on the growth of the archive is made.

A low imprecision of 0–0.5 % or an age uncertainty of approximately 1–2 yr over a period of 200 yr results in minimal relative estimation error and maximal confidence on the similarity peak position for the time series similarity functions $\hat{S}$. If a similarity analysis for real-world data sets covering a time span of 100 000 yr was desired, this would amount to an "allowed" age error of 500 yr at a mean time series resolution of 500 yr, which is a lower than what is usually found (Taylor et al., 2004). Thus, the resolution desired in the analysis is necessarily dependent on age uncertainty – only if that is lower, or comparable, would an analysis of such short time series with full consideration of age uncertainties be feasible. One way to achieve higher certainty could be the incorporation of layer-counted data in the age modeling process, for example, for annually laminated archives (Breitenbach et al., 2012).

The similarity estimators tested show different behavior, dependent on the signal type. The correlation-based estimators perform better for the linear coupling scheme, but fail for the nonlinear processes.

The *gXCF and iXCF* error split is dominated by the age uncertainty as the largest source of error in the linear CAR case. Both have small baseline bias for regular sampling. gXCF estimates coupling strength more effectively, however, for both age uncertainty and irregular sampling contributions of iXCF are significantly larger due to interpolation effects. In the nonlinear coupling scheme there is little difference whether the time series is regular, irregular or age uncertain – the correlation-based methods cannot capture such type of dependencies.

*gMI and iMI* perform badly on the first glance in the linear CAR case, as their baseline bias for regular sampling RMSE is large. However, one needs to take into account that the RMSE is determined by both variance and bias – and that MI estimation, especially using binning estimators, is always associated with a significant positive bias, particularly for short time series. This bias, however, decreases with increasing time series length. If a direct comparison of MI and XC estimates is desired, this bias should be subtracted from the MI estimate prior to scaling it to the correlation scale. In the nonlinear TAR case the Gaussian-kernel-based version has the lowest overall RMSE.

The *ESF*, originally intended for the analysis of event series, performs well and has the lowest total RMSE, followed closely by gXCF, in the linear test case. There, its baseline RMSE dominates the RMSE split, and the RMSE for irregular sampling is *lower* than that for regular sampling. This is similar for the nonlinear processes. One reason for this might be that, for irregularly sampled time series of the same mean observation time distance, the number of observations spaced *closely* together is higher, which might increase the chances to find multiple events spaced closely together, resulting in effective *double-counting* of events. The comparably small contribution from age uncertainty in the linear test indicates that neither the relative nor the absolute observation time distance between the time series are crucially important to the measure. Thus, it is quite a robust similarity measure with respect to age uncertainty and comparable to gXCF for linear coupling and gMI for nonlinear coupling, which both ultimately depend on the notion of simultaneous observations.

Although the irregularity of the time series is rather low (the inter-sampling-time distribution is narrow and close to normally distributed) the estimators that do not require the time series to be sampled regularly perform better than the interpolation-based records, which confirms the previous finding (Rehfeld et al., 2011, 2013) that large sampling irregularity (i.e., the presence of gaps) leads to large interpolation bias, where the adapted estimators gXCF and gMI are particularly suitable. We have applied the similarity estimators to investigate the similarities between the Dandak and Wanxiang cave records. We find that the link strength aptly summarizes the results of the similarity significance tests: the time series are quite likely to be correlated, but age uncertainty blurs the results. There are several other parameters which can have a critical impact on the analysis: the choice of the significance level for link strength estimation, the detrending width and the respective resolution of the time series. The dependence of the results on the detrending parameter (Fig. 9) illustrates the timescale dependence of the analysis: a small detrending width $W$ results in a high-pass filter and very low link strengths, large $W$ yields high similarity on larger timescales. This indicates that the paleoclimatic records are more clearly associated at centennial to multicentennial timescales than at decadal timescales, which are more impacted by age uncertainty. A higher temporal resolution of proxy measurements improves the accuracy of the estimators, particularly for the data-demanding MI estimators. Bootstrapping of the time series to successively lower lengths could be used to test the robustness of the estimators against such effects.

We have only considered five similarity estimators (gXCF, iXCF, gMI, iMI and ESF) here, but this could be expanded for other concepts, for example, based on (cross-)recurrence plots (Romano et al., 2005; Marwan et al., 2007; Marwan, 2002; Lange, 2011), recurrence networks (Feldhoff et al., 2012), convergent cross mapping (Sugihara et al., 2012) or distance measures (Lhermitte et al., 2011). The notion of

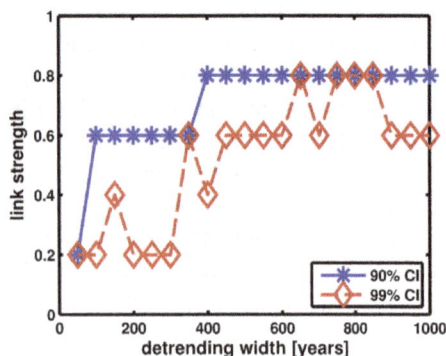

**Fig. 9.** Sensitivity of the link strength result for the original records of Berkelhammer et al. (2010) and Zhang et al. (2008) to changes in the detrending parameter $W$ of a Gaussian-kernel detrending and the significance level in the hypothesis test.

a link strength, instead of XC, MI or ES values, makes it straightforward to extend the analysis to a whole ensemble of time series, be it from age modeling or out of a database of paleoclimate records. If age uncertainty does not impact the cross similarity, the link strength will not drop substantially. The actual value of the link strength can be interpreted in terms of a "degree of confidence": if the value is close to the significance level, a relationship cannot be concluded with confidence. If the link strength is close to one, all the estimators return significant similarity estimates and a similarity can be deduced with certainty.

In the future it could be evaluated whether $p$ values from the surrogate tests can replace the binary thresholding for the link strength metric to improve the sensitivity of the link strength estimate. The ESF alone, however, could be particularly suitable for the analysis of extreme events since it does not place strong restrictions on the time series beyond stationarity, and performs particularly well for irregular time series.

The NESToolbox containing scripts and programs for the similarity analysis of age-uncertain time series in Matlab and the open source software Octave are available with this paper. We also include a function to simulate age uncertainties that arise for archives for which the chronology is based on layer counting, trees, ice cores or laminated sediments, so that these, too, can be investigated using the methods presented in this paper.

## 6 Conclusions

In this paper we have investigated similarity estimators that do not require regular sampling in time and can capture linear (gXCF) and nonlinear (gMI and ESF) relationships. We found that interpolation to regular spacing of the observation times results in worse estimates. By contrast, the adapted estimators are more efficient in the presence of sampling time irregularity and cope with age uncertainty better. Ta-

ble 1 gives a comprehensive overview over the similarity estimators, parameter choices and further references. gXCF and ESF perform particularly well if the relationship is linear, but the correlation estimator fails in the presence of nonlinear coupling, where the ESF and gMI are better suited to infer dependences. The significance of results from different estimators and under varying time series length and sampling can be unified using the concept of a link strength. It combines similarity estimators and significance tests and is given by the relative frequency of positive significance tests and could be especially useful in the analysis of large paleoclimatic data sets where it is infeasible to check each pair of time series for similarity individually. We have shown that age uncertainty is the largest contributor to estimation error for time series similarity, and for a reliable of similarity function shape and coupling structure, the timescale imprecision should be as low as possible. When it exceeds 5 % of the time series length coupling phenomena on timescales close to the sampling resolution can no longer be deduced. While time series irregularity can be well addressed by the use of the adapted estimators, age uncertainty cannot, and should therefore be reduced as much as possible by measuring more ages, improved dating techniques or the use of additional temporal information from layer counting (Breitenbach et al., 2012) where possible. This is, in essence, good news, because the irregular growth of the archives cannot be reversed, but measurement devices can be optimized.

*Acknowledgements.* The authors thank Norbert Marwan, Jobst Heitzig, Bedartha Goswami and Sebastian Breitenbach for helpful comments and discussion, and Franziska Lechleitner for assistance with data pre-processing. We thank Ashish Sinha for providing us with the depth data for the Dandak cave stalagmite and Richard Telford and an anonymous reviewer for their constructive feedback. This work has been financially supported by the Federal Ministry for Education and Research (BMBF) via the Potsdam Research Cluster for Georisk Analysis, Environmental Change and Sustainability (PROGRESS). The NESToolbox containing software tools to handle irregularly sampled data sets can be found on tocsy.pik-potsdam.de/nest.php.

Edited by: K. Mills

## References

Babu, P. and Stoica, P.: Spectral analysis of nonuniformly sampled data – a review, Digit. Signal Process., 20, 359–378, doi:10.1016/j.dsp.2009.06.019, 2010.

Batyrshin, I., Sheremetov, L., and Velasco-Hernandez, J. X.: On axiomatic definition of time series shape association measures, in: Operations Research and Data Mining ORADM 2012 workshop proceedings, edited by: Villa-Vargas, U., Sheremetov, L., and Haasis, H.-D., 1–12, National Polytechnic Institute, Mexico City, 2012.

Berkelhammer, M., Sinha, A., Mudelsee, M., Cheng, H., Edwards, R. L., and Cannariato, K.: Persistent multidecadal power of the

Indian Summer Monsoon, Earth Planet. Sci. Lett., 290, 166–172, doi:10.1016/j.epsl.2009.12.017, 2010.

Breitenbach, S. F. M., Rehfeld, K., Goswami, B., Baldini, J. U. L., Ridley, H. E., Kennett, D. J., Prufer, K. M., Aquino, V. V., Asmerom, Y., Polyak, V. J., Cheng, H., Kurths, J., and Marwan, N.: COnstructing Proxy Records from Age models (COPRA), Clim. Past, 8, 1765–1779, doi:10.5194/cp-8-1765-2012, 2012.

Chatfield, C.: The analysis of time series: an introduction, CRC Press, Florida, US, 6th Edn., 2004.

Cheng, H., Zhang, P. Z., Spötl, C., Edwards, R. L., Cai, Y. J., Zhang, D. Z., Sang, W. C., Tan, M., and An, Z. S.: The climatic cyclicity in semiarid-arid central Asia over the past 500,000 years, Geophys. Res. Lett., 39, 1–5, doi:10.1029/2011GL050202, 2012.

Cover, T. and Thomas, J.: Elements of information theory, John Wiley & Sons, Inc., Hoboken, New Jersey, 2 Edn., 2006.

Donges, J. F., Zou, Y., Marwan, N., and Kurths, J.: Complex networks in climate dynamics, The Eur. Phys. J. Special Top., 174, 157–179, doi:10.1140/epjst/e2009-01098-2, 2009.

Dykoski, C., Edwards, R., Cheng, H., Yuan, D., Cai, Y., Zhang, M., Lin, Y., Qing, J., An, Z., and Revenaugh, J.: A high-resolution, absolute-dated Holocene and deglacial Asian monsoon record from Dongge Cave, China, Earth Planet. Sci. Lett., 233, 71–86, doi:10.1016/j.epsl.2005.01.036, 2005.

Fairchild, I. and Baker, A.: Speleothem Science: from process to past environments, Wiley-Blackwell, 2012.

Feldhoff, J. H., Donner, R. V., Donges, J. F., Marwan, N., and Kurths, J.: Geometric detection of coupling directions by means of inter-system recurrence networks, Phys. Lett. A, 376, 3504–3513, doi:10.1016/j.physleta.2012.10.008, 2012.

Hlinka, J., Hartman, D., Vejmelka, M., Runge, J., Marwan, N., Kurths, J., and Paluš, M.: Reliability of Inference of Directed Climate Networks Using Conditional Mutual Information, Entropy, 15, 2023–2045, doi:10.3390/e15062023, 2013.

Khan, S., Bandyopadhyay, S., Ganguly, A., Saigal, S., Erickson, D., Protopopescu, V., and Ostrouchov, G.: Relative performance of mutual information estimation methods for quantifying the dependence among short and noisy data, Phys. Rev. E, 76, 1–15, doi:10.1103/PhysRevE.76.026209, 2007.

Kraskov, A., Stögbauer, H., and Grassberger, P.: Estimating mutual information, Phys. Rev. E, 69, 1–16, doi:10.1103/PhysRevE.69.066138, 2004.

Kreuz, T., Chicharro, D., Andrzejak, R. G., Haas, J. S., and Abarbanel, H. D. I.: Measuring multiple spike train synchrony., J. Neurosci. Methods, 183, 287–99, doi:10.1016/j.jneumeth.2009.06.039, 2009.

Lange, H.: Recurrence Quantification Analysis in Watershed Ecosystem Research, Int. J. Bifurcat. Chaos, 21, 1113–1125, doi:10.1142/S0218127411028921, 2011.

Lhermitte, S., Verbesselt, J., Verstraeten, W., and Coppin, P.: A comparison of time series similarity measures for classification and change detection of ecosystem dynamics, Remote Sens. Environ., 115, 3129–3152, doi:10.1016/j.rse.2011.06.020, 2011.

Malik, N., Marwan, N., and Kurths, J.: Spatial structures and directionalities in Monsoonal precipitation over South Asia, Nonlin. Processes Geophys., 17, 371–381, doi:10.5194/npg-17-371-2010, 2010.

Malik, N., Bookhagen, B., Marwan, N., and Kurths, J.: Analysis of spatial and temporal extreme monsoonal rainfall over South

Asia using complex networks, Clim. Dynam., 39, 971–987, doi:10.1007/s00382-011-1156-4, 2011.

Marwan, N.: Nonlinear analysis of bivariate data with cross recurrence plots, Phys. Lett. A, 302, 299–307, doi:10.1016/S0375-9601(02)01170-2, 2002.

Marwan, N., Romano, M. C., Thiel, M., and Kurths, J.: Recurrence plots for the analysis of complex systems, Phys. Reports, 438, 237–329, doi:10.1016/j.physrep.2006.11.001, 2007.

Mudelsee, M.: TAUEST: a computer program for estimating persistence in unevenly spaced weather/climate time series, Comput.Geosci., 28, 69–72, doi:10.1016/S0098-3004(01)00041-3, 2002.

Nazareth, D., Soofi, E., and Zhao, H.: Visualizing Attribute Interdependencies Using Mutual Information, Hierarchical Clustering, Multidimensional Scaling, and Self-organizing Maps, 2007 40th Annual Hawaii International Conference on System Sciences (HICSS'07), 53–53, doi:10.1109/HICSS.2007.608, 2007.

Papana, A. and Kugiumtzis, D.: Evaluation of mutual information estimators for time series, Int. J. Bifurcat. Chaos, 19, 4197–4215, doi:10.1142/S0218127409025298, 2009.

Quian Quiroga, R., Kreuz, T., and Grassberger, P.: Event synchronization: A simple and fast method to measure synchronicity and time delay patterns, Phys. Rev. E, 66, 041904, doi:10.1103/PhysRevE.66.041904, 2002.

Rehfeld, K., Marwan, N., Heitzig, J., and Kurths, J.: Comparison of correlation analysis techniques for irregularly sampled time series, Nonlin. Processes Geophys., 18, 389–404, doi:10.5194/npg-18-389-2011, 2011.

Rehfeld, K., Marwan, N., Breitenbach, S. F. M., and Kurths, J.: Late Holocene Asian Summer Monsoon dynamics from small but complex networks of palaeoclimate data, Clim. Dynam., 41, 3–19, doi:10.1007/s00382-012-1448-3, 2013.

Rheinwalt, A., Marwan, N., Kurths, J., Werner, P., and Gerstengarbe, F.-W.: Boundary effects in network measures of spatially embedded networks, (Europhys. Lett.), 100, 28002, doi:10.1209/0295-5075/100/28002, 2012.

Romano, M. C., Thiel, M., Kurths, J., Kiss, I. Z., and Hudson, J. L.: Detection of synchronization for non-phase-coherent and non-stationary data, Europhys. Lett., 71, 466–472, doi:10.1209/epl/i2005-10095-1, 2005.

Roulston, M.: Estimating the errors on measured entropy and mutual information, Phy. D: Nonlinear Phenomena, 125, 285–294, 1999.

Runge, J., Heitzig, J., Marwan, N., and Kurths, J.: Quantifying causal coupling strength: A lag-specific measure for multivariate time series related to transfer entropy, Phys. Rev. E, 86, 061121, doi:10.1103/PhysRevE.86.061121, 2012.

Scargle, J. D.: Studies in astronomical time series analysis. III - Fourier transforms, autocorrelation functions, and cross-correlation functions of unevenly spaced data, The Astrophysical J., 343, 874, doi:10.1086/167757, 1989.

Scholz, D. and Hoffmann, D. L.: StalAge – An algorithm designed for construction of speleothem age models, Quaternary Geochronol., 6, 369–382, doi:10.1016/j.quageo.2011.02.002, 2011.

Schulz, M. and Stattegger, K.: SPECTRUM: spectral analysis of unevenly spaced paleoclimatic time series, Comput. Geosci., 23, 929–945, doi:10.1016/S0098-3004(97)00087-3, 1997.

Sinha, A., Cannariato, K. G., Stott, L. D., Cheng, H., Edwards, R. L., Yadava, M. G., Ramesh, R., and Singh, I. B.: A 900-year (600 to 1500 A.D.) record of the Indian summer monsoon precipitation from the core monsoon zone of India, Geophys. Res. Lett., 34, 1–5, doi:10.1029/2007GL030431, 2007.

Sinha, A., Stott, L., Berkelhammer, M., Cheng, H., Edwards, R. L., Buckley, B., Aldenderfer, M., and Mudelsee, M.: A global context for megadroughts in monsoon Asia during the past millennium, Quaternary Sci. Rev., 30, 47–62, doi:10.1016/j.quascirev. 2010.10.005, 2011.

Stoica, P. and Sandgren, N.: Spectral analysis of irregularly-sampled data: Paralleling the regularly-sampled data approaches, Digit. Signal Process., 16, 712–734, doi:10.1016/j.dsp.2006.08. 012, 2006.

Sugihara, G., May, R., Ye, H., Hsieh, C., Deyle, E., Fogarty, M., and Munch, S.: Detecting causality in complex ecosystems, Science, 338, 496–500, doi:10.1126/science.1227079, 2012.

Taylor, K. C., Alley, R. B., Meese, D. A., Spencer, M. K., Brook, E. J., Dunbar, N. W., Finkel, R. C., Gow, A. J., Kurbatov, A. V., Lamorey, G. W., Mayewski, P. A., Meyerson, E. A., Nishiizumi, K., and Zielinski, G. A.: Dating the Siple Dome (Antarctica) ice core by manual and computer interpretation of annual layering, J. Glaciol., 50, 453–461, doi:10.3189/172756504781829864, 2004.

Telford, R., Heegaard, E., and Birks, H.: All age-depth models are wrong: but how badly?, Quaternary Sci. Rev., 23, 1–5, doi:10. 1016/j.quascirev.2003.11.003, 2004.

Tsay, R.: Testing and modeling threshold autoregressive processes, J. Am. Stat. Assoc., 84, 231–240, 1989.

Webster, J., Brook, G., and Railsback, L.: Stalagmite evidence from Belize indicating significant droughts at the time of Preclassic Abandonment, the Maya Hiatus, and the Classic Maya collapse, Palaeogeogr. Palaeocli. Palaeoecol., 250, 1–17, doi: 10.1016/j.palaeo.2007.02.022, 2007.

Yadava, M., Ramesh, R., and Pant, G.: Past monsoon rainfall variations in peninsular India recorded in a 331-year-old speleothem, The Holocene, 14, 517–524, doi:10.1191/0959683604hl728rp, 2004.

Zhang, J., Chen, F., Holmes, J. A., Li, H., Guo, X., Wang, J., Li, S., Lü, Y., Zhao, Y., and Qiang, M.: Holocene monsoon climate documented by oxygen and carbon isotopes from lake sediments and peat bogs in China: a review and synthesis, Quaternary Sci. Rev., 30, 1973–1987, doi:10.1016/j.quascirev.2011.04.023, 2011.

Zhang, P., Cheng, H., Edwards, R. L., Chen, F., Wang, Y., Yang, X., Liu, J. J. J. J., Tan, M., Wang, X., An, C., Dai, Z., Zhou, J., Zhang, D., Jia, J., Jin, L., and Johnson, K. R.: A test of climate, sun, and culture relationships from an 1810-year Chinese cave record, Science, 322, 940–942, doi:10.1126/science.1163965, 2008.

Zwiers, F. and Storch, H. V.: Regime-dependent autoregressive time series modeling of the Southern Oscillation, J. Climate, 3, 1347–1363, 1990.

# 5

# Ocean biogeochemistry in the warm climate of the late Paleocene

**M. Heinze**[1,2] **and T. Ilyina**[1]

[1]Max Planck Institute for Meteorology, Bundesstrasse 53, 20146 Hamburg, Germany
[2]International Max Planck Research School on Earth System Modelling, Hamburg, Germany

*Correspondence to:* M. Heinze (mathias.heinze@mpimet.mpg.de)

**Abstract.** The late Paleocene is characterized by warm and stable climatic conditions that served as the background climate for the Paleocene–Eocene Thermal Maximum (PETM, $\sim$ 55 million years ago). With respect to feedback processes in the carbon cycle, the ocean biogeochemical background state is of major importance for projecting the climatic response to a carbon perturbation related to the PETM. Therefore, we use the Hamburg Ocean Carbon Cycle model (HAMOCC), embedded in the ocean general circulation model of the Max Planck Institute for Meteorology, MPIOM, to constrain the ocean biogeochemistry of the late Paleocene. We focus on the evaluation of modeled spatial and vertical distributions of the ocean carbon cycle parameters in a long-term warm steady-state ocean, based on a 560 ppm $CO_2$ atmosphere. Model results are discussed in the context of available proxy data and simulations of pre-industrial conditions. Our results illustrate that ocean biogeochemistry is shaped by the warm and sluggish ocean state of the late Paleocene. Primary production is slightly reduced in comparison to the present day; it is intensified along the Equator, especially in the Atlantic. This enhances remineralization of organic matter, resulting in strong oxygen minimum zones and $CaCO_3$ dissolution in intermediate waters. We show that an equilibrium $CO_2$ exchange without increasing total alkalinity concentrations above today's values is achieved. However, consistent with the higher atmospheric $CO_2$, the surface ocean pH and the saturation state with respect to $CaCO_3$ are lower than today. Our results indicate that, under such conditions, the surface ocean carbonate chemistry is expected to be more sensitive to a carbon perturbation (i.e., the PETM) due to lower $CO_3^{2-}$ concentration, whereas the deep ocean calcite sediments would be less vulnerable to dissolution due to the vertically stratified ocean.

## 1 Introduction

The late Paleocene has received interest because of its role as the background climate for the Paleocene–Eocene Thermal Maximum (PETM), which could have been a possible analog for present-day greenhouse warming and ocean acidification (e.g., Zachos et al., 2005; Ridgwell and Zeebe, 2005; Zeebe and Zachos, 2013). The PETM describes a time period of about 170 kyr, which is characterized by an increase in mean surface temperatures of more than 5° (Kennett and Stott, 1991; Dickens et al., 1995; Zachos et al., 2008). During the PETM, the atmospheric $CO_2$ values increased significantly over a relatively short time period of about 10 kyr (Panchuk et al., 2008). Nonetheless, the question about the exact atmospheric $CO_2$ content before the PETM, as well as the maximum values of $CO_2$ during the PETM, still remains unanswered (e.g., Pagani et al., 2006a).

The climate of the late Paleocene was characterized by higher global average temperatures than in the present day, bearing ice-free conditions at the poles (Zachos et al., 2001). The pole-to-equator temperature gradient was smaller, displayed in sea surface temperatures (SST) of more than 30° in the tropics (Pearson et al., 2001) and up to 20° at high latitudes (Sluijs et al., 2006; Lunt et al., 2012). Deep ocean water masses were up to 10° warmer compared to modern values (Kennett and Stott, 1991; Zachos et al., 2008; Tripati and Elderfield, 2005). The warmer climate and the late Paleocene continental configuration influenced global ocean circulation patterns. The main deepwater formation occurred at southern high latitudes, with additional minor regions of deepwater formation in the Northern Hemisphere (Thomas et al., 2003). There is no consensus as to whether the Northern Hemisphere deepwater formation was stronger in the Atlantic or the Pacific (e.g., Bice and Marotzke, 2002; Tripati and Elderfield, 2005; Nunes and Norris, 2006).

Hitherto, the focus of modeling the late Paleocene with complex earth system models (ESM) was set to the physical ocean and the atmospheric system (Huber and Sloan, 2001; Heinemann et al., 2009; Winguth et al., 2010). Studies of the PETM background climate show a wide range of inter-model variability, using prescribed atmospheric $CO_2$ concentrations ranging from $2\times$ to $16\times$ pre-industrial $CO_2$ (Lunt et al., 2012). In previous studies, the late Paleocene ocean biogeochemistry has been addressed exclusively with earth system models of intermediate complexity (EMIC) or box models (e.g., Panchuk et al., 2008; Zeebe et al., 2009; Ridgwell and Schmidt, 2010; Winguth et al., 2012). These modeling studies cover the whole PETM, with the major objective of constraining the absolute amount of the carbon perturbation. Their approach is based on reconstructions of the calcium carbonate compensation depth (CCD). By observing vertical shifts in the $CaCO_3$ dissolution horizons in sediment cores, before and after the peak of the event, rough estimates of the carbon perturbation during the PETM can be obtained. The depth of the pre-PETM CCD is still under discussion for wider geographical areas of the late Paleocene oceans (Zeebe and Zachos, 2013). This leads to quite different estimates of total carbon mass and carbon injection speed into the climate system necessary for obtaining the observed sedimentary $CaCO_3$ dissolution (Dunkley Jones et al., 2010). Hence, the preceding conditions of ocean biogeochemistry are important for a realistic assessment of the PETM itself in order to gain knowledge about ocean biogeochemistry influence on feedback mechanisms for the PETM, e.g., alterations in the carbonate buffer capacity. However, estimates of, for instance, total alkalinity (TA) and dissolved inorganic carbon (DIC) during the late Paleocene, are not well known at present (Dunkley Jones et al., 2010).

In order further to constrain the state of the oceanic part of the carbon cycle during the late Paleocene, we spin up and run the Hamburg Ocean Carbon Cycle model, HAMOCC, and the ocean general circulation model (OGCM) of the Max Planck Institute for Meteorology, MPIOM, under late Paleocene boundary conditions into an equilibrium state. Estimates for late Paleocene atmospheric $CO_2$ concentrations range from 600 to 2800 ppm Pagani et al. (2006a). Based on Heinemann et al. (2009), we use a 560 ppm $CO_2$ late Paleocene atmospheric forcing to achieve a plausible background climate for the PETM. The applied atmospheric $CO_2$ concentrations and late Paleocene boundary conditions cause a new equilibrium climate state, which fits the proxy-record-based SST quite well (Lunt et al., 2012). However, the pre-PETM ocean biogeochemistry is not only affected by modifications in temperature and atmospheric conditions at the ocean–atmosphere boundary (Archer et al., 2004), but also by alterations in the general ocean physical state. Hence, using a state-of-the-art OGCM-based carbon cycle model, we address the following questions. (i) How does the enhanced ver-

tical stratification of the ocean affect the marine carbon cycle? (ii) What is the effect of the higher background steady-state $CO_2$ on the marine carbonate chemistry?

The model used has been applied in a number of previous studies simulating the pre-industrial, modern and future climate/ocean states, for example in the framework of the Climate Model Intercomparison Project (CMIP5; Ilyina et al., 2013). Therefore, here we also use the model output of the CMIP5 experiments for comparison. The model output is based on calculations with the Max Planck Institute for Meteorology-Earth System Model (MPI-ESM) for pre-industrial (1850–1879) climatic conditions.

In Sect. 2 of this paper, we describe the model and give detailed information on the spin-up of HAMOCC under late Paleocene conditions. The general late Paleocene climate state achieved by our simulation is presented in Sect. 3. Section 4 comprises the results of the modeled ocean biogeochemistry, followed by the conclusions in Sect. 5.

## 2 Model description and setup

### 2.1 The ocean biogeochemistry model HAMOCC

For our study, we employ the Hamburg Ocean Carbon Cycle (HAMOCC 5.1) model, which is based on Maier-Reimer (1993) and successive refinements (Maier-Reimer et al., 2005). HAMOCC simulates 18 biogeochemical tracers in the oceanic water column and 12 tracers in the upper 14 cm of the sediment. The tracers are simulated prognostically within a three-dimensional ocean circulation state. HAMOCC is coupled online to the Max Planck Institute ocean model (MPIOM) (Marsland et al., 2003; Jungclaus et al., 2013), which computes tracer advection and mixing. Temperature, pressure and salinity of MPIOM are used to calculate various transformation rates and chemical constants within HAMOCC. The treatment of important biogeochemical processes in HAMOCC, related to this study, is described in some more detail in the following paragraphs. For more complete information on HAMOCC, see Ilyina et al. (2013) and Maier-Reimer et al. (2005).

Air–sea gas exchange is calculated for $O_2$, $CO_2$ and $N_2$. The air–sea $CO_2$ flux is a result of the partial pressure difference between atmosphere and water multiplied by a gas exchange rate and solubility according to Weiss (1974) and Groeger and Mikolajewicz (2011). It is then divided by the actual thickness of the surface layer. The velocity of the gas transfer depends on the Schmidt number and prognostic wind speed at the surface (Wanninkhof, 1992). The oceanic partial pressure of $CO_2$ ($pCO_2$) in the model is prognostically computed as a function of temperature, salinity, DIC, TA, and pressure.

Biological processes are described by an extended NPZD (nutrient–phytoplankton–zooplankton–detritus) type model (Six and Maier-Reimer, 1996). Primary production in

HAMOCC is based on the co-limitation of phosphorous, nitrate and iron, as well as on temperature and radiation. The biogeochemical processes within the model are calculated on the basis of phosphorous. Associated changes between the remaining tracers are calculated using constant stoichiometric ratios (Redfield ratio following Takahashi et al. (1985); $P : N : C : -O_2$ ratio of $1 : 16 : 122 : 172$). Phytoplankton are divided into silicifiers (opal shells) and calcifiers ($CaCO_3$ shells). It is assumed that silicifiers are preferentially produced as long as silicate is available, which is shown by several observational studies (e.g., Lochte et al., 1993). Via prescribed vertical sinking rates, opal, $CaCO_3$ and particulate organic carbon (POC) are transported to depth. During the sinking, the particles undergo remineralization at a constant rate, distributing silicate, DIC, TA and nutrients (while decreasing oxygen) at depth. Remineralization of POC depends on oxygen. If oxygen falls below a concentration of $0.5 \mu mol\,L^{-1}$, organic matter is decomposed by denitrification and sulfate reduction.

The formation of $CaCO_3$ shells consumes DIC and TA in a molar ratio of $1 : 2$. The dissolution of $CaCO_3$ at depth is a function of the calcite saturation state ($\Omega$) of sea water and a dissolution rate constant. $\Omega$ is calculated from $Ca^{2+}$ concentration in sea water, which is kept constant at $1.03 \times 10^{-2}\,kmol\,m^{-3}$, $CO_3^{2-}$ (carbonate ion) concentration and the apparent solubility product of calcite, based on temperature and pressure. Dissolution of opal takes place continuously over the whole water column at a rate of $0.01\,d^{-1}$. While $CaCO_3$ is less soluble in warm waters, the dissolution intensity of opal is positively correlated with temperature (Ragueneau et al., 2000).

The sediment module is based on Heinze and Maier-Reimer (1999) and Heinze et al. (1999). It basically calculates the same tracers as the water column model. The solid components of the sediment comprise opal, $CaCO_3$, organic carbon and chemically inert dust (referred to from here onwards as "clay"). The liquid sediment components (pore water tracer) are DIC, TA, $PO_4$, $O_2$, $N_2$, $NO_3$, $Si(OH)_4$ and Fe. The tracer concentrations within the oceanic bottom layer, and particularly the particle deposition from it, determine the upper boundary for the sediment. The sediment consists of 12 biologically active layers, with increasing thickness and decreasing porosity from top to bottom representing the uppermost 14 cm of the ocean floor. Below the active sediment there is one diagenetically consolidated layer (burial) containing only solid sediment components and representing the bedrock. Major processes simulated in the sediment are vertical diffusion of porewater, decomposition of detritus, as well as dissolution of opal and $CaCO_3$.

## 2.2 Topography and grid

The model setup is based on the interpolation of a late Paleocene $2° \times 2°$ topography (Bice and Marotzke, 2001) onto our $3.5° \times 3.5°$ ocean model grid (Fig. 1). It is used in

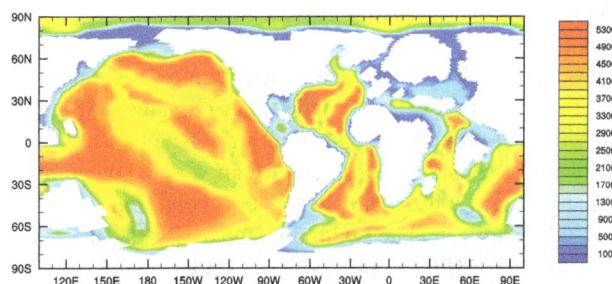

**Figure 1.** Paleocene topography (m depth).

several Paleocene–Eocene climate studies (Panchuk et al., 2008; Roberts et al., 2009; Heinemann et al., 2009; Zeebe, 2012). The main differences to present-day bathymetry lie in the open Central American Seaway, connecting the Atlantic and Pacific, as well as the existence of the Tethys Ocean and its connection to the Arctic Ocean, via the Turgai Strait. Although the Arctic Ocean has an additional link to the surrounding oceans, it lacks, unlike the present-day bathymetry, a deepwater connection. In the Southern Hemisphere, the Drake Passage and the Tasmanian Seaway are already open, but operate just as shallow water connections around Antarctica. The average ocean floor depth amounts to 3135 m (present-day setup: 3700 m) and has its deepest point at 5287 m (present-day setup: 5958 m) in the eastern equatorial Pacific. Vast areas of the Pacific are shallower in depth than in the modern ocean, the Atlantic is narrower than today, and almost the whole Tethys does not exceed depths of 1000 m. It is mainly shaped by extended shelf areas. Taking into account these differences, the model bathymetry yields a 14 % reduced sea water volume compared to today's ocean. Although the missing ice sheets in the late Paleocene setup would suggest an increase in oceanic volume, the provided bathymetry from Bice and Marotzke (2001) results in a reduced ocean volume compared to modern conditions. However, since we adapt the inventories of the ocean biogeochmical tracers (see "Initialization biogeochemistry"), we hold on to the reduced ocean volume bathymetry, since it allows a better comparison of the results to other models using the same bathymetry (e.g., Panchuk et al., 2008; Heinemann et al., 2009).

In the conducted simulations, HAMOCC integrates with a time step of 2.4 h (0.1 days). The horizontal resolution of the ocean model is $3.5° \times 3.5°$, which equals a grid spacing from 70 km around South America to 430 km in the Pacific. The ocean model has 40 vertical layers, with increasing level thicknesses with depth: 9 layers cover the upper 100 m and 23 layers the upper 1000 m of the water column. An orthogonal curvilinear grid is applied, with the poles located over northern Eurasia and South America to achieve the best grid resolution for all ocean regions.

## 2.3  Forcing

The ocean stand-alone model approach requires an atmospheric forcing. We use a late Paleocene climate with an atmospheric $CO_2$ concentration of 560 ppm, which is mimicked by an adequate atmospheric forcing, derived from Heinemann et al. (2009). The atmospheric conditions used in this study were calculated with a coupled climate model using ECHAM5, MPIOM and JSBACH under Paleocene–Eocene boundary conditions (560 ppm $CO_2$). Atmospheric methane and nitrous oxide are set to pre-industrial values. The model shows, after 2300 years of integration, an equilibrium state in atmospheric and oceanic conditions (Heinemann et al., 2009). From the atmospheric model output, we take 30 consecutive years, from which we reproduce a daily mean late Paleocene atmospheric forcing based on the Ocean Model Intercomparison Project (OMIP) forcing used for present-day ocean-model-only setups (Roeske, 2006). The model is then forced using daily heat, freshwater and momentum fluxes in a 30 year cycle.

We initialize the stand-alone ocean model (MPIOM) based on the result of the 2300 year late Paleocene equilibrium run by Heinemann et al. (2009). Additionally, we define a Paleocene climatology for ocean temperature and salinity from the same data. It displays the monthly mean climatological state for the two variables, averaged from daily data over a 30 year period. Sea surface salinity (SSS) and SST (upper 12 m) in our model are relaxed towards this Paleocene-based climatology. Relaxation takes place with a time constant of 180 days; it takes about 3 months till the surface layer is restored completely to the climatology.

## 2.4  Initialization of biogeochemistry

Using the MPIOM physical ocean state, we spin up HAMOCC starting from basin-wide homogeneous distributions of biogeochemical tracers, taking pre-industrial concentrations as a rough orientation for spatial and vertical tracer distributions. We reduce the oceanic carbon inventory (from $\sim 38\,500$ Gt to $\sim 32\,000$ Gt), as well as the different nutrient pools, proportionally to the 14 % reduced ocean volume in the late Paleocene setup. The reduction results in tracer concentrations close to modern values in the water column.

The model is integrated for 3200 years, periodically repeating the forcing (see Sect. 2.3), while the inventories of TA are adjusted to the $CO_2$ level of 560 ppm. Fluxes and tracer distribution stabilize after about 1000 model years within the water column. The distributions of tracers are not restored to any kind of data set, to be consistent with the biological, chemical, and physical dynamics of the model.

Since, for the late Paleocene, the monthly mean dust deposition fields are not available (A. Winguth, personal communication, 2013), we prescribe a spatially homogeneous input of dust at the sea surface. The total amount of annual bio-available iron deposition to the ocean is the same as in the modern ocean setup ($\sim 38 \times 10^7$ kg Fe yr$^{-1}$; Mahowald et al., 2005). Besides the inventory adaptation and homogeneous dust deposition, the ocean chemistry in the late Paleocene simulations is modeled the same as in the modern MPI-ESM (Ilyina et al., 2013).

The weathering fluxes depend on the long-term sedimentation rates. They are used for balancing the water column inventory of the calcite, silicate and OM pools. The annual amount leaving the system through sedimentation is added (globally distributed) at the surface again. The calcite weathering varies from 0 to 900 kmol C s$^{-1}$. Silicate weathering varies from 0 to 650 kmol Si s$^{-1}$, and organic material varies from 0 to 4 kmol C s$^{-1}$ during the spin-up. After establishing an equilibrium state in the sediment, constant weathering fluxes are applied, as shown in Table 1.

## 2.5  Initialization of sediment

An equilibrium sediment state exists for the present-day model configuration of HAMOCC, which is based on long spin-up simulations of approximately 50 kyr (e.g., Heinze et al., 1999). Since an equilibrium sediment state is missing for the Paleocene model configuration of HAMOCC, we initialize the sediment with 100 % clay, while the $CaCO_3$, opal and organic carbon sediment pools start filling from the first year on. However, in matters of computing time, it is not feasible to spin up the sediment module and achieve an equilibrium state within a realistic time frame. To circumvent this problem, we use a computational method to accelerate processes within the sediment module of HAMOCC: we reduce the surface area of the sediment grid boxes (underlying the oceanic bottom layer) by a desired acceleration factor (in this study, 1000), while keeping their vertical extent constant. By using this method, the volume of the sediment grid boxes is effectively reduced. Accordingly, the amount and composition of material that is exported from the oceanic bottom layer and deposited at the sediment surface is distributed faster over the whole sediment column. Since we want to maintain the same dissolution behavior throughout the acceleration process within the sediment, we have to reduce the porewater diffusion by the same factor we apply for the area reduction.

From a modeling perspective, this computational acceleration is acceptable, since the proportion of the single grid cells (100 km × 15 cm) prevents horizontal gradients in the sediment module. Moreover, this method provides a way to accelerate sedimentation processes, while mass conservation is maintained in both, the water column and the sediment module. As soon as the sediment reaches an equilibrium-like state, the sediment surface is extended to its original area and the porewater diffusion is set back.

In our simulation, this sediment acceleration is turned on in year 1350, after having the water column tracers close to

**Table 1.** Globally integrated values of biogeochemical parameters calculated with HAMOCC for the pre-industrial (CMIP5, Ilyina et al., 2013) and the late Paleocene.

| | Pre-industrial (CMIP 5) | Late Paleocene |
|---|---|---|
| Atmospheric $CO_2$ (ppm) | 278 | 560 |
| Ocean volume ($10^{18}\,m^3$) | 1.353 | 1.164 |
| Ocean temperature surface (°C) | 9.6 | 24.8 |
| Ocean temperature at 4000 m depth (°C) | 1.4 | 8.9 |
| Ocean salinity | 34.67 | 34.31 |
| Inventories | | |
| Carbon ($10^3$ Gt) | 38.5 | 32.0 |
| Phosphate ($10^{12}$ kmol) | 2.73 | 2.48 |
| Silicate ($10^{14}$ kmol) | 1.64 | 1.88 |
| Nitrate ($10^{13}$ kmol) | 3.44 | 2.84 |
| Weathering | | |
| Global input of $CaCO_3$ ($Tmol\,yr^{-1}$) | 28.4 | 22.1 |
| Global input of opal ($Tmol\,yr^{-1}$) | 6.5 | 0 |
| Global input of POC ($Tmol\,yr^{-1}$) | 0 | 0.1 |
| Primary production | | |
| Global ($Gt\,C\,yr^{-1}$) | 61.14 | 58.65 |
| Export production | | |
| $CaCO_3$ ($Gt\,C\,yr^{-1}$) | 0.89 | 0.63 |
| Opal ($Tmol\,Si\,yr^{-1}$) | 118.24 | 159.57 |
| POC ($Gt\,C\,yr^{-1}$) | 8.72 | 8.54 |
| Molar export ratio | | |
| C($CaCO_3$) : Si(opal) | 0.63 | 0.33 |
| C($CaCO_3$) : C(POC) | 0.1 | 0.07 |
| Si(opal) : C(POC) | 0.16 | 0.22 |
| N cycle | | |
| $N_2$ fixation ($Tmol\,N\,yr^{-1}$) | 14.89 | 22.33 |
| Denitrification ($Tmol\,N\,yr^{-1}$) | 14.53 | 21.18 |

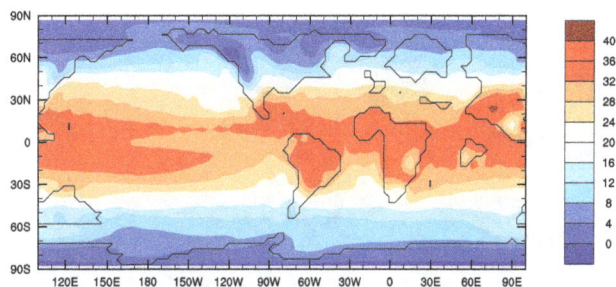

**Figure 2.** Forcing field: annual mean atmospheric temperature (°) at 2 m in height.

latitudes ($-1$ °C at 90° S) are on annual average around 4 °C colder than northern high latitudes (3°C at 90° N).

The prescribed wind forcing displays similar patterns as in the pre-industrial setup (not shown). However, the variability of late Paleocene winds is much stronger than the variability of the pre-industrial state. The highest variability is found in the stormtrack regions at higher latitudes, while around the Equator, winds are comparable to pre-industrial conditions. The atmospheric climate state has been evaluated in an earlier study using the same model version by Heinemann et al. (2009).

The ocean has a mean temperature of 14.7°C (pre-industrial: 5.6°C) and a global annual mean SST of 24.8 °C, which is in agreement with results of other climate models (Lunt et al., 2012). The northern high latitudes reach maximum SST of 12.8 °Cin Northern Hemisphere summer (JJA), but the sea surface of the central Arctic Ocean does not get warmer than 4°C. The southern high latitudes show maximum SST of 17.8 °C in austral summer (DJF). In the presented setup, no sea ice occurs; in each hemisphere's winter, the ocean stays completely ice free. The simulated high-latitude SST are in general agreement with the reconstructions for the Southern Ocean (Thomas et al., 2002), but do not fit the extreme proxy data assumptions of Sluijs et al. (2006) for the Arctic Ocean (for further discussion, see Heinemann et al., 2009). The meridional cross sections of the Pacific and the Atlantic Ocean (Fig. 3) differ from each other in their vertical temperature profiles. While the Atlantic features homogeneous relatively warm temperatures over the largest parts of the water column, the temperature gradient in the Pacific is much more pronounced. Differences in deep sea temperatures reach up to $\sim 5$ °C between the two basins. While deep sea temperatures for the Atlantic (13 °C) are in line with proxies (Tripati and Elderfield, 2005), the deep Pacific (8 °C) seems comparatively cold.

The late Paleocene ocean in our simulations has a mean salinity of 34.31 (pre-industrial: 34.67) and a mean SSS of 33.77. Within the subtropics, regions of elevated SSS around 36 emerge, while higher latitudes show generally lower values (Fig. 3). The Atlantic Ocean has the highest SSS; here, the annual mean surface salinity amounts to 35.06,

an equilibrium state. After approximately 150 years, the net fluxes at the ocean–sediment boundary are strongly reduced (indicating an equilibrium-like state), and the sediment acceleration is turned off. In this way, accumulation of the sediment pool is roughly equivalent to 150 000 years..

## 3 Late Paleocene climate state

The late Paleocene climate state in our simulation is characterized by a global annual mean temperature of 23.6 °C (at 2 m in height), using the atmospheric forcing described above (for a comparison, see Lunt et al., 2012). Maximum annual average temperatures are reached along the Equator over Africa, Asia and South America, with temperatures close to 40 °C (Fig. 2). The absolute annual average heat maximum lies over southern Asia (42.1 °C). Southern high

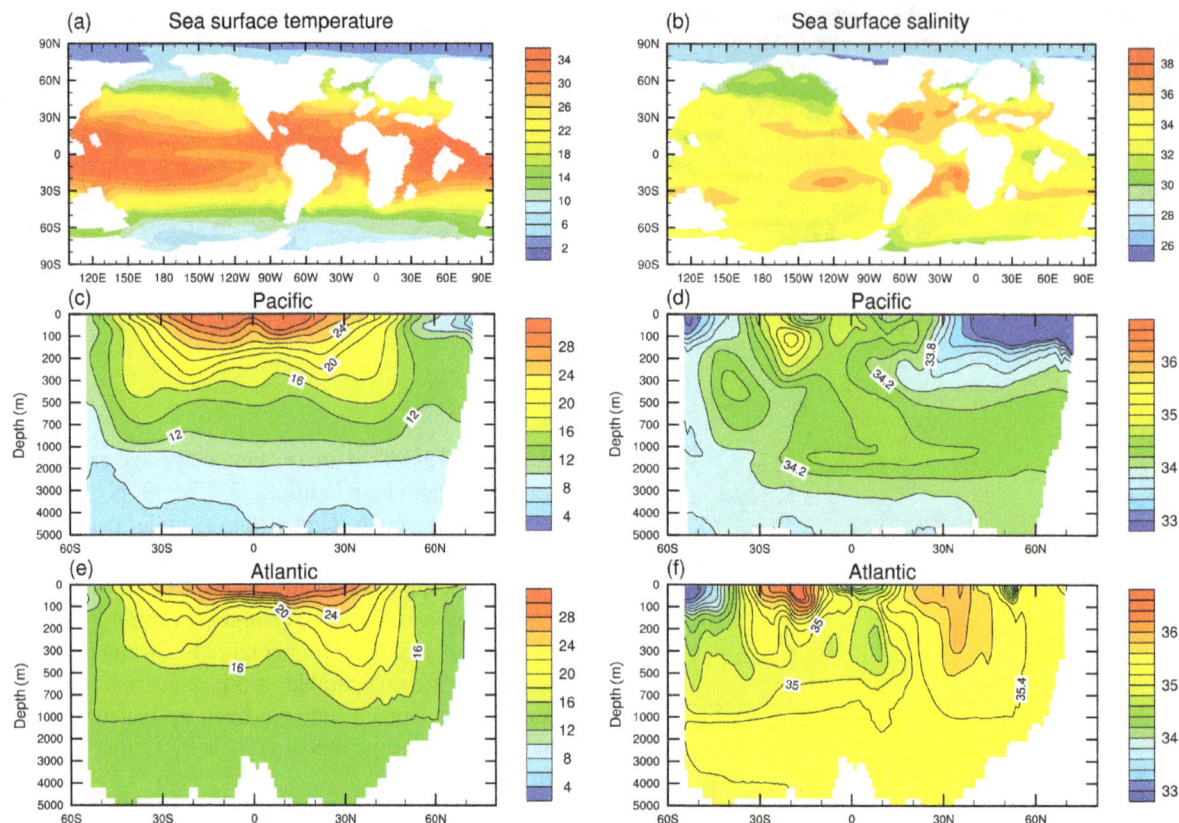

**Figure 3.** Sea surface temperature (°C) (left) and salinity (right) for the surface (**a, b**), Pacific (**c, d**), and Atlantic (**e, f**) averaged meridional crosscut. Note the nonlinear vertical axes, used to zoom in on the upper ocean layers.

and maximum salinities of 36.9 in the North Atlantic and 37.4 in the South Atlantic are reached. The upper 1000 m of the basin reveal the advection of more saline waters from the Tethys around 30° N and less saline waters from the Southern Ocean. At depths greater than 1000 m, the Atlantic basin salinity is characterized by homogeneous distributions, except for the small tongue of slightly less saline Antarctic bottom water (AABW) extending until 20° S (Fig. 3), mirroring the weak vertical mixing and hence strong stratification. The Pacific is much more heterogeneous in terms of vertical salinity distribution. In the northern Pacific, the Arctic inflow of water causes very low surface ocean salinities up to 30° N. The Southern Hemisphere and even parts of the northern deep Pacific are dominated by less saline AABW. The inflow of Atlantic water via the Central American Seaway (strongest net inflow in the uppermost 100 m) causes the highest salinities in the Pacific (Fig. 3). Here, the more saline waters descend, and are then transported northward. This gradient between the southern (low salinity) and northern (high salinity) Pacific shapes the vertical salinity profile. The Arctic Ocean's mean surface salinity amounts to 27.13. This low salinity is interpreted as an effect of freshwater inflow due to an intensified hydrological cycle in a warmer atmosphere and as an effect of the shape of the ocean basin.

Low surface salinities and a poorly ventilated water column are also derived from proxy data (Pagani et al., 2006b; Waddell and Moore, 2008). The bathymetry prevents any deepwater exchange with the surrounding oceans, and hence contributes to a sharper stratification of the Arctic Ocean and its low-salinity cap. There is no additional amplifying mechanism for mixing of Arctic waters due to the absent sea ice formation and consequent brine production.

Salinity and temperature profiles are mainly shaped by the meridional overturning circulation (MOC) (Fig. 4). The large-scale circulation structures are generally similar to modern conditions, but the late Paleocene Atlantic is dominated by just one large-scale circulation cell in our simulation (spreading over nearly the whole basin up to a depth of 4000 m), causing more homogeneous temperature and salinity distributions. The Atlantic Ocean lacks an AABW cell, while the AABW formation and spreading are much more pronounced in the Pacific. Here, formation of deepwater occurs at 70° S. No further overturning takes place in the northern Pacific, as salinity concentrations in the surface ocean are too low, due to the Arctic Ocean water inflow. The late Paleocene climate causes the ocean to be warmer on average (compared to pre-industrial conditions), leading to increased stratification throughout the water column. The

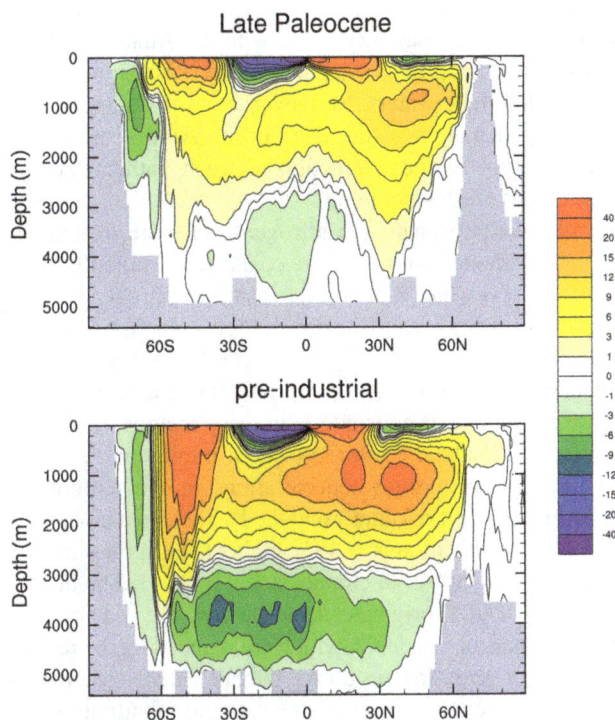

**Figure 4.** Globally averaged meridional overturning circulation (Sv) for the late Paleocene (top) and the pre-industrial (bottom). Positive values correspond to clockwise circulation. Pre-industrial values were calculated within CMIP5 experiments.

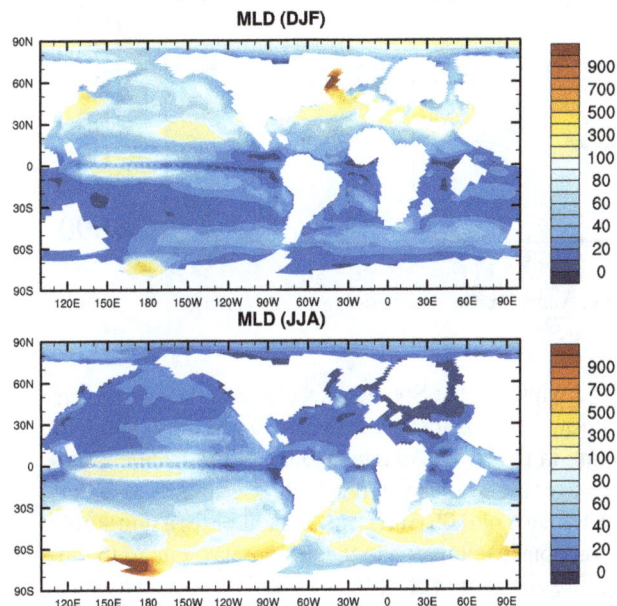

**Figure 5.** Mixed layer depth (m) averaged over boreal winter (DJF) and boreal summer (JJA).

reduced equator-to-pole temperature gradient results in a further slowdown of the MOC. A maximum deepwater formation (for depths greater than 900 m) of $\sim 15$ Sv occurs in the Southern Ocean and the North Atlantic (Fig. 4). Southern Ocean sinking occurs in the Ross Sea, similar to other Paleocene–Eocene simulations (Sijp et al., 2014), whereas the North Atlantic deepwater source is not produced in all models.

In MPIOM, the mixed layer depth (MLD) is defined as the depth where in situ density exceeds surface water density by more than $0.125$ kg m$^3$ (sigma-t criterion). The annual global mean MLD in the late Paleocene setup levels at 52 m in depth. Observations suggest an annual global mean of 65 m for MLD in present-day oceans (de Boyer Montegut et al., 2004). Since the MLD is interpreted as an indicator of the stratification of the ocean, which apparently is stronger in a warmer climate (e.g., Wetzel et al., 2006), the reduced MLD appears to be an expected result of our model simulation. As shown in Fig. 5, the MLD has a strong seasonal signal. In boreal winter, deepwater formation takes place in the North Atlantic. In austral winter, the same deepening of MLD as an effect of convectional processes occurs in the South Pacific.

## 4 Late Paleocene ocean biogeochemistry

### 4.1 Air–sea exchange processes

In this study, we aim at achieving steady-state conditions with respect to ocean biogeochemistry, in accordance with a long-term warm climate. Corresponding to the expectation of an equilibrium climate state, the annual mean $CO_2$ flux at the atmosphere–ocean boundary is balanced around zero. The global annual mean of surface ocean $pCO_2$ is 560 ppm within the late Paleocene model setup. It shows highest values along the Equator and in the eastern boundary currents, along South America in the Pacific, and Africa in the Atlantic (Fig. 6), similar to the present day (Takahashi et al., 2009). While the general spatial distribution of $pCO_2$ is mainly defined by temperature and salinity, the high $pCO_2$ areas in the equatorial and coastal areas result from upwelling of high $pCO_2$ and nutrient-rich waters from mid-ocean depth.

The Atlantic Ocean is the major net emitter of $CO_2$, with an annual outgassing of 0.41 Gt C, while the Pacific Ocean balances its net fluxes around zero over the year. While high oceanic $pCO_2$ is associated with carbon release into the atmosphere, oceanic $CO_2$ uptake occurs in regions with low $pCO_2$ (if $pCO_2$oce $< pCO_2$atm). The model computes the lowest $pCO_2$ around Antarctica, especially close to the Drake Passage, which corresponds to low salinity and TA concentrations in this area. Nevertheless, nearly the whole Southern Ocean ($40–80°$ S) is characterized by $pCO_2$ values below the atmospheric $CO_2$ concentration of 560 ppm. This is consistent with deepwater formation and the deep MLD

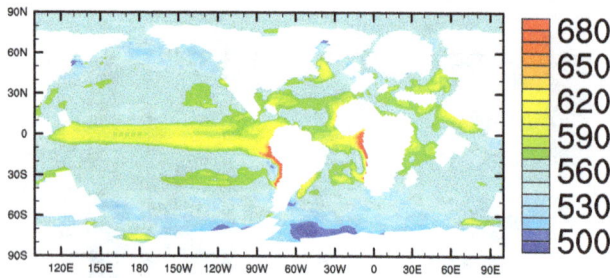

**Figure 6.** Annual mean surface ocean $p\mathrm{CO}_2$ (ppm).

in austral winter in the Southern Ocean. Another prominent zone of low $p\mathrm{CO}_2$ is located between 50 and 80° N, with its maximum in the North Atlantic and the North Pacific. Summarized over the whole year, the Arctic Ocean acts as a $\mathrm{CO}_2$ sink (net uptake of 0.06 Gt C). However, it plays a minor role in carbon uptake compared to the present day, due to the reduction in surface area of 40 % and increased ocean temperature. Instead, the Indian Ocean becomes the major driver in $\mathrm{CO}_2$ net uptake (0.31 Gt C yr$^{-1}$). The southern part of the Indian Ocean is influenced by a deep mixed layer in austral winter (> 200 m), but even parts of the northern Indian Ocean show a MLD of up to 80 m, resulting in rather low $p\mathrm{CO}_2$ values in the Indian Ocean surface waters. Consistent with the solubility effect, which declines with rising temperature (Weiss, 1974), our model shows that the equatorial regions in the late Paleocene acted as a source of $\mathrm{CO}_2$, while the high latitudes operated as a $\mathrm{CO}_2$ sink. This matches the present-day simulations with HAMOCC (Ilyina et al., 2013), but during the late Paleocene, it is mainly the $\mathrm{CO}_2$ uptake of the Indian and Southern oceans that compensates for the $\mathrm{CO}_2$ outgassing in the Atlantic.

Globally, the high ocean temperatures lead to a reduced solubility of $\mathrm{CO}_2$ in the surface ocean, compared to pre-industrial conditions. Moreover, the transfer of $\mathrm{CO}_2$ from the surface to intermediate and deep waters by the oceanic velocity field is reduced by the more sluggish circulation in the late Paleocene compared to the pre-industrial simulation (Fig. 4). The sluggish circulation together with the weak ocean solubility pump would act to reduce the ocean's uptake capacity of atmospheric $\mathrm{CO}_2$ in response to the carbon perturbation during the PETM.

**4.2  Biological production and nutrients**

Phosphate concentrations in the surface ocean show pronounced latitudinal gradients and weaker basin-to-basin differences in the late Paleocene simulation. High phosphate concentrations characterize the surface ocean at the southern high latitudes, the equatorial Pacific and the Atlantic, as well as at Northern Hemisphere mid latitudes (Fig. 7). Strong equatorial upwelling of water masses in the Atlantic and Pacific as well as moderate upwelling along the western coasts and in the Southern Ocean cause a maximum in phosphate

surface concentrations in these regions. All other regions are characterized by phosphate depletion, with the Atlantic and Pacific gyres revealing the lowest phosphate concentrations in subtropical surface waters. Moreover, our simulation reveals a strong depletion in nutrients in the Arctic surface ocean, as a result of the interaction between bathymetry, stratification and freshwater input.

Generally, the phosphate and nitrogen cycles are treated similarly in the model, since they are connected via the Redfield ratio. However, bacterial processes such as nitrogen fixation and denitrification cause deviations between nitrate and phosphate distributions. In oxygen-depleted zones, denitrifying bacteria provide oxygen for remineralization, representing an additional sink for nitrate in these regions. In the late Paleocene simulation, intense oxygen minimum zones (OMZ) lead to low nitrate concentrations in the eastern boundary currents of the Atlantic and the Pacific. On a global average, the denitrification is ∼ 45 % higher than in the simulation for the pre-industrial climate state (Table 1). The increased denitrification originates from the low oxygen concentrations in mid-ocean depth, which are induced by the reduced mixing of water masses during the late Paleocene. Nitrogen fixation occurs in areas where the ratio of nitrate to phosphate is lower than the (constant stoichiometric) value of $R_{\mathrm{N:P}}$. This is primarily in the tropics, the North Pacific and the North Atlantic. Likewise, nitrate fixation is also higher in the late Paleocene than in the pre-industrial climate state.

The gradient in surface ocean nutrient concentrations between low and high latitudes in the Southern Hemisphere resembles the modern one, despite the homogeneous dust/iron concentration prescribed for every grid cell. This indicates that the dust climatology does not produce a strong signal at higher latitudes. Hence, in our late Paleocene setup, the iron limitation is not the major driving mechanism for preventing the surface ocean from complete consumption of nutrients, making the physical conditions at the poles accountable for it.

Surface oxygen concentration decreases from pole to Equator, confirming its strong temperature dependency (Fig. 7). Generally higher SST than under modern conditions lead to slightly lower oxygen concentrations in the surface oceans. Only in the Arctic Ocean does the low salinity counteract the temperature effect, as salinity is inversely related to oxygen solubility in seawater, according to Weiss (1970). This leads to similar oxygen concentrations/solubility like in the present-day setup.

While the physical dynamics and the export production of particulate organic carbon determine the surface concentrations of nutrients, the distribution of nutrients at depth is controlled by remineralization (Maier-Reimer, 1993). Modern nutrient and oxygen concentrations in the Pacific and Atlantic mirror the different age of the deepwater, defined by the global thermohaline circulation. In the late Paleocene simulation, the upper 1000 m of the water column reveal the highest phosphate concentrations in both basins, due to

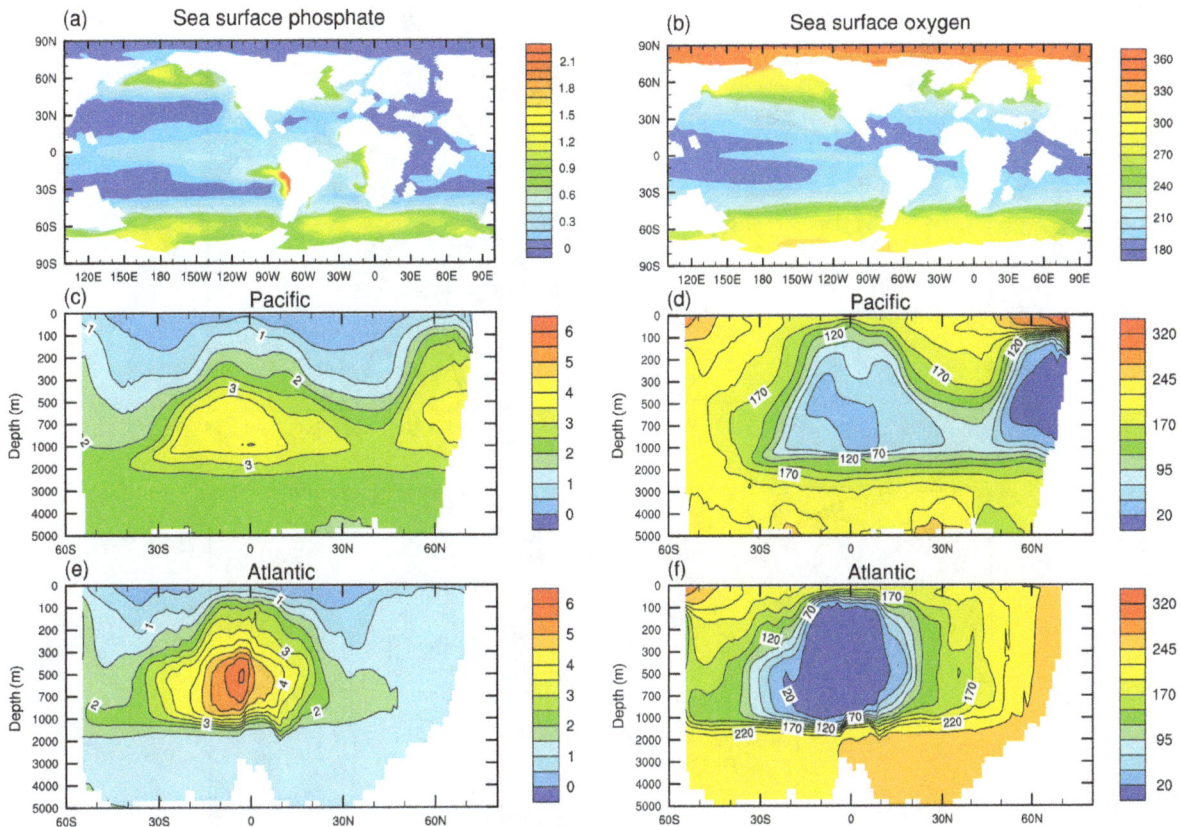

**Figure 7.** Phosphate (left) and oxygen (right) concentrations (both in $\mu$mol L$^{-1}$) for the surface (**a, b**), Pacific (**c, d**), and Atlantic (**e, f**) averaged meridional crosscut. Note the nonlinear vertical axes, used to zoom in on the upper ocean layers.

strong remineralization of organic matter. The phosphate concentrations at depths below 1000 m are rather a product of global ocean circulation, then in situ remineralization. This explains the lower phosphate concentrations in the Atlantic over the Pacific, although OM export rates (in units per area) and oxygen concentrations are higher in the Atlantic. The oxygen concentration in Pacific deepwater is highly increased (Fig. 8) compared to modern conditions, which can be attributed to Southern Ocean deepwater formation and an enhanced exchange of the Pacific and Atlantic through the Central American Seaway.

The OMZ are very pronounced within the upper 1000 m in both basins. Oxygen concentrations are as low as 20 $\mu$mol kg$^{-1}$, on meridional average, along the equatorial Atlantic and the North Pacific. The prominent OMZ are attributed to the existence of productive equatorial zones (Fig. 9) (Norris et al., 2013) and reduced mixing in a more stagnant ocean during the late Paleocene. For the Atlantic, the low oxygen concentrations along the Equator are even intensified on the meridional average, due to an additional high productivity zone along the northern continental margin of South America. In general, the Atlantic shows an increase in OM export (in units per area) of nearly 60 % in comparison to the pre-industrial, although the total export increased by just

13 %, due to the reduced area of the Atlantic in the late Paleocene setup. However, the simulated OMZ might be somewhat overestimated, as illustrated in CMIP5 simulations. The model produces lower than observed oxygen concentrations, spreading over larger areas in the equatorial Pacific and along the western continental margin of Africa (Ilyina et al., 2013), which is typical for other global models as well (Andrews et al., 2013; Cocco et al., 2013).

The annual global primary production amounts to $\sim$ 59 Gt C (Table 1) in the late Paleocene simulation. The coastal upwelling regions along the western continental margins, as well as the equatorial regions of the Pacific and the Atlantic, tend to be the dominant mechanism fueling primary production (Fig. 9). Compared to modern conditions, the production along the eastern boundary currents in the Atlantic and Pacific is less pronounced. However, resulting from the open Central American Seaway, strong production arises along the northern tip of South America. The mid latitudes (nutrient-poor mid latitudinal gyres) and the Arctic Ocean exhibit sparse productivity. On the contrary, the nutrient-rich Southern Ocean is responsible for $\sim$ 11 % of the global primary production.

The production of calcite shells follows the low silicate surface concentrations in the Atlantic, Tethys and Indian

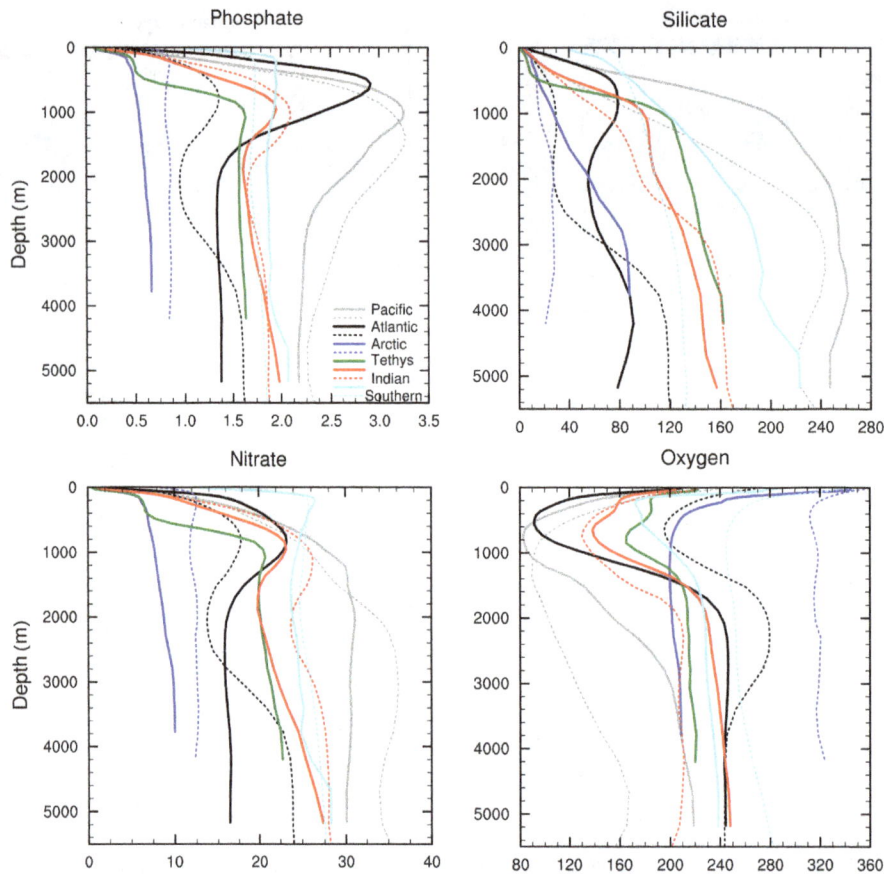

**Figure 8.** Globally averaged vertical profiles of phosphate, silicate, nitrate and oxygen (all in $\mu$mol L$^{-1}$). Solid lines show late Paleocene distributions; dotted lines show pre-industrial concentrations calculated within CMIP5 experiments.

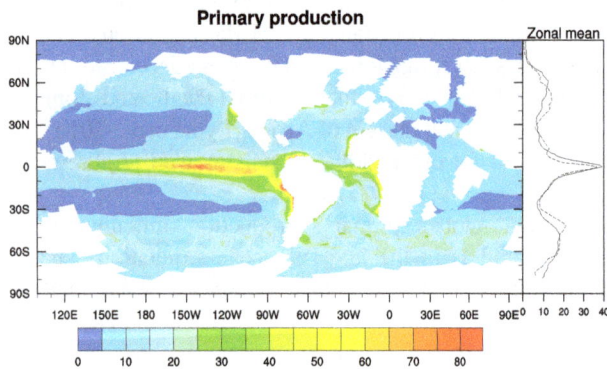

**Figure 9.** Annual mean primary production plotted as map and zonal average (both in mol C m$^{-2}$ yr$^{-1}$). The zonal average plot shows the primary production for the late Paleocene (solid line) and the pre-industrial (dashed line).

oceans, as well as in the western equatorial Pacific. Regions of higher silicate concentrations that correspond to upwelling locations are dominated by the production of opal shells, as implied by our modeling approach.

The export production of CaCO$_3$ is, at 0.63 Gt C yr$^{-1}$, lower than for pre-industrial conditions (see Table 1). This leads to a CaCO$_3$ : POC export ratio of 0.07, contrary to the PETM simulations of Panchuk et al. (2008) and Ridgwell and Schmidt (2010) suggesting a ratio of 0.2. As the CaCO$_3$ : POC rain ratio is an important source of uncertainty, controlling the sedimentary CaCO$_3$ wt % distribution, Panchuk et al. (2008) base their suggestion on an ensemble run using different export ratios. Finally, 0.2 matches best their pre-PETM CCD (3.5–4 km in depth) and CaCO$_3$ sediment distribution. Simulations for the present-day export ratio suggest that the ratio levels more around 0.1 ($\sim$0.06, Sarmiento et al. (2002); $\sim$0.1, CMIP5 runs, MPI-ESM; < 1.4, Ridgwell and Schmidt, 2010). In our simulation, the CaCO$_3$ : POC export ratio is a result of production, remineralization and sinking velocity. Because dissolution of opal is positively correlated with temperature, in a warmer ocean, more silica is available for opal production in the upper ocean (Fig. 8). This indirectly results in less CaCO$_3$ formation in our model setup. The homogeneous dust (iron) climatology can not cause a shift from CaCO$_3$ towards opal-producing skeletons. Thus, it can only lead to an absolute increase in the production of former iron-limiting low-productivity zones,

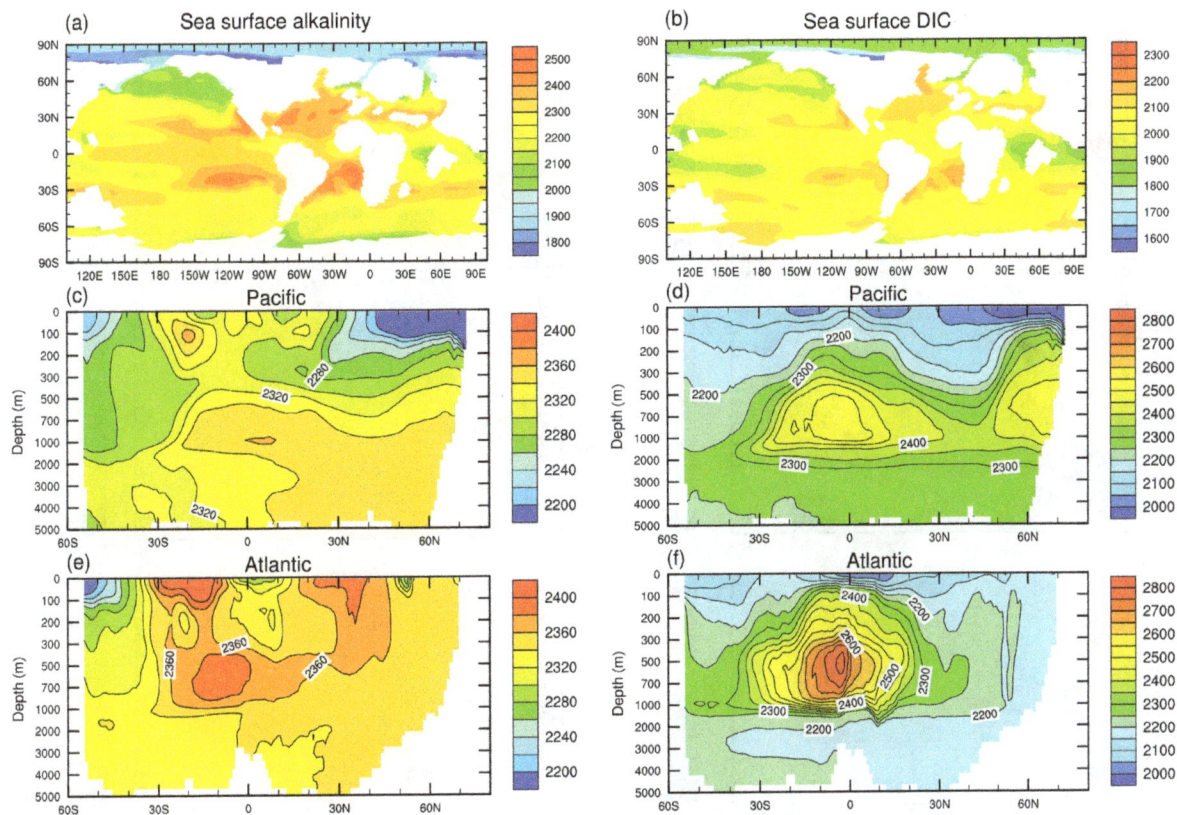

**Figure 10.** TA (left) and DIC (right) concentrations (both in $\mu$mol kg$^{-1}$) for the surface (**a, b**), Pacific (**c, d**), and Atlantic (**e, f**) averaged meridional crosscut. Note the nonlinear vertical axes, used to zoom in on the upper ocean layers.

but can not change the proportions between the two building materials.

An interbasinal comparison reveals increased overall export fluxes in the Atlantic. Opal, CaCO$_3$ and POC exports (in units per area) even exceed the ones in the Pacific. However, in absolute numbers, the Pacific Ocean is the main driver, covering about 50 % of the global ocean surface area; it is responsible for more than 50 % of POC export (4.55 Gt C yr$^{-1}$) globally. The strong depletion in silica in Tethys surface waters (Fig. 8) favors high CaCO$_3$ production. As a consequence, parts of the Tethys Ocean show low surface TA. The Tethys has just a minor impact on the strength of the biological carbon pump, exporting 0.28 Gt C yr$^{-1}$.

### 4.3 Carbonate chemistry

TA concentrations decrease from the Equator towards the poles, with the subtropical gyres showing both, elevated DIC and TA, compared to the surrounding surface waters (Fig. 10). The spatial patterns of surface ocean DIC and TA reproduced by the model are similar to pre-industrial conditions. The highest DIC concentrations, besides the gyres, are located in the Southern Ocean (2100 $\mu$mol kg$^{-1}$), whereas the Arctic Ocean exhibits low DIC concentrations around 1800 $\mu$mol kg$^{-1}$. While the SST mainly determines the sur-

face distribution of DIC, the surface TA rather reflects the structure of salinity (Maier-Reimer, 1993), which is shaped by the precipitation–evaporation gradients, leading to increased TA concentrations in the subtropics. Low salinity in the Arctic Ocean, induced by strong vertical stratification and the low exchange with the surrounding oceans (no deep-water exchange), drives the TA to very low values (< 1800 $\mu$mol kg$^{-1}$). High SST, low TA, and little CO$_2$ uptake in the Arctic Ocean result in generally low DIC concentrations over the whole water column, reducing the Arctic Ocean's carbon storage capacity.

The surface seawater CO$_3^{2-}$ concentration is shaped by the elevated atmospheric CO$_2$ concentration of 560 ppm. Higher atmospheric CO$_2$ concentrations do not have an impact on TA, but cause a shift from CO$_3^{2-}$ to HCO$_3^-$ (bicarbonate). The reduced CO$_3^{2-}$ : DIC ratio, which characterizes the pre-PETM carbonate chemistry, reduces the oceanic buffer capacity for atmospheric CO$_2$ perturbations.

In the vertical profile, the maximum in DIC concentration around the Equator, spreading from 400 to 1000 m in depth (Fig. 10), is related to biological processes. It marks the depth at which intense dissolution and denitrification of the exported particles takes place. The aerobic remineralization of POC releases DIC and consumes oxygen, while at the

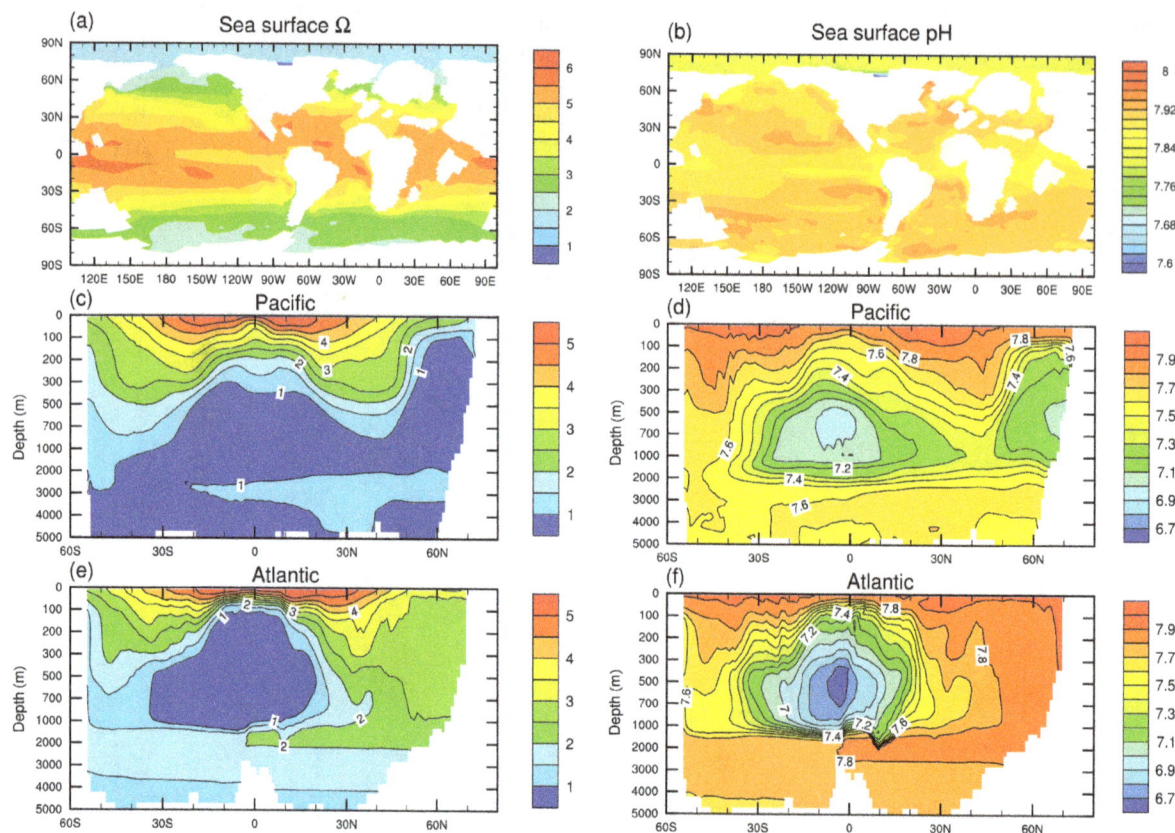

**Figure 11.** $\Omega$ (left) and pH (right) for the surface (**a, b**), Pacific (**c, d**), and Atlantic (**e, f**) averaged meridional crosscuts. Note the nonlinear vertical axes, used to zoom in on the upper ocean layers.

same time, the dissolution of $CaCO_3$ and the denitrification increases the TA at $\sim 1000\,\mathrm{m}$ in depth (Fig. 10). The fact that the subsurface Atlantic (up to 1000 m in depth) exhibits higher TA and DIC concentrations over the Pacific can be explained by the stronger export of $CaCO_3$ and POC (in units per area). In the uppermost layers of the ocean, the difference in $CO_3^{2-}$ concentration between the Atlantic and Pacific is not evident anymore. The atmospheric $CO_2$ concentration of 560 ppm causes low $CO_3^{2-}$ concentrations, reaching up to a depth of 600 m in both oceans. In layers beneath 2000 m, the $CO_3^{2-}$ concentrations within the basins show homogeneous distributions, with the Atlantic $CO_3^{2-}$ concentrations (100–120 $\mu$mol kg$^{-1}$) being nearly twice as high as in the Pacific (60–70 $\mu$mol kg$^{-1}$).

The calculated global average pH in the surface ocean amounts to 7.9, which is in agreement with the estimate by Tyrrell and Zeebe (2004), and close to the suggestion of Ridgwell and Schmidt (2010), that late Paleocene pH surface values were $\sim 0.4$ lower than today. The surface distribution in pH displays a similar pattern to what we know from pre-industrial pH, albeit at lower values. Within the upwelling areas along the Equator and in front of the western continental margins, the pH is lower than at the mid latitudes. The

Arctic Ocean shows particularly low pH, correlated with the very low salinities.

At depth, large parts of the Pacific are undersaturated with respect to $CaCO_3$ (Fig. 11). However, the undersaturation is characterized by a strong gradient in the east–west direction, which is not evident in the zonally averaged $\Omega$ values (Fig. 11). The basin-wide undersaturation in the Pacific starts below a depth of 3700 m. The horizontal gradient in $\Omega$ is even enhanced by inflow of $CO_3^{2-}$-rich water from the Indian Ocean and undersaturated ($\Omega < 1$) near-surface waters in the eastern part of the basin. Undersaturation occurs due to biological respiration processes in intermediate waters, which correlate with very low pH values of up to 6.9, computed for the uppermost 1000 m of the water column within the equatorial Pacific.

The elevated ocean temperatures during the late Paleocene produce a lower $CO_3^{2-}$ saturation concentration, because the saturation concentration depends inversely on temperature. This should lead to an increased $\Omega$, in comparison to the present day, as formerly discussed in Zeebe and Zachos (2007). This effect is strong in the Atlantic, which at depth is on average around 5 °C warmer than the Pacific in our simulation. Here, a reduced calcite saturation concentration counteracts the lower $CO_3^{2-}$ concentrations. The undersaturation with respect to $CO_3^{2-}$ in the equatorial Atlantic between 200 and 1000 m in depth is a result of very low pH in the low-latitude subsurface ocean, caused by aerobe and anaerobe remineralization (Fig. 8). The North Atlantic and the Indian Ocean show no undersaturation with respect to $CaCO_3^{2-}$ at all.

Our approach provides $\Omega$ distributions consistent with the three-dimensional hydrodynamical field and the biogeochemical processes as simulated for the late Paleocene ocean. We perceive a complex pattern of $\Omega$, which includes, e.g., undersaturated waters at shallower depths overlying supersaturated waters in deeper layers, as seen in the Pacific in our simulation. Local undersaturation might be of importance, since 60 to 80 % of the $CaCO_3$ export is dissolved already in the upper 1000 m of the water column, as shown by present-day studies (Feely et al., 2004; Ilyina and Zeebe, 2012). The assumption of a uniform basin-wide saturation horizon (depths where $\Omega = 1$) or lysocline (depths where $\Omega = 0.8$; Ridgwell and Zeebe, 2005) for the late Paleocene oceans (e.g., Panchuk et al., 2008; Zeebe et al., 2009; Cui et al., 2011), as has been made for several EMIC and box model studies, might underestimate these dissolution processes.

## 4.4 Sediment composition

The sediment compartment was initialized with 100 % clay. Hence, the resulting distribution of organic matter, opal and $CaCO_3$ in the sediment reflects water column processes. Large sedimentary opal deposits are located in the equatorial Pacific, along the eastern boundary currents and in the Southern Ocean (Fig. 12). This mirrors the nutrient-rich upwelling areas, zones where high production at the surface takes place. However, it is noticeable that the annual net sedimentation rate of opal is very small on the global scale. Since the opal remineralization rate is positively correlated with temperature, we attribute this effect to the increased deep sea temperature as compared to the pre-industrial setup of the model.

$CaCO_3$ deposits cover wider areas, corresponding to supersaturated ($\Omega > 1$) bottom waters and $CaCO_3$ production at the surface. The present-day apparent predominance of $CaCO_3$ accumulation in the Atlantic and Indian oceans compared to the Pacific is also preserved in the late Paleocene simulations. A greater degree of undersaturation in the deep Pacific exists as a consequence of metabolic $CO_2$ accumulation (Ridgwell and Zeebe, 2005). Unlike today, the Arctic and Southern oceans bear $CaCO_3$ deposits, attributed to a regionally higher $CaCO_3$ : POC rain ratio and less corrosive bottom waters in these regions.

The upper 14 cm of the sediment being affected by dissolution processes contain a total $CaCO_3$ amount of 2049 GtC. This corresponds to a global average $CaCO_3$ wt % of 36. The absolute amount of sedimentary $CaCO_3$ in our simulation is slightly higher than estimates for the present day (1610 GtC,

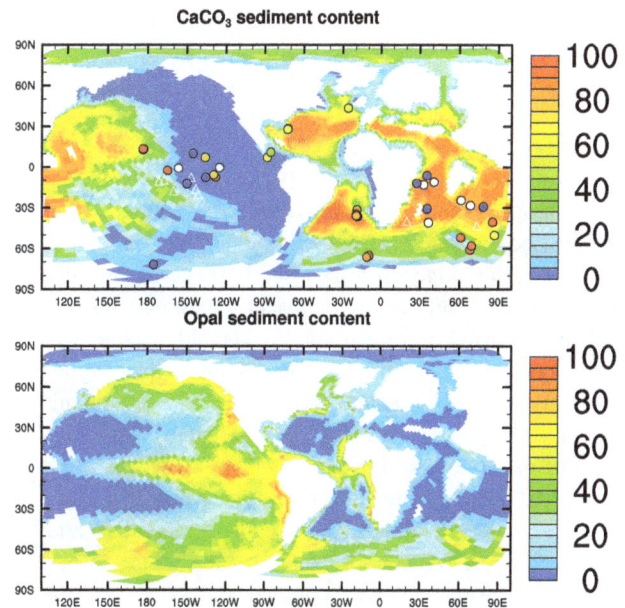

**Figure 12.** Sediment content of $CaCO_3$ and opal (both in wt %), averaged over the last 30 years of the simulation. The $CaCO_3$ plot includes observational data points from Panchuk et al. (2008): colored dots indicate wt %, white dots indicate that $CaCO_3$ is present but wt % is unknown, and triangles indicate a hiatus (due to non-deposition).

CMIP5 simulation with MPI-ESM; 1770 GtC, Archer et al., 1998). Another model study results in a lower total $CaCO_3$ amount of 800 GtC for the present day and a decrease to 620 GtC for a pre-PETM setup (Zeebe, 2012).

Panchuk et al. (2008) evaluate their computed $CaCO_3$ distribution by a compilation of late Paleocene marine sediment cores. We use the same data set to evaluate our results. Our model captures the spatial pattern and the absolute values of the compiled sediment core data relatively well (Fig. 12), yet our model calculates a much smaller $CaCO_3$ : POC rain ratio. Differences between Panchuk et al. (2008) and our estimates persist in the absence of $CaCO_3$ in the central Pacific, as well as in the abundance of $CaCO_3$ in the North Atlantic in our results. This could be due to the different locations for Northern Hemisphere deepwater formation, which is located in the North Pacific in Panchuk et al. (2008), but in the North Atlantic in our simulation. Our model results match the South Atlantic data quite well. The same is true for western and central Pacific sediments and parts of the Indian Ocean. The mismatch in the eastern Pacific and central Indian oceans is mainly due to depth divergences between the cores and the model bathymetry. For instance, the 0 wt % sediment cores in the Indian Ocean display the $CaCO_3$ content between 4000 and 4900 m in depth. The applied bathymetry is several 100 m shallower in these locations. Nevertheless, the model seems to underestimate the $CaCO_3$ abundances in the Atlantic and Indian sectors of the Southern Ocean, while the Pacific sector shows, consistently with the data, very low $CaCO_3$ wt %.

# 5 Summary and conclusions

Using the biogeochemistry model HAMOCC coupled with the MPIOM ocean general circulation model, we establish a steady-state ocean biogeochemistry simulation with late Paleocene boundary conditions. We present spatial and vertical tracer distributions within a warmer climate and display how the late Paleocene ocean physical state influences the biogeochemistry. We provide a general overview of major oceanic carbon cycle components in a high $CO_2$ (560 ppm) world and give estimates of the PETM background climate.

The late Paleocene simulation reveals a strong stratification of water masses, which is displayed by temperature and salinity profiles, as well as the shallow MLD. These conditions are also found in other coupled atmosphere–ocean general circulation models using Eocene boundary conditions (Lunt et al., 2010). The sluggish circulation affects the atmosphere–ocean exchange fluxes of $CO_2$ by shifting its spatial patterns; i.e., uptake in the Indian and Southern Ocean compensates for the $CO_2$ outgassing of the Atlantic Ocean. Moreover, we infer a reduced vertical transfer of carbon from surface to intermediate and deep waters due to the more stagnant circulation in comparison to pre-industrial conditions. Nevertheless, the enhanced thermal vertical stratification is not prominent enough to prevent the supply of nutrients to surface waters and, hence, the global primary production is only slightly decreased.

The intensification of primary production along the Equator causes enhanced remineralization within the upper 1000 m of the Atlantic and Pacific, leading to strong OMZs. The global pattern in oxygen distribution in the intermediate and deep waters is affected by the altered continental configuration. In particular, the open Central American Seaway and deepwater formation in the South Pacific ensure the adjustment between Atlantic and Pacific oxygen concentrations.

The outcomes of the simulation show that an equilibrium $CO_2$ exchange during the late Paleocene can be established without increased concentrations of TA, as assumed by Pearson and Palmer (2000), while the calculated surface ocean pH (lower than today) is in agreement with the results of Tyrrell and Zeebe (2004) and Ridgwell and Schmidt (2010). Low surface TA concentrations are positively influenced by a lower $CaCO_3$ : opal export ratio, since less carbonate leaves the surface ocean. This effect is associated with the temperature-dependent remineralization of opal, which produces an increased silicate/opal turnover rate. The simulated export ratio lies within the range of estimates for modern conditions (Schneider et al., 2008), but falls below the estimates given by the simulations of Panchuk et al. (2008) and Ridgwell and Schmidt (2010) covering the Paleocene–

Eocene period. This difference may be a result of using quite different types of models.

While the deepwater of the Pacific is to a great extent undersaturated with respect to $CaCO_3$, the Atlantic does not show any $CaCO_3$ undersaturation at depths below 1000 m. We claim that the warm temperatures in the deep Atlantic are responsible for producing an overly weak saturation gradient (by decreasing the saturation concentration) compared to other studies (e.g., Zeebe et al., 2009). The surface ocean is characterized by globally lower $CO_3^{2-}$ concentrations than today, because of higher atmospheric $CO_2$ concentrations.

Our results indicate that the late Paleocene climate state produced conditions in the carbonate chemistry that led to different responses of the surface and the deep ocean to a large carbon perturbation during the PETM. In the surface ocean, the lower $CO_3^{2-}$ : DIC ratio implies a reduced carbonate buffer capacity, which would be even further reduced if the $CaCO_3$ export from the surface would have been of the same strength as it is in the modern ocean. In the deep ocean, by contrast, elevated temperatures cause the $CaCO_3$ saturation concentration to decrease (i.e., to increase in $\Omega$). Furthermore, the warm and stratified ocean reduces the transport of $CO_2$ from the surface to greater depths. Both mechanisms lower the vulnerability of the sedimentary $CaCO_3$ to dissolution. In summary, the upper ocean carbonate chemistry must have been more sensitive to a $CO_2$ invasion than under modern conditions, while the deep ocean carbonate sediments were better preserved by the overlaying warm and stratified ocean water.

*Acknowledgements.* We thank H. Haak, who helped with the setup of the model, and M. Heinemann, who provided data for producing a late Paleocene atmospheric forcing. Moreover, we would like to thank K. Panchuk for providing the $CaCO_3$ data set. We wish to thank K. Six for advice and helpful comments on the manuscript. The authors acknowledge the late E. Maier-Reimer for valuable insights on the model.

The service charges for this open access publication have been covered by the Max Planck Society. Finally we would like to thank the two anonymous reviewers which helped to strengthen the manuscript.

Edited by: E. Brook

# References

Andrews, O. D., Bindoff, N. L., Halloran, P. R., Ilyina, T., and Le Quéré, C.: Detecting an external influence on recent changes in oceanic oxygen using an optimal fingerprinting method, Biogeosciences, 10, 1799–1813, doi:10.5194/bg-10-1799-2013, 2013.

Archer, D., Kheshgi, H., and Maier-Reimer, E.: Dynamics of fossil fuel $CO_2$ neutralization by marine $CaCO_3$, Global Biogeochem. Cy., 12, 259–276, doi:10.1029/98GB00744, 1998.

Archer, D., Martin, P., Buffett, B., Brovkin, V., Rahmstorf, S., and Ganopolski, A.: The importance of ocean temperature to global biogeochemistry, Earth Planet. Sc. Lett., 222, 333–348, 2004.

Bice, K. L. and Marotzke, J.: Numerical evidence against reversed thermohaline circulation in the warm Paleocene/Eocene ocean, J. Geophys. Res., 106, 11529–11542, doi:10.1029/2000JC000561, 2001.

Bice, K. L. and Marotzke, J.: Could changing ocean circulation have destabilized methane hydrate at the Paleocene/Eocene boundary?, Paleoceanography, 17, 1018–1033, doi:10.1029/2001PA000678, 2002.

Cocco, V., Joos, F., Steinacher, M., Frölicher, T. L., Bopp, L., Dunne, J., Gehlen, M., Heinze, C., Orr, J., Oschlies, A., Schneider, B., Segschneider, J., and Tjiputra, J.: Oxygen and indicators of stress for marine life in multi-model global warming projections, Biogeosciences, 10, 1849–1868, doi:10.5194/bg-10-1849-2013, 2013.

Cui, Y., Kump, L. R., Ridgwell, A. J., Charles, A. J., Junium, C. K., Diefendorf, A. F., Freeman, K. H., Urban, N. M., and Harding, I. C.: Slow release of fossil carbon during the Palaeocene-Eocene Thermal Maximum, Nature Geosci, 4, 481–485, doi:10.1038/ngeo1179, 2011.

de Boyer Montegut, C., Madec, G., Fischer, A. S., Lazar, A., and Iudicone, D.: Mixed layer depth over the global ocean: An examination of profile data and a profile-based climatology, J. Geophys. Res., 109, C12003, doi:10.1029/2004JC002378, 2004.

Dickens, G. R., O'Neil, J. R., Rea, D. K., and Owen, R. M.: Dissociation of oceanic methane hydrate as a cause of the carbon isotope excursion at the end of the, Paleocene Paleoceanogr., 10, 965–971, 1995.

Dunkley Jones, T., Ridgwell, A., Lunt, D. J., Maslin, M. A., Schmidt, D. N., and Valdes, P. J.: A Palaeogene perspective on climate sensitivity and methane hydrate instability, Philos. T. Roy. Soc. A: Mathematical, Physical and Engineering Sciences, 368, 2395–2415, doi:10.1098/rsta.2010.0053, 2010.

Feely, R. A., Sabine, C. L., Lee, K., Berelson, W., Kleypas, J., Fabry, V. J., and Millero, F. J.: Impact of Anthropogenic $CO_2$ on the $CaCO_3$ System in the Oceans, Science, 305, 362–366, 2004.

Groeger, M. and Mikolajewicz, U.: Note on the $CO_2$ air-sea gas exchange at high temperatures, Ocean Model., 39, 284–290, 2011.

Heinemann, M., Jungclaus, J. H., and Marotzke, J.: Warm Paleocene/Eocene climate as simulated in ECHAM5/MPI-OM, Clim. Past, 5, 785–802, doi:10.5194/cp-5-785-2009, 2009.

Heinze, C. and Maier-Reimer, E.: The Hamburg Oceanic Carbon Cycle Circulation Model version "HAMOCC2s" for longtime integrations, Technical Report 20, Deutsches Klimarechenzenrum, Modellberatungsgruppe, Hamburg, 1999.

Heinze, C., Maier-Reimer, E., Winguth, A. M. E., and Archer, D.: A global oceanic sediment model for long-term climate studies, Global Biogeochem. Cy., 13, 221–250, doi:10.1029/98GB02812, 1999.

Huber, M. and Nof, D.: The ocean circulation in the southern hemisphere and its climatic impacts in the Eocene, Palaeogeogr. Palaeoclimatol. Palaeoecol., 231, 9–28, 2006.

Huber, M. and Sloan, L. C.: Heat transport, deep waters, and thermal gradients: Coupled simulation of an Eocene greenhouse climate, Geophys. Res. Lett, 28, 3481–3484, 2001.

Huber, M., Brinkhuis, H., Stickley, C. E., Döös, K., Sluijs, A., Warnaar, J., Schellenberg, S. A., and Williams, G. L.: Eocene circulation of the Southern Ocean: Was Antarctica kept warm by subtropical waters?, Paleoceanography, 19, PA4026, doi:10.1029/2004PA001014, 2004.

Ilyina, T. and Zeebe, R. E.: Detection and projection of carbonate dissolution in the water column and deep-sea sediments due to ocean acidification, Geophys. Res. Lett., 39, L06606, doi:10.1029/2012GL051272, 2012.

Ilyina, T., Six, K. D., Segschneider, J., Maier-Reimer, E., Li, H., and Núñez-Riboni, I.: The global ocean biogeochemistry model HAMOCC: Model architecture and performance as component of the MPI-earth system model in different CMIP5 experimental realizations, J. Adv. Model. Earth Syst., 5, 287–315, doi:10.1002/jame.20017, 2013.

Jungclaus, J. H., Fischer, N., Haak, H., Lohmann, K., Marotzke, J., Matei, D., Mikolajewicz, U., Notz, D., and von Storch, J. S.: Characteristics of the ocean simulations in the Max Planck Institute Ocean Model (MPIOM) the ocean component of the MPI-Earth system model, J. Adv. Model. Earth Syst., 5, 422–446, doi:10.1002/jame.20023, 2013.

Kennett, J. P. and Stott, L. D.: Abrupt deep-sea warming, palaeoceanographic changes and benthic extinctions at the end of the Palaeocene, Nature, 353, 225–229, doi:10.1038/353225a0, 1991.

Lochte, K., Ducklow, H., Fasham, M., and Stienen, C.: Plankton succession and carbon cycling at 47° N 20° W during the JGOFS North Atlantic Bloom Experiment, Deep Sea Res. II, 40, 91–114, 1993.

Lunt, D. J., Valdes, P. J., Jones, T. D., Ridgwell, A., Haywood, A. M., Schmidt, D. N., Marsh, R., and Maslin, M.: $CO_2$-driven ocean circulation changes as an amplifier of Paleocene-Eocene thermal maximum hydrate destabilization, Geology, 38, 875–878, 2010.

Lunt, D. J., Dunkley Jones, T., Heinemann, M., Huber, M., LeGrande, A., Winguth, A., Loptson, C., Marotzke, J., Tindall, J., Valdes, P., and Winguth, C.: A model-data comparison for a multi-model ensemble of early Eocene atmosphere-ocean simulations: EoMIP, Clim. Past Discuss., 8, 1229–1273, doi:10.5194/cpd-8-1229-2012, 2012.

Mahowald, N. M., Baker, A. R., Bergametti, G., Brooks, N., Duce, R. A., Jickells, T. D., Kubilay, N., Prospero, J. M., and Tegen, I.: Atmospheric global dust cycle and iron inputs to the ocean, Global Biogeochem. Cy., 19, GB4025, doi:10.1029/2004GB002402, 2005.

Maier-Reimer, E.: Geochemical cycles in an ocean general circulation model. Preindustrial tracer distributions, Global Biogeochem. Cy., 7, 645–677, doi:10.1029/93GB01355, 1993.

Maier-Reimer, E., Kriest, I., Segschneider, J., and Wetzel, P.: The HAMburg Ocean Carbon Cycle Model HAMOCC 5.1, Technical Decription Release 1.1, Tech. rep. 14, Reports on Earth System Science, Hamburg, Germany, 2005.

Marsland, S., Haak, H., Jungclaus, J., Latif, M., and Röske, F.: The Max-Planck-Institute global ocean/sea ice model with orthogonal curvilinear coordinates, Ocean Model., 5, 91–127, 2003.

Norris, R. D., Turner, S. K., Hull, P. M., and Ridgwell, A.: Marine Ecosystem Responses to Cenozoic Global Change, Science, 341, 492–498, doi:10.1126/science.1240543, 2013.

Nunes, F. and Norris, R. D.: Abrupt reversal in ocean overturning during the Palaeocene/Eocene warm period, Nature, 439, 60–63, 2006.

Pagani, M., Caldeira, K., Archer, D., and Zachos, J. C.: An Ancient Carbon Mystery, Science, 314, 1556–1557, doi:10.1126/science.1136110, 2006a.

Pagani, M., Pedentchouk, N., Huber, M., Sluijs, A., Schouten, S., Brinkhuis, H., Sinninghe Damste, J. S., Dickens, G. R., Backman, J., Clemens, S., Cronin, T., Eynaud, F., Gattacceca, J., Jakobsson, M., Jordan, R., Kaminski, M., King, J., Koc, N., Martinez, N. C., Matthiessen, J., McInroy, D., Moore, T. C., Moran, K., O'Regan, M., Onodera, J., Palike, H., Rea, B., Rio, D., Sakamoto, T., Smith, D. C., Stein, R., St John, K. E. K., Suto, I., Suzuki, N., Takahashi, K., Watanabe, M., and Yamamoto, M.: Arctic hydrology during global warming at the Palaeocene/Eocene thermal maximum, Nature, 443, 598–598, doi:10.1038/nature05211, 2006b.

Panchuk, K., Ridgwell, A., and Kump, L.: Sedimentary response to Paleocene-Eocene Thermal Maximum carbon release: A model-data comparison, Geology, 36, 315–318, doi:10.1130/G24474A.1, 2008.

Pearson, P. N. and Palmer, M. R.: Atmospheric carbon dioxide concentrations over the past 60 million years, Nature, 406, 695–699, doi:10.1038/35021000, 2000.

Pearson, P. N., Ditchfield, P. W., Singano, J., Harcourt-Brown, K. G., Nicholas, C. J., Olsson, R. K., Shackleton, N. J., and Hall, M. A.: Warm tropical sea surface temperatures in the Late Cretaceous and Eocene epochs, Nature, 413, 481–487, doi:10.1038/35097000, 2001.

Ragueneau, O., Tréguer, P., Leynaert, A., Anderson, R. F., Brzezinski, M. A., DeMaster, D. J., Dugdale, R. C., Dymond, J., Fischer, G., François, R., Heinze, C., Maier-Reimer, E., Martin-Jézéquel, V., Nelson, D. M., and Quéguiner, B.,: A review of the Si cycle in the modern ocean: recent progress and missing gaps in the application of biogenic opal as a paleoproductivity proxy, Global Planet. Change, 26, 317–365, 2000.

Ridgwell, A. and Schmidt, D.: Past constraints on the vulnerability of marine calcifiers to massive carbon dioxide release, Nature Geosci., 3, 196–200, 2010.

Ridgwell, A. and Zeebe, R.: The role of the global carbonate cycle in the regulation and evolution of the Earth system, Earth Planet. Sci. Lett., 234, 299–315, 2005.

Roberts, C. D., LeGrande, A. N., and Tripati, A. K.: Climate sensitivity to Arctic seaway restriction during the early Paleogene, Earth Planet. Sci. Lett., 286, 576–585, 2009.

Roeske, F.: A global heat and freshwater forcing dataset for ocean models, Ocean Modelling, 11, 235–297, 2006.

Sarmiento, J. L., Dunne, J., Gnanadesikan, A., Key, R. M., Matsumoto, K., and Slater, R.: A new estimate of the $CaCO_3$ to organic carbon export ratio, Global Biogeochem. Cy., 16, 54-1–54-12, doi:10.1029/2002GB001919, 2002.

Schneider, B., Bopp, L., Gehlen, M., Segschneider, J., Frölicher, T. L., Cadule, P., Friedlingstein, P., Doney, S. C., Behrenfeld, M. J., and Joos, F.: Climate-induced interannual variability of marine primary and export production in three global coupled climate carbon cycle models, Biogeosciences, 5, 597–614, doi:10.5194/bg-5-597-2008, 2008.

Sijp, W. P., von der Heydt, A. S., Dijkstra, H. A., Flögel, S., Douglas, P. M., and Bijl, P. K.: The role of ocean gateways on cooling climate on long time scales, Global Planet. Change, 119, 1–22, 2014.

Six, K. and Maier-Reimer, E.: Effects of plankton dynamics on seasonal carbon fluxes in an ocean general circulation model, Global Biogeochem. Cy., 10, 559–583, 1996.

Sluijs, A., Schouten, S., Pagani, M., Woltering, M., Brinkhuis, Henk., Damste, J. S., Dickens, G. R., Huber, M., Reichart, G.-J., Stein, R., Matthiessen, J., Lourens, L. J., Pedentchouk, N., Backman, J., and Moran, K.: Subtropical Arctic Ocean temperatures during the Palaeocene/Eocene thermal maximum, Nature, 441, 610–613, 2006.

Takahashi, T., Broecker, W. S., and Langer, S.: Redfield ratio based on chemical data from isopycnal surfaces, J. Geophys. Res.-Oceans, 90, 6907–6924, doi:10.1029/JC090iC04p06907, 1985.

Takahashi, T., Sutherland, S. C., Wanninkhof, R., Sweeney, C., Feely, R. A., Chipman, D. W., Hales, B., Friederich, G., Chavez, F., Watson, A., Bakker, D. C. E., Schuster, U., Metzl, N., Yoshikawa-Inoue, H., Ishii, M., Midorikawa, T., Nojiri, Y., Sabine, C., Olafsson, J., Arnarson, T. S., Tilbrook, B., Johannessen, T., Olsen, A., Bellerby, R., Körtzinger, A., Steinhoff, T., Hoppema, M., de Baar, H. J. W., Wong, C. S., Delille, B., and Bates, N. R.: Climatological mean and decadal change in surface ocean $pCO_2$, and net sea-air $CO_2$ flux over the global oceans, Deep Sea Res. II, 56, 554–577, 2009.

Thomas, D., Zachos, J., Bralower, T., Thomas, E., and Bohaty, S.: Warming the fuel for the fire: Evidence for the thermal dissociation of methane hydrate during the Paleocene-Eocene thermal maximum, Geology, 30, 1067–1070, 2002.

Thomas, D. J., Bralower, T. J., and Jones, C. E.: Neodymium isotopic reconstruction of late Paleocene early Eocene thermohaline circulation, Earth Planet. Sc. Lett., 209, 309–322, 2003.

Tripati, A. and Elderfield, H.: Deep-Sea Temperature and Circulation Changes at the Paleocene-Eocene Thermal Maximum, Science, 308, 1894–1898, doi:10.1126/science.1109202, 2005.

Tyrrell, T. and Zeebe, R. E.: History of carbonate ion concentration over the last 100 million years, Geochim. Cosmochim. Acta, 68, 3521–3530, doi:10.1016/j.gca.2004.02.018, 2004.

Waddell, L. M. and Moore, T. C.: Salinity of the Eocene Arctic Ocean from oxygen isotope analysis of fish bone carbonate, Paleoceanography, 23, PA1S12, doi:10.1029/2007PA001451, 2008.

Wanninkhof, R.: Relationship Between Wind Speed and Gas Exchange Over the Ocean, J. Geophys. Res., 97, 7373–7382, doi:10.1029/92JC00188, 1992.

Weiss, R.: The solubility of nitrogen, oxygen and argon in water and seawater, Deep Sea Research and Oceanographic Abstracts, 17, 721–735, 1970.

Weiss, R.: Carbon dioxide in water and seawater: the solubility of a non-ideal gas, Mar. Chem., 2, 203–215, 1974.

Wetzel, P., Maier-Reimer, E., Botzet, M., Jungclaus, J., Keenlyside, N., and Latif, M.: Effects of ocean biology on the penetrative radiation in a coupled climate model, J/ Climate, 19, 3973–3987, 2006.

Winguth, A., Shellito, C., Shields, C., and Winguth, C.: Climate Response at the Paleocene–Eocene Thermal Maximum to Greenhouse Gas Forcing – A Model Study with CCSM3, J. Climate, 23, 2562–2584, doi:10.1175/2009JCLI3113.1, 2010.

Winguth, A. M., Thomas, E., and Winguth, C.: Global decline in ocean ventilation, oxygenation, and productivity during the Paleocene-Eocene Thermal Maximum: Implications for the benthic extinction, Geology, 40, 163–266, 2012.

Zachos, J., Pagani, M., Sloan, L., Thomas, E., and Billups, K.: Trends, rhythms, and aberrations in global climate 65 Ma to present, Science, 292, 686–693, 2001.

Zachos, J. C., Röhl, U., Schellenberg, S. A., Sluijs, A., Hodell, D. A., Kelly, D. C., Thomas, E., Nicolo, M., Raffi, I., Lourens, L. J., McCarren, H., and Kroon, D.: Rapid Acidification of the Ocean during the Paleocene-Eocene Thermal Maximum, Science, 308, 1611–1615, 2005.

Zachos, J. C., Dickens, G. R., and Zeebe, R. E.: An early Cenozoic perspective on greenhouse warming and carbon-cycle dynamics, Nature, 451, 279–283, doi:10.1038/nature06588, 2008.

Zeebe, R. E.: LOSCAR: Long-term Ocean-atmosphere-Sediment CArbon cycle Reservoir Model v2.0.4, Geosci. Model Dev., 5, 149–166, doi:10.5194/gmd-5-149-2012, 2012.

Zeebe, R. E. and Zachos, J. C.: Reversed deep-sea carbonate ion basin gradient during Paleocene-Eocene thermal maximum, Paleoceanography, 22, PA3201, doi:10.1029/2006PA001395, 2007.

Zeebe, R. E. and Zachos, J. C.: Long-term legacy of massive carbon input to the Earth system: Anthropocene versus Eocene, Philos. T. Roy. Soc. A, 371, 20120006, doi:10.1098/rsta.2012.0006, 2013.

Zeebe, R. E., Zachos, J. C., and Dickens, G. R.: Carbon dioxide forcing alone insufficient to explain Palaeocene-Eocene Thermal Maximum warming, Nature Geosci., 2, 576–580, doi:10.1038/ngeo578, 2009.

# 6

# Thenardite after mirabilite deposits as a cool climate indicator in the geological record: lower Miocene of central Spain

**M. J. Herrero**[1], **J. I. Escavy**[1], **and B. C. Schreiber**[2]

[1]Departamento de Petrología y Geoquímica, Fac. Ciencias Geológicas, Universidad Complutense Madrid, C/Jose Antonio Novais 2, 28040 Madrid, Spain
[2]Department of Earth and Space Sciences, University of Washington, Seattle, WA 98195, USA

*Correspondence to:* J. I. Escavy (jiescavy@ucm.es)

**Abstract.** Salt deposits are commonly used as indicators of different paleoclimates and sedimentary environments, as well as being geological resources of great economic interest. Ordinarily, the presence of salt deposits is related to warm and arid environmental conditions, but there are salts, like mirabilite, that form by cooling and a concentration mechanism based on cooling and/or freezing. The diagenetic transformation of mirabilite into thenardite in the upper part of the lower Miocene unit of the Tajo basin (Spain) resulted in the largest reserves of this important industrial mineral in Europe. This unit was formed in a time period ($\sim 18.4$ Ma) that, in other basins of the Iberian Peninsula, is characterized by the existence of particular mammal assemblages appropriate to a relatively cool and arid climate. Determining the origin of the thenardite deposits as related to the diagenetic alteration of a pre-existing mirabilite permits the establishment and characterization of the sedimentary environment where it was formed and also suggests use as a possible analog with comparable deposits from extreme conditions such as Antarctica or Mars.

## 1 Introduction

Salt deposits are natural chemical deposits that have significant economic, scientific and social implications (Herrero et al., 2013; Warren, 2006). They constitute or contain valuable geological resources such as industrial minerals and building materials, and they are both a source and cap rock of hydrocarbons, etc. (Warren, 2010). It is commonly accepted that most salt deposits are formed under arid environmental conditions, being that most salts are produced in hot arid climates. For some saline deposits, this is the case, but there are many examples in current settings being produced under arid but cool conditions (Dort and Dort, 1970; Last, 1994; Socki et al., 2012; Stankevich et al., 1990; Zheng et al., 2000), and a few are not really evaporites, although they may appear so, superficially, but are pressure and thermal hydrothermal release precipitates (Chaboureau et al., 2012; Hovland et al., 2006).

Saline deposits have proved to be very useful in the study of paleoclimatology and sedimentology (Babel and Schreiber, 2014; Escavy et al., 2012; Fan-Wei et al., 2013; Kendall, 1992; Lowenstein et al., 1999; Minghui et al., 2010; Rouchy and Blanc-Valleron, 2009; Schreiber and El Tabakh, 2000; Warren, 2010). These studies are focused on the relationship between the main periods of the Earth's history with salt deposit formation and climate, which is one of the key factors involved in their deposition. Particularly, continental deposits in arid closed basins may record very accurately the changes in paleoclimate, being the most important factors controlling these changes the water inflow-outflow ratio, temperatures, wind patterns, storm records, and evaporation rates (Lowenstein et al., 1999). Comparison of salt deposits from lakes with marine records permits the development of land–sea correlations in the perspective of global reconstructions of environmental and climatic changes (Magny and Combourieu Nebout, 2013).

Most studies point to evaporative concentration as the main mechanism controlling precipitation of salts. Evaporitic salts precipitate after salt saturation of brines, and indicate hydrological systems in which evaporative water loss is greater than water gain. An alternative way to concentrate brine is by cooling–freezing processes that remove water from it through ice formation. These two concentration mechanisms lead to two different salt formation pathways: evaporative and "frigid" concentration (Strakhov, 1970). In addition, precipitation of certain salts occurs due to their reduction in solubility with temperature decrease (positive temperature coefficient of solution). The resultant minerals, like epsomite, sylvite and hexahydrite, are called cryophile salts (Stewart, 1963) or cryophilic salts (Sánchez-Moral et al., 2002). The main difference between salt deposits formed under cool or hot temperatures is the resulting mineral assemblage (Zheng et al., 2000), indicating the high dependence of the resultant mineralogy on the mechanism of brine concentration.

Sodium sulfate minerals appear to be highly dependent on temperature range (Dort and Dort, 1970), the anhydrous phase being the most common, $Na_2SO_4$ (thenardite), and two hydrated forms, $Na_2SO_4 \times 7H_2O$ (sodium sulfate heptahydrate) and $Na_2SO_4 \times 10H_2O$ (mirabilite). Both thenardite and mirabilite occur extensively in nature, while the heptahydrate is metastable and does not form or become preserved as natural deposits (Dort and Dort, 1970). Attempts to classify evaporitic minerals by their temperature of formation was proposed by Zheng et al. (2000), with mirabilite ($Na_2SO_4 \times 10H_2O$) being the typical product of cool periods, bloedite ($Na_2Mg(SO_4)_2 \times 4H_2O$) for slightly warm phases, and thenardite ($Na_2SO_4$) being formed under warm conditions. This work has been the first attempt to classify minerals by their temperature of formation, but the precipitation of saline minerals should always be related to the environmental and geological conditions of the salt deposit because their temperature of formation may vary from one setting to another.

Mirabilite, therefore, is the most common evaporitic mineral crystallizing under cool temperatures (Nai'ang et al., 2012; Wang et al., 2003; Zheng et al., 2000). An example is the mirabilite layers from the Huahai lake (China), where they precipitated under mean temperatures around 11 °C lower than current ones, during the Quaternary Younger Dryas event (Nai'ang et al., 2012). Thenardite, however, is the most common sodium sulfate found in ancient deposits (Garrett, 2001), occurring mostly in Neogene continental endorheic settings (Warren, 2010). As a primary mineral, it forms either by direct precipitation from warm brines in shallow lakes (Last, 1994), either in/or near the surface of playas as capillary efflorescent crusts by evaporative concentration (Jones, 1965). It commonly occurs as thin layers interbedded with other evaporitic minerals forming salt assemblages (Garrett, 2001). In Lake Beida (Egypt), thenardite occurs as a 50 cm thick crust together with halite (NaCl), trona ($Na_3(CO_3)(HCO_3) \times 2(H_2O)$) and burkeite ($Na_6(CO_3)(SO_4)_2$) (Shortland, 2004).

Therefore, both mirabilite and thenardite precipitate in modern lacustrine systems, whereas only thenardite appears as the prevalent sodium phase in the geological record (Ortí et al., 2002). Mirabilite is a very reactive mineral due to its low melting point and high solubility (Garrett, 2001), this being the main reason for the lack of this mineral in ancient deposits. When the conditions where mirabilite has accumulated change (increase in temperature, evaporation rate, burial, interaction with concentrated salt solutions, etc.), it melts, dissolves, or is transformed into more stable minerals such as thenardite, astrakanite (bloedite), glauberite ($Na_2Ca(SO_4)_2$) or burkeite (Garrett, 2001).

The Oligocene–lower Miocene sequence of the Tajo basin contains a 100 to 650 m thick succession of evaporitic materials (Calvo et al., 1989) that include one of the major thenardite deposits of the world (Garrett, 2001). This paper presents the results of the analysis of this thenardite deposit and establishes its secondary origin as a transformation phase after mirabilite. As a result, we postulate that the thenardite level within the lower Miocene lacustrine sequence is a cool paleotemperature indicator. Therefore, we have been able to identify a decrease in temperature and precipitation regime in the lower Miocene geological record of the Iberian Peninsula during which there also was a significant change in faunal diversity, coincident with the Mi-1a event of Miller et al. (1991) that took place at 18.4 Ma that is well documented on a worldwide scale (Zachos et al., 2001).

## 2　Study site

The Tajo basin, located in the central part of the Iberian Peninsula (Fig. 1), was formed during the Cenozoic by several basement uplifts (De Vicente et al., 1996). Growth strata related to syntectonic alluvial deposits appear in the margins of the basin and pass into lacustrine and palustrine deposits towards the center (Calvo et al., 1989). Ordoñez and García del Cura (1994) defined four main units in the Neogene of the Tajo basin: a Lower or Saline Unit, an Intermediate or Middle Unit, and Upper Miocene Units and the Pliocene Unit. Based on the study of cores from several drill holes, they divided the Lower Unit into a Lower Saline Subunit that occupies a broad area of the Tajo basin, and the Upper Saline Subunit that contains the thenardite deposits, which is restricted to the area south of the Tajo River.

During the lower Miocene (23.2–16.2 Ma), the Lower Unit was formed by syntectonic coarse alluvial detrital deposits located close to the tectonically active margins of the basin, gradually passing into finer clastic sediments (sandstones and shales) and wide saline lake systems that occupied the basin center (Calvo et al., 1996). The saline deposits, a succession up to 500 m thick, are composed of alternating anhydrite, halite and glauberite beds with some thin layers of fine interbedded detritic sediments. This unit grades laterally into shale beds with abundant calcium sulfate nodules (of both gypsum and anhydrite), followed by

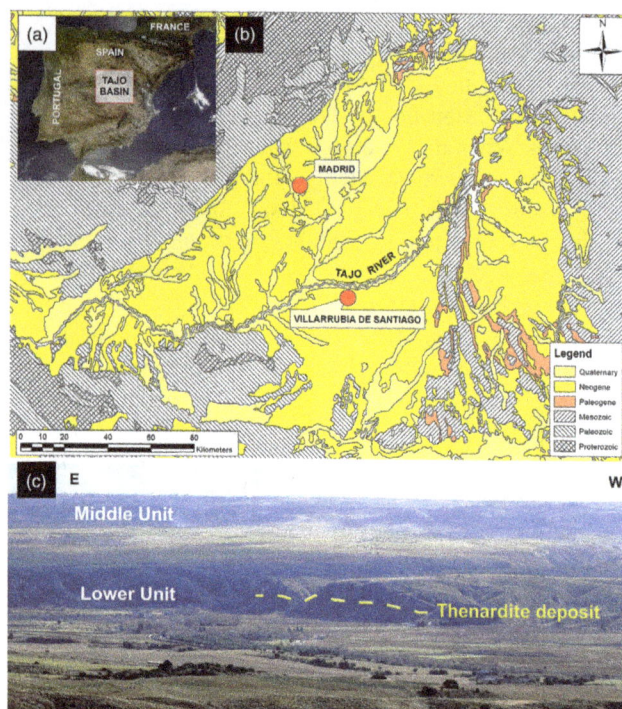

Figure 1. (a) Location of the Tajo basin in the central part of the Iberian Peninsula. (b) Simplified geological map of the Tajo basin and surrounding mountain belts (modified from the Spanish Geological Map, scale 1 : 50 000, IGME, 2013). The thenardite deposits appear near the village of Villarrubia de Santiago. (c) General view of the upper part of the Lower Unit and the base of the Middle Unit of the Miocene sequence of the Tajo basin in the Villarrubia de Santiago area. The thenardite deposit is laterally continuous for tens of kilometers, although, due to the high solubility of the thenardite, it is not easily identified at all locations.

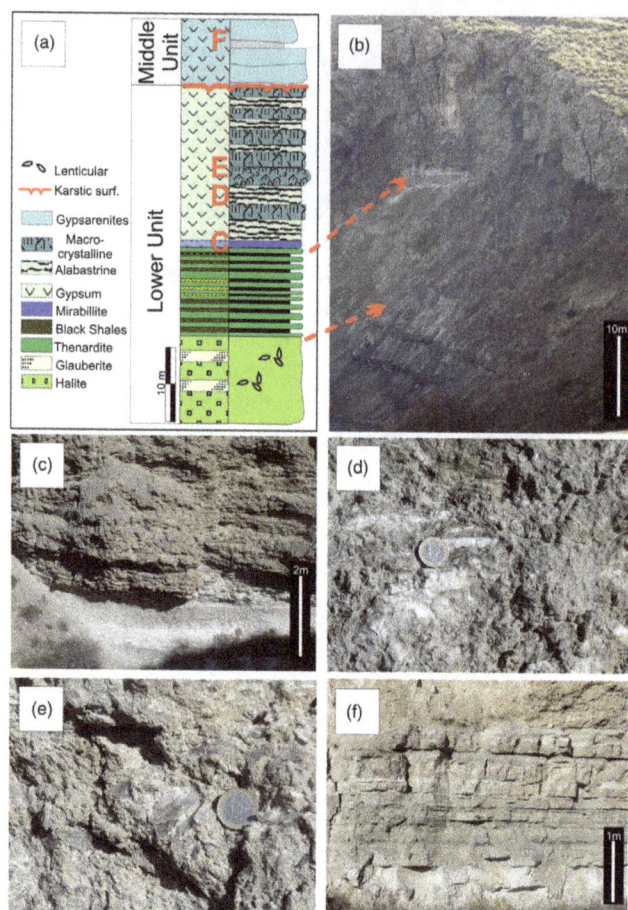

Figure 2. (a) Stratigraphic section of the upper part of the Lower Unit and the lowermost part of the Middle Unit of the Miocene in the Tajo basin. (b) General view of the Lower Unit outcropping along the current Tajo River (south bank). (c) Outcrop view of the contact between the upper part of the thenardite body and the overlaying unit with secondary gypsum. This secondary gypsum appears as two main lithofacies: (d) alabastrine gypsum, and (e) macro-crystalline gypsum (the coin for scale is 2.2 cm in diameter). (f) Outcrop of the detritic gypsum beds that comprise the lower part of the Middle Unit.

coarsersiliciclasticdeposits that correspond to alluvial fans formed at the foot of the surrounding mountain ranges. To the south of the Tajo River valley, the upper part of the Lower Unit (Fig. 2a, b) contains a massive deposit of thenardite, 7–12 m thick (Fig. 2b, c) (Ortí et al., 1979). This thenardite overlies a massive halite and decimeter-thick beds of glauberite. Thus, unlike the broadly distributed glauberite, the deposit of thenardite is restricted to a relatively small area in the central part of the basin (Ordóñez et al., 1991). Above the thenardite layer, there is a layer of mirabilite (several cm thick), a product of recent hydration of the thenardite by meteoric water (Ortí et al., 1979). At the top of the Lower Unit, there is a 10–20 m thick alternation of layers (tens of cm) composed of secondary gypsum, both alabastrine (Fig. 2d) and macro-crystalline (Fig. 2e), interbedded with shales and marls. This unit has been interpreted as a weathering cover, a product of the replacement of glauberite and anhydrite by gypsum (Ordoñez and García del Cura, 1994). The top of the Lower Unit is established at a paleokarstic surface (Cañaveras et al., 1996).

Higher up in the sequence, the Miocene Middle Unit is mostly composed of primary gypsum forming tabular beds up to 1 m thick (Fig. 2f). This unit passes upwards into abundant carbonate bodies and is characterized by the absence of evaporitic minerals such as halite or glauberite. During this period, there was a significant progradation of siliciclastic sediments that reached the central part of the basin. Sedimentation during the Miocene Upper Unit was dominated by carbonates and marls (Calvo et al., 1996).

## 3 Materials and analytical methods

### 3.1 Sampling

A total of 30 samples have been collected from different levels of the evaporitic sequence, both from the subsurface mine walls and from drilled cores provided by SAMCA, the company that owns the sodium sulfate deposit. The thenardite samples were collected from the mine face, while the halite–glauberite samples have been obtained from exploration drill cores. In order to avoid the alterations produced during drilling (by the interaction with drilling fluids, the pressure and heat transmitted by the coring bits, etc.), only the centermost part of the core has been used. Special care has been taken during sample preparation in order to prevent mineral alteration, avoiding high temperatures ($> 25\,^{\circ}$C) and exposure to humid conditions.

### 3.2 Optical microscopy and SEM analysis

For the petrographic study, rock samples were cut with an oil-refrigerated 1 mm thick diamond disc saw (Struers Discoplan-TS). A low viscosity oil was used (rhenus GP 5M) for cutting, grinding and polishing. When cutting samples in the disc saw, extra rock was included at the cut ($\sim 1$ mm) and was removed by hand grinding, thus eliminating any possible alteration of the samples. Grinding to the final thickness has been done using emery papers with different grit sizes, impregnated with oil. Unconsolidated samples were previously indurated with a resin under vacuum (Struers Epofix resin). The thin sections were glued on 4.8 cm $\times$ 2.8 cm glass slides using LOCTITE 358, and cured afterwards under ultraviolet light.

Petrographic characterization was performed using a Zeiss West Germany Optical 316 Microscope (OM) at the Department of Petrology and Geochemistry of Complutense University of Madrid (UCM). By analyzing the doubly polished plates, it has been possible to undertake a petrographic characterization of fluid inclusions, determining the moment of formation in relation to the crystal growth.

Textural characterization of the samples was completed by scanning electron microscopy observations performed using a JEOL 6.400 instrument working at 20 kV 320 microscopy (SEM), at the CAI Geological Techniques Laboratory (UCM).

### 3.3 X-Ray diffraction analysis

To obtain the whole rock mineralogy by X-ray diffraction, a portion of 20 of the 30 samples was ground in an agate mortar at low rotation speed (avoiding high temperatures). A Bruker D8 Advance diffractometer equipped with a Sol-X detector was used. The mineralogical composition of crystalline phases was estimated following Chun's (1975) method and using Bruker software (EVA). The XRD analysis was performed at the Geological Techniques Laboratory (UCM).

### 3.4 Low-temperature scanning electron microscopy (LTSEM)

This technique was used to assess the chemical composition of fluid inclusions and to establish qualitatively and quantitatively the elemental characterization of the host minerals and the fluid inclusion fluids (Ayora et al., 1994). Low-temperature scanning electron microscopy (LTSEM or Cryo-SEM) was performed in 20 fluid inclusions from 11 samples using small pieces of thenardite and halite that were cut, mounted, and mechanically fixed onto a specimen holder at room temperature. The instrument used was a CT 1500 Cryotrans system (Oxford Instruments) mounted on a Zeiss 960 SEM. This study was done at the Spanish Institute of Agricultural Sciences (ICA) of the CSIC.

## 4 Results

### 4.1 Mineralogy

The lower part of the Lower Unit sequence, below the thenardite deposit (Fig. 3a), is characterized by evaporitic layers composed of a mixture of glauberite (45.8 %) and halite (41.7 %) (Fig. 3b), with a minor content of polyhalite (7.8 %), dolomite (2.1 %), and clay minerals (1.8 %) (Table 1). This mineral assemblage is common in evaporitic Neogene continental basins of the Iberian Peninsula such as those of the Zaragoza (Salvany et al., 2007) or Lerín gypsum formations (Salvany and Ortí, 1994), both in the Ebro basin (Spain). The relative proportions of halite and glauberite are variable, with halite ranging from 30 to 51 % and glauberite from 23 to 59 %. Glauberite crystals (Fig. 3b), with sizes between 1 mm and 10 cm, occur either forming banded or nodular layers with abundant structures indicating fluid escape, or as irregular masses or nodules accompanying halite crystals in the halite-rich horizons (Fig. 3b). No stratification or competitive growth typical of primary halite formation is found. Therefore, a secondary origin of these halite crystals can be inferred.

Higher up in the sequence, a sharp change in mineralogy takes place, passing upward into a fairly pure and thick sodium sulfate body mainly composed of thenardite (96.5 %) with a minor content of glauberite (2.5 %) and anhydrite (1.0 %) (Table 1). Thenardite (Fig. 3c) occurs as cm-sized subeuhedral to anhedral crystals, with sizes from 1 mm to several cm, forming aggregates. Crystal color is also variable, ranging from blue to clear and transparent. When crystals have a high volume of fluid inclusions, they have a cloudy aspect. Thenardite layers usually present abundant fluid escape structures (Ortí et al., 1979).

### 4.2 Fluid inclusion analysis

Fluid inclusions are abundant within the thenardite and halite crystals, whereas they are very scarce in the

**Table 1.** XRD mineralogical composition of the samples.

| Sample ID | Thenardite | Glauberite | Halite | Polyhalite | Dolomite | Anhydrite | Clay Min. |
|---|---|---|---|---|---|---|---|
| | | | Halite–glauberite layer | | | | |
| 522116–01 | – | 45.7 | 30.2 | 19.6 | 2.6 | – | 0.0 |
| 522116–02 | – | 23.5 | 51.1 | 12.1 | 5.9 | – | 7.4 |
| 522116–03 | – | 54.9 | 45.1 | 0.0 | 0.0 | – | 0.0 |
| 522116–04 | – | 59.4 | 40.6 | 0.0 | 0.0 | – | 0.0 |
| 522116–10 | – | 51.5 | 47.5 | 0.0 | 0.0 | – | 1.0 |
| Mean | – | 47.0 | 42.9 | 6.3 | 1.7 | – | 1.7 |
| | | | Thenardite layer | | | | |
| 522116–05 | 100.0 | 0.0 | – | – | – | – | – |
| 522116–06 | 91.3 | 5.5 | – | – | – | 3.2 | – |
| 522116–07 | 99.0 | 1.0 | – | – | – | – | – |
| 522116–08 | 95.8 | 3.2 | – | – | – | 1.0 | – |
| 522116–09 | 99.0 | 1.0 | – | – | – | – | – |
| 522116–11 | 95.0 | 3.6 | – | – | – | 1.4 | – |
| 522116–12 | 91.3 | 5.5 | – | – | – | 3.2 | – |
| 522116–13 | 99.5 | 0.5 | – | – | – | – | – |
| 522116–14 | 99.0 | 1.0 | – | – | – | – | – |
| 522116–15 | 94.6 | 3.8 | – | – | – | 1.6 | – |
| 522116–16 | 92.9 | 5.0 | – | – | – | 2.1 | – |
| 522116–17 | 100.0 | – | – | – | – | – | – |
| 522116–18 | 94.5 | 4.4 | – | – | – | 1.1 | – |
| 522116–19 | 98.2 | 1.0 | – | – | – | 0.8 | – |
| 522116–20 | 97.3 | 1.7 | – | – | – | 1 | – |
| Mean | 96.5 | 2.5 | – | – | – | 1.0 | – |

glauberitecrystals. Most of them are primary fluid inclusions that were formed during the growth of the crystals. Therefore, the brine trapped in the primary fluid inclusions is the same from which these minerals precipitated. In the case of diagenetic minerals, fluid inclusions show the conditions of recrystallization rather than the conditions of formation of the precursor mineral (Goldstein and Reynolds, 1994). There appear to be few primary inclusions in the form of two-phase inclusions (containing gases or solids), most of them being single-phase aqueous liquid inclusions at room temperature.

In this study, only primary fluid inclusions have been analyzed, established as primary by their relationship with the crystals' growth zonation (Fig. 3d), mainly because voids that house these fluid inclusions are crystallographically regular (mimic crystal terminations) (Goldstein and Reynolds, 1994). Some sparse secondary fluid inclusions have been found aligned or associated with fractures.

The primary fluid inclusion chemical composition has been analyzed by Cryo-SEM. Fluid aqueous inclusions analyzed (15 analyses) in the thenardite crystals (Fig. 3e, f) have shown that, systematically, the only elements found in the brine are Na and S (Fig. 3g). The composition obtained by analyzing the fluid inclusions from the halite crystals (five analyses) is Na and Cl, with trace contents of Ca (Fig. 3h).

## 5  Discussion

### 5.1  Dates and climate during the thenardite formation

The formation of the many Cenozoic lacustrine systems in Spain was mainly controlled by the tectonic activity that affected the Iberian microplate and by changes in the paleogeography and paleoclimatic conditions of the western Mediterranean–eastern Atlantic zone (De Vicente et al., 1996). The base of the Lower Unit of the Miocene (the Lower Saline Subunit) of the Tajo basin is at the Oligocene–Miocene boundary ($\sim$23 Ma) and ends at the top of the Burdigalian stage ($\sim$16 Ma) (Calvo et al., 1993). Paleoclimatic curves have been obtained through the study of mammal associations (Calvo et al., 1993; Daams and Freudenthal, 1988; Van der Meulen and Daams, 1992), and show that this period was warm and humid and became relatively more arid towards its end. Nevertheless, within this unit, the temperature and humidity curves for northern–central Spain (Van der Meulen and Daams, 1992) show the existence of a stage where both temperature and humidity were reduced. The thenardite of this study appears within the sequence that corresponds to this time period. Previous authors (Calvo et al., 1996; Ordóñez et al., 1991) have interpreted the thenardite layer as the result of thermal evaporative concentration when

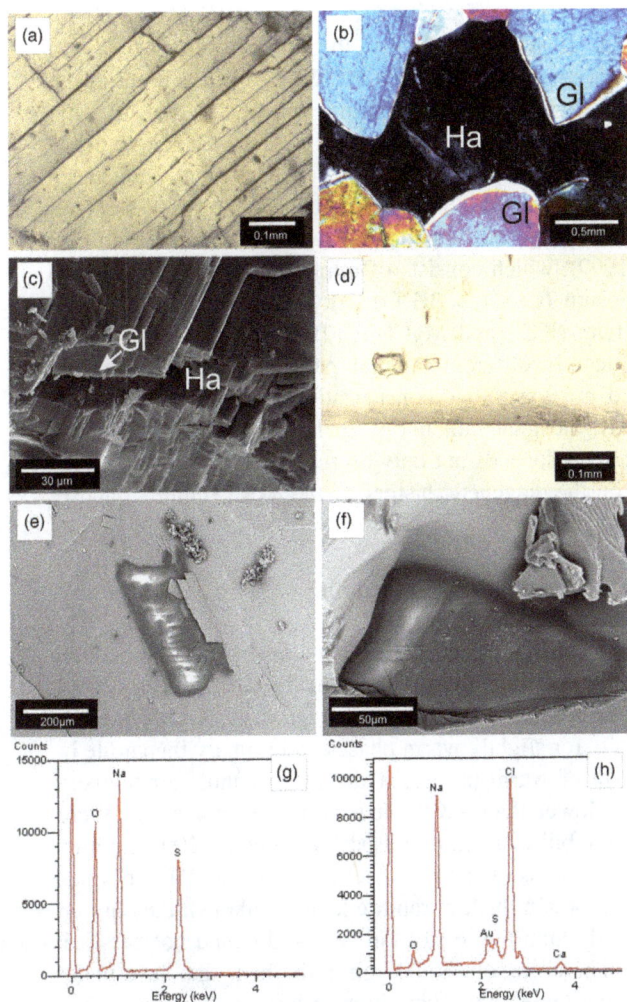

**Figure 3. (a)** Thenardite crystal under a thin section (crossed nicols) with a splintery fracture along cleavage planes. **(b)** Photomicrograph of idiomorphic crystals of glauberite (Gl) cemented by halite (Ha) (crossed nicols). **(c)** SEM image of a thenardite crystal showing splintery fractures along cleavage planes. **(d)** Photomicrograph of primary fluid inclusions in a thenardite crystal mimicking the thenardite crystal termination. **(e)** and **(f)**: frozen fluid inclusion within thenardite crystals studied by Cryo-SEM SEM. **(g)** EDX spectrum of a fluid inclusion in a thenardite crystal with sodium and sulfate as the only ions present, analyzed by Cryo-SEM. **(h)** Cryo-SEM EDX spectrum of a fluid inclusion in the halite, containing sodium, chlorine and low quantities of sulfate and calcium ions.

the lake water volume was reduced, although they indicated that the environments required to follow this brine concentration path do not fit with the temperature and humidity curves proposed for that time span in other parts of the Iberian Peninsula. This difference in environmental conditions has been explained as the establishment of a microclimate in this area, placed in a "rain shadow" region that resulted from the uplift of the surrounding mountain belts (Ordóñez et al., 1991) and also the existence of highly concentrated brines

sourced by recycling of older evaporites (Calvo et al., 1996). Ordoñez and García del Cura (1994) suggested, as one of the options for the formation of the thenardite deposit, that mirabilite formed within these lakes could have been transformed into thenardite during early diagenesis.

## 5.2 Evaporative concentration versus frigid precipitation: mineralogical criteria

Salt precipitation from a given aqueous solution undersaturated with respect to a given mineral can be achieved in three different ways: (1) removal of the solvent (water) at more or less constant temperature by evaporation (evaporative concentration); (2) removal of water by freezing, called frigid concentration, producing cryogenic salts according to Strakhov (1970); and (3) change in temperature at constant salinity (or total concentration) producing the precipitation of cryophilic salts, according to Borchert and Muir (1964). By the first two mechanisms, there is an increase in the concentration of all the dissolved species, leading to the formation of a brine. The third mechanism, related to changes in mineral solubility with temperature, only modifies the concentration of the dissolved species that constitute the precipitating mineral. The second mechanism (freezing) compulsorily implies the third one (solubility change with $T$), and therefore they should be able to happen together in natural environments.

These distinct pathways of brine concentration result in two different pathways of salt formation: by evaporation of the solvent (evaporative concentration) or by cooling/freezing (frigid concentration). Nevertheless, the resulting mineralogy is obviously also dependent on the ion content of the mother brine (Eugster and Hardie, 1978; Hardie and Eugster, 1970).

When a brine is concentrated by evaporation, the salt content increases, reaching saturation and precipitating progressively from less soluble to more soluble minerals. If evaporation continues, at the eutonic point all the remaining salts precipitate simultaneously. In natural conditions with natural brines the eutonic point is reached at temperatures above 32 °C, with salinities between 35 and 40 % (Strakhov, 1970).

The freezing process concentrates the brine in the same way as by evaporation, by removing $H_2O$ from the solution, but in this case by formation of ice, leading to a concentrated residual brine as well as the progressive precipitation of saline minerals (Stark et al., 2003). Freezing ends when the eutectic or cryohydric point is reached, at the point when all compounds (including $H_2O$) pass to the solid state (Mullin, 2001). Depending on the initial mineralization and composition of the brine, the eutectic point is reached between −21 and −54 °C (Marion et al., 1999; Strakhov, 1970). The liquid brines, called cryobrines, are those that reach the eutectic point at temperatures below 0 °C (Möhlmann and Thomsen, 2011), and such brines exist in the Earth's polar regions

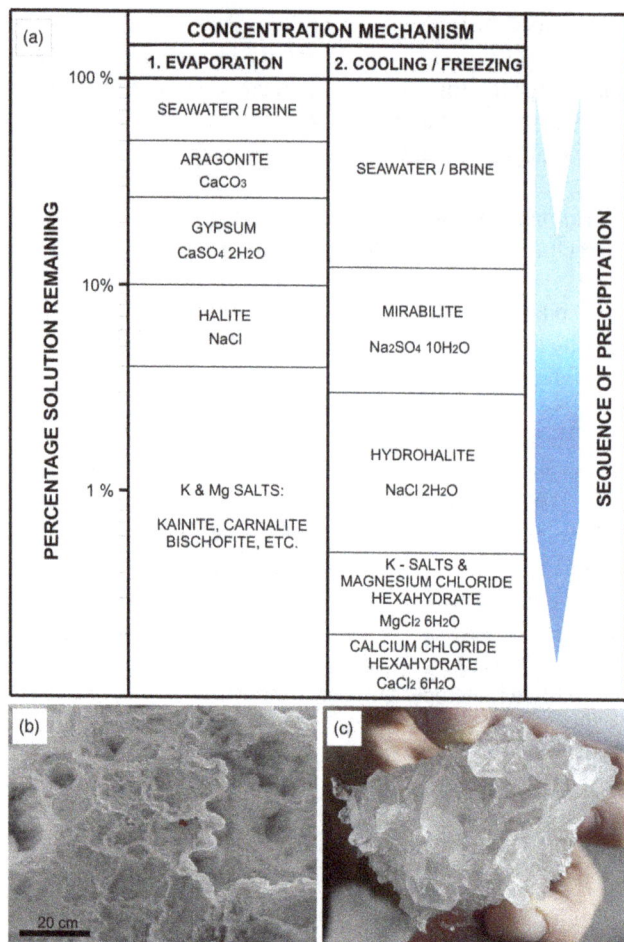

**Figure 4. (a)** Mineral precipitation sequences from seawater depending on the concentration mechanism. Arrow shows the sense of precipitation. Left scale (logarithmic) shows the percentage of remaining brine during the concentration process. Evaporative concentration sequence defined by Orti (2010) and frigid concentration by Dort and Dort (1969). **(b)** and **(c)**: mirabilite precipitation in a pond near a $Na_2SO_4$-rich water spring in Belorado (Burgos, Spain). Photographs taken early in the morning after 3 days of continuous cool temperatures (30 November 2011). General view of the mirabilite pond **(b)**. See coin as scale (diameter 1.8 cm). Detail of the mirabilite crystals **(c)**.

(Garrett, 2001) and probably on Mars (Peterson et al., 2007). The minerals formed under these conditions are called cryogenic (Babel and Schreiber, 2014; Brasier, 2011).

Sodium salts precipitate in nature by both mechanisms: (1) concentration of the brine by solar-driven evaporation like in the Quaternary playas of the USA, and (2) by brine cooling and freezing, like in Kara Bogaz Gol in Turkmenistan, and the Great Plains of Canada (Last, 1994; Warren, 2010).

The different results obtained from the same brine, by using evaporative or frigid concentrations, can be illustrated with the different resultant mineral paragenesis obtained from the seawater (Fig. 4a). Path 1 is the result of evaporative

concentration: calcite ($CaCO_3$) – gypsum ($CaSO_4 \times 2H_2O$) – halite (NaCl) and, finally, the bittern salts (K and Mg salts) (Harvie et al., 1980; Ortí, 2010); path 2 is the result of frigid concentration: mirabilite ($Na_2SO_4 \times 10H_2O$) – hydrohalite ($NaCl \times 2H_2O$) – sylvite (KCl) and $MgCl_2 \times 6H_2O$ – $CaCl_2 \times 6H_2O$ (Dort and Dort, 1970). Recently, a new sequence of precipitation for frigid concentrations of seawater has been proposed, named the Gitterman pathway (Marion et al., 1999), which consists of mirabilite ($Na_2SO_4 \times 10H_2O$) – gypsum ($CaSO_4 \times 2H_2O$) – hydrohalite ($NaCl \times 2H_2O$) – sylvite (KCl) and $MgCl_2 \times 12H_2O$, which also offers a significantly different mineral paragenesis to the one obtained by evaporative concentration from seawater. Nevertheless, the final mineralogy that is found in the geological record depends not only on the primary mineralogy, but also on the diagenetic history of the rock (Schreiber and El Tabakh, 2000).

### 5.2.1 Precipitation of mirabilite

Attempts to classify evaporitic minerals by their temperature of formation have been carried out by Zheng et al. (2000), with mirabilite being the typical product of cool periods, bloedite for slightly warm phases and primary thenardite indicative of warm phases. If the mean annual air temperature is lower than $-3\,°C$, it is possible for the newly created mirabilite layers to persist (Wang et al., 2003). With at least 7 months of mean temperatures below $0\,°C$, it is possible to obtain thick mirabilite layers, although they are unstable through the rest of the year and would not persist in time. Therefore, only thick mirabilite beds that have undergone the temperature conditions necessary for the precipitation and preservation of this mineral can be indicators of sustained cool periods (Minghui et al., 2010).

The sodium sulfate solubility curve shows a rapid decrease when the temperature drops (Dort and Dort, 1970), resulting in mirabilite crystallization from concentrated brines during cool temperature periods like in glacial periods or during fall and winter at high latitudes. Thick beds of mirabilite are common in modern Canadian playa lakes (Last, 1994, 1984). Mirabilite may naturally crystallize even from diluted brines such as seawater if temperature drops severely (Garrett, 2001). This is the case in the McMurdo area, Antarctica, where average air temperatures are about $-20\,°C$. Mirabilite precipitates at higher temperatures from more concentrated brines (Garrett, 2001). In Ebeity Lake in Siberia (Russia), mirabilite starts forming at the end of the summer when the brine temperature drops below $19\,°C$ (Strakhov, 1970). When the environmental temperature of $0\,°C$ is reached, 70 % of the mirabilite has already precipitated, and almost all the mirabilite has formed when temperature reaches $-15\,°C$. If the brine continues freezing, hydrohalite precipitates at $-21.8\,°C$ (Strakhov, 1970). We have recorded mirabilite current precipitation in Burgos (northern Spain) at a height of 820 m a.s.l. under night-time temperatures between $-2$ and

0 °C. At this location, mirabilite precipitates in ponds as microterraces (Fig. 4b, c) that form from emergent groundwater that flows through Cenozoic glauberite deposits.

### 5.2.2 Precipitation of thenardite

From the range of sodium sulfate minerals, thenardite is the most commonly found in ancient deposits (Garrett, 2001). Thenardite can be formed as a primary mineral by direct precipitation from warm brines in shallow lakes (Last, 1994), or as capillary efflorescent crusts in playas with sulfate-rich waters (Jones, 1965). Primary thenardite normally occurs together with many other evaporitic minerals forming salts assemblages (Garrett, 2001), usually as layered evaporite deposits like in Lake Beida, Egypt (Shortland, 2004). According to Lowenstein and Hardie (1985), layered evaporites can accumulate in (1) ephemeral saline pans, (2) shallow perennial lagoons or lakes, and (3) deep perennial basins. Evaporitic sediments occurring in saline pans consist of centimeter-scale crystalline salt levels, alternating with millimeter- to centimeter-scale detrital siliciclastic-rich muds. Thenardite precipitating in modern saline pans appears to be associated with halite, gypsum, mirabilite, epsomite and trona (Lowenstein and Hardie, 1985), and there are no thick deposits of pure thenardite described in the literature as being formed as a primary deposit. Lake Beida (Egypt) contains the purest primary thenardite deposit in the world, reaching 60 % of thenardite at some locations, with variable contents of halite (up to 60 %), sodium carbonate (trona, up to 14 %), sodium bicarbonate (nahcolite, up to 16 %) and minor amounts of other K, Ca and Mg salts (Nakhla et al., 1985).

Instead, thenardite can be formed by the transformation of other minerals during diagenesis (secondary thenardite), the most common case being the mirabilite dehydration by increasing temperature, evaporation rate, burial, or by interaction with NaCl-concentrated brines (Last, 1994).

### 5.2.3 Transformation: mirabilite to thenardite

Mirabilite is a very reactive mineral due to its low melting point and high solubility (Garrett, 2001). This high reactivity is the main reason for the lack of this mineral in ancient deposits, because when the conditions of formation of mirabilite change, it usually dissolves or is transformed into other more stable minerals such as thenardite (Garrett, 2001). The mirabilite-to-thenardite transformation commonly takes place at about 32.4 °C, but in the presence of NaCl, this transition occurs at approximately 18 °C and drops down to 16 °C if $Mg^{2+}$ is present (Charykova et al., 1992). The impact of additional ions within the solution in the transition temperatures in the sodium sulfate system is due to the double salt effect (Warren, 2010). The transformation of mirabilite into thenardite may occur soon after deposition (in the early diagenesis) or later, when the change in the condi-

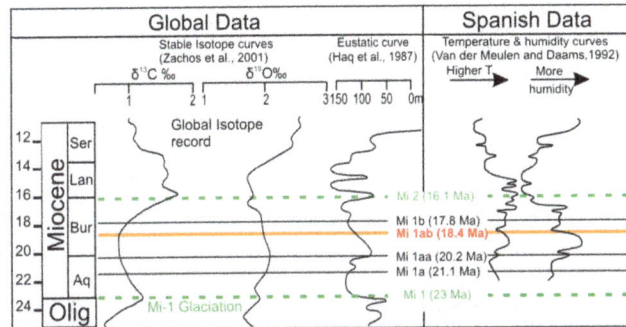

**Figure 5.** Correlation of the Oligocene–Miocene of the global deep-sea carbon and oxigen isotope curves of Zachos et al. (2001), with the Haq et al. sea-level curve (Haq et al., 1987). The main significant ages of the Miocene oxigen isotope events (Mi events) are shown (Miller et al., 1991). The green Mi 1 event corresponds to the Oligocene–Miocene glaciation produced during the Oligocene–Miocene limit. The red line corresponds to the Mi-1ab, the time at which the mirabilite deposits (thenardite precursor) of the Tajo basin were formed. The absolute ages are relative to the USGS Chronostratigraphic Chart (2013).

tions makes mirabilite unstable, for example by compaction during burial (Garrett, 2001). An example of an early diagenesis transformation is found in Kuchuk Lake (Russia), where mirabilite precipitates during the cool winters, and, during summers, the level of the lake drops due to evaporation and becomes NaCl saturated, producing the transformation into thenardite (Stankevich et al., 1990). Mirabilite can also be transformed into thenardite directly by heating upon burial (Warren, 2010). The water of crystallization of mirabilite, which escapes during the transformation into thenardite, produces fluid escape structures within the sedimentary sequence. Part of this water may also be trapped as fluid inclusions within the emerging thenardite crystals.

### 5.2.4 Sedimentology and diagenesis

There are several characteristics that indicate that mirabilite is the precursor mineralogy of the thenardite beds from the lower Miocene deposits of the Tajo basin. Textural features (Ordoñez and García del Cura, 1994; Ortí et al., 1979), mineral assemblage and fluid inclusion chemistry presented in this study suggest this mineral progression. This information, combined with paleontological evidence, is indicative of the existence of cool and arid environmental conditions at the time of formation, and therefore, it can be correlated with a time period having these characteristics.

The thenardite deposit of the Tajo basin commonly occurs as large crystals in thick and fairly pure layers that present fluid escape structures. The thenardite deposit appears as interbedded layers of pure thenardite (cm to m thick) with thin intercalations of black shales (Ortí et al., 1979), similar to the sequence described in Lake Kuchuk in the Volga region of Russia (Stankevich et al., 1990). No textural characteristics such as dissolution (flooding stage), crystal growth (saline lake stage), or syndepositional diagenetic growth features (a desiccation stage) (Lowenstein and Hardie, 1985) have been found in the Tajo basin thenardite deposit, which would indicate a primary thenardite origin within a salt-pan environment. Instead, fluid escape structures appear that are indicative of the fluids produced during the mirabilite dehydration and transformation to thenardite.

### 5.2.5  Fluid inclusions

Primary fluid inclusions within evaporitic minerals contain and preserve samples of the brine where the crystals were growing. In our case, the chemistry of the fluid inclusions is mainly chlorine and sodium within the halite fluid inclusions, and sulfur and sodium within the thenardite ones. This is not surprising because, in aqueous fluid inclusions occurring in very soluble minerals, like the ones under study, the chemistry of the aqueous solution will very rapidly come to equilibrium with the surrounding mineral. Therefore, the fluid inclusion chemistry will contain a relevant amount of the ions forming the hosting minerals. The most common mineral precursor for diagenetic thenardite is the original hydrated sodium sulfate (mirabilite) (Dort and Dort, 1970). Part of the water of crystallization escapes, but another part may be trapped in the fluid inclusions that are formed during the growth process of the resulting thenardite crystals. The chemical composition of such aqueous inclusions should be exclusively water and ions from the hosting mineral (in this case, sodium and sulfate).

Previous studies of fluid inclusions in other primary salts of the Cenozoic sequences of the Tajo basin show a broad range of cations defining a mother brine rich in $Ca^{2+}$, $Na^+$, $Mg^{2+}$ and $K^+$ (Ayllón-Quevedo et al., 2007). The fluid inclusions, within the thenardite crystals, exclusively contain the same ions as the host mineral (sodium and sulfate), highlighting the lack of any trace of $K^+$, which would be the last ion to combine in this kind of brine. This is evidence of thenardite being a diagenetic product (secondary mineral) formed after a precursor mineralogy. A similar mechanism could explain the chemistry of the halite fluid inclusions, in this case being produced by the dehydration of hydrohalite, another salt formed in severe cool environments, although the origin of the halite in the Tajo basin sequence is still under study.

Consequently, based on the textural patterns of the thenardite crystals, the internal arrangements of the sedimentary structures and the ionic content of the fluid inclusions, the thenardite deposits of the Tajo basin are clearly a diagenetic product of a precursor mirabilite, which had to be formed under cool temperature conditions.

### 5.3  Cool and arid climate indicator

The Lower Unit of the Miocene in the Tajo basin ($\sim$ 23–16 Ma) (Alberdi et al., 1984; Calvo et al., 1993) is subdivided into different stages based on mammal associations (Daams et al., 1997). The time span of the lower Miocene sequence of another Iberian basin (the Calatayud–Teruel basin, 200 km to the northeast of the Tajo basin) corresponds to Zone A ($\sim$ 22–18 Ma) (Van der Meulen and Daams, 1992) established on the basis of particular stages of evolution of rodents and other species (Daams et al., 1997). The fauna from the younger part of this unit is characterized by the existence of a particular *Gliridae* (a dormouse) that lived in forest or open forest environments, as well as other rodent taxocenoses, which are dominated by *Eomyids* of the genus *Ligermimys*. Zone A is thought to be humid, although there is a change to drier and relatively cooler conditions towards the top of the zone ($\sim$ 18.4 Ma). At this moment, the number of specimens decreases significantly, and a higher percentage of *Peridyromys murinus* (46 %) appears, a species that shows abundance at higher latitudes because of its greater tolerance to lower temperatures than other species such as *Mycrodyromys* (present at 3 %), a thermophile taxon that disappears during cooling events (Daams et al., 1997). During this same time interval, in other parts of Europe, a noticeable increase in mesothermic plants and high-elevation conifers has been documented, interpreted as a result of climate cooling possibly caused by Antarctic glaciations or by uplift of surrounding mountains (Kuhlemann and Kempf, 2002; Utesche et al., 2000), a process even favored by the progressive movement of Eurasia towards northern latitudes as a result of the northward collision of Africa.

Among other characteristics, it is of great importance to point out that, during the upper part of the lower Miocene, there is a marked fauna turnover, with the appearance of new mammals such as *Anchitherium*, the first *Proboscideans*, etc. (Morales and Nieto, 1997). The existence of turnover cycles in rodent faunas from Spain (periods of 2.4 to 2.5 and 1 Ma) appears related to low-frequency modulations of Milankovitch-controlled climate oscillation (Van Daam et al., 2006). The Earth's climate and its evolution, studied by the analysis of deep-sea sediment cores, experience gradual trends of warming and cooling, with cycles showing $10^4$ to $10^6$ years of rhythmic or periodic cyclicality explained as related to variations in orbital parameters such as eccentricity, obliquity and precession that affect the distribution and amount of incident solar energy (Zachos et al., 2001). Obliquity nodes and eccentricity minima are associated with ice sheet expansion in Antarctica that altered precipitation regimes together with cooling and aridity. These climatic changes produce perturbations in terrestrial biota through reduced food availability (Kuhlemann and Kempf, 2002; Utesche et al., 2000; Van Daam et al., 2006).

The Oligocene–Miocene boundary ($\sim$ 23 Ma) corresponds to a brief ($\sim$ 200 kyr), but deep, Antarctic glacial maximum, referred to as Mi-1 (Fig. 5), followed by a series of intermittent but smaller phases of glaciation (Mi events) where maximum ice volume took place on a scale of over 100 kyr on the East Antarctic continent (Mawbey and Lear, 2013). The

Mi-1 event was accompanied by a series of accelerated rates of turnover and speciation in certain groups of biota, such as the extinction of Caribbean corals at this boundary. This limit is accompanied by sharp positive carbon isotope excursions that suggest perturbations of the global carbon cycle (Fig. 5). Correlating the $\delta^{18}O$ and $\delta^{13}C$ values of deep-sea sediment cores with sea-level calibrations has shown that, during the early Miocene, the ice volume ranged from 50 to 125 % of the present day volume (Pekar and DeConto, 1996). Tectonic changes such as the opening of the Drake Passage may have modified portions of the planet's ocean circulation system, promoting synchronous global cooling trends (Coxall et al., 2005). The cold water from the southern Atlantic and abyssal Pacific basins (Lear et al., 2004) mixed with a warm deep-water mass located in the Atlantic and Indian oceans (Billups et al., 2002; Wright and Miller, 1996). Wright and Colling (1995) estimated that, during these glacial periods, there was a temperature gradient of up to 6 °C, larger than observed today ($\sim$ 3–4 °C). The influence of this temperature drop on a global scale could have had some influence on the precipitation of cryophilic and even cryogenic salts from salt-concentrated brines during these particular moments at the Iberian Peninsula latitudes.

Hence, in the upper part of the sedimentary sequence of the Lower Unit, the presence of a higher proportion of species with high tolerance to cool climatic conditions, and the lowering of the individual count and species variety, indicate the existence of a climatic change into a cool and arid period within the Iberian Peninsula ($\sim$ 18.4 Ma). This age appears to coincide with a global Mi-1ab event (Miller et al., 1991) that represented an interval of ice expansion, at least in East Antarctica. The global low temperature and arid conditions of the environment could have been magnified in this area by its continental character and the regional uplift of the surrounding mountains that left this area at a higher altitude and within a "rain shadow" region. In addition, recycling of ancient saline formations provides concentrated brines that promote the precipitation of mirabilite at even higher temperatures. Higher up in the sequence, the gypsum deposits formed by evaporative concentration of the saline brines as a result of the climate warming indicated the temperature curves that show the trend towards the Miocene optimum (Zachos et al., 2001).

## 6    Conclusions

The appearance of thick, pure thenardite beds in the geological record can be used as a paleoclimate indicator of cool and arid periods. By fieldwork analysis and laboratory techniques, we have described a way to establish the diagenetic character of the thenardite deposits formed after a mirabilite precursor, a salt that is well known to form under cool and arid weather conditions. Mirabilite deposits require a sustained period of time to develop, with a fairly continuous,

persistent period of a cool climate because it is normally formed during a frigid-concentration process. This mechanism of formation has led to the development of a typical salt paragenesis, and its fingerprint is recorded within the geochemistry of its fluid inclusions.

The establishment of the age of this unit, based on mammal assemblages, has permitted us to determine the existence of a relatively cool and dry period from a lacustrine record that correlates with an Antarctic ice expansion "Mi" event (Mi-1ab that took place at $\sim$ 18.4 Ma) determined from marine deposits and established on a global scale by isotope studies. This period represents a moment of the expansion of, at least, the East Antarctic ice sheet. This expansion has been interpreted as being related to changes in the Earth's orbital parameters such as obliquity and eccentricity that even control the turnover cycles of different biotas, as appears to be the case in the Iberian Peninsula. Therefore, the correlation of terrestrial and marine records contributes to a more precise knowledge of environmental and climatic changes on a global scale.

Hence, the lacustrine deposits of the upper part of the Lower Unit of the Tajo Miocene succession do not require a regressive sequence of a lacustrine system due to the reduction of water by desiccation alone (due to intense evaporation). Instead, the mirabilite was formed in a lake with high $Na^+$ and $SO_4^{2-}$ saturated waters. At a time period where temperature was subject to a significant decrease and aridity became a key factor ($\sim$ 18.4 Ma), the brines were concentrated by a cooling–freezing mechanism that led to the formation of thick well-differentiated mirabilite layers, which later were diagenetically transformed to thenardite.

*Acknowledgements.* We would like to thank the SAMCA company, and especially Francisco Gonzalo and Carlos Lasala, for providing the necessary samples (drilling cores and mine samples) employed in this research and for their encouragement. Special thanks to M. E. Arribas, M. A. Alvarez, the editor Y. Godderis and the reviewers C. Monnin and S. Bourquin for their comments and suggestions that have led to improvements in the manuscript. This study was financed by the Fundación General de la Universidad Complutense de Madrid (projects 396/2009-4153239 and 139/2014-4155418).

Edited by: Y. Godderis

## References

Alberdi, M. T., Hoyos, M., Junco, F., López-Martínez, N., Morales, J., Sesé, C., and Soria, D.: Biostratigraphy and sedimentary evolution of continental Neogene in the Madrid area, Paléobiol. Continent., 14, 47–68, 1984.

Ayllón-Quevedo, F., Souza-Egipsy, V., Sanz-Montero, M. E., and Rodriguez-Aranda, J. P.: Fluid inclusion analysis of twinned selenite gypsum beds from the Miocene of the Madrid basin

(Spain), Implication on dolomite bioformation., Sediment. Geol., 201, 212–230, 2007.

Ayora, C., García-Veigas, J., and Pueyo Mur, J. J.: X-ray micro-analysis of fluid inclusions and its application to the geochemical modelling of evaporite basins, Geochim. Cosmochim. Ac., 58, 43–55, 1994.

Babel, M. and Schreiber, B. C.: Geochemistry of Evaporites and Evolution of Seawater. In: Treatise on Geochemistry (Second Edition), edited by: Turekian, K. and Holland, H., Elsevier, Oxford, 2014.

Billups, K., Channell, J. E. T., and Zachos, J.: Late Oligocene to early Miocene geochronology and paleoceanography from the subantarctic South Atlantic, Paleoceanography, 17, 4.1–4.11, 2002.

Borchert, H. and Muir, R. O.: Salt Deposits. The Origin, Metamorphism and Deformation of Evaporites, D. Van Nostrand Company, London, 1964.

Brasier, A. T.: Searching for travertines, calcretes and speleothemes in deep time: Processes, appearances, predictions and the impact of plants, Earth-Science Reviews, 104, 213–239, 2011.

Calvo, J. P., Daams, R., Morales, J., López-Martínez, N., Agustí, J., Anadón, P., Armenteros, I., Cabrera, L., Civis, J., Corrochano, A., Díaz-Molina, M., Elizaga, E., Hoyos, M., Martín-Suarez, E., Martínez, J., Moissenet, E., Muñoz, A., Pérez-García, A., Pérez-González, A., Portero, J. M., Robles, F., Santisteban, C., Torres, T., Van der Meulen, A., Vera, J. A., and Mein, P.: Up-to-date Spanish continental Neogene synthesis and paleoclimatic interpretation, Revista de la Sociedad Geológica de España, 6, 29–40, 1993.

Calvo, J. P., Alonso-Zarza, A. M., García del Cura, M. A., Ordoñez, S., Rodriguez-Aranda, J. P., and Sanz-Montero, M. E.: Sedimentary evolution of lake systems through the Miocene of the Madrid Basin: paleoclimatic and paleohydrological constraints, in: Tertiary basins of Spain, the stratigraphic record of crustal kinematics, edited by: Friend, P. F. and Dabrio, C. J., Cambridge University Press, Cambridge, 1996.

Calvo, J. P., Ordoñez, S., García del Cura, M. A., Hoyos, M., Alonso-Zarza, A. M., Sanz, E., and Rodriguez, J. P.: Sedimentología de los complejos lacustres miocenos de la Cuenca de Madrid, Acta Geol. Hisp., 24, 281–298, 1989.

Cañaveras, J. C., Sánchez-Moral, S., Calvo, J. P., Hoyos, M., and Ordóñez, S.: Dedolomites associated with karstic features, an example of early dedolomitization in lacustrine sequences from the Tertiary Madrid Basin, Central Spain., Carb. Evaporit., 11, 85–103, 1996.

Chaboureau, A.-C., Donnadieu, Y., Sepulchre, P., Robin, C., Guillocheau, F., and Rohais, S.: The Aptian evaporites of the South Atlantic: a climatic paradox?, Clim. Past, 8, 1047–1058, doi:10.5194/cp-8-1047-2012, 2012.

Charykova, M. V., Kurilenko, V. V., and Charykov, N. A.: Temperatures of formation of certain salts in sulfate-type brines, J. Appl. Chem. USSR, 65-1, 1037–1040, 1992.

Chun, F. H.: Quantitative interpretation of X-ray diffraction patterns of mixtures, III. simultaneous determination of a set of reference intensities, J. Appl. Chrystallogr., 8, 17–19, 1975.

Coxall, H. K., Wilson, P. A., Palike, H., Lear, C. H., and Backman, J.: Rapid stepwise onset of Antarctic glaciation and deeper calcite compensation in the Pacific Ocean, Nature, 433, 53–57, 2005.

Daams, R. and Freudenthal, M.: Synopsis of the Dutch-Spanish collaboration program in the Aragonian type area, 1975–1986, in: Biostratigraphy and paleoecology of the Neogene micromammalian faunas from the Calatayud-Teruel Basin (Spain), Scripta Geológica, Spec. Issue, 1, 3–18, 1988.

Daams, R., Álvarez-Sierra, M. A., Van der Meulen, A., and Peláez-Campomanes, P.: Los micromamíferos como indicadores de paleoclimas y evolución de las cuencas continentales, in: Registros fósiles e Historia de la Tierra, edited by: Aguirrre, E., Morales, J., and Soria, D., Editorial Complutense, Madrid, 1997.

De Vicente, G., González-Casado, J. M., Muñoz-Martín, A., Giner, J. L., and Rodríguez-Pascua, M. A.: Structure and Tertiary evolution of the Madrid basin, in: Tertiary basins of Spain, the stratigraphic record of crustal kinematics, edited by: Friend, P. F. and Dabrio, C. J., Cambridge University Press, Cambridge, 1996.

Dort, W. J. and Dort, D. S.: Low Temperature Origin of Sodium Sulfate Deposits, Particularly in Antarctica, in: Third Symposium on Salt, 1, Ohio Geological Society, 1970.

Escavy, J. I., Herrero, M. J., and Arribas, M. E.: Gypsum resources of Spain: Temporal and spatial distribution, Ore Geol. Rev., 49, 72–84, 2012.

Eugster, H. P. and Hardie, L. A.: Saline Lakes ,in: Lakes, Chemistry, Geology, Physics, edited by: Lerman, A., Springer Verlag, 237–293, 1978.

Fan-Wei, M., Pei, N., Xun-Lai, Y., Chuan-Ming, Z., Chun-He, Y., and Yin-Ping, L.: Choosing the best ancient analogue for projected future temperatures: A case using data from fluid inclusions of middle-late Eocene halites, J. Asian Earth Sci., 67/68, 46–50, 2013.

Garrett, D.: Sodium Sulfate. Handbook of Deposits, Processing, and Use, Academic Press, San Diego, 2001.

Goldstein, R. H. and Reynolds, T. J.: Systematics of Fluid Inclusions in Diagenetic Minerals, Society for Sedimentary Geology, Tulsa, USA, 1994.

Haq, B. U., Hardenbol, J., and Vail, P. R.: Chronology of fluctuating sea levels since the Triassic 250 million years ago to present, Science, 235, 1156–1167, 1987.

Hardie, L. A. and Eugster, H. P.: The evolution of closed-basin brines, Mineral. Soc. Am. Spec. Paper 3, 1970, 273–290, 1970.

Harvie, C. E., Weare, J. H., Hardie, L. A., and Eugster, H. P.: Evaporation of sea-water: calculated mineral sequences, Science, 208, 498–500, 1980.

Herrero, M. J., Escavy, J. I., and Bustillo, M.: The Spanish building crisis and its effect in the gypsum quarry production (1998–2012), Resour. Pol., 38, 123–129, 2013.

Hovland, M., Rueslatten, H., Johnsen, H. K., Kvamme, B., and Kuznetsova, T.: Salt formation associated with sub-surface boiling and supercritical water, Mar. Petrol. Geol., 23, 855–869, 2006.

Jones, B.: The Hydrology and Mineralogy of Deep Springs Lake, Inyo County, California, United States Government, Washington, 1965.

Kendall, A. C.: Evaporites, in: Facies Models, Respons to sea level changes, Walker, R. G. and James, N. P., Geological Association of Canada, St John's, 1992.

Kuhlemann, J. and Kempf, O.: Post-Eocene evolution of the North Alpine Foreland, Sediment. Geol., 152, 45–78, 2002.

Last, W. M.: Modern sedimentology and hydrology of Lake Manitoba, Canada, Environ. Geol., 5, 177–190, 1984.

Last, W. M.: Deep-water evaporite mineral formation in lakes of
western canada, in: Sedimentology and geochemistry of modern and ancient saline lakes, edited by: Renant, R. and Last, W.,
SEPM, Tulsa, Oklahoma, 51–59, 1994.

Lear, C. H., Rosenthal, Y., Coxall, H. K., and Wilson, P.
A.: Late Eocene to early Miocene ice sheet dynamics and
the global carbon cycle, Paleoceanography, 19, PA4015,
doi:10.1029/2004PA001039, 2004.

Lowenstein, T. K. and Hardie, L. A.: Criteria for the recognition of
salt-pan evaporites, Sedimentology, 32, 627–644, 1985.

Lowenstein, T. K., Jianren, L., Brown, C., Roberts, S. M., Teh-
Lung, K., Shangde, L., and Wembo, Y.: 200 k.y. paleoclimate
record from Death Valley salt core, Geology, 27, 3–6, 1999.

Magny, M. and Combourieu Nebout, N.: Holocene changes in environment and climate in the central Mediterranean as reflected
by lake and marine records, Climate of the Past, 9, 1447–1454,
doi:10.5194/cp-9-1447-2013, 2013.

Marion, G. M., Farren, R. E., and Komrowski, A. J.: Alternative
pathways for seawater freezing, Cold Reg. Sci. Technol., 29,
259–266, 1999.

Mawbey, E. M. and Lear, C. H.: Carbon cycle feedbacks during the
Oligocene-Miocene transient glaciation, Geology, 41, 963–966,
2013.

Miller, K. G., Wright, J. D., and Fairbanks, R. G.: Unlocking the Ice
House: Oligocene-Miocene oxygen isotopes, eustasy, and margin erosion, J. Geophys. Res., 96, 6829–6848, 1991.

Minghui, L., Xiaomin, F., Chaulou, Y., Shaopeng, G., Weilin, Z.,
and Galy, A.: Evaporite minerals and geochemistry of the upper
400 m sediments in a core from the Western Qaidam Basin, Tibet,
Quaternary Internat., 218, 176–189, 2010.

Möhlmann, D. and Thomsen, K.: Properties of cryobrines on Mars,
Icarus, 212, 123–130, 2011.

Morales, J. and Nieto, M.: El registro terciario y cuaternario de los
mamíferos de España, in: Registros fósiles e Historia de la Tierra,
edited by: Calvo, J. P. and Morales, J., Editorial Complutense,
Madrid, 297–322, 1997.

Mullin, J. W.: Crystallization, 4th Edition, Butlerworth-Heinemann,
Oxford, 2001.

Nai'ang, W., Zhuolun, L., Yu, L., Hongyi, C., and Rong, H.:
Younger Dryas event recorded by the mirabilite deposition in
Huahai lake, Hexi Corridor, NW China, Quaternary Interna-
tional, 250, 93–99, 2012.

Nakhla, F. M., Saleh, S. A., and Gad, N. L.: Mineralogy, chemistry
and paragenesis of the thenardite ($Na_2SO_4$), in: Applied Mineralogy, The Metallurgical Society of AIME, New York, 1985.

Ordóñez, S., Calvo, J. P., García del Cura, M. A., Alonso-Zarza, A.
M., and Hoyos, M.: Sedimentology of sodium sulphate deposits
and special clays from the Tertiary Madrid Basin (Spain), in: Lacustrine Facies Analysis, edited by: Anadón, P., Cabrera, L., and
Kelts, K., Special Publications of the International Association
of Sedimentologists, Wiley, 1991.

Ordoñez, S. and García del Cura, M. A.: Deposition and diagenesis
of sodium-calcium sulfate salts in the tertiary saline lakes of the
Madrid Basin, Spain, in: Sedimentology and Geochemistry of
Modern and Ancient Saline Lakes, SEPM Spec. Publ., 50, 229–
238, 1994.

Ortí, F.: Evaporitas: Introducción a la sedimentología evaporítica,
in: Sedimentología, edited by: Arche, A., 675–770, 2010.

Ortí, F., Pueyo, J. J., and San Miguel, A.: Petrogénesis del
yacimiento de sales sódicas de Villarubia de Santiago, Toledo
(Terciario continental de la Cuenca del Tajo), Boletín Geológico
y Minero, T. XC, 347–373, 1979.

Ortí, F., Gündogan, I., and Helvaci, C.: Sodium sulphate deposits
of Neogene age: the Kirmir Formation, Beypazari Basin, Turkey,
Sediment. Geol., 146, 305–333, 2002.

Pekar, S. F. and DeConto, R. M.: High-resolution ice-volume estimates for the early Miocene: Evidence for a dynamic ice sheet in
Antarctica, Palaeogeography, Palaeoclimatology, Palaeoecology,
231, 101–109, 1996.

Peterson, R. C., Nelson, W., Madu, B., and Shurvell, H. F.: Merid-
ianiite: A new mineral species observed on Earth and predicted
to exist on Mars, Am. Mineral., 92, 1756–1759, 2007.

Rouchy, J. M. and Blanc-Valleron, M. M.: Les évaporites: materi-
aux singuliers, millieux extrèmes, Vuibert, Société géologique de
France, Paris, 2009.

Salvany, J. M. and Ortí, F.: Miocene glauberite deposits of Al-
canadre, Ebro Basin, Spain: Sedimentary and diagenetic processes, in: Sedimentology and Geochemistry of Modern and Ancient Saline Lakes, SEPM Spec. Publ. 50, edited by: Renault, R.
W. and Last, W. M., Tulsa, EEUU, 203–215, 1994.

Salvany, J. M., García-Veigas, J., and Ortí, F.: Glauberite-halite association of the Zaragoza Gypsum Formation (Lower Miocene,
Ebro Basin, NE Spain, Sedimentology, 54, 443–467, 2007.

Sánchez-Moral, S., Ordóñez, S., Benavente, D., and García del
Cura, M. A.: The water balance equations in saline playa lakes:
comparison between experimental and recent data from Quero
Playa Lake (central Spain), Sediment. Geol., 148, 221–234,
2002.

Schreiber, B. C. and El Tabakh, M.: Deposition and early alteration
of evaporites, Sedimentology, 47, 215–238, 2000.

Shortland, A. J.: Evaporites of the Wadi Natrun: Seasonal and annual variation and its implication for ancient exploitation, Ar-
chaemetry, 46, 497–516, 2004.

Socki, R. A., Sun, T., Niles, P. B., Harvey, R. P., Bish, D. L., and
Tonui, E.: Anctarctic Mirabilite Mounds as Mars Analogs: The
Lewis Cliffs Ice Tongue Revisited, The Woodlands, Texas2012,
2012.

Stankevich, E. F., Batalin, Y. V., and Sinyavskii, E. I.: Sedimentation
and dissolution of salts during fluctuations of the brine level in a
self-sedimenting lake, Geol. Geofiz, 3, 35–41, 1990.

Stark, S. C., O'Grady, B. V., Burton, H. R., and Carpenter, P. D.:
Frigidly concentrated seawater and the evolution of Antarctic
saline lakes, Australian J. Chem., 56, 181–186, 2003.

Stewart, F. H.: Marine Evaporites, in: Data of Geochemistry, edited
by: Fleischer, M., US Geological Survey, Washington, 1963.

Strakhov, N. M.: Principles of lithogenesis, Plenum Publishing Corporation, New York, 1970.

Utesche, T., Mosbrugger, V., and Ashraf, A.: Terrestrial Climate
Evolution in Northwest Germany Over the Last 25 Million Years,
PALAIOS, 15, 430–449, 2000.

Van Daam, J. A., Abdul-Aziz, H., Álvarez-Sierra, M. A., Hilgen,
F. J., Van de Hoek Ostende, L. W., Lourens, L. J., Mein, P., Van
der Meulen, A., and Pelaez-Campomanes, P.: Long-period astronomical forcing of mammal turnover, Nature, 443, 687–691,
2006.

Van der Meulen, A. and Daams, R.: Evolution of Early-Middle
Miocene rodent faunas in relation to long-term palaeoenviron-

mental changes, Palaeogeography, Palaeoclimatology, Palaeoecology, 93, 227–253, 1992.

Wang, N., Zhang, J., Cheng, H., Guo, J., and Zhao, Q.: The age of formation of the mirabilite and sand wedges in the Hexi Corridor and their paleoclimatic interpretation, Chinese Sci. Bull., 48, 1439–1445, 2003.

Warren, J.: Evaporites: Sediments, Resources and Hydrocarbons, Springer, Berlin, 2006.

Warren, J.: Evaporites through Time: Tectonic, climatic and eustatic controls in marine and nonmarine deposits, Elsevier, 2010.

Wright, J. D. and Colling, A.: Seawater: its composition, properties, and behaviour, Open University Press and Elsevier, Oxford, 1995.

Wright, J. D. and Miller, K. G.: Control of North Atlantic deep water circulation by the Greenland-Scotland Ridge, Paleoceanography, 11, 157–170, 1996.

Zachos, J., Pagani, M., Sloan, L., Thomas, E., and Billups, K.: Trends, Rhythms, and Aberrations in Global Climate 65 Ma to Present, Science, 292, 686–693, 2001.

Zheng, M., Zhao, Y., and Liu, J.: Palaeoclimatic Indicators of China's Quaternary Saline Lake Sediments and Hydrochemistry, Acta Geologica Sinica, 74, 259–265, 2000.

# 7

# Persistent millennial-scale link between Greenland climate and northern Pacific Oxygen Minimum Zone under interglacial conditions

O. Cartapanis[1,*], K. Tachikawa[1], O. E. Romero[2,**], and E. Bard[1]

[1]Aix-Marseille Université, CNRS, IRD, Collège de France, CEREGE UM34, 13545 Aix en Provence, France
[2]Instituto Andaluz de Cs. de la Tierra (CSIC-UGR), Ave. de las Palmeras 4, 18100 Armilla-Granada, Spain
[*]now at: McGill University, Department of Earth and Planetary Sciences, 3450 University Street, Montreal, H3A 0E8, Quebec, Canada
[**]now at: MARUM, Center for Marine Environmental Sciences, University of Bremen, Leobener Str., 28359 Bremen, Germany

*Correspondence to:* O. Cartapanis (olivier.cartapanis@mcgill.ca)

**Abstract.** The intensity and/or extent of the northeastern Pacific Oxygen Minimum Zone (OMZ) varied in-phase with the Northern Hemisphere high latitude climate on millennial timescales during the last glacial period, indicating the occurrence of atmospheric and oceanic connections under glacial conditions. While millennial variability was reported for both the Greenland and the northern Atlantic Ocean during the last interglacial period, the climatic connections with the northeastern Pacific OMZ has not yet been observed under warm interglacial conditions. Here we present a new geochemical dataset, spanning the past 120 ka, for major components (terrigenous fraction, marine organic matter, biogenic opal, and carbonates) generated by X-ray fluorescence scanning alongside with biological productivity and redox sensitive trace element content (Mo, Ni, Cd) of sediment core MD02-2508 at 23° N, retrieved from the northern limit of the modern OMZ. Based on elemental ratios Si/Ti (proxy for opal), Cd/Al and Ni/Al, we suggest that biological productivity was high during the last interglacial (MIS5). Highly resolved opal reconstruction presents millennial variability corresponding to all the Dansgaard-Oeschger interstadial events over the last interglacial, while the Mo/Al ratio indicates reduced oxygenation during these events. Extremely high opal content during warm interstadials suggests high diatom productivity. Despite the different climatic and oceanic background between glacial and interglacial periods, rapid variability in

the northeastern Pacific OMZ seems to be tightly related to Northern Hemisphere high latitude climate via atmospheric and possibly oceanic processes.

## 1 Introduction

Oxygen Minimum Zones (OMZ) develop in areas where oxygen consumption by organic matter degradation in the water column and on the seafloor outmatches lateral/vertical advection/diffusion of dissolved oxygen. Thus, intermediate depth OMZ variations are mainly determined by (1) variations of the oxygen content in source intermediate water masses, (2) changes in rate of intermediate water formation and advection, and/or (3) oxygen consumption occurring during the flow of intermediate water masses toward the OMZ and within the OMZ itself (Karstensen et al., 2008; Paulmier and Ruiz-Pino, 2009; Stramma et al., 2010a). Oxygen consumption is related to the degradation of organic matter as it sinks through the water column and settles on the seafloor (Karstensen et al., 2008).

During the past decades, the scientific community has put considerable effort in understanding the past dynamics of the OMZ, owing to its potential role in the carbon cycle and greenhouse gas Emissions (Paulmier et al., 2008, 2011), and the global marine nitrogen cycle (Galbraith et al., 2004). The modern expansion of the OMZ, probably in relation to anthropogenic climate change (Stramma et al., 2008, 2010b), points out the importance of understanding the factors that influence OMZ dynamics in an abrupt climate change context, as it could behave as an internal feedback on the global climate (Altabet et al., 2002; Pichevin et al., 2007).

The variability of the mid-latitude north Pacific OMZ during the last glacial period has been extensively studied (Hendy and Pedersen, 2005; Dean, 2007; Hendy, 2010; Cartapanis et al., 2011, 2012). Most of these studies revealed a strong link between oxygen deficit in the northern Pacific and Northern Hemisphere high latitude climate on millennial timescales, with enhanced oxygen depletion and higher biological productivity during warm Dansgaard-Oeschger (DO) interstadial events, and enhanced oxygenation during Heinrich events (HE). Both biological productivity and oceanic ventilation are suspected to have influenced past OMZ variations (Hendy and Kennett, 2000; Hendy et al., 2004; Cartapanis et al., 2011, 2012). In particular, modelling studies indicate that the disruption of North Atlantic Deep Water formation during HE was associated with an increase of the oxygenation in the northern Pacific. North Pacific Intermediate Waters (NPIW) formation/oxygenation increased, enhancing oxygen advection towards the OMZ, while a reduction of the deep upwelling of nutrient rich waters reduced productivity and associated oxygen consumption at low latitudes (Schmittner, 2005; Schmittner et al., 2007).

Climate variability during the last interglacial (Marine Isotopic Stage = MIS5) is particularly interesting because of climatic similarities with the Holocene. However, productivity and ventilation variations during MIS5 have been poorly documented so far. Except for MIS5e, MIS5 is characterised by lower sea level than during full interglacial periods (e.g. Holocene and MIS5e), but higher sea level than during glacial (e.g. MIS2, 3 and 4) (Lambeck et al., 2002; Waelbroeck et al., 2002; Cutler et al., 2003; Hu et al., 2010; Grant et al., 2012). This suggests the existence of a small but significant ice cap over Northern Hemisphere high latitude continents (Bonelli et al., 2009; Ganopolski et al., 2010). During MIS5, DO (Johnsen et al., 2001) and HE-like events (massive iceberg discharge in the North Atlantic) (McManus et al., 1994; Eynaud et al., 2000; Oppo et al., 2006) occurred together with rapid high amplitude sea level changes (Grant et al., 2012).

A few studies in the northeastern Pacific suggested higher productivity (Ganeshram and Pedersen, 1998; Kienast et al., 2002) and lower oxygenation at intermediate depth (Nameroff et al., 2004) during MIS5 as compared to glacial periods (MIS2 to MIS4). However, the millennial scale variations of productivity and oxygenation in North Pacific OMZ over the last interglacial remain unknown because of scarce high-resolution records.

Siliceous organisms such as diatoms represent a significant proportion of the production in the Pacific, but the mechanisms responsible for spatial and temporal variations of diatoms productivity (e.g. wind driven upwelling, nutrient con-

tent of water masses, biological competition for nutrients use) remain controversial and are likely related to changes in atmospheric and oceanic circulation. While several studies displayed millennial and/or glacial/interglacial changes of the opal production at low latitude in the eastern Pacific (Ganeshram and Pedersen, 1998; Kienast et al., 2006; Dubois et al., 2010; Pichevin et al., 2010; Arellano-Torres et al., 2011; Dubois et al., 2011; Romero et al., 2011a; Cartapanis et al., 2012), high temporal resolution opal records in northeastern Pacific are still limited.

Here we present a new geochemical dataset spanning the last 120 ka from the northern edge of the modern North Pacific OMZ. High-resolution X-ray fluorescence (XRF) measurements for major sediment components (marine organic carbon, opal, carbonates and terrigenous fraction), were performed on MD02-2508 core (MD08, northeastern coastal Pacific), in order to monitor the past productivity variations. Trace element response to productivity and bottom water oxygenation were measured by ICP-MS in an attempt to reconstruct past variability of the OMZ and to determine the factors driving its variations. High sedimentation rate at site MD08 during MIS5 (around $35\,\text{cm}\,\text{ka}^{-1}$) provides a unique opportunity to reconstruct the millennial scale variability of productivity and ventilation during the last interglacial period, and the mechanisms involved in linking high latitude climate variations to lower latitude climate, oceanic conditions, and biogeochemical cycles.

## 2  Materials and method

Core MD08 (23°27.91′ N, 111°35.74′ W, 606 m water depth, Fig. 1) was retrieved from the Baja California margin by the R.V. Marion-Dufresne during the coring campaign IMAGES MD126-MONA. Detailed analytical procedures are shown in a previous study (Cartapanis et al., 2011). In order to trace the past OMZ dynamics, we measured Br and Si / Ti, representative of the marine organic matter (Mayer et al., 2007) and opal content, respectively, and Ca and Ti, representative of the carbonates and the terrigenous sediment fractions respectively. XRF measurements were performed using ITRAX XRF core scanner (Cox Analytical Systems) with both Mo (30 kV and 45 mA) and Cr (35 kV and 35 mA) X-ray sources, with a counting time of 15 s. XRF measurement resolution was increased in the presence of lamination (from 0.5 cm to $200\,\mu\text{m}$) and is equivalent to 2 yr on average. Discrete bulk sediment samples were dissolved by acid digestion and elemental concentrations (Cd, Mo, Ni, Ti, Ca, and Al) were analysed by ICP-MS (Agilent 7500ce). Trace elements contents (Cd, Mo, Ni) are normalized against Al to estimate authigenic enrichment. Temporal resolution of these measurements is better than 500 yr. While Cd and Ni are closely associated with biogenic sinking particles, Mo accumulation in the sediment is instead associated with redox condition at bottom water/sediment interface (Nameroff et al.,

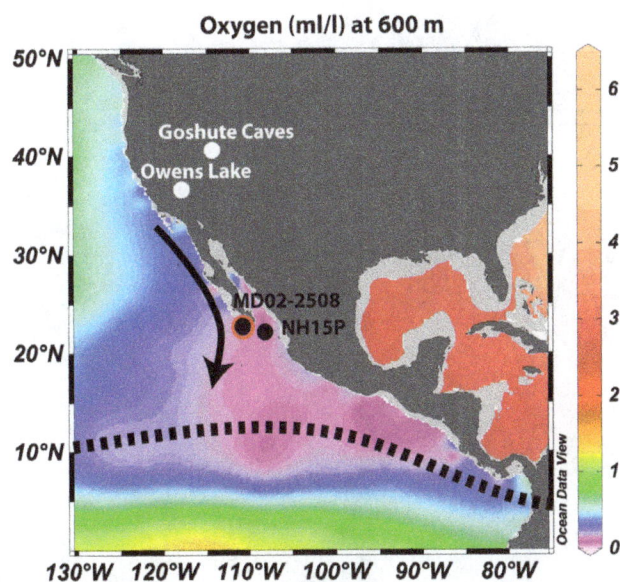

**Oxygen (ml/l) at 600 m**

**Fig. 1.** Dissolved oxygen content at 600 m in northeastern Pacific (World Ocean Atlas 09, Garcia et al., 2010) and positions of the records discussed in the text. Cores MD02-2508 (this study); NH15P (Ganeshram and Pedersen, 1998; Nameroff et al., 2004); Goshute Cave (Denniston et al., 2007); Owens Lake (Li et al., 2004). Black arrow indicates dominant winds during upwelling season and black dotted line represents ITCZ position during boreal summer. Figure 1 was generated using the Ocean Data View software (http://odv.awi.de).

**Table 1.** Tie points (corrected depth in cm, age in year B.P.), method and references used to build the age model of MD02-2508 core in this study.

| Corrected depth (cm) | Age (yr B.P.) | Method | Reference |
|---|---|---|---|
| 0 | 0 | NGRIP visual correlation | Cartapanis et al. (2011) |
| 500 | 11 450 | NGRIP visual correlation | Cartapanis et al. (2011) |
| 592 | 14 550 | NGRIP visual correlation | Cartapanis et al. (2011) |
| 892 | 28 500 | NGRIP visual correlation | Cartapanis et al. (2011) |
| 980 | 33 500 | NGRIP visual correlation | Cartapanis et al. (2011) |
| 1014 | 35 300 | NGRIP visual correlation | Cartapanis et al. (2011) |
| 1082 | 38 250 | NGRIP visual correlation | Cartapanis et al. (2011) |
| 1168 | 41 700 | NGRIP visual correlation | Cartapanis et al. (2011) |
| 1218 | 43 600 | NGRIP visual correlation | Cartapanis et al. (2011) |
| 1306 | 47 300 | NGRIP visual correlation | Cartapanis et al. (2011) |
| 1514 | 54 950 | NGRIP visual correlation | Cartapanis et al. (2011) |
| 1656 | 59 100 | NGRIP visual correlation | Cartapanis et al. (2011) |
| 3689 | 117 000 | Blake magnetic excursion | Blanchet et al. (2007) |

2002; Tribovillard et al., 2006; Cartapanis et al., 2011). We thus use Cd / Al and Ni / Al as productivity indicators, and Mo / Al as an oxygenation indicator. The terrigenous fraction of the sediment was calculated using Ti content in the mean Upper Continental Crust (UPC, 4100 µg g$^{-1}$, McLennan, 2001). Total organic carbon (TOC) was determined using a FISONS NA 1500 elemental analyser after carbonates were removed with 1M HCl. All the elemental analyses and organic carbon determination were carried out at CEREGE (Aix en Provence, France).

Samples for the analysis of opal were freeze-dried and ground in an agate mortar. Opal content was determined by the sequential leaching technique of DeMaster (1981) and by molybdate-blue spectrophotometry, with further modifications by Müller and Schneider (1993). Opal measurements were performed at MARUM (Bremen, Germany).

The age model of the core is based on previously published studies. Given a known relationship between sediment physical property variation and high latitude climate variability at southern Baja California margin (Ortiz et al., 2004; Dean et al., 2006; Dean, 2007; Marchitto et al., 2007), we used a visual correlation of wet bulk density obtained from the GEOTEK logger and lightness obtained by spectrophotometry (Beaufort et al., 2002) to the NGRIP isotopic oxygen record (Johnsen et al., 2001) for the past 70 ka (see details in Leduc (2007), Cartapanis et al. (2011) and in Table 1). One

additional tie point was added for the Blake magnetic excursion (115–122 ka) identified in MD08 core by Blanchet et al. (2007) (see crosses at the top of Fig. 2 and Table 1).

## 3 Study area

The intertropical Pacific hosts the most extended modern OMZ (Paulmier and Ruiz-Pino, 2009, see also Fig. 1). Oxygen depletion is even more pronounced in the northern Pacific, because of restrained NPIW formation in the Okhotsk sea (Talley, 1991; Shcherbina et al., 2003) as compared to the southern Pacific where Antarctic Intermediate Water (AAIW) ventilates the intermediate depth more efficiently. MD08 core was retrieved off the Baja California margin, at the southern limit of the NPIW in the eastern Pacific (Fig. 1, see details in Cartapanis et al., 2011). Further south, intermediate depths are occupied by the Equatorial Pacific Intermediate Waters, mainly derived from AAIW and Pacific Deep Waters (Bostock et al., 2010). At the surface, the California Current (CC) that transports relatively fresh and cold Subarctic Water (SW) along eastern Pacific coast is diverted westward and slowly sinks in the water column around 25° N, slightly north of MD08 core (Auad et al., 2011).

Presently, the productivity of diatoms and small phytoplankton (as defined by Moore et al., 2004, and which include Coccolithophoridae), is thought to be limited by nitrate content in surface waters (Moore et al., 2004). Wind driven upwelling activity related to atmospheric circulation is an important driver of the modern productivity along northeastern Pacific shores (Lynn and Simpson, 1987; Thomas et al., 2001; Zaytsev et al., 2003; Pennington et al., 2006). During summer, the North Pacific high-pressure cell strengthens and the continental low deepens, resulting in a high east-west pressure gradient inducing equatorward winds. Modern productivity off northeastern Pacific shores is consequently higher during summer months when alongshore southward

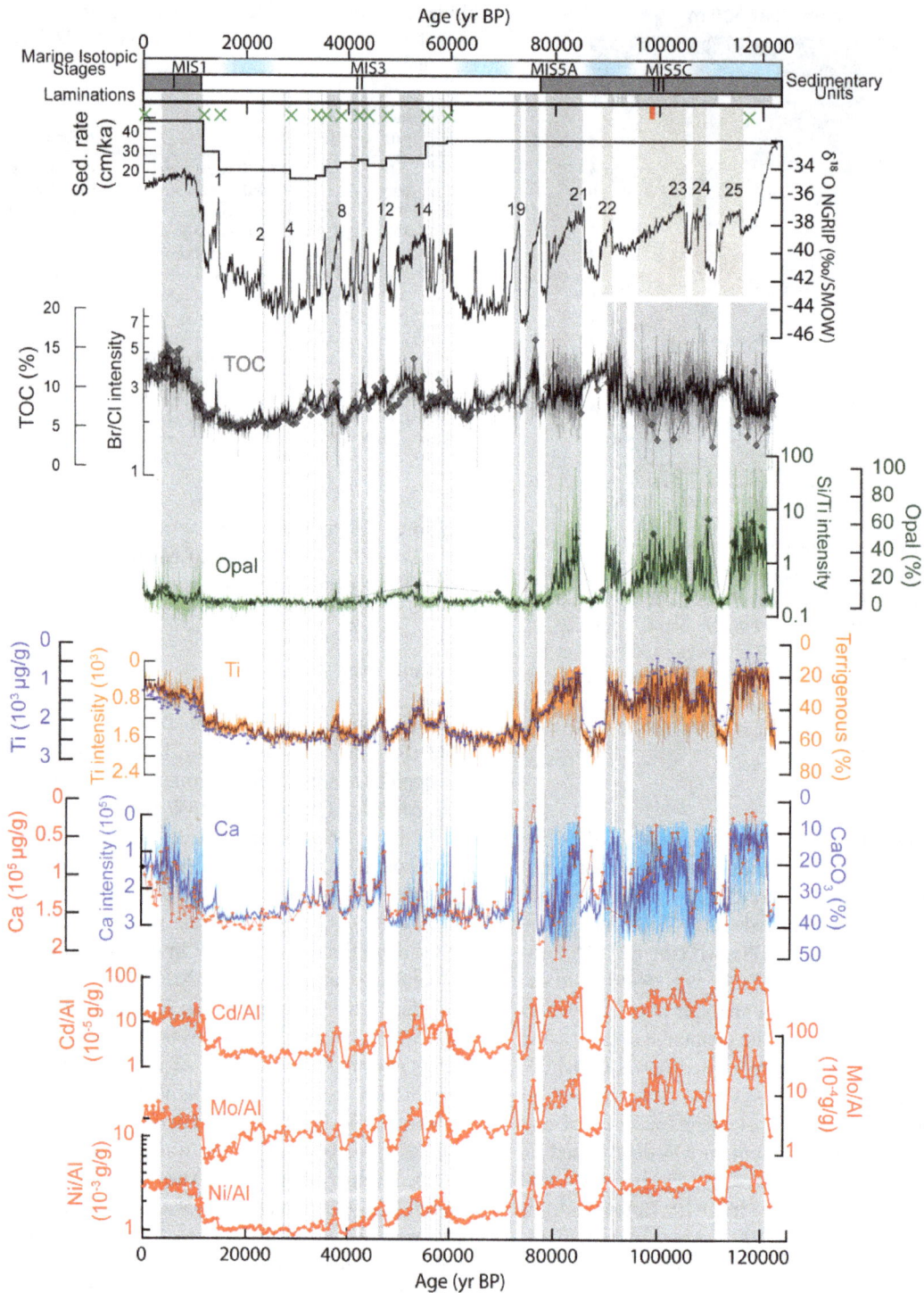

**Fig. 2.** The sedimentation rate in MD08 core; the isotopic composition of NGRIP ice core ($\delta^{18}$O‰/SMOW) (NGRIP Members, 2004); the Br/Cl based marine organic matter high resolution record for MD02-2508 core calibrated using TOC measurements on the same core (the TOC data were first reported by Blanchet et al. (2007), and 40 additional TOC measurements were added from 120 to 60 ka). Si/Ti record calibrated using opal measurements (green diamonds). Ca measurements (XRF scan and ICP-MS), converted into carbonate content considering that all the Ca belongs to the carbonate phase. Ti XRF and ICP-MS records converted into terrigenous matter content calculated using mean Ti content in the upper continental crust (McLennan, 2001). Note that Ca and Ti scales are reversed. Cd/Al, Mo/Al, and Ni/Al measured using ICP-MS on log scale. Green crosses at the top of the figure indicate tie points used to build the age model. Marine isotopic stages and sedimentary units as described in part 3 are indicated. Grey vertical bars highlight the laminated layers. Red tick at the top of the figure indicates the position of the sample displayed in Fig. 3.

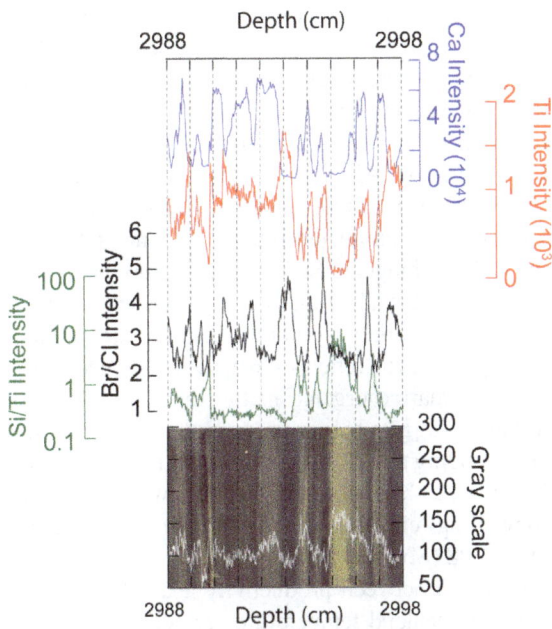

**Fig. 3.** The XRF intensities for Ca, Ti, Br/Cl, and Si/Ti (spatial resolution of 200 $\mu$m) of a 10 cm long sample of MD02-2508, representing roughly 300 yr of sediment deposition, and the corresponding photography of the core with gray scale analyses. Contrast and lightness of the image were slightly enhanced, and gray scale measurements were performed using ImageJ software. The position of the sample within the record is indicated with a red tick at the top of Fig. 2. Vertical dashed lines show 1 cm intervals.

winds are enhanced (Fig. 1, Thomas et al., 2001). Despite the influence of local bottom topography and angular orientation of the coastline, upwelling is stronger during early summer at MD08 core (Zaytsev et al., 2003; Cartapanis et al., 2011). Upwelled waters originate from up to 300 m depth along the California coast (van Geen and Husby, 1996), at the mixing zone between SW and NPIW. Thus, NPIW nutrient inventory related to tidal mixing at high northern latitude in the Pacific (Sarmiento et al., 2004), might influence surface productivity. High local productivity is thus associated with warm condition on seasonal timescales when the Intertropical Convergence Zone (ITCZ) is situated on its northern position.

## 4 Results

Highly resolved XRF Si/Ti (log-scale) is well correlated ($R^2 = 0.93$) to opal content obtained by chemical leaching of bulk sediments (Fig. 2). Br is contained in marine organic matter, but also in pore water (Ziegler et al., 2008). Since pore water content in opal-rich sediment is strongly affected by opal concentration, we used Br/Cl ratio as indicator of marine organic matter to correct pore water contribution using Cl (Croudace et al., 2006). Log Br/Cl is well

**Table 2.** Minimal, maximal, and mean values for Al content ($\mu$g g$^{-1}$), Cd/Al, Mo/Al, and Ni/Al in MD08 core. Mean values in Upper Continental Crust (McLennan, 2001) and mean enrichment factors (((Mean Element/Al)$_{MD08}$/(Element/Al)$_{UPC}$) are displayed at the bottom of the table.

|                      | Al ($\mu$g g$^{-1}$) | Cd/Al (g g$^{-1}$) | Mo/Al (g g$^{-1}$) | Ni/Al (g g$^{-1}$) |
|----------------------|---------|-----------------------|-----------------------|-----------------------|
| Min                  | 3200    | $2.77 \times 10^{-5}$ | $1.32 \times 10^{-4}$ | $1.38 \times 10^{-3}$ |
| Max                  | 51813   | $4.21 \times 10^{-3}$ | $2.00 \times 10^{-2}$ | $8.49 \times 10^{-3}$ |
| Mean                 | 32343   | $3.97 \times 10^{-4}$ | $1.30 \times 10^{-3}$ | $3.41 \times 10^{-3}$ |
| UPC                  | 80400   | $1.22 \times 10^{-6}$ | $1.87 \times 10^{-5}$ | $5.47 \times 10^{-4}$ |
| Enrichment Factor    |         | 326                   | 70                    | 6                     |

correlated ($R^2 = 0.68$) to TOC content (Fig. 2). The MD08 sediment column is divided in three different units (Fig. 2). The first unit (Unit I) that corresponds to the Holocene consists of laminated high TOC ($\approx 13\%$), relatively low carbonates ($\approx 20\%$) and terrigenous compounds ($\approx 45\%$) sapropelic mud (Fig. 2). The second unit (Unit II) was deposited during MIS2, 3 and 4, and is composed of light homogenous calcareous clay (carbonates $\approx 35\%$, terrigenous $\approx 60\%$, and TOC $\approx 7\%$). Interbeds of dark laminated sapropelic mud similar to Unit I (Fig. 2) were deposited during interstadial events (Cartapanis et al., 2011). Opal content within Units I and II is low (0 to 10 %) but is higher within laminated intervals (up to 20 %, Fig. 2). The third unit (Unit III) corresponds to MIS5 and mainly consists of laminated sapropelic diatom ooze (Fig. 3) with some homogenous layers composed of calcareous clays, similar to Unit II non-laminated sediment (Fig. 2). The millimetric to centimetric laminations in the Unit III are composed of light colour biogenic remains, and dark coloured organic matter mixed with terrigenous sediment (Fig. 3). Opal content rises up to 40 % in MIS5 laminations and reaches over 80 % in particular laminations (Fig. 2). Micro sedimentologic analyses suggest that high opal content laminae in Unit III were deposited abruptly, and could correspond to algal blooms during upwelling events or in the frontal zone between water masses of different densities (Murdmaa et al., 2010).

Trace element (Cd, Mo, Ni) to Al ratios show extremely high values relative to the so-called UPC values (McLennan, 2001, Table 2): The mean Ni/Al ratio is six times higher than in UPC, seventy times higher for Mo/Al, and more than three hundreds times higher for Cd/Al (Table 2). Even the lowest element to Al ratios, which occur mainly in Unit II, are higher than the corresponding value in UPC. The results indicate that significant parts of these elements are not related to the input of terrigenous minerals, but rather to in-situ biogenic and authigenic enrichments. The highest values for elemental ratios and enrichments in MD08 core occur within laminated intervals, and are even higher within sedimentary Unit III as compared to Unit I (Fig. 2).

## 5 Discussion

A previous study of the MD08 core revealed the strong link between productivity in surface waters and OMZ variability during the last glacial period (MIS2 and 3), in relation to changes in atmospheric circulation over Baja California driven by Northern Hemisphere high latitude climate variability (Cartapanis et al., 2011). High productivity may have led to reduced bottom water oxygenation at site MD08 during interstadials. Changes in the wind patterns were proposed as mechanisms for changes in the strength, duration, and frequency of upwelling events, leading to higher productivity during interstadials events of MIS3. Higher bottom water oxygenation during Heinrich events was suggested to be produced by changes in oxygen advection by NPIW, formed in the northern Pacific. In light of the new datasets on MD08 core, that now cover the past 120 ka, we will focus our discussion on the oceanic and climatic millennial scale variability of the last interglacial period (MIS5).

### 5.1 Millennial-scale productivity variations during MIS5

The high-resolution records of opal, terrigenous, and carbonate fractions in the MD08 core display striking similarities with the NGRIP $\delta^{18}$O record on millennial timescales across the last interglacial period, despite limited age constraint earlier than 60 ka (Fig. 2). The MD08 record shows higher opal content within laminated intervals that likely correspond to DO interstades 25 to 19 (Figs. 2 and 4). The opal content is extremely high during interstadials 25 to 21 (more than 40 %), while this enrichment is attenuated for interstades 20 and 19 (Fig. 4). In contrast, TOC remains rather stable during interstadials 25 to 21 (around 8 %), and increases up to 12 % for interstades 20 and 19 (Fig. 2). Carbonate and terrigenous fractions show low values during Greenland interstades 25 to 19, most likely due to dilution by opal and TOC as well as carbonate dissolution (Figs. 2 and 3).

We roughly estimated opal mass accumulation rate using gamma density obtained on board (Beaufort et al., 2002) and a constant sedimentation rate from 59.1 to 117 ka (Table 1) to examine whether temporal variation of opal accumulation rate is different from that of opal concentration. The maximum range of wet bulk density is 1.1 to 1.8 g cm$^{-3}$ for core MD08, which is much smaller than the change of opal content from several % to nearly 50 % during MIS5. Consequently, the opal concentration and opal accumulation rates have very similar variability during MIS5.

The increase of opal concentration in bulk sediments could be produced by decline of carbonates and terrigenous fractions. If the dilution effect were the main cause of opal concentration changes, the sedimentation rate of high opal periods would be lower than for periods of low opal content. Alternatively, since we assume a constant sedimentation rate from 117 to 59.1 ka (Table 1), the periods of high opal con-

centration should be systematically shorter than interstadials of NGRIP record. Nonetheless, such a trend is not found in the records (Fig. 4). Consequently, opal variations reflect at least partly changes in the opal flux to the sea floor and/or the preservation of this component in sediments.

Both Cd/Al and Ni/Al, closely related to biogenic productivity of surface waters (Nameroff et al., 2002), show increased values within laminated intervals. Elemental ratios (Cd/Al, Ni/Al and biogenic Si to organic carbon, Si$_{bio}$/C, Sect. 5.2, and Figs. 2 and 5) that are not affected by dilution, display similar variability to opal content, further suggesting that opal content reflects changes in the productivity. In summary, marine organic matter, opal, Cd/Al, and Ni/Al variations suggest higher productivity during the intervals corresponding to Greenland warm DO interstadials of MIS5 (Fig. 4). This observation is consistent with the occurrence of upwelling-related diatom blooms in the laminated intervals of MD08 (Murdmaa et al., 2010).

The good match between productivity and oxygenations at MD08 with Greenland temperature records (Figs. 2 and 3) suggest strong connections between Northern Hemisphere high latitude climate and northeastern Pacific OMZ on the millennial-scale during MIS5 via atmospheric and/or oceanic pathways. In the following section, we discuss possible processes responsible for high opal content during MIS5 interstadials compared to Holocene and MIS3 interstadials.

### 5.2 Atmospheric circulation impact on productivity

Despite different climatic settings, high latitude cold Northern Hemisphere intervals (MIS2 and MIS4 as well as stadials of MIS3 and MIS5) were generally less productive than warmer interglacial and DO interstadial periods at site MD08. Considering that modern productivity over site MD08 is driven by wind, changes in local wind fields may have influenced productivity during MIS5 (Fig. 4). During cold events simulated by fresh water perturbation under pre-industrial conditions, low temperatures in the northern Pacific enhance the Aleutian Low (Mikolajewicz et al., 1997; Okumura et al., 2009). Cold temperatures over the North American continent favour the establishment of anticyclonic circulation (Kutzbach and Wright Jr., 1985; Romanova et al., 2006). These mechanisms could have reduced the southward wind intensity over the studied site during MIS5 stadials. Substantial millennial-scale changes in the volume, surface extent, and height of the North American ice sheet suggested by high amplitude sea level changes (Fig. 5) (Lambeck and Chappell, 2001; Siddall et al., 2007; Grant et al., 2012; Medina-Elizalde, 2013), could have been responsible for variations in atmospheric circulation (Romanova et al., 2006).

Continental records indicate millennial-scale variability in atmospheric circulation over the North American continent during MIS5. A multiproxy study in Lake Owens (southeastern California) suggests reduced rainfall and/or warmer

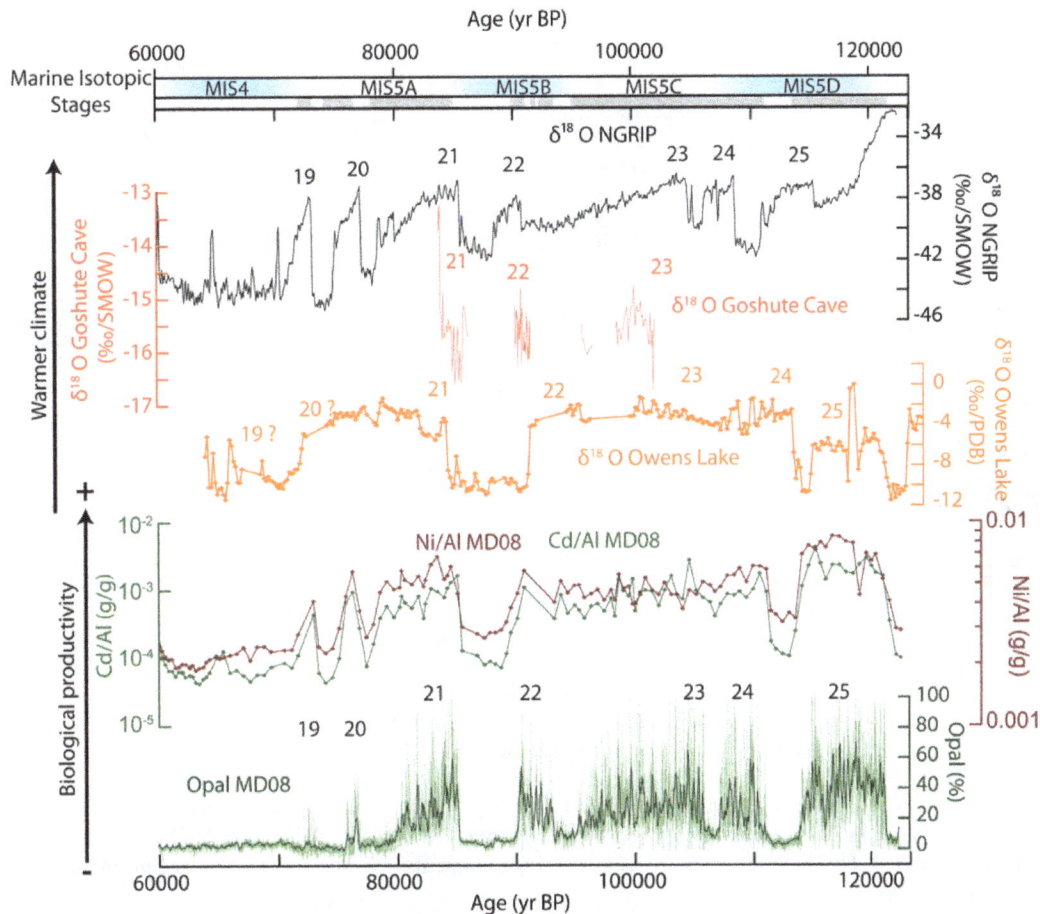

**Fig. 4.** The isotopic composition of NGRIP ice core ($\delta^{18}$O‰/SMOW) (NGRIP Members, 2004) ; the isotopic composition of Owens Lake sediment ($\delta^{18}$O‰/PDB) (Li et al., 2004), the isotopic composition of Goshute cave stalagmite ($\delta^{18}$O‰/PDB) (Denniston et al., 2007); and the opal content, Cd / Al, and Ni / Al in MD02-2508 core.

conditions during interstadial events of MIS5 (e.g. $\delta^{18}$O of Owens Lake sediment, Figs. 1 and 4, (Li et al., 2004)), consistent with variations of isotopic composition of Goshute cave stalagmite (NE Nevada, (Denniston et al., 2007), Figs. 1 and 4). Previous studies in the southwestern United States over the last glacial interpreted millennial-scale speleothem $\delta^{18}$O variations to reflect latitudinal migrations of storm tracks and rainfall associated to the Polar Jet Stream and the ITCZ, implying atmospheric circulation changes, at least, in the Northern Hemisphere (Asmerom et al., 2010; Wagner et al., 2010). During warm interstadials, both polar jet and Northern Hemisphere summer ITCZ were proposed to have shifted northward (Asmerom et al., 2010; Wagner et al., 2010; Wang et al., 2012). By affecting upwelling intensity, duration or frequency, this atmospheric circulation change may have triggered productivity variability at MD08 (Cartapanis et al., 2011), as well as in the Panama Basin (Romero et al., 2011a; Cartapanis et al., 2012). Based on the fact that millennial-scale ITCZ latitudinal shifts existed during MIS5a

(Peterson et al., 2000), we infer the similar processes could have operated over the whole MIS5.

## 5.3 Oceanic circulation impact on oxygenation and productivity

High Mo / Al during MIS5 interstadials suggest reduced oxygenation of bottom water (Figs. 2 and 5). The high correlation between Mn and terrigenous element content ($R^2 = 0.97$ between Mn and Ti) indicates that Mn oxides were dissolved in the sediment, and that bottom waters had never been oxic, even during stadial events (not shown). Changes in oceanic circulation could have impacted the OMZ dynamic by changing ventilation. Modelling studies indicate that reduction of North Atlantic Deep Water formation during Heinrich events and associated global oceanic circulation modifications could have increased the NPIW formation (Saenko et al., 2004; Timmermann et al., 2005; Schmittner et al., 2007; Okazaki et al., 2010), which was confirmed by observational studies for the last glacial period (Okazaki et al., 2010; Cartapanis et al., 2011; Rella et al., 2012). Oceanic circulation

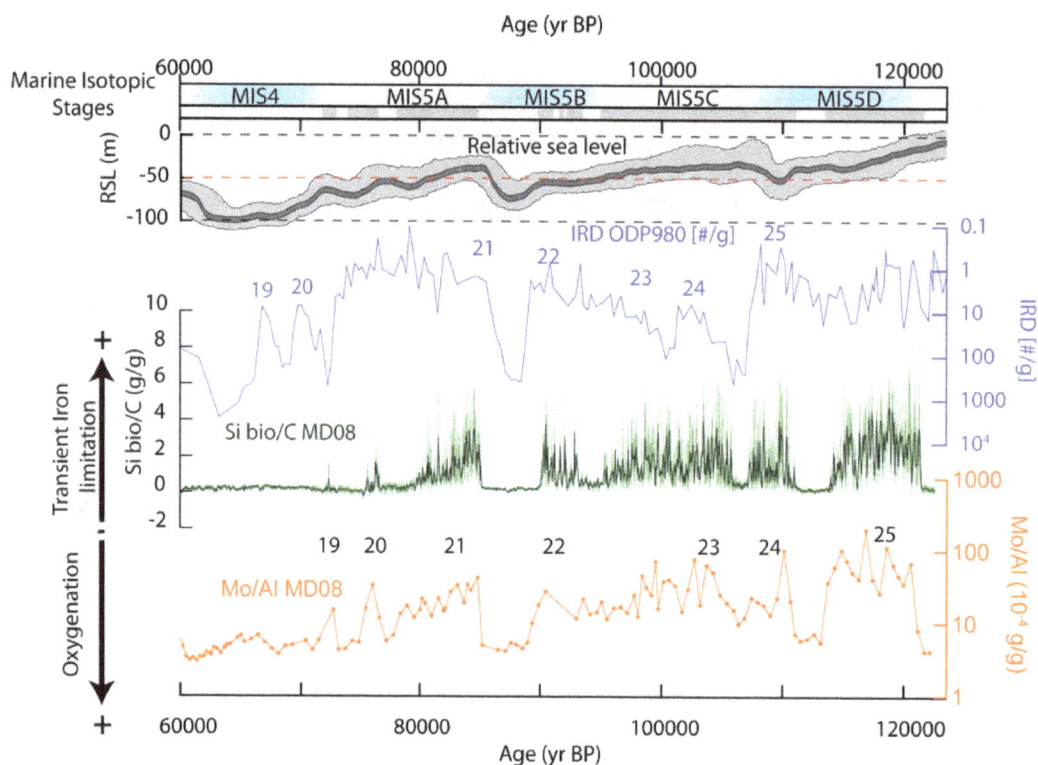

**Fig. 5.** The global relative sea level (Grant et al., 2012). Confidence interval of 95 % (light grey) and probability maximum (dark grey) are indicated. The 50 m depth below modern sea level, corresponding to modern Bering Strait depth is highlighted in red. The Ice Rafted Debris content in North Atlantic ODP site 980 core plotted on an inverse logarithmic scale (Oppo et al., 2006), and the biogenic Si to C ratio ($Si_{bio}$ / C) and Mo / Al in MD02-2508 core.

variations could have modified the nutrient content of the upwelled water over site MD08, affecting biological productivity and oxygen consumption in the OMZ (Schmittner, 2005; Schmittner et al., 2007). Considering that Heinrich-like events probably occurred during the MIS5 (Fig. 5) (McManus et al., 1994; Eynaud et al., 2000; Oppo et al., 2006), changes in the ventilation and/or the nutrient inventory at Baja California cannot be ruled out. Numerical simulations evaluated the relative impact of productivity and ventilation on oxygen level in the Pacific during a simulated HE. It was shown that oxygenation changes at high latitude in the northern Pacific during HE were mainly driven by the advection of oxygen through NPIW (Schmittner et al., 2007). In contrast, at low latitudes in the Pacific, the change in biological productivity, related to the reduction of the upwelling of deep nutrient rich water was the main factor for the increased oxygenation of the intermediate water (Schmittner et al., 2007). MD08 site is situated at 23° N, close to the modern NPIW southern limit, such as both oxygen advection and consumption could play a role in regulating the local intermediate depth oxygenation.

Silicic acid and nitrate inventory in subsurface waters of the northern Pacific is partly related to the upward advection of nutrient-rich Pacific Deep Water toward NPIW at high northern latitudes (Sarmiento et al., 2004). The enhanced NPIW formation during HE and stadials related to oceanic circulation changes (Saenko et al., 2004; Timmermann et al., 2005; Schmittner et al., 2007; Okazaki et al., 2010), possibly reduced the upward advection of deep, old, oxygen poor and nutrient rich water toward NPIW (Sarmiento et al., 2004; Rella et al., 2012). Considering that NPIW is probably the source for upwelling water in the northeastern Pacific (van Geen and Husby, 1996), stronger NPIW production could have lowered the productivity at site MD08, reduced oxygen consumption in NPIW, and enhanced oxygen advection toward the OMZ. This scenario would explain the correlation between productivity and bottom water oxygenation proxies (Fig. 5) along the NE Pacific margin as both oxygen advection and productivity (thus oxygen consumption) would have been related to NPIW formation. Alternatively, atmospheric connections (Harada et al., 2006; Clement and Peterson, 2008) altered surface hydrological conditions in NPIW formation zones (such as Okhotsk and Bering Seas), which affected the NPIW formation and/or oxygenation (Sarmiento et al., 2004; Schmittner et al., 2007; Rella et al., 2012), and thus the ventilation and/or the nutrient availability at MD08 core site during stadials/interstadials and HE.

A fundamental difference in oceanic circulation conditions between the MIS5 and the last glacial period is related to sea level, which affected the communication between the Arctic Ocean and the northern Pacific through the Bering Strait (De Boer and Nof, 2004). Modelling studies indicate that the impact of Heinrich-like events on oceanic circulation over both northern Atlantic and Pacific oceans might have been reduced during interglacial periods (De Boer and Nof, 2004; Hu et al., 2012). During sea level high stands, any increase of the freshwater inputs in the northern Atlantic was partially diverted toward the northern Pacific through the still open Bering Strait, thus weakening the impact of HE on the North Atlantic Deep Water (De Boer and Nof, 2004; Hu et al., 2012). At the same time, freshwater influx into the northern Pacific probably reduced NPIW formation. Moreover, global reorganization of oceanic circulation, that favoured NPIW formation during glacial age HEs (Saenko et al., 2004; Timmermann et al., 2005), would have been weakened during MIS5. Consequently, Heinrich-like events during MIS5 should have had reduced impact on the northeastern Pacific OMZ as compared to glacial periods. In spite of this, no obvious differences in millennial-scale response of redox sensitive trace elements for glacial and interglacial periods in our records are seen (Fig. 2). This suggests that productivity is the main factor controlling millennial-scale variations of the bottom water oxygenation at core MD08 site throughout the MIS5. Productivity changes could have been related to either atmospheric or oceanic processes. This scenario, however, depends on the sea level changes during MIS5 stadial/Heinrich-like events. The present depth of the Bering Strait is around 50 m, close to the reconstructed (Fig. 5) and simulated sea level during MIS5 (Lambeck et al., 2002; Waelbroeck et al., 2002; Cutler et al., 2003; Hu et al., 2010; Grant et al., 2012). This suggests that Arctic/Pacific water exchanges were probably limited like the present state. In order to address the Bering Strait influence on the past North Pacific climate and oceanic circulation, and furthermore, the global climate and oceanic circulation variations on both glacial/interglacial and millennial timescales (Okumura et al., 2009; Hu et al., 2012), it will be necessary to develop a well dated record of the opening and closure of the Bering Strait.

## 5.4   High biogenic opal content in MIS5 sediments

One of the most striking features of the 120 ka MD08 record is the extraordinary high opal content during the MIS5 (Fig. 2). The sediment deposited during MIS5 interstadial events has been identified as diatom-rich laminated sediment (Murdmaa et al., 2010). The formation of this type of sediment structure may occur during major diatom blooms within active frontal zones, which favor the aggregation of diatoms and the rapid sinking of frustules downward (Kemp et al., 1996; Shimada et al., 2008; Romero et al., 2011b). Changes in the seasonal and latitudinal insolation pattern

during MIS5 as compared to the Holocene could be responsible for changes in the atmospheric circulation (Montero-Serrano et al., 2011; Herold et al., 2012), wind driven upwelling, and productivity. However, high opal production extends at least over 40 ka (120–80 ka), during periods of different changes in the intensity and the latitudinal distribution of insolation. This suggests that insolation variation was not likely responsible for the buildup of mat deposits in Baja California.

Since site MD08 is located south of the modern limit of Subarctic Water transported by the California Current, the southward extension of the latter toward site MD08 (ca. 23° N) might have contributed to the build-up of diatom-rich laminations during MIS5. The nearby site NH15P (ca. 22° N, 106° W, 420 m water depth, Fig. 1 Ganeshram and Pedersen, 1998) shows similar variability in opal content with increased values during interglacial periods. It is worth of noting that there is no obvious difference between Holocene and MIS5 opal content at site NH15P, which is consistent with the hypothesis that the southern limit of California Current was situated above MD08 during MIS5 whereas site NH15P was beyond of the influence of California Current for both MIS5 and Holocene.

Despite the fact that nitrate availability was thought to be the limiting factor for diatom production in surface waters overlying MD08 (Moore et al., 2004), shipboard seawater incubation experiments demonstrated that diatom production within upwelling zones along California margins was rather limited by iron (Hutchins and Bruland, 1998; Firme et al., 2003; Moore et al., 2013). Intense productivity could have led to total Fe utilisation during upwelling events thus to transient Fe limitation at the end of an upwelling event (Hutchins and Bruland, 1998; Firme et al., 2003). Iron limitation can lead to reduced cellular organic matter production and increased silicification (high cellular Si : C and Si / N ratio) of diatoms (Baines et al., 2010, 2011; Brzezinski et al., 2011), high Si:C ratio in sinking biogenic particles, high opal flux (Hutchins and Bruland, 1998; Firme et al., 2003; Pichevin et al., 2012), and better frustule preservation in sediments. This scenario is consistent with the unvarying TOC contents in MD08 core across stadial/interstadials DO intervals of MIS5 (Fig. 2), and high biogenic Si to C ratio ($Si_{bio}$ / C, Fig. 5). Moreover, terrigenous element content during MIS5 is low in MD08 core, possibly because of reduced aeolian and fluvial inputs, suggesting decreased iron inputs to the surface waters. Dissolved Fe in upwelling areas also depends on interactions between upwelled waters and underlying shelf sediments (Elrod et al., 2004; Bruland et al., 2005), such as reduced Fe content in the sediment could have further enhanced Fe depletion in upwelled waters.

As suggested by (Romero et al., 2011b), change in nutrient availability and supply in the photic zone might have affected the buildup of diatoms mats deposits. In this scenario, changes in the intensity of processes described in Sect. 5.2 (wind driven upwelling) and Sect. 5.3 (oceanic circulation impact on nutrient availability) between Holocene and MIS5 could be responsible for higher diatom productivity during MIS5.

Regardless of the mechanism responsible (California Current position, nutrient inventory at subsurface or nutrient supply to the photic zone), we suggest that the climatic/oceanic setting might have led to high opal production during MIS5. Global ice volume (Lambeck et al., 2002; Waelbroeck et al., 2002; Cutler et al., 2003; Hu et al., 2010; Grant et al., 2012) suggests the existence of a small ice sheet over the North American continent during MIS5d to MIS5a (Bonelli et al., 2009; Ganopolski et al., 2010). Leaving aside millennial-scale variability, the presence of this ice sheet (absent during the Holocene and MIS5e) could have been responsible for high MIS5 opal content in MD08 as compared to Holocene, by either modifying local wind stress, California Current extension, or nutrient content of upwelling waters. In order to determine whether these specific features (higher opal during MIS5 than during Holocene) reflect regional or global processes rather than a local non-linear response to climatic variations, more records in upwelling zones of northeastern Pacific for the MIS5 are required.

## 6 Conclusions

We studied core MD02-2508, retrieved off the Baja California margin, at present northern limit of the northeastern Pacific Oxygen Minimum Zone (OMZ), using high resolution XRF measurements (Si/Ti as a biogenic opal indicator, Ca as a biogenic carbonate indicator, Br as a marine organic carbon indicator, and Ti as a terrigenous fraction indicator) and ICP-MS trace elements measurements (Mo, Ni, Cd), for the last interglacial period (MIS5). The data allowed us to better constrain the millennial scale variability of past biological productivity and oxygenation at the core site over the studied period.

Opal, Ni/Al, and Cd/Al increased during the DO interstadial events identified in Greenland ice cores over the MIS5, strongly suggesting an increase in biological productivity. A high Mo/Al ratio within interstadials events indicates that the OMZ strengthened during these intervals.

Despite differences in the climatic and oceanic conditions, productivity and oxygenation during MIS5 displayed millennial-scale variability in relation to high northern latitude climates that are comparable to those over the last glacial period. Enhanced communication between the Pacific and Arctic Oceans during MIS5 due to higher sea level does not seem to have strongly modified the productivity and bottom water ventilation response. This suggests that atmospheric teleconnection was the major link between the mid-latitude productivity and ventilation variations, and the high latitude Northern Hemisphere climate. Changes in atmospheric circulation, such as latitudinal shift of the polar jet stream, the storm track, and the ITCZ location during DO-

H events of the past 120 ka probably affected wind driven upwelling and hence biological productivity. Alternatively, changes in the oceanic circulation could have modified productivity via changes in the deep upwelling of nutrient rich water, if this process was not affected by changes in Arctic-Pacific communication through the Bering strait.

Oxygenation variations at the core site were more likely related to changes in the biological productivity, but also possibly to changes of North Pacific Intermediate Water production because of atmospheric changes at Northern Hemisphere high latitudes in the Pacific, that affected oxygen inputs and nutrient availability in subsurface waters.

While several mechanisms (changes in California Current southward extension, modification of nutrient availability in surface water through changes of atmospheric and oceanic circulation) could explain the high opal content and Si/C during MIS5, we suggest that the presence of an ice sheet over the northern part of northern America was probably indirectly involved in the build up of diatom ooze deposits at the Baja California.

*Acknowledgements.* We thank Luc Beaufort (chief scientist), Yvon Balut (IPEV), and the team of the *IMAGESVIII-MONA* cruise (MD126 North American Margin) on board of *Marion Dufresne*. We also thank Marta Garcia for technical assistance with ICP-MS measurements, Frauke Rostek for additional TOC measurements, Genevieve Elsworth for English polishing, Eric Galbraith for discussion and support, and A. Schmittner and L. Pichevin for constructive comments during the review of the manuscript. We are grateful to MESR, the Collège de France and McGill University for providing salary support to O. Cartapanis. Paleoclimate work at CEREGE is supported by grants from the Collège de France, the Comer Science and Education Foundation, the CNRS (LEFE-EVE MISLOLA), and the European Community (Project Past4Future).

Edited by: E. Brook

## References

Altabet, M. A., Higginson, M. J., and Murray, D. W.: The effect of millennial-scale changes in Arabian Sea denitrification on atmospheric $CO_2$, Nature, 415, 159–162, 2002.

Arellano-Torres, E., Pichevin, L. E., and Ganeshram, R. S.: High-resolution opal records from the eastern tropical Pacific provide evidence for silicic acid leakage from HNLC regions during glacial periods, Quaternary Sci. Rev., 30, 1112–1121, 2011.

Asmerom, Y., Polyak, V. J., and Burns, S. J.: Variable winter moisture in the southwestern United States linked to rapid glacial climate shifts, Nat. Geosci., 3, 114–117, 2010.

Auad, G., Roemmich, D., and Gilson, J.: The California Current System in relation to the Northeast Pacific Ocean circulation, Prog. Oceanogr., 91, 576–592, 2011.

Baines, S. B., Twining, B. S., Brzezinski, M. A., Nelson, D. M., and Fisher, N. S.: Causes and biogeochemical implications of regional differences in silicification of marine diatoms, Global Biogeochem. Cy., 24, GB4031, doi:10.1029/2010GB003856, 2010.

Baines, S. B., Twining, B. S., Vogt, S., Balch, W. M., Fisher, N. S., and Nelson, D. M.: Elemental composition of equatorial Pacific diatoms exposed to additions of silicic acid and iron, Deep Sea Res. II, 58, 512–523, doi:10.1016/j.dsr2.2010.08.003, 2011.

Beaufort, L. et al. (Members of the scientific party): MD126-IMAGES VIII Marges Ouest Nord Américaines MONA Cruise Report, Institut Paul Emile Victor, Plouzané, France, 2002.

Blanchet, C. L., Thouveny, N., Vidal, L., Leduc, G., Tachikawa, K., Bard, E., and Beaufort, L.: Terrigenous input response to glacial/interglacial climatic variations over southern Baja California: a rock magnetic approach, Quaternary Sci. Rev., 26, 3118–3133, 2007.

Bonelli, S., Charbit, S., Kageyama, M., Woillez, M.-N., Ramstein, G., Dumas, C., and Quiquet, A.: Investigating the evolution of major Northern Hemisphere ice sheets during the last glacial-interglacial cycle, Clim. Past, 5, 329–345, doi:10.5194/cp-5-329-2009, 2009.

Bostock, H. C., Opdyke, B. N., and Williams, M. J. M.: Characterising the intermediate depth waters of the Pacific Ocean using [delta]13C and other geochemical tracers, Deep-Sea Res. I, 57, 847–859, doi:10.1016/j.dsr.2010.04.005, 2010.

Bruland, K. W., Rue, E. L., Smith, G. J., and DiTullio, G. R.: Iron, macronutrients and diatom blooms in the Peru upwelling regime: brown and blue waters of Peru, Mar. Chem., 93, 81–103, doi:10.1016/j.marchem.2004.06.011, 2005.

Brzezinski, M. A., Baines, S. B., Balch, W. M., Beucher, C. P., Chai, F., Dugdale, R. C., Krause, J. W., Landry, M. R., Marchi, A., Measures, C. I., Nelson, D. M., Parker, A. E., Poulton, A. J., Selph, K. E., Strutton, P. G., Taylor, A. G., and Twining, B. S.: Co-limitation of diatoms by iron and silicic acid in the equatorial Pacific, Deep Sea Res. II, 58, 493–511, doi:10.1016/j.dsr2.2010.08.005, 2011.

Cartapanis, O., Tachikawa, K., and Bard, E.: Northeastern Pacific oxygen minimum zone variability over the past 70 kyr: Impact of biological production and oceanic ventilation, Paleoceanography, 26, PA4208, doi:10.1029/2011pa002126, 2011.

Cartapanis, O., Tachikawa, K., and Bard, E.: Latitudinal variations in intermediate depth ventilation and biological production over northeastern Pacific Oxygen Minimum Zones during the last 60ka, Quaternary Sci. Rev., 53, 24–38, 2012.

Clement, A. C. and Peterson, L. C.: Mechanisms of abrupt climate change of the last glacial period, Rev. Geophys., 46, 39, Rg4002, doi:10.1029/2006rg000204, 2008.

Croudace, I. W., Rindby, A., and Rothwell, R. G.: ITRAX: description and evaluation of a new multi-function X-ray core scanner, New Techniques in Sediment Core Analysis, Geological Society, London, Special Publications, 267, 51–63, 2006.

Cutler, K. B., Edwards, R. L., Taylor, F. W., Cheng, H., Adkins, J., Gallup, C. D., Cutler, P. M., Burr, G. S., and Bloom, A. L.: Rapid sea-level fall and deep-ocean temperature change since the last interglacial period, Earth. Planet. Sci. Lett., 206, 253–271, doi:10.1016/S0012-821X(02)01107-X, 2003.

De Boer, A. M. and Nof, D.: The Bering Strait's grip on the northern hemisphere climate, Deep-Sea Res. I, 51, 1347–1366, 2004.

Dean, W. E.: Sediment geochemical records of productivity and oxygen depletion along the margin of western North America during the past 60,000 years: teleconnections with Greenland Ice and the Cariaco Basin, Quaternary Sci. Rev., 26, 98–114, doi:10.1016/j.quascirev.2006.08.006, 2007.

Dean, W. E., Zheng, Y., Ortiz, J. D., and van Geen, A.: Sediment Cd and Mo accumulation in the oxygen-minimum zone off western Baja California linked to global climate over the past 52 kyr, Paleoceanography, 21, PA4209, doi:10.1029/2005pa001239, 2006.

DeMaster, D. J.: The supply and accumulation of silica in the marine environment, Geochim. Cosmochim. Acta, 45, 1715–1732, doi:10.1016/0016-7037(81)90006-5, 1981.

Denniston, R. F., Asmerom, Y., Polyak, V., Dorale, J. A., Carpenter, S. J., Trodick, C., Hoye, B., and Gonzalez, L. A.: Synchronous millennial-scale climatic changes in the Great Basin and the North Atlantic during the last interglacial, Geology, 35, 619–622, doi:10.1130/g23445a.1, 2007.

Dubois, N., Kienast, M., Kienast, S., Calvert, S. E., Francois, R., and Anderson, R. F.: Sedimentary opal records in the eastern equatorial Pacific: It is not all about leakage, Global Biogeochem. Cy., 24, GB4020, doi:10.1029/2010gb003821, 2010.

Dubois, N., Kienast, M., Kienast, M., Normandeau, C., Calvert, S. E., Herbert, T. D., and Mix, A. C.: Millennial-scale variations in hydrography and biogeochemistry in the Eastern Equatorial Pacific over the last 100 kyr, Quaternary Sci. Rev., 30, 210–223, 2011.

Elrod, V. A., Berelson, W. M., Coale, K. H., and Johnson, K. S.: The flux of iron from continental shelf sediments: A missing source for global budgets, Geophys. Res. Lett., 31, L12307, doi:10.1029/2004GL020216, 2004.

Eynaud, F., Turon, J. L., Sanchez-Goni, M. F., and Gendreau, S.: Dinoflagellate cyst evidence of Heinrich-like events off Portugal during the Marine Isotopic Stage 5, Mar. Micropaleontol., 40, 9–21, 2000.

Firme, G. F., Rue, E. L., Weeks, D. A., Bruland, K. W., and Hutchins, D. A.: Spatial and temporal variability in phytoplankton iron limitation along the California coast and consequences for Si, N, and C biogeochemistry, Global Biogeochem. Cy., 17, 1016, doi:10.1029/2001GB001824, 2003.

Galbraith, E. D., Kienast, M., Pedersen, T. F., and Calvert, S. E.: Glacial-interglacial modulation of the marine nitrogen cycle by high-latitude O-2 supply to the global thermocline, Paleoceanography, 19, PA4007, doi:10.1029/2003pa001000, 2004.

Ganeshram, R. S. and Pedersen, T. F.: Glacial-interglacial variability in upwelling and bioproductivity off NW Mexico: Implications for quaternary paleoclimate, Paleoceanography, 13, 634–645, 1998.

Ganopolski, A., Calov, R., and Claussen, M.: Simulation of the last glacial cycle with a coupled climate ice-sheet model of intermediate complexity, Clim. Past, 6, 229–244, doi:10.5194/cp-6-229-2010, 2010.

Garcia, H. E., Locarnini, R. A., Boyer, T. P., Antonov, J. I., Baranova, O. K., Zweng, M. M., and Johnson, D. R.: World Ocean Atlas 2009, Volume 3: Dissolved Oxygen, Apparent Oxygen Utilization, and Oxygen Saturation. S. Levitus, (Ed.) NOAA Atlas NESDIS 70, U.S. Government Printing Office, Washington, D.C., 344 pp., 2010.

Grant, K. M., Rohling, E. J., Bar-Matthews, M., Ayalon, A., Medina-Elizalde, M., Ramsey, C. B., Satow, C., and Roberts, A. P.: Rapid coupling between ice volume and polar temperature over the past 150,000 years, Nature, 491, 744–747, doi:10.1038/nature11593, 2012.

Harada, N., Ahagon, N., Sakamoto, T., Uchida, M., Ikehara, M., and Shibata, Y.: Rapid fluctuation of alkenone temperature in the southwestern Okhotsk Sea during the past 120 ky, Global Planet. Change, 53, 29–46, 2006.

Hendy, I. L.: The paleoclimatic response of the Southern Californian Margin to the rapid climate change of the last 60 ka: A regional overview, Quaternerary Int., 215, 62–73, doi:10.1016/j.quaint.2009.06.009, 2010.

Hendy, I. L. and Kennett, J. P.: Dansgaard-Oeschger cycles and the California Current System: Planktonic foraminiferal response to rapid climate change in Santa Barbara Basin, Ocean Drilling Program hole 893A, Paleoceanography, 15, 30–42, 2000.

Hendy, I. L. and Pedersen, T. F.: Is pore water oxygen content decoupled from productivity on the California Margin? Trace element results from Ocean Drilling Program Hole 1017E, San Lucia slope, California, Paleoceanography, 20, PA4026, doi:10.1029/2004pa001123, 2005.

Hendy, I. L., Pedersen, T. F., Kennett, J. P., and Tada, R.: Intermittent existence of a southern Californian upwelling cell during submillennial climate change of the last 60 kyr, Paleoceanography, 19, PA3007, doi:10.1029/2003pa000965, 2004.

Herold, N., Yin, Q. Z., Karami, M. P., and Berger, A.: Modelling the climatic diversity of the warm interglacials, Quaternary Sci. Rev., 56, 126–141, doi:10.1016/j.quascirev.2012.08.020, 2012.

Hu, A., Meehl, G. A., Otto-Bliesner, B. L., Waelbroeck, C., Han, W., Loutre, M.-F., Lambeck, K., Mitrovica, J. X., and Rosenbloom, N.: Influence of Bering Strait flow and North Atlantic circulation on glacial sea-level changes, Nat. Geosci., 3, 118–121, doi:10.1038/ngeo729, 2010.

Hu, A., Meehl, G. A., Han, W., Abe-Ouchi, A., Morrill, C., Okazaki, Y., and Chikamoto, M. O.: The Pacific-Atlantic seesaw and the Bering Strait, Geophys. Res. Lett., 39, L03702, doi:10.1029/2011gl050567, 2012.

Hutchins, D. A. and Bruland, K. W.: Iron-limited diatom growth and Si:N uptake ratios in a coastal upwelling regime, Nature, 393, 561–564, 1998.

Johnsen, S. J., Dahl-Jensen, D., Gundestrup, N., Steffensen, J. P., Clausen, H. B., Miller, H., Masson-Delmotte, V., Sveinbjörnsdottir, A. E., and White, J.: Oxygen isotope and palaeotemperature records from six Greenland ice-core stations: Camp Century, Dye-3, GRIP, GISP2, Renland and NorthGRIP, J. Quaternary Sci., 16, 299–307, 2001.

Karstensen, J., Stramma, L., and Visbeck, M.: Oxygen minimum zones in the eastern tropical Atlantic and Pacific oceans, Prog. Oceanogr., 77, 331–350, doi:10.1016/j.pocean.2007.05.009, 2008.

Kemp, A. E. S., Baldauf, J. G., and Pearce, R. B.: Origins and palaeoceangraphic significance of laminated daitom ooze from the Eastern Equatorial Pacific Ocean, Geological Society, London, Special Publications, 116, 243–252, doi:10.1144/gsl.sp.1996.116.01.19, 1996.

Kienast, S. S., Calvert, S. E., and Pedersen, T. F.: Nitrogen isotope and productivity variations along the northeast Pacific margin over the last 120 kyr: Surface and subsurface paleoceanography, Paleoceanography, 17, doi:10.1029/2001pa000650, 2002.

Kienast, S. S., Kienast, M., Jaccard, S., Calvert, S. E., and Francois, R.: Testing the silica leakage hypothesis with sedimentary opal records from the eastern equatorial Pacific over the last 150 kyrs, Geophys. Res. Lett., 33, L15607, doi:10.1029/2006gl026651, 2006.

Kutzbach, J. E. and Wright Jr., H. E.: Simulation of the climate of 18,000 years BP: Results for the North American/North Atlantic/European sector and comparison with the geologic record of North America, Quaternary Sci. Rev., 4, 147–187, 1985.

Lambeck, K. and Chappell, J.: Sea level change through the last glacial cycle, Science, 292, 679–686, 2001.

Lambeck, K., Esat, T. M., and Potter, E.-K.: Links between climate and sea levels for the past three million years, Nature, 419, 199–206, 2002.

Leduc, G.: Temporal variations of hydrological changes in the eastern pacific. Geochemical, isotopic and micropaleontological approaches, Ph-D, Aix-Marseille III, 2007.

Li, H.-C., Bischoff, J. L., Ku, T.-L., and Zhu, Z.-Y.: Climate and hydrology of the Last Interglaciation (MIS 5) in Owens Basin, California: isotopic and geochemical evidence from core OL-92, Quaternary Sci. Rev., 23, 49–63, 2004.

Lynn, R. J. and Simpson, J. J.: The California Current System: The Seasonal Variability of its Physical Characteristics, J. Geophys. Res., 92, 12947–12966, doi:10.1029/JC092iC12p12947, 1987.

Marchitto, T. M., Lehman, S. J., Ortiz, J. D., Fluckiger, J., and van Geen, A.: Marine radiocarbon evidence for the mechanism of deglacial atmospheric $CO_2$ rise, Science, 316, 1456–1459, doi:10.1126/science.1138679, 2007.

Mayer, L. M., Schick, L. L., Allison, M. A., Ruttenberg, K. C., and Bentley, S. J.: Marine vs. terrigenous organic matter in Louisiana coastal sediments: The uses of bromine:organic carbon ratios, Mar. Chem., 107, 244–254, 2007.

McLennan, S. M.: Relationships between the trace element composition of sedimentary rocks and upper continental crust, Geochem. Geophys. Geosyst., 2, 1021, doi:10.1029/2000GC000109, 2001.

McManus, J. F., Bond, G. C., Broecker, W. S., Johnsen, S., Labeyrie, L., and Higgins, S.: High-resolution climate records from the North Atlantic during the last interglacial, Nature, 371, 326–329, 1994.

Medina-Elizalde, M.: A global compilation of coral sea-level benchmarks: Implications and new challenges, Earth. Planet. Sci. Lett., 362, 310–318, doi:10.1016/j.epsl.2012.12.001, 2013.

Mikolajewicz, U., Crowley, T. J., Schiller, A., and Voss, R.: Modelling teleconnections between the North Atlantic and North Pacific during the Younger Dryas, Nature, 387, 384–387, 1997.

Montero-Serrano, J.-C., Bout-Roumazeilles, V., Carlson, A. E., Tribovillard, N., Bory, A., Meunier, G., Sionneau, T., Flower, B. P., Martinez, P., Billy, I., and Riboulleau, A.: Contrasting rainfall patterns over North America during the Holocene and Last Interglacial as recorded by sediments of the northern Gulf of Mexico, Geophys. Res. Lett., 38, L14709, doi:10.1029/2011gl048194, 2011.

Moore, C. M., Mills, M. M., Arrigo, K. R., Berman-Frank, I., Bopp, L., Boyd, P. W., Galbraith, E. D., Geider, R. J., Guieu, C., Jaccard, S. L., Jickells, T. D., La Roche, J., Lenton, T. M., Mahowald, N. M., Maranon, E., Marinov, I., Moore, J. K., Nakatsuka, T., Oschlies, A., Saito, M. A., Thingstad, T. F., Tsuda, A., and Ulloa, O.: Processes and patterns of oceanic nutrient limitation, Nat. Geosci., 6, 701–710, doi:10.1038/ngeo1765, 2013.

Moore, J. K., Doney, S. C., and Lindsay, K.: Upper ocean ecosystem dynamics and iron cycling in a global three-dimensional model, Global Biogeochem. Cy., 18, GB4028, doi:10.1029/2004gb002220, 2004.

Müller, P. J. and Schneider, R.: An automated leaching method for the determination of opal in sediments and particulate matter, Deep-Sea Res. Part I Oceanogr. Res. Pap., 40, 425–444, doi:10.1016/0967-0637(93)90140-X, 1993.

Murdmaa, I. O., Kazarina, G. H., Beaufort, L., Ivanova, E. V., Emelyanov, E. M., Kravtsov, V. A., Alekhina, G. N., and Vasileva, V. E.: Upper Quaternary Laminated Sapropelic Sediments from the Continental Slope of Baja California, Lithol. Mineral Resour., 45,, 154–171, doi:10.1134/S0024490210020057, 2010.

Nameroff, T. J., Balistrieri, L. S., and Murray, J. W.: Suboxic trace metal geochemistry in the eastern tropical North Pacific, Geochim. Cosmochim. Acta, 66, 1139–1158, 2002.

Nameroff, T. J., Calvert, S. E., and Murray, J. W.: Glacial-interglacial variability in the eastern tropical North Pacific oxygen minimum zone recorded by redox-sensitive trace metals, Paleoceanography, 19, PA1010, doi:10.1029/2003pa000912, 2004.

NGRIP Members: High-resolution record of Northern Hemisphere climate extending into the last interglacial period, Nature, 431, 147–151, doi:10.1038/nature02805, 2004.

Okazaki, Y., Timmermann, A., Menviel, L., Harada, N., Abe-Ouchi, A., Chikamoto, M. O., Mouchet, A., and Asahi, H.: Deepwater Formation in the North Pacific During the Last Glacial Termination, Science, 329, 200–204, doi:10.1126/science.1190612, 2010.

Okumura, Y. M., Deser, C., Hu, A., Timmermann, A., and Xie, S. P.: North Pacific Climate Response to Freshwater Forcing in the Subarctic North Atlantic: Oceanic and Atmospheric Pathways, J. Climate, 22, 1424–1445, doi:10.1175/2008jcli2511.1, 2009.

Oppo, D. W., McManus, J. F., and Cullen, J. L.: Evolution and demise of the Last Interglacial warmth in the subpolar North Atlantic, Quaternary Sci. Rev., 25, 3268–3277, 2006.

Ortiz, J. D., O'Connell, S. B., DelViscio, J., Dean, W., Carriquiry, J. D., Marchitto, T., Zheng, Y., and van Geen, A.: Enhanced marine productivity off western North America during warm climate intervals of the past 52 ky, Geology, 32, 521–524, 2004.

Paulmier, A. and Ruiz-Pino, D.: Oxygen minimum zones (OMZs) in the modern ocean, Prog. Oceanogr., 80, 113–128, 2009.

Paulmier, A., Ruiz-Pino, D., and Garcon, V.: The oxygen minimum zone (OMZ) off Chile as intense source of $CO_2$ and $N_2O$, Cont. Shelf Res., 28, 2746–2756, 2008.

Paulmier, A., Ruiz-Pino, D., and Garçon, V.: $CO_2$ maximum in the oxygen minimum zone (OMZ), Biogeosciences, 8, 239–252, doi:10.5194/bg-8-239-2011, 2011.

Pennington, J. T., Mahoney, K. L., Kuwahara, V. S., Kolber, D. D., Calienes, R., and Chavez, F. P.: Primary production in the eastern tropical Pacific: A review, Prog. Oceanogr., 69, 285–317, 2006.

Peterson, L. C., Haug, G. H., Hughen, K. A., and Rohl, U.: Rapid changes in the hydrologic cycle of the tropical Atlantic during the last glacial, Science, 290, 1947–1951, 2000.

Pichevin, L., Bard, E., Martinez, P., and Billy, I.: Evidence of ventilation changes in the Arabian Sea during the late Quaternary: Implication for denitrification and nitrous oxide emission, Global Biogeochem. Cy., 21, GB4008, doi:10.1029/2006gb002852, 2007.

Pichevin, L., Ganeshram, R., Francavilla, S., Arellano-Torres, E., Pedersen, T., and Beaufort, L.: Interhemispheric leakage of isotopically heavy nitrate in the eastern tropical Pacific during the last glacial period, Paleoceanography, 25, 1204, doi:10.1029/2009pa001754, 2010.

Pichevin, L., Ganeshram, R. S., Reynolds, B. C., Prahl, F., Pedersen, T. F., Thunell, R., and McClymont, E. L.: Silicic acid biogeochemistry in the Gulf of California: Insights from sedimentary Si isotopes, Paleoceanography, 27, doi:10.1029/2011PA002237, 2012.

Rella, S. F., Tada, R., Nagashima, K., Ikehara, M., Itaki, T., Ohkushi, K. I., Sakamoto, T., Harada, N., and Uchida, M.: Abrupt changes of intermediate water properties on the northeastern slope of the Bering Sea during the last glacial and deglacial period, Paleoceanography, 27, PA3203, doi:10.1029/2011pa002205, 2012.

Romanova, V., Lohmann, G., Grosfeld, K., and Butzin, M.: The relative role of oceanic heat transport and orography on glacial climate, Quaternary Sci. Rev., 25, 832–845, 2006.

Romero, O. E., Leduc, G., Vidal, L., and Fischer, G.: Millennial variability and long-term changes of the diatom production in the eastern equatorial Pacific during the last glacial cycle, Paleoceanography, 26, PA2212, doi:10.1029/2010pa002099, 2011a.

Romero, O. E., Swann, G. E. A., Hodell, D. A., Helmke, P., Rey, D., and Rubio, B.: A highly productive Subarctic Atlantic during the Last Interglacial and the role of diatoms, Geology, 39, 1015–1018, doi:10.1130/g32454.1, 2011b.

Saenko, O. A., Schmittner, A., and Weaver, A. J.: The Atlantic-Pacific seesaw, J. Climate, 17, 2033–2038, 2004.

Sarmiento, J. L., Gruber, N., Brzezinski, M. A., and Dunne, J. P.: High-latitude controls of thermocline nutrients and low latitude biological productivity, Nature, 479, 56–60, 2004.

Schmittner, A.: Decline of the marine ecosystem caused by a reduction in the Atlantic overturning circulation, Nature, 434, 628–633, 2005.

Schmittner, A., Galbraith, E. D., Hostetler, S. W., Pedersen, T. F., and Zhang, R.: Large fluctuations of dissolved oxygen in the Indian and Pacific oceans during Dansgaard-Oeschger oscillations caused by variations of North Atlantic Deep Water subduction, Paleoceanography, 22, Pa3207 doi:10.1029/2006pa001384, 2007.

Shcherbina, A. Y., Talley, L. D., and Rudnick, D. L.: Direct observations of North Pacific ventilation: Brine rejection in the Okhotsk Sea, Science, 302, 1952–1955, 2003.

Shimada, C., Sato, T., Toyoshima, S., Yamasaki, M., and Tanimura, Y.: Paleoecological significance of laminated diatomaceous oozes during the middle-to-late Pleistocene, North Atlantic Ocean (IODP Site U1304), Mar. Micropaleontol., 69, 139–150, 2008.

Siddall, M., Chappell, J., and Potter, E. K.: 7. Eustatic sea level during past interglacials, in: Developments in Quaternary Sciences, edited by: Frank Sirocko, M. C. M. F. S. G. and Thomas, L., Elsevier, 75–92, 2007.

Stramma, L., Johnson, G. C., Sprintall, J., and Mohrholz, V.: Expanding oxygen-minimum zones in the tropical oceans, Science, 320, 655–658, doi:10.1126/science.1153847, 2008.

Stramma, L., Johnson, G. C., Firing, E., and Schmidtko, S.: Eastern Pacific oxygen minimum zones: Supply paths and multidecadal changes, J. Geophys. Res.-Ocean., 115, C09011, doi:10.1029/2009jc005976, 2010a.

Stramma, L., Schmidtko, S., Levin, L. A., and Johnson, G. C.: Ocean oxygen minima expansions and their biological impacts, Deep-Sea Res. Part I-Oceanogr. Res. Pap., 57, 587–595, doi:10.1016/j.dsr.2010.01.005, 2010b.

Talley, L. D.: An Okhotsk sea-water anomaly-Implication for ventilation in the north Pacific, Deep-Sea Res. I, 38, S171–S190, 1991.

Thomas, A. C., Carr, M. E., and Strub, P. T.: Chlorophyll variability in eastern boundary currents, Geophys. Res. Lett., 28, 3421–3424, 2001.

Timmermann, A., Krebs, U., Justino, F., Goosse, H., and Ivanochko, T.: Mechanisms for millennial-scale global synchronization during the last glacial period, Paleoceanography, 20, PA4008, doi:10.1029/2004pa001090, 2005.

Tribovillard, N., Algeo, T. J., Lyons, T., and Riboulleau, A.: Trace metals as paleoredox and paleoproductivity proxies: An update, Chem. Geol., 232, 12–32, doi:10.1016/j.chemgeo.2006.02.012, 2006.

van Geen, A. and Husby, D. M.: Cadmium in the California current system: Tracer of past and present upwelling, J. Geophys. Res.-Ocean., 101, 3489–3507, 1996.

Waelbroeck, C., Labeyrie, L., Michel, E., Duplessy, J. C., McManus, J. F., Lambeck, K., Balbon, E., and Labracherie, M.: Sea-level and deep water temperature changes derived from benthic foraminifera isotopic records, Quaternary Sci. Rev., 21, 295–305, 2002.

Wagner, J. D. M., Cole, J. E., Beck, J. W., Patchett, P. J., Henderson, G. M., and Barnett, H. R.: Moisture variability in the southwestern United States linked to abrupt glacial climate change, Nat. Geosci., 3, 110–113, 2010.

Wang, H., Stumpf, A. J., Miao, X., and Lowell, T. V.: Atmospheric changes in North America during the last deglaciation from dune-wetland records in the Midwestern United States, Quaternary Sci. Rev., 58, 124–134, doi:10.1016/j.quascirev.2012.10.018, 2012.

Zaytsev, O., Cervantes-Duarte, R., Montante, O., and Gallegos-Garcia, A.: Coastal upwelling activity on the pacific shelf of the Baja California Peninsula, J. Oceanogr., 59, 489–502, 2003.

Ziegler, M., Jilbert, T., de Lange, G. J., Lourens, L. J., and Reichart, G. J.: Bromine counts from XRF scanning as an estimate of the marine organic carbon content of sediment cores, Geochem. Geophys. Geosyst., 9, 5009, doi:10.1029/2007gc001932, 2008.

# 8

# What controls the isotopic composition of Greenland surface snow?

H. C. Steen-Larsen[1,2,3], V. Masson-Delmotte[1], M. Hirabayashi[4], R. Winkler[1], K. Satow[4], F. Prié[1], N. Bayou[2], E. Brun[5], K. M. Cuffey[6], D. Dahl-Jensen[3], M. Dumont[7], M. Guillevic[1,3], S. Kipfstuhl[8], A. Landais[1], T. Popp[3], C. Risi[9], K. Steffen[2,10], B. Stenni[11], and A. E. Sveinbjörnsdottír[12]

[1]Laboratoire des Sciences du Climat et de l'Environnement, UMR8212, CEA-CNRS-UVSQ/IPSL, Gif-sur-Yvette, France
[2]Cooperative Institute for Research in Environmental Sciences, University of Colorado, Boulder, USA
[3]Centre for Ice and Climate, Niels Bohr Institute, University of Copenhagen, Copenhagen, Denmark
[4]National Institute of Polar Research, Tokyo, Japan
[5]Meteo-France – CNRS, CNRM-GAME UMR 3589, GMGEC, Toulouse, France
[6]Department of Geography, Center for Atmospheric Sciences, 507 McCone Hall, University of California, Berkeley, CA 94720-4740, USA
[7]Meteo-France – CNRS, CNRM-GAME UMR 3589, CEN, Grenoble, France
[8]Alfred Wegener Institute for Polar and Marine Research, Bremerhaven, Germany
[9]Laboratoire de Météorologie Dynamique, Jussieu, Paris, France
[10]ETH, Swiss Federal Institute of Technology, Zurich, Switzerland
[11]Department of Geological, Environmental and Marine Sciences, University of Trieste, Trieste, Italy
[12]Institute of Earth Sciences, University of Iceland, Reykjavik, Iceland

*Correspondence to:* H. C. Steen-Larsen (hanschr@gfy.ku.dk)

**Abstract.** Water stable isotopes in Greenland ice core data provide key paleoclimatic information, and have been compared with precipitation isotopic composition simulated by isotopically enabled atmospheric models. However, post-depositional processes linked with snow metamorphism remain poorly documented. For this purpose, monitoring of the isotopic composition ($\delta^{18}$O, $\delta$D) of near-surface water vapor, precipitation and samples of the top (0.5 cm) snow surface has been conducted during two summers (2011–2012) at NEEM, NW Greenland. The samples also include a subset of $^{17}$O-excess measurements over 4 days, and the measurements span the 2012 Greenland heat wave. Our observations are consistent with calculations assuming isotopic equilibrium between surface snow and water vapor. We observe a strong correlation between near-surface vapor $\delta^{18}$O and air temperature (0.85 ± 0.11‰ °C$^{-1}$ ($R = 0.76$) for 2012). The correlation with air temperature is not observed in precipitation data or surface snow data. Deuterium excess (d-excess) is strongly anti-correlated with $\delta^{18}$O with a stronger slope for vapor than for precipitation and snow surface data. During nine 1–5-day periods between precipitation events, our data demonstrate parallel changes of $\delta^{18}$O and d-excess in surface snow and near-surface vapor. The changes in $\delta^{18}$O of the vapor are similar or larger than those of the snow $\delta^{18}$O. It is estimated using the CROCUS snow model that 6 to 20 % of the surface snow mass is exchanged with the atmosphere. In our data, the sign of surface snow isotopic changes is not related to the sign or magnitude of sublimation or deposition. Comparisons with atmospheric models show that day-to-day variations in near-surface vapor isotopic composition are driven by synoptic variations and changes in air mass trajectories and distillation histories. We suggest that, in between precipitation events, changes in the surface snow isotopic composition are driven by these changes in near-surface vapor isotopic composition. This is consistent with an estimated 60 % mass turnover of surface snow per day driven by snow recrystallization processes under NEEM summer surface snow temperature gradients. Our findings have implications for ice core data interpretation and model–data comparisons, and call for further process studies.

# 1 Introduction

Ice cores drilled in central Greenland, with limited summer melt, provide direct archives of past precipitation. Water stable isotope ($\delta^{18}$O and/or $\delta$D) measurements have been conducted along numerous shallow and deep ice cores in order to characterize past Greenland climate variability, offering seasonal records during the past millennia (Vinther et al., 2010; Ortega et al., 2014) and recently extending back to the last interglacial period (NEEM Community members, 2013).

The processes controlling water stable isotopes in mid- to high-latitude vapor and precipitation are, based on modern data and modeling, relatively well understood. Theoretical calculations of Rayleigh distillation show an expected $\delta^{18}$O–condensation temperature slope for Greenland precipitation of $0.96\,‰\,°C^{-1}$ (Johnsen et al., 2001), coherent with the modern spatial gradient of $0.8\,‰\,°C^{-1}$ of near-surface air temperature established from coastal precipitation data together with shallow ice core data (Sjolte et al., 2011). This isotope–temperature relationship (isotope thermometer) (Johnsen et al., 2001) has been central to the use of ice core water isotope records to reconstruct past Greenland climate variations. However, the comparison of water stable isotope measurements with past temperatures inferred either from the inversion of borehole temperature data (Dahl-Jensen et al., 1998) or from the fingerprint of firn air fractionation in ice core air $\delta^{15}$N has revealed that (i) for a given site, the isotope–temperature relationship varies through time (e.g., Guillevic et al., 2013; Landais et al., 2004; Kindler et al., 2013; Severinghaus and Brook, 1999), and (ii) for a given stadial–interstadial event, the isotope–temperature relationship varies between sites (Guillevic et al., 2013). The reported temporal isotope–temperature relationships vary between 0.3 and $0.6\,‰\,°C^{-1}$.

Differences in the estimated isotope–surface-temperature relationship have been suggested to arise from (i) precipitation intermittency and the covariance between precipitation and temperature (Persson et al., 2011), (ii) changes in relationships between surface and condensation temperature linked with changes in boundary layer dynamics, and (iii) changes in moisture sources and distillation along air mass trajectories (Masson-Delmotte et al., 2005b). Simulations of water stable isotopes within regional or general circulation atmospheric models (GCM) have been used to explore the drivers of changes in isotope–temperature relationships, and evaluate models against ice core data. Such studies have confirmed the importance of precipitation seasonality for glacial–interglacial changes and highlighted the role of changes in atmospheric circulation and moisture sources (Cuffey and Steig, 1998; Jouzel et al., 1997; Krinner et al., 1997).

Second-order parameters such as the deuterium excess (d-excess; d-excess $= \delta$D$-8 \times \delta^{18}$O) and more recently the $^{17}$O-excess are expected, based on modeling, to preserve the signature of the moisture source. The reason for this is the iso-topic composition of source moisture being controlled by kinetic effects at evaporation related to wind speed, sea surface temperature and relative humidity (for d-excess) or relative humidity (for $^{17}$O-excess) (Merlivat and Jouzel, 1979; Landais et al., 2012; Johnsen et al., 1989). Measurements of d-excess in Greenland ice cores have therefore been used to infer present and past evaporation conditions and locate the main moisture sources (Johnsen et al., 1989; Masson-Delmotte et al., 2005b; Steen-Larsen et al., 2011). A few measurements of $^{17}$O-excess conducted at the seasonal scale show a seasonal cycle in anti-correlation with respect to Greenland temperature and $\delta^{18}$O (Landais et al., 2012). On stadial–interstadial timescales, existing ice core records have revealed an anti-correlation of d-excess with $\delta^{18}$O, reflecting the impact of changes in cloud condensation temperature on (i) the ratio of equilibrium fractionation for $\delta$D and $\delta^{18}$O (Merlivat and Nief, 1967; Ellehoj et al., 2013) and (ii) kinetic fractionation on ice crystals (Jouzel and Merlivat, 1984). However, ice cores have also depicted specific d-excess signals spanning both present and past decadal to millennial and orbital timescales, interpreted as reflecting changes in moisture source conditions (Steffensen et al., 2008; Masson-Delmotte et al., 2005a; Steen-Larsen et al., 2011). It is, however, difficult to simulate the observed Greenland ice core d-excess shifts for glacial–interglacial transitions (Werner et al., 2001; Risi et al., 2010).

The interpretation of the ice core data and the comparison with atmospheric model results implicitly rely on the assumption that the snowfall precipitation signal is perfectly preserved in the snow–ice matrix. However, post-deposition processes associated with wind scouring and firn isotopic diffusion are known to introduce a "post-deposition noise" in the surface snow. Comparisons of isotopic records obtained from nearby shallow ice cores have allowed for estimation of a "signal-to-noise" ratio with respect to a common "climate" signal (Fisher and Koerner, 1994, 1988; White et al., 1997; Steen-Larsen et al., 2011). Diffusion lengths of typically 7–10 cm have been diagnosed in Greenland ice cores based on the loss of magnitude of seasonal cycles in shallow ice cores (Johnsen et al., 2000), and statistical methods have been used to "backdiffuse" ice core signals for the purpose of identifying seasonal cycles for dating ice cores or for the correction of loss of amplitude for winter and summer water stable isotope signals (Johnsen, 1977). In parallel, numerical snow models have been developed in order to represent the surface snow metamorphism, a process associated with vapor–snow mass exchanges in the upper centimeters of the firn. Snow metamorphism affects changes in grain size, surface albedo, and more generally the surface snow energy budget and mass balance (Vionnet et al., 2012; Brun et al., 2011). Snow models are growingly incorporated in atmospheric–land-surface models or used for the coupling between atmospheric and ice sheet models (Rae et al., 2012), but none of them is yet equipped with the explicit modeling of water stable isotopes.

The motivation for our study is to investigate the impacts of post-deposition processes on (i) the isotope–temperature relationships, (ii) the d-excess vs. $\delta^{18}O$ relationships and (iii) the surface snow isotopic composition in between precipitation events.

For this purpose, a surface-water-isotope-monitoring program has been established at the NEEM site, NW Greenland, with the goal of improving the interpretation of the NEEM deep ice core through a better understanding of the processes controlling the water isotopic composition measured in the ice core record at the event scale. In summer 2008, this program combined event and sub-event precipitation sampling for water stable isotope analysis, together with shallow ice core data, and water vapor monitoring using cryogenic trapping (Steen-Larsen et al., 2011). These first measurements showed parallel isotopic variations between vapor and snowfall. The near-surface water vapor isotopic composition was found to predominantly be close to isotopic equilibrium with surface snow (Steen-Larsen et al., 2011). The resolution of water vapor isotope observations was subsequently strongly improved thanks to continuous, in situ measurements using cavity ring-down spectrometers (CRDS) during summer 2010 (Steen-Larsen et al., 2013). The day-to-day variability of the near-surface atmospheric water vapor $\delta^{18}O$ was in good agreement with the results from an atmospheric general circulation model, LMDZiso, nudged to atmospheric analyses. While the model did not capture the magnitude of vapor d-excess variations observed at NEEM, it showed that high d-excess events coincided with inflows of moisture originating from the Arctic (north of 70° N). This finding demonstrated that large-scale atmospheric circulation changes drive day-to-day variations of NEEM near-surface water vapor. During clear-sky days, CRDS measurements conducted at heights from 1 to 13 m showed a strong diurnal variability in humidity and $\delta^{18}O$, interpreted to reflect the fact that the snow surface acts as a moisture source (sink) during the warming (cooling) phase.

Altogether these preliminary findings have qualitatively evidenced interactions between the atmospheric water vapor and the snow surface, which has motivated further observations. Here, we report new data acquired at NEEM in summer 2011 and spring–summer 2012, which include a systematic monitoring of surface snow and water vapor, precipitation (only for 2011), and the first $^{17}O$-excess measurements simultaneously conducted during 4 days on water vapor and surface snow. The CROCUS snow model (Vionnet et al., 2012) has been adapted to the NEEM site and used to calculate the snow–air net mass exchange.

This manuscript is organized in the following way. Section 2 describes the NEEM site, sampling strategy and analytical methods, as well as the set up for the CROCUS model and the basis for vapor-surface snow equilibrium calculations. Section 3 describes the results of the new isotopic composition measurements and reports the $\delta^{18}O$–temperature and d-excess–$\delta^{18}O$ relationships, as well as the $^{17}O$-excess

results. Section 4 is finally devoted to the discussion of our results and the comparison with CROCUS calculations, in order to qualitatively understand the processes controlling the isotopic composition of surface snow.

## 2 Methods

### 2.1 NEEM site description

The sampling and measurements were carried out as part of the international deep drilling program conducted at NEEM, NW Greenland (77.45° N, 51.05° W; 2484 m a.s.l.), from 2007 to 2012, providing climatic and glaciological information back to the last interglacial period (NEEM Community members, 2013; Steen-Larsen et al., 2011). An automatic weather station (AWS) was installed at the NEEM site in 2006 to supply meteorological observations. Air temperature and relative humidity (post-corrected with respect to ice) were measured using a Campbell Scientific HMP45C ($\pm 0.1$ °C and $\pm 5\% < 90\%$ RH and $\pm 10\% > 90\%$ RH), wind direction and speed using an RM Young propeller-type vane ($\pm 5°$ and $\pm 0.1$ m s$^{-1}$), and station pressure using Vaisala PTB101B ($\pm 0.1$ mb) (Steffen and Box, 2001; Steffen et al., 1996). The estimated mean summer (JJA) temperature at NEEM is $\sim -11 \pm 5$ °C ($1\sigma$ based on 3-hourly observations during the summers 2006–2011; the $1\sigma$ on the mean summer temperatures 2006–2011 is $\sim 1$ °C), but the summer of 2012 was found to be significantly warmer ($\sim -7.5$ °C) than average. The annual mean accumulation rate from 1964 to 2005 is estimated to 20 cm a$^{-1}$ (water equivalent), with a large part (between a factor of 2.5 and 4.5) of precipitation occurring in JJA compared to DJF (Steen-Larsen et al., 2011). During the field campaigns of 2009–2011, a thermistor string was installed in the top snowpack between the surface and two meters depth. Extra thermistors were installed in the top snow layer throughout the season. The resistances of the thermistors were recorded using a Pico Technology 24 bit data logger. Occurrence of precipitation was recorded in the daily field report managed by the field leader.

### 2.2 CRDS-analyzer measurements

A tent was installed at the edge of the clean-air sector in the southwest corner of the NEEM camp $\sim 50$ m from the nearest building. Inside this tent we installed a temperature-regulated box able to control the temperature to within 0.2 °C. A commercial laser-based spectrometer from Picarro Inc. (product number L1102-i) was installed inside this box. The detailed sampling and post-calibration procedure is given in Steen-Larsen et al. (2013). The inlet tubes were placed in insulation material and heated to above 50 °C. Bottles with holes in the bottom were placed at the beginning of the inlet to prevent snow from getting inside the tubes. The inlet tubes consisted of 0.25 in. outer diameter copper tubes. Two pumps were installed to increase the flow speed in the inlet tubes, thereby minimizing the resident time of the air inside the tubes. The flow speed of the tube being sampled was $\sim 5\,\mathrm{L\,min^{-1}}$, while the flow speed through each of the tubes not being sampled was $\sim 2\,\mathrm{L\,min^{-1}}$. To correct for the humidity dependence on the measured isotope signal, a humidity–isotope response curve was calibrated in the beginning of each measurement campaign according to the description in Steen-Larsen et al. (2013). The isotopic measurements were converted to the VSMOW-SLAP scale by measuring standards of known isotopic composition. To correct for drifts, water vapor with a known isotopic composition was measured every 6 h. The introduction lines were installed to sample air from five levels at $\sim 1, 3, 7, 10$ and 13 m above the snow surface during the 2011 campaign and two levels at $\sim 20\,\mathrm{cm}$ and 3 m above the snow surface during the 2012 campaign. For the 2011 campaign, each level was measured for 15 min, of which the first 5 min were discarded to rule out memory effects of the inlet tubes. For the 2012 campaign the 3 m level was measured continuously for 45 min of every hour, with the other 15 min used for measuring the 20 cm level. Values are therefore reported with an hourly resolution for the 2011 data, and 15 min resolution for 2012 data, and with similar accuracy and precision as reported by Steen-Larsen et al. (2013) (Table 1). The measurements of the drift standard with known isotopic composition used to correct the drift of the instrument are shown in the Supplement Figs. S1 and S2. We do not expect that the data gaps due to drift calibrations produce any significant effect on the analysis. When the vertical gradients of water stable isotopes were investigated in Steen-Larsen et al. (2013), it was shown that they only depict diurnal gradients, and that water vapor isotopes covary at the day-to-day scale at different heights. This was also verified for our new measurements (not shown), and here we will only describe and discuss the measurements obtained at 3 m height (Table 1). Note that the 2011 sampling was initiated on day 185 (5 July) (noon UTC on 1 January is day 0.5), while the 2012 sampling covers a longer time period, starting on day 141 (21 May). In both years, measurements stopped at day 216 (August) due to the closure of the summer camp. Hereafter, vapor measurements are reported as $\delta^{18}O_v$ and d-excess$_v$.

### 2.3  Precipitation and snow surface samples

A white table with $\sim 20\,\mathrm{cm}$ high sides covering an area of $\sim 0.7\,\mathrm{m^2}$ made out of opaque Plexiglas was installed on the edge of the clean-air sector next to the atmospheric vapor station in order to collect precipitation. The table was installed at a height of $\sim 1.5\,\mathrm{m}$ to limit the collection of blowing snow. Precipitation was collected in 2011 on event and sub-event basis as reported in Steen-Larsen et al. (2011), leading to 41 samples (Table 1). We have discarded two outliers with very low d-excess$_p$ of $\sim -10\permil$.

A designated area (5 × 5 m) was marked from which the snow surface samples were collected. The snow surface sam-

ples were collected from the top $\sim 0.5\,\mathrm{cm}$ of the snow surface. All samples were collected within the designated area but never from any previously sampled place. The surface snow was sampled every 12 h (a few (25) samples were only collected every 24 h in the beginning of 2012) by collecting the surface from a 15 cm × 15 cm area. Altogether, 51 samples were collected in 2011, and 122 samples in 2012.

Snowfall and surface snow samples were melted in sealed plastic bags before being transferred to a vial, which was kept frozen until being measured. The precipitation and snow surface samples were measured using a Picarro Inc. liquid analyzer at the Laboratoire des Sciences du Climat et de l'Environnement (LSCE), Gif-Sur-Yvette, and the Centre for Ice and Climate (CIC), Copenhagen (see Table 1). Hereafter, precipitation measurements are reported as $\delta^{18}O_p$ and d-excess$_p$, and surface snow measurements as $\delta^{18}O_s$ and d-excess$_s$.

### 2.4  $^{17}O$-excess of snow surface and atmospheric water vapor

During the period from 11 to 14 July 2011, a specific sampling was conducted to explore day-to-day variations in $^{17}O$-excess of snow surface and atmospheric water vapor. Using a cryogenic trapping system similar to Steen-Larsen et al. (2011), water vapor from both 1 and 10 m above the snow surface was collected with the aim of measuring $^{17}O$-excess. Vapor trapping was conducted over 6 h, leading to two samples per level per day. Because the results are very similar at the two heights, we only report the 10 m data here. $^{17}O$-excess measurements were also conducted on the corresponding subset of surface snow samples. The $^{17}O$-excess of the snow surface and atmospheric water vapor samples were measured at Laboratoire des Sciences du Climat et de l'Environnement (LSCE) using the fluorination technique (Barkan and Luz, 2005; Landais et al., 2012) (Table 1). The same notation for $\delta^{18}O$ and d-excess is used for $^{17}O$-excess to report vapor data ($^{17}O$-excess$_v$) and surface snow data ($^{17}O$-excess$_s$).

### 2.5  Calculation of equilibrium between surface snow and water vapor isotopic composition

In order to investigate the relationship between near-surface water vapor and snow, we use the observed air temperature (see Sect. 2.1) and vapor isotopic composition (see Sect. 2.2) (integrated 12 h back) to estimate the expected isotopic composition of the snow surface at equilibrium with the water vapor. The calculation is performed (i) using the fractionation coefficients for liquid water extrapolated below 0 °C (Majoube, 1971), then (ii) using the fractionation coefficients for ice (Majoube, 1970) (Merlivat and Nief, 1967) and (iii) new fractionation coefficients for ice determined by Ellehoj et al. (2013). We justify the calculations assuming liquid

**Table 1.** Overview of data collected during the 2011 and 2012 field campaign.

| Sample type | Time period | Isotopic measurement type | Number of samples | Precision and accuracy | Measurement type | Place of measurement |
|---|---|---|---|---|---|---|
| Picarro water vapor isotope | Day 185 to day 216 of 2011 | $\delta D_v$, $\delta^{18}O_v$ continuously from 3 m above snow surface interrupted by calibration 30 min every 6 h | ~ hourly resolution | $\delta^{18}O_v = 0.23‰$ $\delta D_v = 1.4‰$ (Steen-Larsen et al., 2013) | Picarro CRDS analyzer | NEEM |
| Picarro water vapor isotope | Day 141 to day 216 of 2012 | $\delta D_v$, $\delta^{18}O_v$ continuously from 3 m above snow surface interrupted by calibration 30 min every 6 h | ~ 15 min resolution | $\delta^{18}O_v = 0.23‰$ $\delta D_v = 1.4‰$ (Steen-Larsen et al., 2013) | Picarro CRDS analyzer | NEEM |
| Cryogenic collected vapor | Day 191 to 194 of 2012 | $^{17}O$-excess from 6 h trapping of 10 m height vapor (2 samples per day) | 7 samples | 6 ppm (Landais et al., 2012) | IRMS using a flourination line | LSCE |
| Precipitation samples | Day 189 to day 210 of 2011 | $\delta D_p$, $\delta^{18}O_p$ Sub-event resolution | 41 samples | $\delta^{18}O_p = 0.1‰$ $\delta D_p = 1.0‰$ | Picarro CRDS analyzer | LSCE |
| Snow surface samples | Day 188 to day 215 of 2011 | $\delta D_s$, $\delta^{18}O_s$ every 12 h | 51 samples | $\delta^{18}O_s = 0.1‰$ $\delta D_s = 1.0‰$ | Picarro CRDS analyzer | LSCE |
| Snow surface samples | Day 191 to day 194 of 2011 | $^{17}O$-excess every 12 h | 7 samples | 6 ppm | IRMS using a flourination line | LSCE |
| Snow surface samples | Day 143 to day 214 of 2012 | $\delta D_s$, $\delta^{18}O_s$ every 24 h (day 143–165) every 12 h (day 166–214) | 122 samples | $\delta^{18}O_s = 0.1‰$ $\delta D_s = 1.0‰$ | Picarro CRDS analyzer | CIC |

water by the fact that the summer temperatures at NEEM are close to 0 °C (Fig. 1). Results are reported in Sect. 3.8.

## 2.6 LMDZiso

The measurements are compared with simulations using the atmospheric general circulation model LMDZ (Hourdin et al., 2006). It is enabled with water isotopes (Risi et al., 2010) and the simulated water isotopic distribution has been validated at various timescales globally (Risi et al., 2010, 2013) and over Greenland (Steen-Larsen et al., 2011; Ortega et al., 2014). The details of isotopic implementation are given in Risi et al. (2010).

LMDZ is used here with a resolution of 2.5° in latitude, 3.75° in longitude and 39 vertical levels. The first atmospheric layer is 60 m thick. The simulated winds are nudged towards those of the ECMWF operational analyses, allowing the model to reproduce the day-to-day large-scale atmospheric conditions (Risi et al., 2010; Yoshimura et al., 2008).

Snow is represented as a single, vertically homogeneous layer. The mass fluxes into the layer are snowfall and frost (deposition) and the mass fluxes out of the layer are melt, sublimation "and model-layer overflow". Melt and sublimation are assumed not to fractionate. Frost is formed in equilibrium with the water vapor of the first atmospheric layer. "Model-layer overflow" occurs whenever the snow height exceeds a maximum capacity of 3 m. The "model-layer overflow" is carried out in the simulation by simply removing

from the calculations any part of the homogeneous snow layer that exceeds 3 m. In practice, at NEEM, the snow layer is almost always at its maximum capacity. Therefore, the isotopic composition of the snow varies very little at the daily scale in the model, since the snow height variations through snowfall or through frost are several orders of magnitudes smaller than the total snow layer height.

## 2.7 CROCUS

We run the detailed snowpack model CROCUS (Vionnet et al., 2012) to drive the energy and mass fluxes between the atmosphere and the snow surface as well as the snow grain metamorphism. The meteorological forcing required by the model was extracted from ERA-Interim (Dee et al., 2011) and projected onto NEEM as in Brun et al. (2013). The precipitation rate was multiplied by a factor of 2 in order to match the annual accumulation record at NEEM. This adjustment does not significantly change the simulated heat and vapor surface fluxes. ERA-Interim air temperature and humidity were compared with observations from the GC-NET (Greenland Climate Network) station at NEEM (Steffen and Box, 2001; Steen-Larsen et al., 2011). A strong linear correlation of ERA-Interim temperature vs. observations is found with < 1 °C deviation to the 1 : 1 line for the range −20 to 0 °C ($R = 0.95$). Similar strong linear correlation between ERA-Interim air humidity and observation is found with < 100 ppmv deviation to the 1 : 1 lines for the range

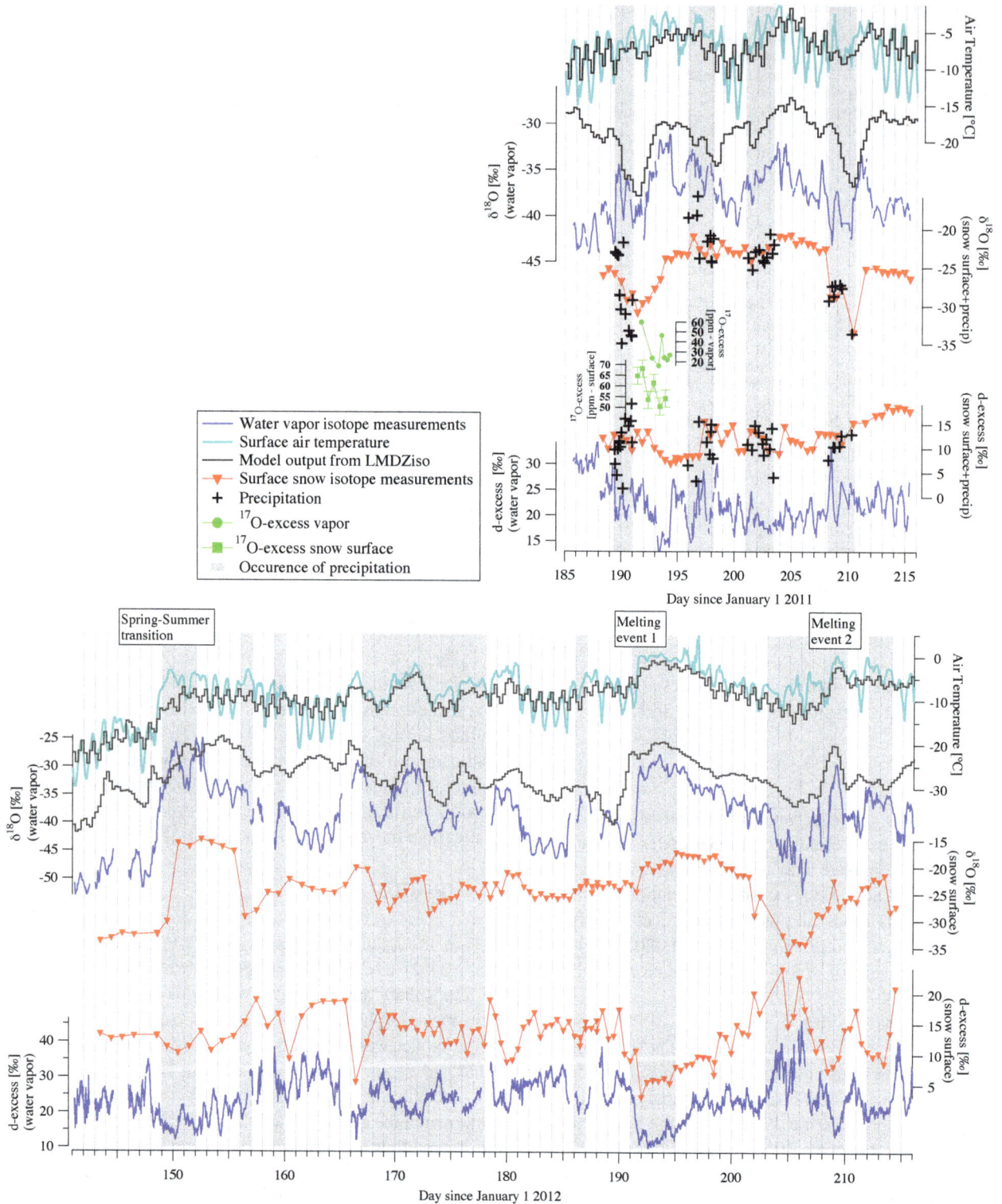

**Fig. 1.** The observed water vapor isotope signal during the 2011 (top panel) and 2012 (bottom panel) field season at NEEM (shown in blue) together with the concomitant variability in the surface snow isotopes (show in red). Air temperature from local GC-Net station shown in cyan. Model outputs from LMDZiso shown with solid dark-gray line. Occurrence of precipitation indicated by vertical gray columns. Precipitation samples collected in 2011 on sub-event shown with black crosses in top panel.

2000 to 8000 ppmv ($R = 0.94$). An evaluation of the snow surface temperature simulated with CROCUS was also carried out. As part of the snow surface monitoring program we installed thermistor probes from the surface and to a depth of $\sim 2$ m. However due to solar heating of the top thermistors and uncertainty in vertical position we use surface temperature estimates from MODIS MOD11_L2 from TERRA and MYD11_L2 from AQUA (Wan, 2009) for clear-sky days and find a very good agreement to surface temperature estimates from CROCUS (see Fig. S3 in Supplement). This performance makes it possible to reliably calculate the vapor fluxes exchanged between the atmosphere and the surface snow layers.

## 3 Results

### 3.1 NEEM climate during summer 2011 and spring–summer 2012

The mean air temperature during summer 2011 (day 185–216) and spring–summer 2012 (day 140–216) is $-6.9\,°C$ and $-6.8\,°C$, respectively (see Fig. 1). Diurnal cycles have a magnitude of $\sim 5$–$10\,°C$ during clear-sky days, but with a reduced magnitude during precipitation events, caused by the clouds leading to a increase in downwelling longwave radiation. Daily averaged temperatures (Fig. 1) show day-to-day variations associated with changes in the large-scale atmospheric circulation, well captured by the LMDZiso nudged simulation (Fig. 1). The summer near-surface air temperatures reached levels close to or slightly above $0\,°C$ for several episodes during both 2011 and 2012. Particularly remarkable are two events in 2012 around day 191–196 (10–15 July) and around day 209 (28 July), during which melting of the surface layer was observed (indicated in Fig. 1). As reported in Steen-Larsen et al. (2011) based on satellite microwave data, starting in 1987, melt has previously (since 1987) occurred in summer 2005 (24 and 25 July 2005). This 2005 two-day event was likely caused by the same cyclonic event rather than from two separate events as in 2012. The 2012 measurements span the spring–summer transition around day 147–150. This transition was not recorded in the 2010 or 2011 data sets, as measurements were not initiated before the summer period had started. During the transition over these days the air temperature increased by $\sim 20$–$25\,°C$ (minimum to maximum)

### 3.2 Variability of $\delta^{18}O_v$ and d-excess$_v$ during summer 2011 and spring–summer 2012

We describe the 2011 and 2012 vapor $\delta^{18}O_v$ and d-excess$_v$ data, report the mean value and range of variations (Table 2), and highlight two remarkable events recorded in spring and summer 2012.

Figure 1 shows that $\delta^{18}O_v$ varied between $-44$ and $-32\,‰$, with an average of $\sim -37\,‰$ during July to early

August 2011. The mid-May to early August 2012 data vary between $-53$ and $-25\,‰$, with an average of $\sim -38\,‰$. Altogether, the range of variation and the mean value are representative of summer values at NEEM as reported earlier (Steen-Larsen et al., 2011, 2013), with the exception of the strongly depleted values in May 2012, which captures seasonal variations associated with major warming from spring to summer.

During days with clear-sky conditions, $\sim 4\,‰$ diurnal variations of $\delta^{18}O_v$ occur in phase with the near-surface air temperature and humidity diurnal cycle, confirming the results obtained in 2010 (Steen-Larsen et al., 2013). These diurnal cycles are not further investigated here, as we will subsequently focus on day-to-day variations.

Both in 2011 and 2012, synoptic variations (large-scale cyclonic variations) in $\delta^{18}O_v$ of typically 4–10 ‰ occur within 1 to 3 days with parallel variations in near-surface air temperatures and humidity. We note that significant variations in $\delta^{18}O_v$ occur during precipitation events (gray-shaded areas of Fig. 1) and in between precipitation events. As also shown for summer 2010 (Steen-Larsen et al., 2013), the day-to-day variability of temperature and $\delta^{18}O_v$ is well captured by LMDZiso (Fig. 1). This confirms that such changes in the $\delta^{18}O_v$ are driven by changes in large-scale circulation, since only the large-scale winds are nudged in this simulation. We do not further investigate the comparison between LMDZiso and our data, as this will be the focus of a separate multi-model–data paper currently in preparation.

Two remarkable $\delta^{18}O_v$ events are observed during the 2012 campaign. The largest event ("spring 2012 event") with an increase of $\sim 25\,‰$ occurs from day 147 to 150, reflecting the transition from spring to summer. The fastest event ("summer 2012 event") is a sharp increase in $\delta^{18}O_v$ by $\sim 12\,‰$ over 6 h associated with $\sim 11\,°C$ warming on day 191. This event reflects the advection of warm air associated with an atmospheric river (Newell and Zhu, 1994), which led to a record heat wave and melting on 98 % of the surface of the Greenland Ice Sheet (Nghiem et al., 2012).

During summer 2011, d-excess$_v$ varies between $\sim 13$ and $\sim 34\,‰$, with an average of $\sim 22\,‰$, while during spring–summer 2012 a larger range of variations was obtained from $\sim 10$ to $\sim 46\,‰$, with a similar average of $\sim 23\,‰$. The range and mean value are comparable to observations from 2010 reported in Steen-Larsen et al. (2013), where a mean value of $\sim 26\,‰$ was observed together with observations of large d-excess values ($> 40\,‰$). Data from 2010, 2011 and 2012 all show a general anti-correlation between $\delta^{18}O_v$ and d-excess$_v$ for synoptic events.

Based on the anomaly of the spring signal in 2012, we disregard this period when investigating $\delta^{18}O$–temperature (Sect. 3.6) or d-excess–$\delta^{18}O$ (Sect. 3.7) relationships.

**Table 2.** Overview of the distribution of the observed $\delta^{18}O$ and d-excess together with the slope and correlation for d-excess vs. $\delta^{18}O$ and $\delta^{18}O$ vs. air temperature for the atmospheric water vapor, precipitation samples, and snow surface samples.

| Sample type | Year | Measurement type | Minimum | Maximum | Average | Slope d-excess vs. $\delta^{18}O$ | $R$ d-excess vs. $\delta^{18}O$ | Slope $\delta^{18}O$ vs. Air temp. | $R$ $\delta^{18}O$ vs. Air temp. |
|---|---|---|---|---|---|---|---|---|---|
| Atmospheric water vapor | 2011 | $\delta^{18}O$ | −44 | −32 | −37 | −1.14 | −0.76 | 0.81 | 0.77 |
| | | d-excess | 13 | 34 | 22 | | | | |
| | 2012 | $\delta^{18}O$ | −53 | −25 | −38 | −1.03 | −0.86 | 0.85 | 0.76 |
| | | d-excess | 10 | 46 | 23 | | | | |
| Precipitation samples | 2011 | $\delta^{18}O$ | −35 | −18 | −26 | −0.47 | −0.57 | 0.64 | 0.24 |
| | | d-excess | 2 | 20 | 11 | | | | |
| Snow surface samples | 2011 | $\delta^{18}O$ | −33 | −21 | −25 | −0.31 | −0.28 | − | 0.17 |
| | | d-excess | 7 | 19 | 12 | | | | |
| | 2012 | $\delta^{18}O$ | −36 | −15 | −24 | −0.44 | −0.53 | 0.31 | 0.32 |
| | | d-excess | 3 | 24 | 13 | | | | |

## 3.3 Variability of $\delta^{18}O_p$ and d-excess$_p$ during summer 2011

The precipitation samples collected during summer 2011 cover four precipitation events (Fig. 1). The $\delta^{18}O_p$ and d-excess$_p$ of the samples vary respectively between $\sim$ −35 and $\sim$ −18‰ with a mean value (arithmetic mean) of $\sim$ −26‰ and between $\sim$ 2 and $\sim$ 20‰ with a mean value of $\sim$ 11‰ (Table 2). No systematic pattern of change can be observed during precipitation events regarding the trends in the isotopic composition of precipitation or vapor.

## 3.4 Variability of $\delta^{18}O_s$ and d-excess$_s$ during summer 2011 and spring–summer 2012

During summer 2011, $\delta^{18}O_s$ varies between $\sim$ −33 and $\sim$ −21‰ with an average value of $\sim$ −25‰. The range of values is smaller compared to $\delta^{18}O_p$, and the mean value is 1‰ more enriched than the average $\delta^{18}O_p$. A larger range of variations is observed in spring–summer 2012, with $\delta^{18}O_s$ varying between $\sim$ −36 and $\sim$ −15‰ with an average value of −24‰ (a similar value is obtained when averaging over the same data collection period in 2011).

For d-excess$_s$, the summer 2011 is characterized by values between $\sim$ 7 and $\sim$ 19‰ with a mean value of $\sim$ 12‰. The range is smaller than depicted in the precipitation data (consistent with different sampling durations), and the mean value is identical. A larger range of variations is observed in spring–summer 2012, from $\sim$ 3 and $\sim$ 24‰ with an average value of $\sim$ 13‰ (averaging over the same period as 2011 yields a value of $\sim$ 12‰).

The isotopic composition of the surface snow varies during precipitation events, consistent with the magnitude of the precipitation isotopic composition changes (Fig. 1, summer 2011). While this is expected from new snow deposition, variations in the isotopic composition of the snow surface are easily observed in between precipitation events. System-

atic comparisons between these evolutions of surface snow and vapor isotopic composition are conducted and discussed in Sect. 4.

## 3.5 $^{17}O$-excess data

Between days 191 and 194 of 2011, larger variations (from 16 to 60 ppm) are recorded in $^{17}O$-excess$_v$ than in $^{17}O$-excess$_s$ (50 to 68 ppm) (Table 2). This finding confirms similar results obtained from samples collected during days 224–228 of 2008, showing a larger range of variability in vapor (10–70 ppm) compared to precipitation data (20–40 ppm) (Landais et al., 2012). These reported values are within the range observed in shallow cores at NEEM (Landais et al., 2012). We also conclude from this small data set that changes in $^{17}O$-excess of water vapor and surface snow occur in between precipitation events, and show qualitatively parallel trends, albeit with larger amplitude in the vapor than the snow surface.

## 3.6 $\delta^{18}O$–temperature relationships in vapor, precipitation and surface snow

The summer $\delta^{18}O_v$ data exhibit a strong correlation with near-surface air temperature (Fig. 2, Table 2), with a slope of $\sim 0.81 \pm 0.13$‰ $°C^{-1}$ ($R = 0.77$) in 2011 (from daily mean values), very similar to the slope observed in 2012 summer observations, $\sim 0.85 \pm 0.11$‰ $°C^{-1}$ ($R = 0.76$) for 2012. When including the observations before the spring–summer transition results in a slope of $\sim 0.86$‰ $°C^{-1}$.

A very low correlation is found between $\delta^{18}O_p$ and temperature, with a slope of $\sim 0.64 \pm 0.49$‰ $°C^{-1}$ ($R = 0.24$). In order to explore the relationship between isotopic composition of surface snow and temperature, we use the mean air temperature averaged over 12 h prior to collection of the snow surface sample. The result does not depend on the integration time (tested over durations of 1 to 48 h). No

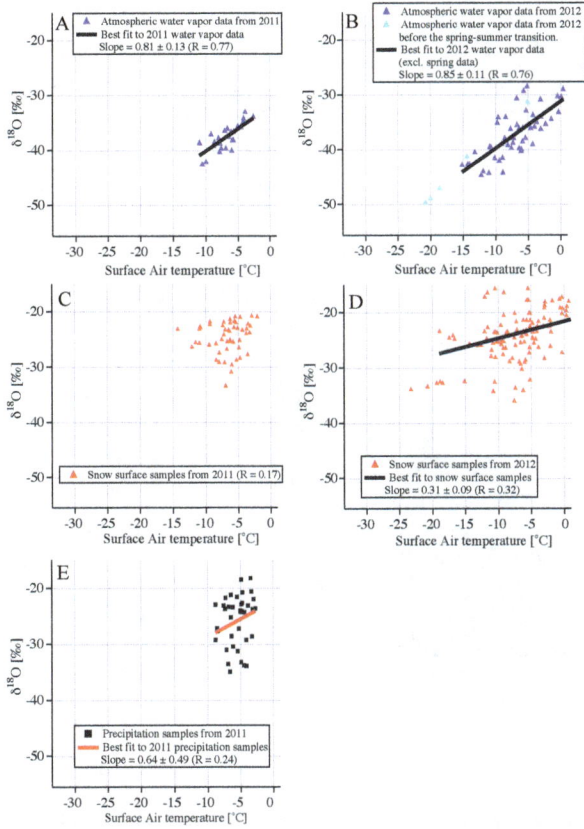

**Fig. 2.** The $\delta^{18}O$ vs. surface air temperature for the observed mean daily water vapor isotopes (**A**: 2011; **B**: 2012), surface snow samples (**C**: 2011; **D**: 2012), and precipitation samples (**E**: 2011).

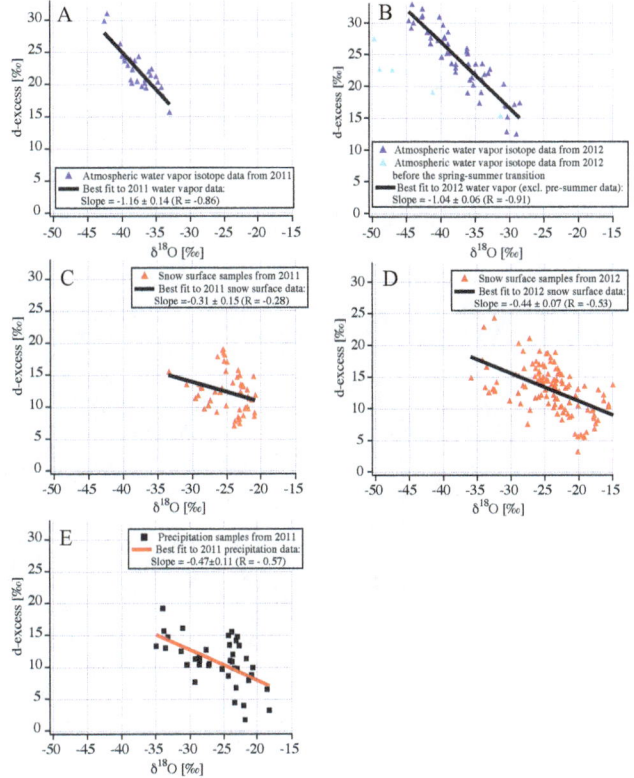

**Fig. 3.** The d-excess vs. $\delta^{18}O$ for the observed mean daily water vapor isotopes (**A**: 2011; **B**: 2012), surface snow samples (**C**: 2011; **D**: 2012), and precipitation samples (**E**: 2011).

significant correlation is detected between $\delta^{18}O_s$ and air temperature during 2011. In summer 2012, a significant correlation is associated with a slope of $\sim 0.31 \pm 0.09\,‰\,°C^{-1}$ ($R = 0.32$) (Fig. 2, Table 2) due to the presence of surface snow samples collected before the spring–summer transition.

### 3.7   d-excess–$\delta^{18}O$ relationships in vapor, precipitation and surface snow in summer 2011 and 2012

For summer vapor data (excluding the 2012 spring–summer transition), daily mean d-excess$_v$ is anti-correlated with $\delta^{18}O_v$, with slopes of $\sim -1.16 \pm 0.14\,‰$ per permil in 2011 ($R = -0.86$) and $\sim -1.04 \pm 0.06\,‰$ per permil in 2012 ($R = -0.91$) (Fig. 3). Similar slopes are obtained from hourly data. By contrast, a weaker relationship associated with a smaller slope emerges from 2011 surface snow data ($\sim -0.31 \pm 0.15\,‰$ per ‰, $R = -0.28$) as well as 2012 surface snow data ($\sim -0.44 \pm 0.07\,‰$ per permil, $R = -0.53$). The latter is closer to the relationship obtained from precipitation data both from our 2011 data set ($\sim -0.47 \pm 0.11\,‰$ per permil, $R = -0.57$) and reported from 2008 precipitation data by Steen-Larsen et al., 2011 ($-0.43 \pm 0.14$, $R = -0.58$).

**Table 3.** Mean summer surface snow isotopic composition from observations and the calculation of equilibrium with surface water vapor at surface air temperature.

| Summer | Observed mean values | Equilibrium with water (Majoube, 1971) | Equilibrium with ice (Majoube, 1970; Merlivat and Nief, 1967) | Equilibrium with ice (Ellehoj et al., 2013) |
|---|---|---|---|---|
| 2011 | | | | |
| $\delta^{18}O_s$: | $-25$ | $-25\,‰$ | $-21\,‰$ | $-22\,‰$ |
| d-excess$_s$ | 12 | $16\,‰$ | $1\,‰$ | $10\,‰$ |
| 2012 | | | | |
| $\delta^{18}O_s$: | $-24$ | $\sim -26\,‰$ | $-22\,‰$ | $-23\,‰$ |
| d-excess$_s$ | 13 | $\sim 16\,‰$ | $1\,‰$ | $10\,‰$ |

### 3.8   Comparison between measured surface snow isotopic composition and values calculated from isotopic equilibrium with near-surface water vapor

To study the relationship between the surface snow and water vapor isotopic composition, we compare the calculated surface snow isotopic composition, using the method described in Sect. 2.7, to the observed data in Fig. 4, with the mean summer isotopic values summarized in Table 3.

The observed $\delta^{18}O_s$ variations lie in between the theoretical calculations of a surface snow at equilibrium with

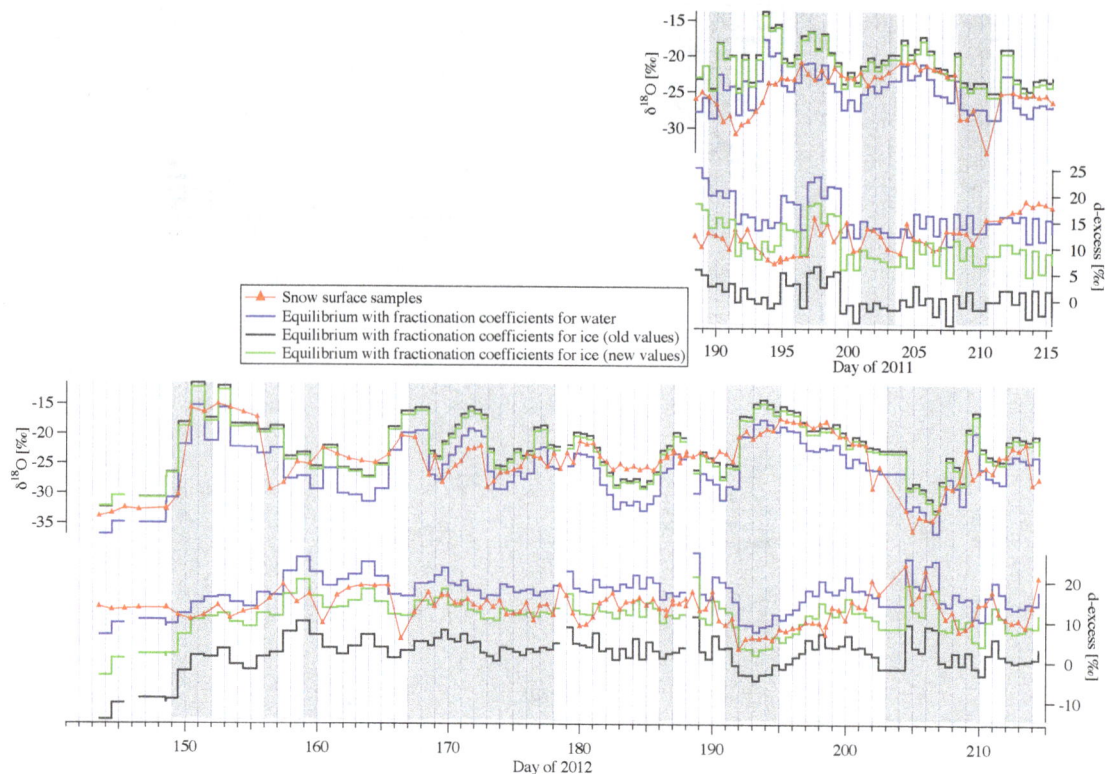

**Fig. 4.** $\delta^{18}$O and d-excess of the condensate calculated based on the air temperature and observed atmospheric water vapor isotopic composition under the assumptions of isotopic equilibrium. Different fractionation coefficients were used assuming liquid water (Majoube, 1971), ice (Majoube, 1970; Merlivat and Nief, 1967), and ice but with new fractionation coefficients estimated by Ellehoj et al. (2013).

vapor using liquid water and ice. During precipitation events, a better agreement is obtained using fractionation with water, while between precipitation events a better agreement is found for calculations of fractionation with ice in between precipitation events.

The observed d-excess$_s$ data also lie between the results obtained when using fractionation factors for liquid water and ice, but a much better fit is obtained when using the new ice fractionation coefficients (Ellehoj et al., 2013). Calculations with the old ice fractionation coefficients led to values much lower than observed.

From this comparison, the hypothesis of equilibrium between vapor and surface snow cannot be rejected. We note that equilibrium between near-surface vapor and precipitation was also suggested from 2008 $\delta^{18}$O, d-excess and $^{17}$O-excess data (Steen-Larsen et al., 2011; Landais et al., 2012).

## 4 Discussion

### 4.1 Robustness of findings when comparing different years:

1. We find similar mean temperature, $\delta^{18}$O and d-excess values in vapor and surface snow for 2011 and 2012.

2. For different years a comparable relationship is found between d-excess vs. $\delta^{18}$O, and $\delta^{18}$O vs. temperature in vapor (2008, 2010, 2011, 2012), precipitation (2008, 2011), and surface snow (2011 and 2012).

3. We conclude that different $\delta^{18}$O-temperature and d-excess-$\delta^{18}$O slopes exist for vapor and surface snow.

4. Our observations indicate that surface snow is at equilibrium with near-surface water vapor (or vice versa).

5. The good agreement between LMDZiso simulations of day-to-day variations in $\delta^{18}$O$_v$ indicates that water vapor isotope variations are driven by large-scale circulation features.

We will therefore subsequently assume that the same processes are at play for several summers despite different meteorological conditions. We seek to investigate what drives the observed variation of surface snow isotopic composition in between snowfall events by comparing magnitudes of change in surface snow isotopes with simultaneous change in water vapor isotope (Sect. 4.2) and investigating possible processes that can explain the observed variation (Sect. 4.3).

**Table 4.** Changes in surface snow and water vapor isotopic composition in between precipitation events. The value is calculated by comparing the mean over the specific days.

| Day begin | Day end | Change in snow surface $\delta^{18}O_s$ [‰] | Change in water vapor $\delta^{18}O_v$ [‰] | Change in snow surface d-excess$_s$ [‰] | Change in water vapor d-excess$_v$ [‰] |
|---|---|---|---|---|---|
| 2011 | | | | | |
| 191 | 194 | ∼ +7.0 | ∼ +8.0 | ∼ −6 | ∼ −10 |
| 199 | 200 | ∼ −1.3 | ∼ −1.3 | – | – |
| 204 | 207 | ∼ +2.4 | ∼ +5.0 | ∼ −5 | – |
| 211 | 214 | ∼ −0.5 | ∼ −3.5 | ∼ +3 | – |
| 2012 | | | | | |
| 153 | 155 | ∼ −2 | ∼ −2 | ∼ +2 | ∼ +2 |
| 160 | 163/164 | ∼ −4 | ∼ −4 | ∼ +10 | ∼ +7 |
| 163/164 | 166 | ∼ +5 | ∼ +12 | ∼ −13 | ∼ −13 |
| 180 | 183 | ∼ −4 | ∼ −8.5 | ∼ +8 | ∼ +7 |
| 196 | 201 | ∼ −4 | ∼ −8 | ∼ +7 | ∼ +6 |

## 4.2 Magnitude of isotopic changes in surface snow and near-surface water vapor

We focus on changes in snow surface isotope values occurring in between precipitation events. We have (between day 191 and 194 of 2011) concomitant $\delta^{18}O$, d-excess and $^{17}O$-excess data for one period. During this period, $\delta^{18}O_s$ increases by ∼ 7‰ in parallel with a ∼ 8‰ increase in $\delta^{18}O_v$. Similarly, d-excess$_s$ decreases by ∼ 6‰, while d-excess$_v$ decreases by ∼ 10‰. Finally, $^{17}O$-excess$_s$ increases by ∼ 15 ppm, while $^{17}O$-excess$_v$ increases by ∼ 40 ppm. During this event, the isotopic composition of the surface snow varies in parallel with that of water vapor, albeit with an attenuated magnitude.

In order to test whether this finding is valid for all periods in between precipitation, Table 3 shows a systematic investigation of the sign and magnitude of changes in $\delta^{18}O_s$, $\delta^{18}O_v$, d-excess$_s$ and d-excess$_v$. Note that no lead or lag time could be identified for the simultaneous changes in the surface snow and water vapor isotopic composition seen in Fig. 1 and documented in Table 4, due to the time resolution of the surface snow measurements.

Table 4 demonstrates that changes in snow surface and water vapor isotopes systematically occur in the same direction, with the exception of two periods (2011, days 204–207 and 211–214) where significant changes are detected in d-excess$_s$ but not in d-excess$_v$. Magnitudes of observed changes in $\delta^{18}O_s$ are either similar (2 situations out of 9) or smaller (7 cases out of 9) than changes in the $\delta^{18}O_v$. Results are more ambiguous for d-excess$_s$, which is possibly due to the magnitude of the signal with respect to the larger analytical uncertainty associated with the water vapor measurements.

From this coevolution the following question arises: are the changes in the snow surface isotopes caused by changes in the water vapor isotopic composition, or are the water vapor isotope changes caused by changes in the snow surface isotopic composition?

## 4.3 Possible causes for change in snow surface and water vapor isotopes

We will only discuss changes in between precipitation events. If changes in the snow surface isotopes lead the changes in the water vapor isotopic composition, we need to explain why the isotopic composition of the snow surface is modified. Our first hypothesis is that the surface snow isotopic composition is affected by the isotopic composition of the firn below the top layer, itself reflecting the isotopic composition, at the time of campaign, of spring or winter snowfall. If this were the case, one would expect the surface snow isotopic composition to show a systematic decrease towards more depleted spring values. However, our observations depict 3 out of 9 cases where $\delta^{18}O_v$ is increasing, and 6 cases out of 9 when it is decreasing. A second hypothesis could be that, for a given $\delta^{18}O_s$, changes in near-surface air temperature affect the isotopic composition of the water vapor formed at equilibrium with this snow. However, a 10 °C warming only leads to a change of ∼ 2‰ in $\delta^{18}O_v$ under the assumption of isotopic equilibrium. This mechanism is therefore unable to account for the magnitude of $\delta^{18}O_s$ changes (Table 4). Finally, we cannot identify any mechanism that could explain why changes in $\delta^{18}O_s$ (alone) can lead to a larger change $\delta^{18}O_v$. We now investigate how mass fluxes between surface snow and atmosphere may explain this finding.

In order to guide our discussion of exchanges between the snow surface and the atmosphere, we use the mean daily vapor mass flux calculated by CROCUS (Fig. 5) to estimate the daily vapor mass flux from the snow surface to the atmosphere. The mass of the snow surface layer (top 0.5 cm)

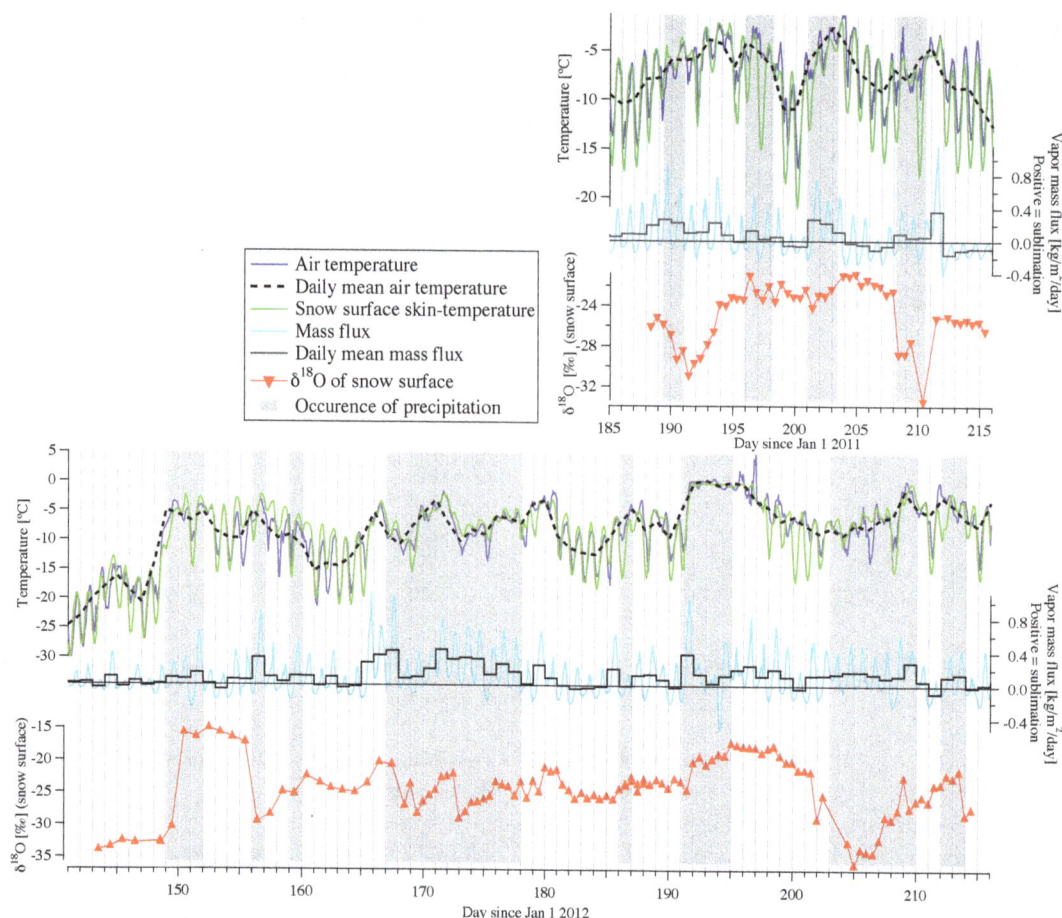

**Fig. 5.** Top and bottom panel show data from 2011 and 2012, respectively. The air temperature and snow surface skin temperature modeled by CROCUS are shown in blue and green, respectively (Daily mean air temperature shown in black). The estimated vapor mass flux from CROCUS with hourly resolution and daily mean values are shown in cyan and black, respectively. The variability in the surface snow $\delta^{18}O_s$ value is shown below.

is estimated to be $\sim 1.7\,\mathrm{kg\,m^{-2}}$, assuming a surface density of $340\,\mathrm{kg\,m^{-3}}$ (Steen-Larsen et al., 2011). For summer 2011 and 2012, the estimated flux varies between $\sim -0.15$ and $\sim 0.40\,\mathrm{kg\,m^{-2}\,day^{-1}}$ (20 % of our surface samples), with a mean value of $\sim 0.1\,\mathrm{kg\,m^{-2}\,day^{-1}}$ (6 % of our surface samples). The largest values are identified during precipitation events. In between precipitation events, the average mass flux corresponds to $\sim 10\,\%$ of the mass of the top layer each day.

Sublimation is normally assumed not to change the isotopic composition of the snow (Dansgaard, 1964; Town et al., 2008; Neumann and Waddington, 2004). However to test this hypothesis we will now investigate whether there is a simple fingerprint of sublimation in the changes of the isotopic composition of the snow surface. We note that CROCUS simulates net sublimation (positive mass flux) between days 191 and 196 of 2011, which corresponds to a negative trend of $\delta^{18}O_s$, and between days 152 and 156 (positive trend of $\delta^{18}O_s$), 160 and 166 (negative then positive trend of $\delta^{18}O_s$), and between 195 and 202 of 2012 (negative trend of $\delta^{18}O_s$).

We are therefore not able to identify any systematic relationship between sublimation and trends in $\delta^{18}O_s$.

Another hypothesis is that sampling of progressively older snow, as the snow surface is removed by sublimation, may cause the observed change in the snow surface isotopes. This hypothesis can be tested by comparing either earlier precipitation isotope data (for summer 2011) or by comparing earlier surface snow isotope data during precipitation events. The changes in d-excess$_s$ appear to contradict this hypothesis. On day 194 of 2011, d-excess$_s$ reaches a level lower than observed during the previous precipitation event. Similarly, the drop in d-excess$_s$ observed on day 165 of 2012 is lower than earlier values measured in the snow surface. The d-excess$_s$ maximum measured on day 202 of 2012 is higher than earlier snow measurements. During the same period (days 195 to 202, 2012), the observed d-excess$_s$ increase is opposite to what would be expected based on the preceding precipitation event. We therefore do not find support for the hypothesis that sublimation causes sampling of the snow surface to be similar to playing a tape recorder in reverse.

We also hypothesize that surface snow isotopic composition reflects surface hoar formation. During NEEM field campaigns, such surface hoar formation was observed during clear-sky nights in response to condensation of water vapor in the air on to the cold snow surface. Similarly, deposition (negative mass flux from surface to atmosphere) is also simulated by CROCUS on days with a clear diurnal cycle. In these cases the simulated diurnal amount of sublimation is much larger than the simulated diurnal amount of condensation. It is therefore likely that the night frost would vanish throughout the day. The ice self-diffusion (diffusion inside the ice matrix) has a value of $\sim 5 \times 10^{-8}\,\mathrm{m^2\,yr^{-1}}$ at $-20\,^\circ\mathrm{C}$ (Whillans and Grootes, 1985) and the characteristic time for diffusion in a grain in the surface layer is about $10^{-2}\,\mathrm{yr}$ (Waddington et al., 2002). We therefore do not expect the snow surface to take up a significant isotopic signal from the surface hoar formed during the night. We do notice that, during clear-sky days, the diurnal cycle of sublimation/condensation is reflected in the water vapor diurnal cycles of both $\delta^{18}O_v$ and d-excess$_v$ as also observed by Steen-Larsen et al. (2013).

At the day-to-day scale, our final hypothesis is that changes in snow surface isotopic composition are driven by synoptic changes in the atmospheric water vapor isotopic composition. First, the day-to-day variability of $\delta^{18}O_v$ is well captured by the LMDZiso atmospheric general circulation model, nudged to ECMWF operational analysis wind fields (Fig. 1). We therefore conclude that changes in the snow surface isotopes do not drive the variability in the water vapor isotopes, which instead is driven by changes in large-scale winds and moisture advection. The fact that changes in $\delta^{18}O_v$ in between precipitation events are always greater than or equal to the changes in $\delta^{18}O_s$ (Table 3) supports the hypothesis that the snow surface isotopic composition takes up part of the atmospheric water vapor signal.

So far, we do not have data available to constrain the nature of the process at play. For surface snow subjected to a temperature gradient of $0.5\,^\circ\mathrm{C\,cm^{-1}}$, Pinzer et al. (2012) showed using X-ray-computed tomography that the characteristic residence time for a snow crystal to stay in place before being sublimated is 2–3 days. This corresponds to a mass turnover time of $\sim 60\,\%$ of total ice mass per day (Pinzer et al., 2012). Based on model outputs from CROCUS and our in situ temperature observations, such a temperature gradient occurs in the top 5 cm of the NEEM snowpack on clear-sky days. Larger temperature gradients in the top 1–2 cm of the snowpack cannot be excluded. We speculate that wind pumping (Clarke and Waddington, 1991; Neumann and Waddington, 2004) can cause a continuous replacement of the interstitial water vapor in the top snow layer. Due to the continuous recrystallization described by Pinzer et al. (2012), we hypothesize that this process leads to an imprint of changes in near-surface water vapor isotopic composition into surface snow. Further investigations including controlled laboratory experiments and isotopic modeling are needed to understand how metamorphism processes can impact the $\delta^{18}O$–temperature and d-excess–$\delta^{18}O$ relationships.

## 5 Conclusions

During the two warm summers of 2011 and 2012 at NEEM, continuous measurements of near-surface water vapor combined with isotopic measurements of snow samples collected every 12–24 h from the top 0.5 cm of the snow surface reveal parallel variations in between precipitation events, with larger variations in the vapor $\delta^{18}O_v$ than in the snow surface $\delta^{18}O_s$. We also report positive correlations between $\delta^{18}O$ and temperature and negative correlations between d-excess and $\delta^{18}O$ in the vapor data, but weaker correlations as well as different slopes in precipitation and surface snow data, a finding confirmed by results obtained from data sets acquired during earlier years at NEEM. We note a decoupling between vapor $\delta^{18}O$ and temperature during the warmest days such as the summer 2012 heat wave. Changes in near-surface air temperature during the summer only account for a tiny fraction of the variability in precipitation and surface snow isotopic composition. This finding is consistent with investigations of shallow ice core data, which show much weaker relationships between Greenland summer $\delta^{18}O$ and temperature records than during winter (Vinther et al., 2010). Simple isotopic calculations show that the hypothesis of equilibrium (Steen-Larsen et al., 2011) between near-surface vapor and surface snow cannot be ruled out when considering fractionation assuming liquid water or fractionation assuming ice when using newly estimated fractionation coefficient coefficients (Ellehoj et al., 2013).

The most surprising result from our work is the fact that surface snow and near-surface vapor evolve in tandem in between precipitation events, with similar or larger changes in vapor $\delta^{18}O$ than in surface snow. The fact that an atmospheric general circulation model nudged to large-scale operational analysis wind fields is able to capture day-to-day variations in vapor isotopic composition shows that the near-surface water vapor isotopic composition is controlled by large-scale changes in air mass trajectories and distillation paths.

In an earlier work (Steen-Larsen et al., 2011), we had reported a diurnal cycle observed in near-surface air temperature, humidity and water vapor isotopic composition at the NEEM site during clear-sky days. The qualitative comparison with CROCUS mass flux calculations confirms that this cycle is very likely driven by diurnal cycle of condensation (hoar deposition at night and sublimation at day time). Further quantitative simulations would be required to implement water stable isotopes in a modeling framework including the surface snow and atmospheric boundary layer. For the NEEM site, 3 yr of continuous water vapor data (2010–2012) are available for the evaluation of such simulations.

On day-to-day scales, no systematic relationship is observed between the CROCUS surface–atmosphere mass flux and our isotopic trend. We also show that sublimation does not make the surface reveal earlier precipitation and snow surface isotope values similar to a tape recorder played in reverse. We therefore suggest that day-to-day variations of surface snow isotopic composition in between precipitation events are caused by an uptake of the atmospheric water vapor isotopic signal driven by the continuous replacement of interstitial water vapor with atmospheric water vapor. This isotopic signal of the interstitial water vapor is then transferred into the surface snow due to snow metamorphism associated with a strong temperature gradient in the upper centimeters of the snow surface. Laboratory experiments conducted by injecting an isotopically known vapor into a snow disk of known isotopic composition and different temperature gradients are needed in order to quantify the magnitude and rates of changes of the isotopic processes occurring during controlled snow metamorphism. Such laboratory experiments would also allow for validating the implementation of water stable isotopes in snow models such as CROCUS. Similar monitoring frameworks in different places could investigate the validity of our findings for other sites. If our interpretation of the observed signals is correct, changes in surface snow isotopic composition are expected to be significant (i) if large day-to-day surface changes in water vapor occur in between precipitation events, (ii) wind pumping is efficient and (iii) snow metamorphism is enhanced by large temperature gradients in the upper first centimeters of the snow. Due to prolonged periods of time between precipitation events, central East Antarctica would be an excellent site for a case study, albeit with the challenge to accurately measure the isotopic composition of water vapor at very low concentrations, even in summer, when metamorphism is expected to be larger.

Our hypothesis that the surface snow isotopic composition is affected by isotopic exchanges with the atmospheric water vapor in between precipitation events also has implications for the interpretation of ice core records. Indeed, classically, ice core stable isotope records are interpreted as reflecting precipitation-weighted signals, and compared to observations (e.g., station data, atmospheric reanalysis) and atmospheric model results for precipitation (using $\delta^{18}O$ precipitation data, or precipitation-weighted temperature), ignoring such snow–vapor exchanges. Recording a surface climate signal even when no precipitation is deposited suggests a more continuous archiving process than previously thought, but makes the comparison with atmospheric simulations more challenging. It has long been known that processes in the surface snow attenuate the signal associated with each precipitation event. Mathematical calculations of so-called isotopic diffusion require being reconciled with the physical understanding of the processes at play. Our findings also challenge the use of purely statistical back-diffusion calculations in order to restore the full magnitude of seasonal variations, a method classically applied for identifying seasonal cycles in damped isotopic signals.

*Acknowledgements.* This paper is dedicated to the memory of our friend and mentor Sigfús J. Johnsen (1940–2013) for being an unlimited source of inspiration and encouragement. NEEM is directed and organized by the Centre for Ice and Climate at the Niels Bohr Institute and US NSF, Office of Polar Programs. It is supported by funding agencies and institutions in Belgium (FNRS-CFB and FWO), Canada (GSC), China (CAS), Denmark (FIST), France (IPEV, INSU/CNRS and ANR VMC NEEM), Germany (AWI), Iceland (RannIs), Japan (NIPR), Korea (KOPRI), the Netherlands (NWO/ALW), Sweden (VR), Switzerland (SNF), the UK (NERC) and the USA (US NSF, Office of Polar Programs). The numerical simulations with LMDZiso were performed on the NEC-SX8 of the IDRIS/CNRS computing center. The work was supported by the Danish Council for Independent Research – Natural Sciences grant number 09-072689 and 10-092850, by the French Agence Nationale de la Recherche (grants VMC NEEM and CEPS GREEN-LAND), and by the AXA Research Fund. We also acknowledge the MODIS mission scientists and associated NASA personnel for the production of the data used in this research effort.

The authors thank the editor and two anonymous reviewers for valuable suggestions during the review process, which improved the final version of this paper.

Edited by: E. Wolff

## References

Barkan, E. and Luz, B.: High precision measurements of $O_{17}/O_{16}$ and $O_{18}/O_{16}$ ratios in $H_2O$, Rapid Commun. Mass Sp., 19, 3737–3742, doi:10.1002/rcm.2250, 2005.

Brun, E., Six, D., Picard, G., Vionnet, V., Arnaud, L., Bazile, E., Boone, A., Bouchard, A., lie, Genthon, C., Guidard, V., Le Moigne, P., Rabier, F., and Seity, Y.: Snow/atmosphere coupled simulation at dome c, antarctica, J. Glaciol., 57, 721–736, 2011.

Brun, E., Vionnet, V., Boone, A., Decharme, B., Peings, Y., Valette, R. M., Karbou, F., and Morin, S.: Simulation of northern eurasian local snow depth, mass, and density using a detailed snowpack model and meteorological reanalyses, J. Hydrometeorol., 14, 203–219, doi:10.1175/jhm-d-12-012.1, 2013.

Clarke, G. K. C. and Waddington, E. D.: A three-dimensional theory of wind pumping, J. Glaciol., 37, 89–96, 1991.

Cuffey, K. M. and Steig, E. J.: Isotopic diffusion in polar firn: implications for interpretation of seasonal climate parameters in ice-core records, with emphasis on central greenland, J. Glaciol., 44, 273–284, 1998.

Dahl-Jensen, D., Mosegaard, K., Gundestrup, N., Clow, G. D., Johnsen, S. J., Hansen, A. W., and Balling, N.: Past temperatures directly from the greenland ice sheet, Science, 282, 268–271, 1998.

Dansgaard, W.: Stable isotopes in precipitation, Tellus, 16, 436–468, 1964.

Dee, D. P., Uppala, S. M., Simmons, A. J., Berrisford, P., Poli, P., Kobayashi, S., Andrae, U., Balmaseda, M. A., Balsamo, G., Bauer, P., Bechtold, P., Beljaars, A. C. M., van de Berg, L., Bidlot, J., Bormann, N., Delsol, C., Dragani, R., Fuentes, M., Geer, A. J., Haimberger, L., Healy, S. B., Hersbach, H., Hólm, E. V., Isaksen, L., Kållberg, P., Köhler, M., Matricardi, M., McNally, A. P., Monge-Sanz, B. M., Morcrette, J. J., Park, B. K., Peubey, C., de Rosnay, P., Tavolato, C., Thépaut, J. N., and Vitart, F.: The era-interim reanalysis: configuration and performance of the data assimilation system, Q. J. Roy. Meteor. Soc., 137, 553–597, doi:10.1002/qj.828, 2011.

Ellehoj, M. D., Steen-Larsen, H. C., Johnsen, S. J., and Madsen, M. B.: Ice-vapor equilibrium fractionation factor of hydrogen and oxygen isotopes: experimental investigations and implications for stable water isotope studies, Rapid Commun. Mass Sp., 27, 2149–2158, doi:10.1002/rcm.6668, 2013.

Fisher, D. A. and Koerner, R. M.: The effects of wind on d($^{18}$O) and accumulation give an inferred record of seasonal d amplitude from the agassiz ice cap, ellesmere island, Canada, Ann. Glaciol., 10, 34–37, 1988.

Fisher, D. A. and Koerner, R. M.: Signal and noise in four ice-core records from the agassiz ice cap, ellesmere island, canada: Details of the last millennium for stable isotopes, melt and solid conductivity, The Holocene, 4, 113–120, 1994.

Guillevic, M., Bazin, L., Landais, A., Kindler, P., Orsi, A., Masson-Delmotte, V., Blunier, T., Buchardt, S. L., Capron, E., Leuenberger, M., Martinerie, P., Prié, F., and Vinther, B. M.: Spatial gradients of temperature, accumulation and $\delta^{18}$O-ice in Greenland over a series of Dansgaard–Oeschger events, Clim. Past, 9, 1029–1051, doi:10.5194/cp-9-1029-2013, 2013.

Hourdin, F., Musat, I., Bony, S., Braconnot, P., Codron, F., Dufresne, J. L., Fairhead, L., Filiberti, M. A., Friedlingstein, P., Grandpeix, J. Y., Krinner, G., Levan, P., Li, Z. X., and Lott, F.: The lmdz4 general circulation model: Climate performance and sensitivity to parametrized physics with emphasis on tropical convection, Clim. Dynam., 27, 787–813, doi:10.1007/s00382-006-0158-0, 2006.

Johnsen, S. J.: Stable isotope homogenization of polar firn and ice, in: Proc. of Symp. on isotopes and impurities in snow and ice, i.U.G.G. Xvi, general assembly, grenoble aug. September 1975, Iahs-aish publ. 118, Washington DC, 210–219, 1977.

Johnsen, S. J., Dansgaard, W., and White, J. W. C.: The origin of arctic precipitation under present and glacial conditions, Tellus B, 41, 452–468, 1989.

Johnsen, S. J., Clausen, H. B., Cuffey, K. M., Hoffmann, G., Schwander, J., and Creyts, T.: Diffusion of stable isotopes in polar firn and ice: the isotope effect in firn diffusion, in: Physics of Ice Core Records, edited by: Hondoh, T., Hokkaido University Press, Sapporo, 121–140, 2000.

Johnsen, S. J., Dahl-Jensen, D., Gundestrup, N., Steffensen, J. P., Clausen, H. B., Miller, H., Masson-Delmotte, V., Sveinbjörnsdottir, A. E., and White, J.: Oxygen isotope and palaeotemperature records from six greenland ice-core stations: Camp century, dye-3, grip, gisp2, renland and northgrip, J. Quaternary Sci., 16, 299–307, 2001.

Jouzel, J. and Merlivat, L.: Deuterium and oxygen 18 in precipitation: modeling of the isotopic effects during snow formation, J. Geophys. Res., 89, 11749–11757, 1984.

Jouzel, J., Alley, R. B., Cuffey, K. M., Dansgaard, W., Grootes, P., Hoffmann, G., Johnsen, S. J., Koster, R. D., Peel, D., Shuman, C. A., Stievenard, M., Stuiver, M., and White, J.: Validity of the temperature reconstruction from water isotopes in ice cores, J. Geophys. Res., 102, 26471–26487, 1997.

Kindler, P., Guillevic, M., Baumgartner, M., Schwander, J., Landais, A., and Leuenberger, M.: NGRIP temperature reconstruction from 10 to 120 kyr b2k, Clim. Past Discuss., 9, 4099–4143, doi:10.5194/cpd-9-4099-2013, 2013.

Krinner, G., Genthon, C., and Jouzel, J.: Gcm analysis of local influences on ice core d signals, Geophys. Res. Lett., 24, 2825–2828, 1997.

Landais, A., Caillon, N., Severinghaus, J., Barnola, J.-M., Goujon, C. L., Jouzel, J., and Masson-Delmotte, V. R.: Isotopic measurements of air trapped in ice to quantify temperature changes, C. R. Geosci., 336, 963–970, doi:10.1016/j.crte.2004.03.013, 2004.

Landais, A., Steen-Larsen, H. C., Guillevic, M., Masson-Delmotte, V., Vinther, B., and Winkler, R.: Triple isotopic composition of oxygen in surface snow and water vapor at neem (greenland), Geochim. Cosmochim. Ac., 77, 304–316, 2012.

Majoube, M.: Fractionation factor of $^{18}$O between water vapour and ice, Nature, 226, p. 1242, 1970.

Majoube, M.: Fractionnement en oxygène 18 et en deutérium entre l'eau et sa vapeur, J. Climate Phys., 68, 1423–1436, 1971.

Masson-Delmotte, V., Jouzel, J., Landais, A., Stievenard, M., Johnsen, S. J., White, J. W. C., Werner, M., Sveinbjornsdottir, A., and Fuhrer, K.: Grip deuterium excess reveals rapid and orbital-scale changes in greenland moisture origin, Science, 309, 118–121, 2005a.

Masson-Delmotte, V., Landais, A., Stievenard, M., Cattani, O., Falourd, S., Jouzel, J., Johnsen, S. J., Dahl-Jensen, D., Sveinbjornsdottir, A., White, J. W. C., Popp, T., and Fisher, H.: Holocene climatic changes in greenland: Different deuterium excess signals at greenland ice core project (grip) and northgrip, J. Geophys. Res., 110, D14102, doi:10.1029/2004JD005575, 2005b.

Merlivat, L. and Jouzel, J.: Global climatic interpretation of the deuterium-oxygen 18 relationship for precipitation, J. Geophys. Res., 84, 5029–5033, 1979.

Merlivat, L. and Nief, G.: Fractionnement isotopique lors des changements d'état solide-vapeur et liquide-vapeur de l'eau à des températues inférieures à 0 °C, Tellus, 1, 122–127, 1967.

NEEM Community members: Eemian interglacial reconstructed from a greenland folded ice core, Nature, 493, 489–494, 2013.

Neumann, T. A. and Waddington, E. D.: Effects of firn ventilation on isotopic exchange, J. Glaciol., 50, 183–194, 2004.

Newell, R. E. and Zhu, Y.: Tropospheric River: A one-year record and a possible application to ice core data, Geophys. Res. Lett., 21, 113–116, doi:10.1029/93GL03113, 1994.

Nghiem, S. V., Hall, D. K., Mote, T. L., Tedesco, M., Albert, M. R., Keegan, K., Shuman, C. A., DiGirolamo, N. E., and Neumann, G.: The extreme melt across the greenland ice sheet in 2012, Geophys. Res. Lett., 39, L20502, doi:10.1029/2012gl053611, 2012.

Ortega, P., Swingedouw, D., Masson-Delmotte, V., Risi, C., Vinther, B., Yiou, P., Vautard, R., and Yoshimura, K.: Characterizing atmospheric circulation signals in greenland ice cores: insights from weather regime approach, Clim. Dynam., accepted, doi:10.1007/s00382-014-2074-z, 2014.

Persson, A., Langen, P. L., Ditlevsen, P., and Vinther, B. M.: The influence of precipitation weighting on interannual variability of stable water isotopes in greenland, J. Geophys. Res.-Atmos., 116, D20120, doi:10.1029/2010jd015517, 2011.

Pinzer, B. R., Schneebeli, M., and Kaempfer, T. U.: Vapor flux and recrystallization during dry snow metamorphism under a steady temperature gradient as observed by time-lapse microtomography, The Cryosphere, 6, 1141–1155, doi:10.5194/tc-6-1141-2012, 2012.

Rae, J. G. L., Alalgeirsdottir, G., Edwards, T. L., Fettweis, X., Gregory, J. M., Hewitt, H. T., Lowe, J. A., Lucas-Picher, P., Mottram, R. H., Payne, A. J., Ridley, J. K., Shannon, S. R., van de Berg, W. J., van de Wal, R. S. W., and van den Broeke, M. R.: Greenland ice sheet surface mass balance: evaluating simulations and making projections with regional climate models, The Cryosphere, 6, 1275–1294, doi:10.5194/tc-6-1275-2012, 2012.

Risi, C., Bony, S., Vimeux, F., and Jouzel, J.: Water stable isotopes in the lmdz4 general circulation model: model evaluation for present day and past climates and applications to climatic interpretations of tropical isotopic records, J. Geophys. Res., 115, D12118, doi:10.1029/2009JD013255, 2010.

Risi, C., Noone, D., Frankenberg, C., and Worden, J.: Role of continental recycling in intraseasonal variations of continental moisture as deduced from model simulations and water vapor isotopic measurements, Water Resour. Res., 49, 4136–4156, doi:10.1002/wrcr.20312, 2013.

Severinghaus, J. P. and Brook, E. J.: Abrupt climate change at the end of the last glacial period inferred from trapped air in polar ice, Science, 286, 930–943, 1999.

Sjolte, J., Hoffmann, G., Johnsen, S. J., Vinther, B. M., Masson-Delmotte, V., and Sturm, C.: Modeling the water isotopes in greenland precipitation 1959–2001 with the meso-scale model remo-iso, J. Geophys. Res., 116, D18105, doi:10.1029/2010jd015287, 2011.

Steen-Larsen, H. C., Masson-Delmotte, V., Sjolte, J., Johnsen, S. J., Vinther, B. M., Breon, F. M., Clausen, H. B., Dahl-Jensen, D., Falourd, S., Fettweis, X., Gallee, H., Jouzel, J., Kageyama, M., Lerche, H., Minster, B., Picard, G., Punge, H. J., Risi, C., Salas, D., Schwander, J., Steffen, K., Sveinbjornsdottir, A. E., Svensson, A., and White, J.: Understanding the climatic signal in the water stable isotope records from the neem shallow firn/ice cores in northwest greenland, J. Geophys. Res.-Atmos., 116, D06108, doi:10.1029/2010jd014311, 2011.

Steen-Larsen, H. C., Johnsen, S. J., Masson-Delmotte, V., Stenni, B., Risi, C., Sodemann, H., Balslev-Clausen, D., Blunier, T., Dahl-Jensen, D., Ellehøj, M. D., Falourd, S., Grindsted, A., Gkinis, V., Jouzel, J., Popp, T., Sheldon, S., Simonsen, S. B., Sjolte, J., Steffensen, J. P., Sperlich, P., Sveinbjörnsdóttir, A. E., Vinther, B. M., and White, J. W. C.: Continuous monitoring of summer surface water vapor isotopic composition above the Greenland Ice Sheet, Atmos. Chem. Phys., 13, 4815–4828, doi:10.5194/acp-13-4815-2013, 2013.

Steffen, K. and Box, J.: Surface climatology of the greenland ice sheet: Greenland climate network 1995–1999, J. Geophys. Res., 106, 33951–33964, 2001.

Steffen, K., Box, J. E., and Abdalati, W.: Greenland climate network: Gc-net, CRREL 96-27 Special Report on glaciers, Ice Sheets and Volcanoes, trib. to M. Meier, 98–103, 1996.

Steffensen, J. P., Andersen, K. K., Bigler, M., Clausen, H. B., Dahl-Jensen, D., Fischer, H., Goto-Azuma, K., Hansson, M., Johnsen, S. J., Jouzel, J., Masson-Delmotte, V., Popp, T., Rasmussen, S. O., Rothlisberger, R., Ruth, U., Stauffer, B., Siggaard-Andersen, M. L., Sveinbjornsdottir, A. E., Svensson, A., and White, J. W. C.: High-resolution greenland ice core data show abrupt climate change happens in few years, Science, 321, 680–684, doi:10.1126/science.1157707, 2008.

Town, M. S., Warren, S. G., Walden, V. P., and Waddington, E. D.: Effect of atmospheric water vapor on modification of stable isotopes in near-surface snow on ice sheets, J. Geophys. Res.-Atmos., 113, D24303, doi:10.1029/2008jd009852, 2008.

Vinther, B. M., Jones, P. D., Briffa, K. R., Clausen, H. B., Andersen, K. K., Dahl-Jensen, D., and Johnsen, S. J.: Climatic signals in multiple highly resolved stable isotope records from greenland, Quaternary Sci. Rev., 29, 522–538, 2010.

Vionnet, V., Brun, E., Morin, S., Boone, A., Faroux, S., Le Moigne, P., Martin, E., and Willemet, J.-M.: The detailed snowpack scheme Crocus and its implementation in SURFEX v7.2, Geosci. Model Dev., 5, 773–791, doi:10.5194/gmd-5-773-2012, 2012.

Waddington, E. D., Steig, E. J., and Newmann, T. A.: Using characteristic times to assess whether stable isotopes in polar snow can be reversibly deposited, Ann. Glaciol., 35, 118–124, 2002.

Wan: Collection-5 modis land surface temperature products user's guide, available at: www.icess.ucsb.edu/modis/LstUsrGuide/MODIS_LST_products_Users_guide_C5.pdf (last access: 1 October 2013), 2009.

Werner, M., Heimann, M., and Hoffmann, G.: Isotopic composition and origin of polar precipitation in present and glacial climate simulations, Tellus B, 53, 53–71, 2001.

Whillans, I. M. and Grootes, P. M.: Isotopic diffusion in cold snow and firn, J. Geophys. Res., 90, 3910–3918, 1985.

White, J. W. C., Barlow, L. K., Fisher, D., Grootes, P. M., Jouzel, J., Johnsen, S. J., Stuiver, M., and Clausen, H.: The climate signal in the stable isotopes from summit, greenland: Results of comparisons with modern climate observations., J. Geophys. Res., 102, 26425–26439, 1997.

Yoshimura, K., Kanamitsu, M., Noone, D., and Oki, T.: Historical isotope simulation using reanalysis atmospheric data, J. Geophys. Res.-Atmos., 113, D19108, doi:10.1029/2008jd010074, 2008.

# 9

# On-line and off-line data assimilation in palaeoclimatology: a case study

A. Matsikaris[1,2], M. Widmann[1], and J. Jungclaus[2]

[1]University of Birmingham, Edgbaston, Birmingham B15 2TT, UK
[2]Max Planck Institute for Meteorology, Hamburg, Germany

*Correspondence to:* A. Matsikaris (axm368@bham.ac.uk)

**Abstract.** Different ensemble-based data assimilation (DA) approaches for palaeoclimate reconstructions have been recently undertaken, but no systematic comparison among them has been attempted. We compare an off-line and an on-line ensemble-based method, with the testing period being the 17th century, which led into the Maunder Minimum. We use a low-resolution version of Max Planck Institute for Meteorology Earth System Model (MPI-ESM) to assimilate the Past Global Changes (PAGES) 2k continental temperature reconstructions. In the off-line approach, the ensemble for the entire simulation period is generated first and then the ensemble is used in combination with the empirical information to produce the analysis. In contrast, in the on-line approach, the ensembles are generated sequentially for sub-periods based on the analysis of previous sub-periods. Both schemes perform better than the simulations without DA. The on-line method would be expected to perform better if the assimilation led to states of the slow components of the climate system that are close to reality and the system had sufficient memory to propagate this information forward in time. In our comparison, which is based on analysing correlations and differences between the analysis and the proxy-based reconstructions, we find similar skill for both methods on the continental and hemispheric scales. This indicates either a lack of control of the slow components in our setup or a lack of skill in the information propagation on decadal timescales. Additional experiments are however needed to check whether the conclusions reached in this particular setup are valid in other cases. Although the performance of the two schemes is similar and the on-line method is more difficult to implement, the temporal consistency of the analysis in the on-line method makes it in general preferable.

## 1 Introduction

Reconstructing the climate of the past is crucial for quantifying and understanding natural climate change, which in turn is essential for detecting anthropogenic climate change, as well as for the validation of climate models that are used to provide future climate projections. As the instrumental meteorological records are too short to estimate low-frequency variability, reconstructions based on climate proxy data or numerical simulations are used for this purpose. However, both approaches are associated with substantial uncertainties. In principle, the best state estimates can be expected by employing data assimilation (DA) techniques, which systematically combine the empirical information from proxy data with the representation of the processes that govern the climate system given by climate models. Although DA is a very mature field in numerical weather prediction, the specific problem in palaeoclimatology is different and the methods cannot be directly transferred (e.g. Widmann et al., 2010; Hakim et al., 2013). DA is an emerging research area and can be considered as one of the key challenges in palaeoclimatology.

There are two types of proxy-based reconstructions, those for large-scale, e.g. continental or hemispheric averages (e.g. Crowley and Lowery, 2000; Moberg et al., 2005; Mann et al., 2008; Ljungqvist, 2010; PAGES 2K Consortium, 2013) and spatial field reconstructions (e.g. Briffa et al., 1994; Luterbacher et al., 2004; Jones and Mann, 2004; Xoplaki et al., 2005; Mann et al., 2009). Proxy-based estimates of climate variability contain considerable errors: different proxies usually represent different seasons, different statistical methods used in the reconstructions lead to different results, and non-climatic factors influence the proxies (e.g. Jansen et al., 2007; Jones and Mann, 2004). Moreover, the poor spatial coverage of the climate proxies leads to errors in hemispheric or continental means and even larger errors in full-field reconstructions. The climate states provided by standard model simulations are spatially complete and provide an independent estimate which can be checked for consistency with the proxies, on both large and regional scales. However, the simulations also have errors, e.g. systematic model biases and errors in the climate forcings or in the response to them. Additionally, interannual-to-decadal temperature variations have a large random, non-forced component, and thus agreement of simulations and observations is very unlikely on these timescales. The forcings do not precisely determine the temporal evolution of the climate, particularly on regional scales. Ensemble simulations are indispensable in order to better assess the internal variability for periods within the last millennium (Jungclaus et al., 2010).

Data assimilation combines the two previous methods to find estimates that are both consistent with the empirical knowledge and with the dynamical understanding of the climate system, providing complete spatial fields. It uses the empirical data after the construction of the model to either estimate, correct or select the system state (e.g. Hakim et al., 2013; Bronnimann et al., 2013), or to systematically improve some model parameters (e.g. Annan et al., 2013). Here, we consider the case of state estimation, where DA aims to capture the real-world random, non-forced variability in a simulation and to provide information for variables for which no empirical estimates exist.

Attempts to assimilate proxy data into models include different approaches, such as the selection of ensemble members, forcing singular vectors, and pattern nudging (e.g. Widmann et al., 2010). Ensemble member selection techniques, like the one implemented here, are based on the selection of simulations from an ensemble that are closest to the empirical evidence on climate. A general advantage of these techniques is that they are easy and straightforward to implement, and they are the most frequently used methods by the community. Goosse et al. (2006) were the first to use this method for palaeoclimate research, employing a simplified global 3-D climate model. An updated version was employed by Goosse et al. (2010), using a more advanced 3-D Earth-System Model of Intermediate Complexity (EMIC), along with a set of 56 proxy series derived from a comprehensive compilation of Mann et al. (2008). In the first study, the best model analog was selected by comparing the simulations with proxy-based temperature reconstructions after the completion of the simulations, an approach called off-line DA. In the second study, a new ensemble was generated at each step of the assimilation procedure, starting from the best simulation selected for the previous period, an approach called on-line DA. The revised method offered dynamical consistency between best model analogs of different periods, while the former benefited from its computational simplicity. Both

methods showed positive reconstruction skill, particularly at the regional scale in areas with high data coverage. The on-line method was also employed by Crespin et al. (2009) to analyse the 15th century Arctic warming. The novelty of the current manuscript is the focus on the comparison of the on-line and off-line approaches.

In addition to the above methods, where a single simulation having the best fit to the data is chosen during the assimilation ("degenerate particle filter"), another approach employs weights for each member of the ensemble, calculated after the comparison with the proxies and generating a probabilistic posterior distribution ("particle filter"). The technique was applied by Annan and Hargreaves (2012), who performed off-line assimilation based on a simple likelihood weighting algorithm, implementing thus all the DA after the completion of the ensemble integration. In the particle filter methods (both in the on-line and off-line techniques), more than one member proceeds to the next assimilation step after the first filtering. The most unlikely ensemble members (particles) are being discarded and the highly likely particles are being copied proportionally to their likelihood. The same "probabilistic posterior distributions" technique was used by Goosse et al. (2012). The outcomes of the approach led to distributions with larger overlaps with the proxy-based reconstruction. The method has also been used by Mairesse et al. (2013) to reconstruct the climate of the mid-Holocene (6 kyr BP).

Other ensemble-based DA approaches include the use of the Kalman filter and the explicit treatment of time-averaged observations. The off-line approach of DA was advanced by Bhend et al. (2012), through the assimilation of proxy data into a high-resolution general circulation model (GCM). The ensemble square root filter (EnSRF), a variant of the ensemble Kalman filter, was used to update the ensembles with climate proxy information. The use of an atmosphere-only GCM rather than a coupled atmosphere–ocean GCM left no possibility for information propagation over long timescales; therefore, the DA was performed off-line. In other words, an on-line DA scheme would not have benefited the reconstruction skill, apart from leading to temporal consistency of the analysis. Dirren and Hakim (2005) examined the case where only time-averaged observations are available. Their algorithm constitutes a natural extension of the ensemble Kalman filter, and reduces to the ensemble Kalman filter in the limit of zero time averaging (Dirren and Hakim, 2005). Huntley and Hakim (2010) applied the new algorithm to test the method in a simple atmospheric model. Similarly, Pendergrass et al. (2012) tested two idealized models, which captured adequate climate variability related to the palaeo-proxies. In order to identify initial conditions, an ensemble Kalman filter technique was applied to the two models. Another computationally inexpensive DA method, adapted for past climates, was presented by Steiger et al. (2014), requiring only a static ensemble of climatologically plausible states.

An advantage of the on-line compared to the off-line ensemble-based DA methods is the temporal consistency of the simulated states. The off-line approach on the other hand is computationally less complicated and can also be computationally cheaper if one uses simulations that already exist. The question we address in this paper is whether the on-line reconstruction is closer to the proxy-based reconstructions compared to the off-line version. This depends on the memory of the slow components of the climate system, such as the ocean. If these propagate the information contained in the assimilated proxy data forward in time on decadal timescales, and this information is correct, the on-line approach is expected to perform better. If, on the other hand, the chaotic nature of the system dominates and the predictability of the system is limited, or the simulated ocean states are unrealistic, the computationally easier off-line method would be sufficient. The experiment design with decadal assimilation is motivated by a number of reasons. Firstly, since we aimed for a complete Northern Hemisphere reconstruction, the 10-year resolution of the North American proxy reconstructions did not allow us to use annually resolved proxy data for the assimilation. Additionally, the annually resolved proxies include substantial noise, which is cancelled out with the decadal averaging. Finally, in a climate change context, the yearly changes are in general of less interest compared to the decadal variability. GCMs exhibit up to decadal predictability in the North Atlantic (e.g. Branstator et al., 2012; Hawkins and Sutton, 2009a) and the ocean predictability can in turn lead to atmospheric predictability. The extent of decadal predictability and the relevant mechanism behind are not yet clear and many studies have recently been performed on these topics (e.g. Hawkins and Sutton, 2009a, b; Keenlyside and Ba, 2010).

In this paper, we compare two ensemble-based DA approaches, an off-line and an on-line method, to reconstruct the climate for the period AD 1600–1700. This is a period for which many proxy studies and model simulations exist, and which is interesting due to the large temperature variations exhibited in the transition to the prolonged cold period of the Maunder Minimum (about AD 1645–1715). We employ ensemble simulations with the Max Planck Institute for Meteorology's Earth System Model (MPI-ESM), and specifically a low-resolution version of the MPI Coupled Model Intercomparison Project (CMIP) Phase 5 model. The proxy temperature reconstructions of the PAGES 2K project are used in our assimilation (PAGES 2K Consortium, 2013). The structure of the paper is as follows: in Sect. 2, we review the model characteristics and the proxy data sets used, and give the details of our methodology. Section 3 gives the results of the validation of the off-line and the on-line DA approaches and a comparison of them, discusses their limitations and includes a significance test of the results. Finally, in Sect. 4, we summarize, draw conclusions, and discuss the benefits of each approach.

## 2 Experimental design

### 2.1 Model simulations

We used the Max Planck Institute for Meteorology Earth System Model (MPI-ESM), comprising of the general circulation models ECHAM6 (European Centre Hamburg Model) (Stevens et al., 2013) for the atmosphere and MPIOM (Max Planck Institute Ocean Model) (Marsland et al., 2003) for the ocean. ECHAM6 was run at T31 horizontal resolution ($3.75° \times 3.75°$), with 31 vertical levels, resolving the atmosphere up to 10 hPa. MPIOM was run at a horizontal resolution of 3.0° (GR30) and 40 vertical levels. The OASIS3 coupler was used to couple the ocean and the atmosphere daily without flux corrections. The land surface model was JSBACH (Jena Scheme for Biosphere-Atmosphere Coupling in Hamburg; Raddatz et al., 2007) and no ocean biogeochemistry model was employed. The model is a low-resolution version of the model used for the Coupled Model Intercomparison Project (CMIP) Phase 5 simulations.

The simulations described here are based on a simulation covering the last millennium (AD 850–1849) following the "past1000" protocol of the Paleo Model Intercomparison Project Phase 3 (Schmidt et al., 2011). Prescribed external forcing factors are reconstructed variations of total solar irradiance (Vieira et al., 2011), volcanic aerosols (Crowley and Unterman, 2012), concentrations of the most important greenhouse gases (Schmidt et al., 2011), and anthropogenic land-cover changes (Pongratz et al., 2008). The past1000 simulation has been started after a 700-year-long spin-up with constant AD 850 boundary conditions.

The high computational cost restricted us to running 10 ensemble members for each experiment. This choice is consistent with Bhend et al. (2012), who found that ensembles of size 10 or more can be successful in finding a simulation moderately close to the proxies, and that considerable skill in regions close to the assimilated data can be found for ensembles of 15 members or more, while larger sizes are needed for areas further away. The ensemble members have been generated by slightly varying values of an atmospheric diffusion parameter. The method leads to a fast divergence of the different simulations and an adequate ensemble spread, not only in surface variables like the 2 m or sea-surface temperature, but also in deeper ocean variables, such as the Atlantic meridional overturning circulation (AMOC). Figure 1 shows the AMOC time series of the ensemble spread at 26.5°, for the first 100 days after the initialization of the ensemble in year AD 1600, illustrating the fast growth of the ensemble spread in ocean variables. The selected ensemble generation method does not directly introduce any disturbance in the ocean, which may limit the capability of the assimilation scheme. For this reason, a different way of generating ensembles was also tested, namely the lagged-ocean initialization method, generating the ensemble members by using different ocean initial conditions, based on different dates close

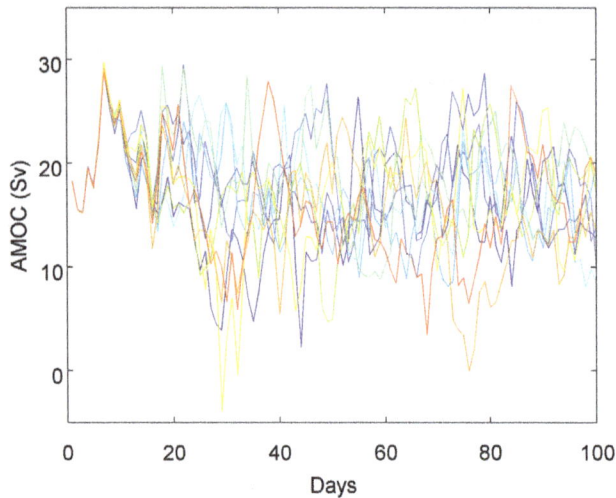

**Figure 1.** 26.5° AMOC time series of the ensemble spread for the first 100 days after the initialization of the ensemble in year AD 1600, measured in Sverdrups ($1\,\mathrm{Sv} = 10^6\,\mathrm{m}^3\,\mathrm{s}^{-1}$).

to the original starting date of the generation. The similarity in the output of the two methods however, and the fact that the lagged-ocean initialization is more complicated, led us to choose the atmosphere-only disturbance.

## 2.2  Proxy data sets

For our assimilation procedure, we used the "2k Network" of the International Geosphere–Biosphere Programme Past Global Changes (PAGES) proxy data sets. The PAGES project used a global set of proxy records and produced temperature reconstructions for seven continental-scale regions (PAGES 2K Consortium, 2013). The data set covers different periods during the last millennium for each continent, and specifically the years AD 167–2005 for Antarctica, AD 1–2000 for the Arctic, AD 800–1989 for Asia, AD 1001–2001 for Australasia, AD 1–2003 for Europe, AD 480–1974 for North America, and AD 857–1995 for South America. It has been produced by nine regional working groups, who identified the best proxy climate records for the temperature reconstruction within their region, using criteria they had established a priori.

Here, we assimilate the reconstructions for the period AD 1600–1700, which led into the Maunder Minimum. The Maunder Minimum (AD 1645–1715) was characterized by a large reduction in the number of sunspots and hence a reduction in solar radiation, and corresponds to the middle part of the Little Ice Age. Volcanic forcing likely had a role in this cooling as well. The PAGES 2K reconstructions exhibit a cooling in all the continents except Antarctica for this period, being in agreement with previous studies.

The techniques followed by the majority of the groups were either the "composite plus scale" (CPS) approach for the adjustment of the mean and variance of a predictor com-

posite to an instrumental target (e.g. Mann et al., 2008, 2009), or regression-based techniques for the predictors, including principal component pre-filters or distance weighting (PAGES 2K Consortium, 2013). The data set of individual proxies consists of 511 time series that include ice cores, tree rings, pollen, speleothems, corals, lake and marine sediments, and historical documents of changes in biological or physical processes. The reconstructions have annual resolution, apart from North America, which is resolved in 10- and 30-year periods.

## 2.3  Selection of the best ensemble members

We simulated the period AD 1600–1700 using the standard forcings for this period. The initial conditions were taken as the last day of the year AD 1599 from a transient forced simulation starting in AD 850. We performed ensemble experiments of 100 years' duration. In the off-line experiment, in the first year (AD 1600), the ten ensemble members used slightly different values of an atmospheric diffusion parameter. For each member, the simulation period was divided into 10-year intervals, and the decadal means of the 2 m temperature were calculated for each of the Northern Hemisphere continents. Using a root mean square (rms) error-based cost function, the model outputs were compared to the proxy-based continental temperature reconstructions, averaged over the respective 10-year periods. The ensemble member that minimized the cost function in each decade was selected as the best simulation for that period. The same process was followed for all the decades within the analysis period so that in the end we obtained the analysis by merging the best members of each decade.

The selection of the "optimal" simulation of the ensemble for each decade of the simulation period was done after the calculation of the following cost function:

$$\mathrm{CF}(t) = \sqrt{\sum_{i=1}^{k} \left(T_{\mathrm{mod}}^i(t) - T_{\mathrm{prx}}^i(t)\right)^2},\qquad(1)$$

where $i$ are the Northern Hemisphere continents, namely the Arctic, Asia, Europe, and North America, $T_{\mathrm{mod}}^i(t)$ is the standardized modelled decadal mean of the temperatures in each Northern Hemisphere continent and $T_{\mathrm{prx}}^i(t)$ is the standardized proxy-based reconstruction for the decadal mean of the temperatures in each Northern Hemisphere continent. The algorithm filters out the ensemble members that are considered poor representations of the actual state by throwing away the ones that are less consistent with the proxies and promoting the best fitting member. We include only the data of the Northern Hemisphere in the cost function, in an effort to reduce the degrees of freedom of the system and make it easier to find good analogues with our small ensemble size. Moreover, the Southern Hemisphere is affected by bigger uncertainties and is reconstructed by less dense proxy networks.

The cost function is based on standardized simulated and proxy-based temperatures in order to remove systematic biases in means and variances between the model and the proxy-based reconstructions, and to ensure that continental temperatures with differing variance contribute equally to the analysis. The standardized model and proxy time series were calculated by subtracting the AD 850–1850 means of the model output and the proxies from the AD 1600–1700 raw model output and proxies respectively, and then dividing by the respective standard deviations, based on the decadal averages for the AD 850–1850 period. The data sets were not weighted according to the size of the different regions as we consider all continents to be equally important. We also decided against weighting on the base of the errors of the proxy data sets as the different methods followed by each of the PAGES 2K groups make the errors not directly comparable. Moreover, the errors of the continental reconstructions are of similar order and thus error weighting would only have a small effect.

In the on-line experiment, a 10-member ensemble was generated for the first year of the analysis period, by introducing small perturbations in the atmospheric diffusion field. Simulations with 10 years' duration were run. Using the same cost function as the one used in the off-line experiment, the temperature decadal means of the model outputs were compared to the PAGES 2K continental proxy reconstructions. In contrast to the off-line method, the selected member for that period, i.e. the one that minimized the cost function, was used as the initial condition for the subsequent simulation. A new ensemble consisting of 10 members was performed for the second decade, starting from the previous best member's final conditions and having slightly varying values of the atmospheric diffusivity parameter in the different members. The same procedure was repeated until the year AD 1700.

The comparison of the two experiments is based on the proximity to the proxy-based reconstructions. We note however that it is not the aim of DA to exactly reproduce the assimilated empirical information since these have errors. Ideally, a validation of different DA methods would be based on a comparison with the true and spatially complete temperature field, but as this is not available, a validation based on proximity to the assimilated information is a useful first step to investigate whether the on-line and off-line approaches perform differently.

In order of have a good chance of finding a close analogue of an atmospheric state, one requires a large number of ensemble members, if the state space has a high dimension. Van Den Dool (1994) showed that to find an accurate analogue for daily data over a large area, such as the Northern Hemisphere, one needs daily data from a period of about $10^{30}$ years. According to Van Den Dool (1994), using a shorter library, like the current libraries of only 10–100 years of data, analogues can be found only in just 2 or 3 degrees of freedom (e.g. Bretherton et al., 1999). In our case, by using only the continental averages of the Northern Hemisphere as targets for the assimilation process, we have a low number of degrees of freedom for our cost function (less than 3). This makes the detection of a good analogue much more likely with our small ensemble size of 10 members.

## 3   Results

The performance of the two schemes was assessed by computing the correlation and the rms error for each Northern Hemisphere (NH) continent between the simulated and the proxy-based reconstructions of the 2 m air temperatures. We also investigated whether there is information propagation on decadal timescales in the model by comparing the standard deviation of the ensembles during the sub-periods in the on-line and off-line cases. An additional significance test to evaluate the role of the sampling effects that may affect many of the aspects discussed in the study was also conducted.

### 3.1   Comparison of the two DA schemes

Despite the fact that the cost function for the selection of the best members was based on standardized data, we demonstrate the performance of the two schemes using the non-standardized, but unbiased model output (absolute anomalies). This is because the latter represents the actual assimilated temperatures that come out of the model, which can be compared with other studies. Starting with the off-line DA scheme, the validation shows a clear improvement of the simulated reconstruction for the period under consideration, presenting higher correlations between model and proxies for all the continents of the Northern Hemisphere and lower root mean square errors for the analysis compared to the individual members. The on-line DA scheme was also successful, improving the skill of the analysis time series compared to the individual members. However, the scheme presented very similar correlations between the DA analysis and the proxy-based reconstructions with the ones found with the off-line approach, and no major improvements to the rms errors, both on the continental and hemispheric scales.

Figure 2 shows the Northern Hemisphere continents' decadal mean temperature anomalies w.r.t. the AD 850–1850 mean for the 17th century, for the on-line and off-line ensemble members, the on-line and off-line DA analysis, and the proxy-based reconstructions. The figure displays the ensemble spreads as shadings, but a more detailed investigation shows that the DA analysis for all the NH continents is closer to the proxies than any of the individual ensemble members in both schemes. This result is not trivial as the cost function only minimizes the rms error with respect to all NH continents. Even better agreement is exhibited by the direct average of the four Northern Hemisphere continents and the Northern Hemisphere mean for both DA schemes, as illustrated in Fig. 3. The direct average of the four NH conti-

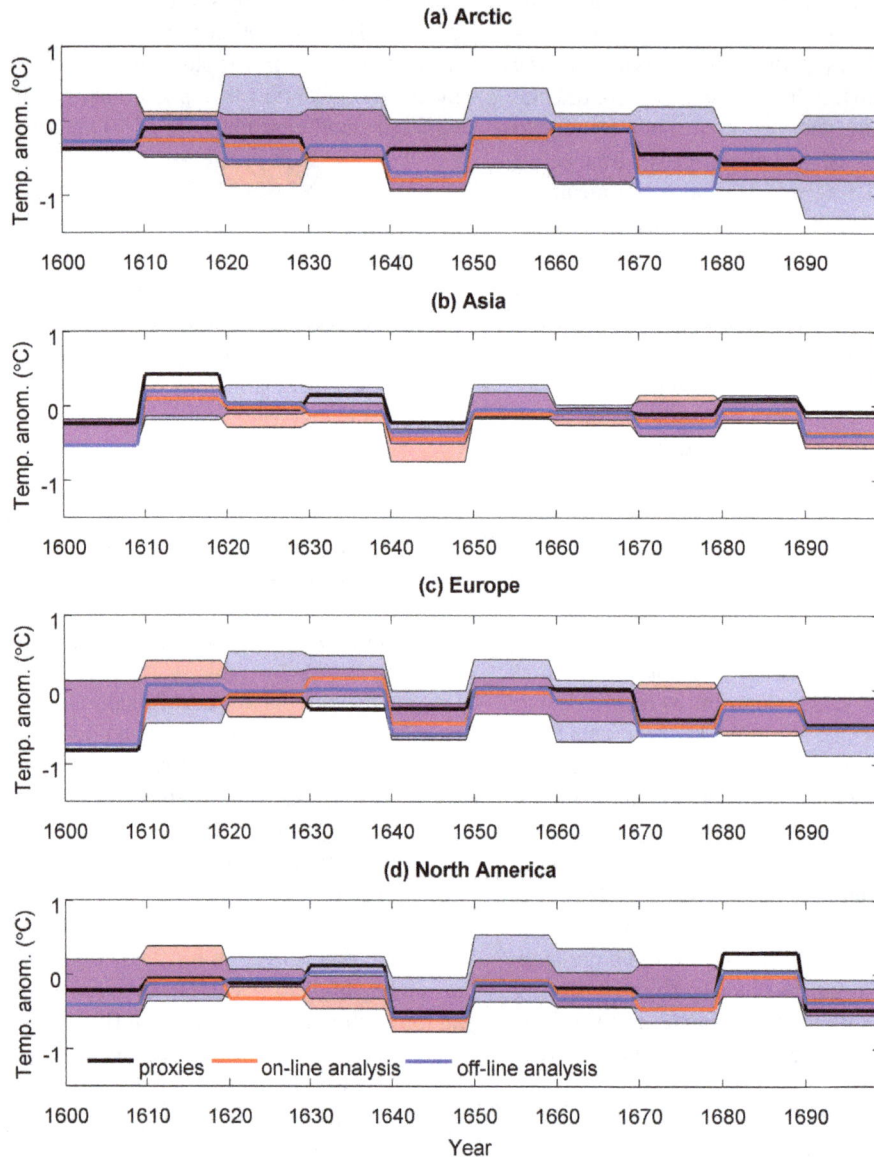

**Figure 2.** Continental decadal mean temperature anomalies w.r.t. the AD 850–1850 mean in the Northern Hemisphere for the 17th century, for the on-line (red shading) and off-line (blue shading) ensemble members, the on-line (red line) and off-line DA analysis (blue line), and the proxy-based reconstructions (black line).

nental temperatures in the simulations makes use of the same sea-land masks and seasonal representativity as the ones employed by the proxy reconstructions. Hence, it is directly comparable to the proxy data sets, which are only available as continental means. The NH mean on the other hand is the true spatial average temperature of the whole Northern Hemisphere. We show this time series as it is the usual mean temperature given in most climate studies, despite the fact that in our comparison it not the direct equivalent of the proxy-based reconstructions (the proxy time series in the two cases are the same).

The correlations in the off-line experiment between the analysis and the proxies are relatively high for all the NH

continents (0.56 for the Arctic, 0.78 for Asia, 0.79 for Europe, and 0.89 for North America). Since the cost function includes all the NH continents, the correlation is highest for the Northern Hemisphere direct average (0.94), while the correlation for the Northern Hemisphere mean is also high (0.92). These values are much higher than the correlations of the individual members with the proxies, and also higher than the correlation of the ensemble mean with the proxies (0.73 for the NH direct average). The ensemble mean has a higher ratio of forced to random variability and thus a higher correlation with the proxy-based reconstructions than the individual members, but because of the fact that the random components of the individual members partly cancel each other out,

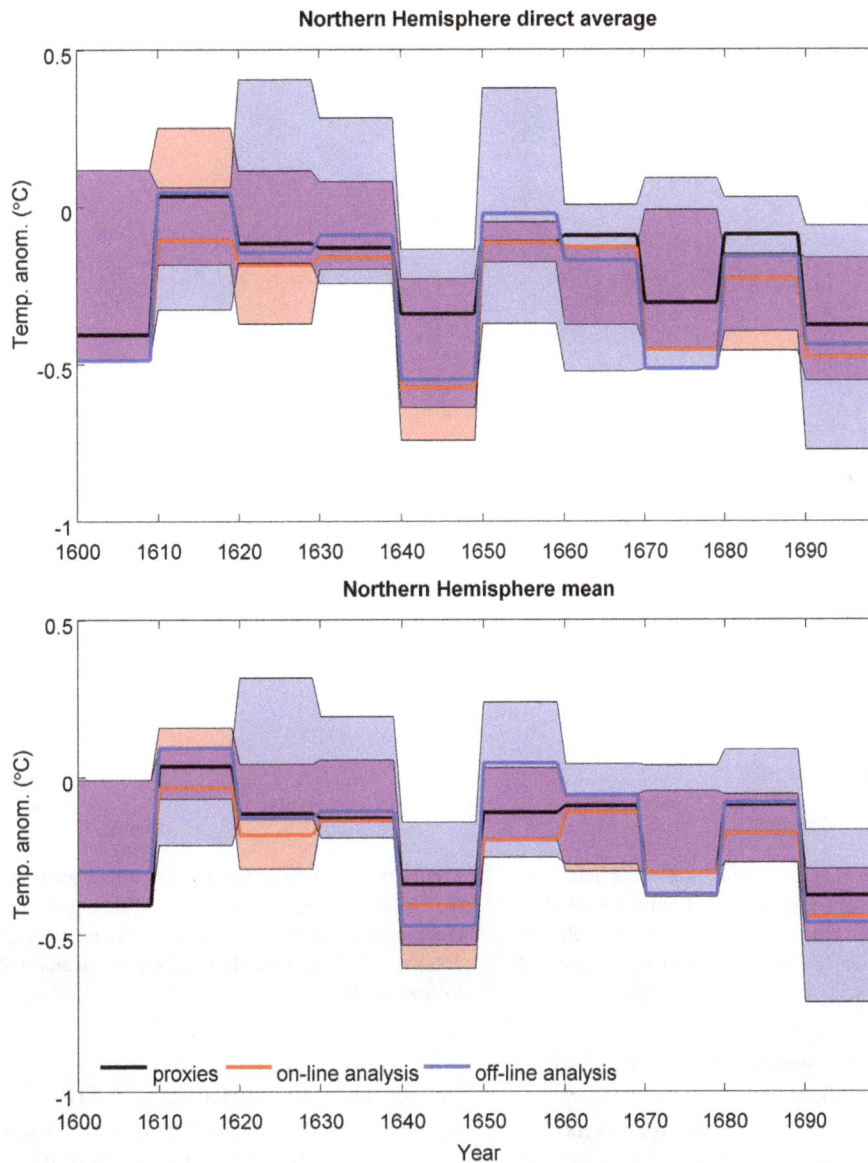

**Figure 3.** Direct average of the four Northern Hemisphere continental temperatures (anomalies w.r.t. the AD 850–1850 mean) and NH mean for the 17th century, for the on-line (red shading) and off-line (blue shading) ensemble members, the on-line (red line) and off-line DA analysis (blue line), and the proxy-based reconstructions (black line).

the total variance of the ensemble mean is much lower than the individual members. Similarly, the validation of the absolute anomalies in the on-line experiment reveal high correlations between analysis and proxies for all the NH continents (0.79 for the Arctic, 0.76 for Asia, 0.79 for Europe, and 0.81 for North America). The correlation is again the highest for the Northern Hemisphere direct average (0.93), and the Northern Hemisphere mean (0.92). The above values are again higher than the correlations of any individual member with the proxies, as well as higher than the correlation of the ensemble mean with the proxies (0.67).

The rms error of the simulated time series for each continent provides a quantification of the local agreement between the model and the proxy-based reconstructions. It is calculated based on the decadal mean differences of the model and the proxy time series for each continent. Figure 4 shows the rms errors for the individual members, the ensemble mean, and the analysis of the four Northern Hemisphere continents in the two DA schemes. In both experiments, the rms errors are either minimal or among the lowest for the analysis compared to all other members. The result is even more evident when considering the rms errors for the direct average and the mean of the Northern Hemisphere (Fig. 5). The fact that the rms error of the ensemble mean is lower than the error of most of the individual members in the two experiments might either indicate the influence of forcings or can be simply due

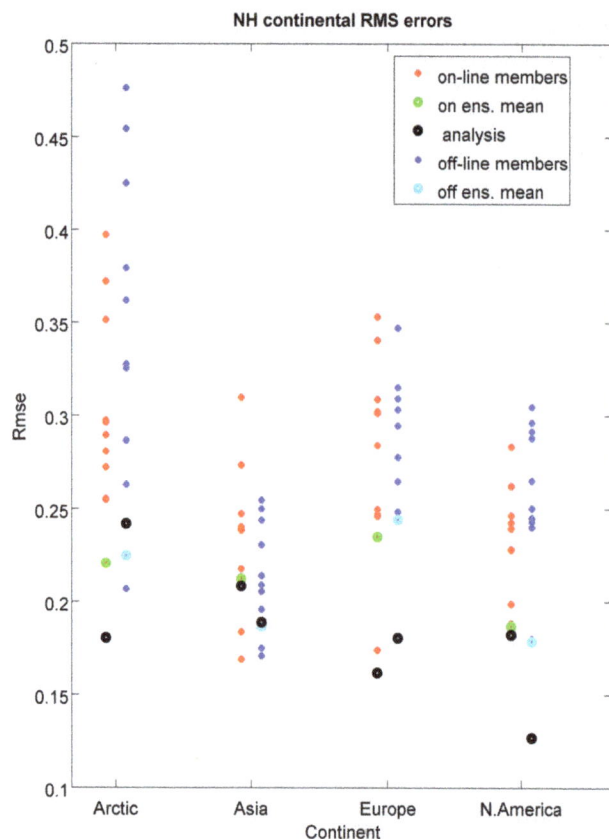

**Figure 4.** The rms errors for the four Northern Hemisphere continents for the 17th century, for the on-line (red dots) and off-line (blue dots) ensemble members, the on-line (green dots) and off-line (cyan dots) ensemble means, and the two analyses (black dots).

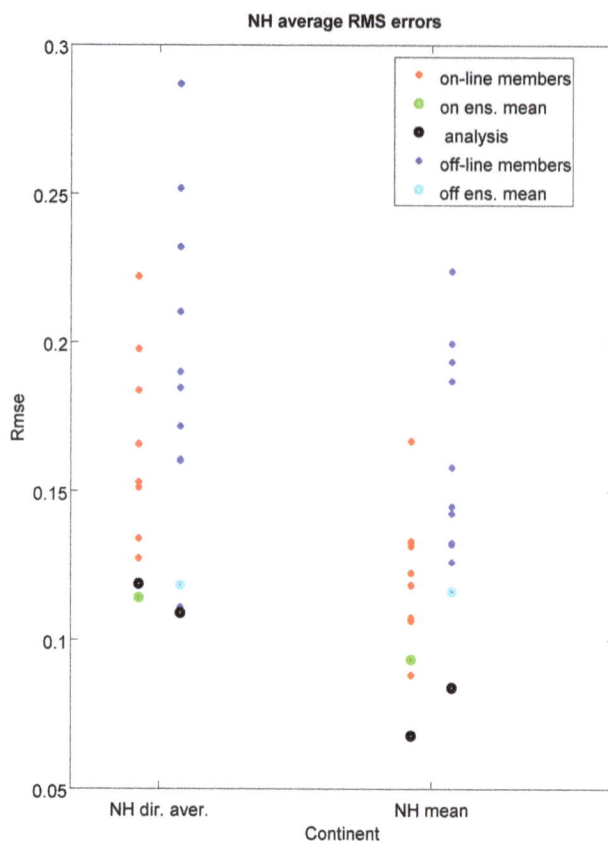

**Figure 5.** The rms errors for the direct average and the mean of the Northern Hemisphere for the 17th century, for the on-line (red dots) and off-line (blue dots) ensemble members, the on-line (green dots) and off-line (cyan dots) ensemble means, and the two analyses (black dots).

to the lower variance of the ensemble mean compared to the individual members, which might bring it closer to the proxies. However, a better estimate can be obtained from the DA analysis, which indicates that some of the internal variability has been successfully captured by the assimilation schemes. The rms errors between the analysis and the proxies in the on-line DA scheme are 0.18 for the Arctic, 0.21 for Asia, 0.16 for Europe, and 0.18 for North America. The rms error for the direct average of the four Northern Hemisphere continents is 0.12, insignificantly different from the off-line one (0.11).

The assessment of the performance of the two DA schemes using the standardized data produced correlations and rms errors very similar to the ones found when using the absolute anomalies as presented above. For the Southern Hemisphere, it is more meaningful to assess the performance of the method using the standardized data as the rms error only has a meaning with this approach. Not using the standardized outputs in this case would result in non-comparable scales because of the different standard deviations between model and proxies. In contrast to the good skill of the two schemes in the Northern Hemisphere, the agreement between the anal-

ysis for the Southern Hemisphere (SH) and the proxy-based reconstructions is not good, as expected from the fact that SH data are not included in the cost function.

The construction of our cost function on the basis of decadal mean temperatures of the NH means that the analysis is not expected to be more skilful than the individual members when considering the 100-year average. The absolute differences between simulated and reconstructed 17th century average temperatures, for the on-line and off-line ensemble members, the on-line and off-line ensemble means, and the two analyses are presented in Fig. 6, and indeed do not exhibit the best agreement between the analysis and the proxy-based reconstructions in all the regions, although this is the case in some continents.

### 3.2 Random sampling effects

Sampling effects may affect many of the aspects discussed in the study due to the limited ensemble size and the relatively short time period analysed. Therefore, sampling uncertainty should be more thoroughly addressed where possible. We ap-

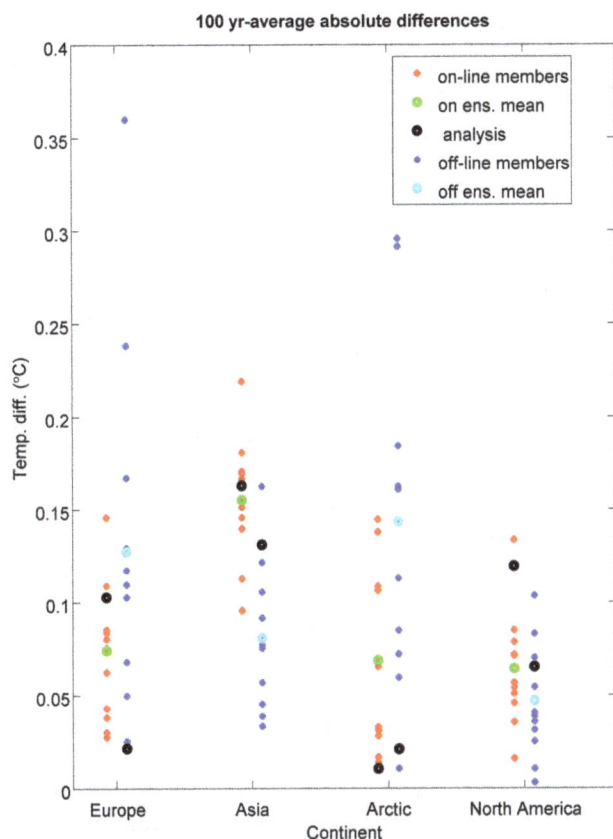

**Figure 6.** Absolute differences between simulated and reconstructed 17th century average temperatures, for the on-line (red dots) and off-line (blue dots) ensemble members, the on-line (green dots) and off-line (cyan dots) ensemble means, and the two analyses (black dots).

**Table 1.** Best cost functions for the off-line and the on-line DA schemes, for the decades 1 (1600–1609) to 10 (1690–1699). The respective ensemble mean (EM) cost functions are also shown.

| Decade | Off-line Best | On-line Best | Off-line EM | On-line EM |
|---|---|---|---|---|
| 1 | 1.47 | 1.47 | 2.53 | 2.53 |
| 2 | 1.44 | 1.55 | 1.96 | 2.00 |
| 3 | 0.51 | 0.45 | 1.31 | 0.86 |
| 4 | 1.71 | 2.10 | 2.39 | 2.49 |
| 5 | 0.72 | 0.60 | 1.57 | 1.14 |
| 6 | 1.04 | 0.50 | 1.65 | 0.95 |
| 7 | 0.53 | 0.49 | 1.22 | 0.97 |
| 8 | 0.62 | 0.38 | 1.66 | 1.62 |
| 9 | 1.72 | 0.66 | 2.28 | 2.21 |
| 10 | 1.46 | 1.45 | 1.97 | 1.93 |

**Table 2.** Northern Hemisphere correlations between simulations and proxy-based reconstructions for the analysis and the ensemble mean of the two data assimilation schemes.

| | Arctic | Asia | Europe | N. America | NH dir. aver. |
|---|---|---|---|---|---|
| Off-line DA analysis | 0.56 | 0.78 | 0.79 | 0.89 | 0.94 |
| On-line DA analysis | 0.79 | 0.76 | 0.79 | 0.81 | 0.93 |
| Off-line DA EM | 0.32 | 0.55 | 0.58 | 0.66 | 0.73 |
| On-line DA EM | 0.07 | 0.67 | 0.38 | 0.64 | 0.67 |

plied a resampling method to illustrate the distribution of the skill metrics (correlation and rms error) when randomly sampling a best model in the off-line method.

Initially, we calculated the correlations between model and proxy-based reconstructions for the NH direct average for 100 random analyses in the off-line experiment after randomly selecting one member as the best for each of the 10 decades. The mean correlation of the randomly sampled distribution with the proxies was 0.48 (with a standard deviation of 0.21), ranging between negative values and 0.8. These correlations are very low compared to the value of 0.94 from the off-line DA analysis. For the NH mean, the mean correlation of the randomly sampled analyses was 0.63 (with a standard deviation of 0.15). It is noteworthy that the correlations from the random analyses are not centred around zero, due to the presence of the forcings.

The same resampling experiment was performed for the rms error of the NH direct average. The mean rms error was 0.62 (with a standard deviation of 0.13), ranging between 0.3 and 0.95. On the other hand, the rms error found for the off-line DA analysis was only 0.11, falling well outside the above

range. Similarly, for the NH mean, the mean rms error of the random analyses was 0.51 (with a standard deviation of 0.10). The above results reveal that the DA analysis performs much better and is clearly outside the range of the randomly sampled distribution. The skill of the DA analysis is significantly different from the skill obtained from the random sampling.

### 3.3 Discussion

As previously noted, both DA schemes perform better than the simulations without DA, but there is not much difference in performance between them. In 7 out of the 10 decades of the testing period, a lower cost function for the best member and the ensemble mean is found when using the on-line method, but the differences to the off-line approach are very small (Table 1). The respective ensemble mean (EM) cost functions are also shown in the table and are substantially larger than in the DA cases. Tables 2 and 3 summarize the Northern Hemisphere correlations and rms errors respectively, between simulations and proxy-based reconstructions for the analysis and the ensemble mean of the two data assimilation schemes. The correlations and the rms errors, on the continental scale and the hemispheric averages of the NH, are very close to each other. None of the two analyses can be deemed as better in following the proxy-based reconstruction. The similarity of the two analyses can also be seen in Fig. 7, which shows the 2 m mean temperature for the two analyses (anomalies w.r.t. the AD 1961–1990 mean) and the

**Table 3.** Northern Hemisphere rms errors between simulations and proxy-based reconstructions for the analysis and the ensemble mean of the two data assimilation schemes.

| | Arctic | Asia | Europe | N. America | NH dir. aver. |
|---|---|---|---|---|---|
| Off-line DA analysis | 0.24 | 0.19 | 0.18 | 0.13 | 0.11 |
| On-line DA analysis | 0.18 | 0.21 | 0.16 | 0.18 | 0.12 |
| Off-line DA EM | 0.23 | 0.19 | 0.24 | 0.18 | 0.12 |
| On-line DA EM | 0.22 | 0.21 | 0.24 | 0.19 | 0.11 |

**Table 4.** Standard deviations of the ensemble spreads for the Northern Hemisphere of the two data assimilation schemes, calculated for all the years and for the final year of each decade.

| | Arctic | Asia | Europe | N. America | NH dir. aver. |
|---|---|---|---|---|---|
| Off-line all years | 0.48 | 0.28 | 0.50 | 0.41 | 0.30 |
| On-line all years | 0.42 | 0.28 | 0.46 | 0.37 | 0.25 |
| Off-line last year | 0.49 | 0.32 | 0.49 | 0.40 | 0.31 |
| On-line last year | 0.47 | 0.30 | 0.49 | 0.38 | 0.28 |

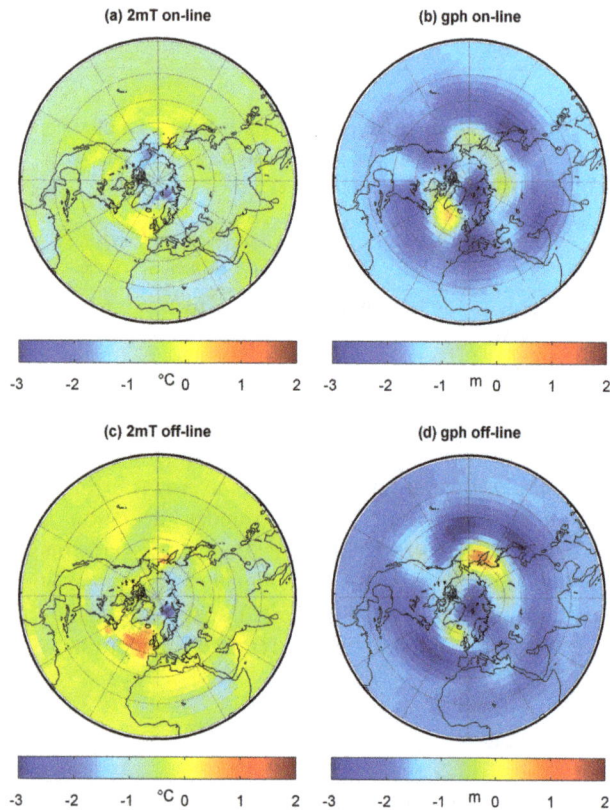

**Figure 7.** Analyses of the on-line and off-line DA methods for the 2 m mean temperature (anomalies w.r.t. the AD 1961–1990 mean) and 500 hPa geopotential height (anomalies w.r.t. the AD 1961–1990 mean) of the decade AD 1640–1649.

500 hPa geopotential height (anomalies w.r.t. the AD 1961–1990 mean) for the decade AD 1640–1649. Similar patterns can be seen, e.g. cool Barents Sea and warm NW Atlantic.

There are three potential reasons for the fact that the on-line method does not perform better than the off-line method: (i) there might be no information propagation on decadal timescales in the model, (ii) the simulated information propagation might be not skilful, i.e. different from reality, or (iii) the ocean initial conditions used at the start of each decade in the on-line DA might be not sufficiently close to reality. A possible insufficient control of the ocean state would

affect only the on-line method, as the off-line method is an a posteriori selection for which the ocean state is irrelevant.

While it is difficult and beyond the scope of this study to test whether the second and the third factors contribute to the similarity of skill of the two DA methods, we have assessed in a simple way whether there is any information propagation during the decadal sub-periods used in our DA. In the on-line assimilation, all ensemble members are initialized with the same ocean state at the beginning of each decade. Therefore, if there is information propagation, one would expect less spread in the on-line ensemble than in the off-line one. We tested this by calculating the standard deviation of the ensemble spreads for the on-line and off-line methods for the different continents. The results are shown in Table 4. For the NH direct average, we computed the standard deviation of the ensemble spreads for the whole period (for every year of the simulation period), as well as for the final year of each decade, and then computed the mean of these standard deviations. The standard deviations were 0.25 for the on-line compared to 0.30 for the off-line ensemble in the yearly test, and 0.28 compared to 0.31 respectively for the final year test. For the NH mean, the differences were a bit smaller. The standard deviations were 0.19 for the on-line compared to 0.23 for the off-line in the yearly test, and 0.22 compared to 0.23 respectively for the final year test. All the results show that the members are slightly closer together in the on-line experiment, a fact which is also in agreement with Figs. 2 and 3. It can also be noted that the all-year ensemble spread for the on-line method is consistently smaller than the respective last year spread. The different spreads in the two DA approaches is evidence of the influence of the initialization during the entire decadal assimilation time step. The smaller spread in the on-line ensemble compared to the off-line one, which starts from different ocean initial states, hints at information propagation. However, we note that it is not clear from this analysis whether the information propagation is strong enough to lead to substantially higher skill of the on-line DA method.

As mentioned above the question whether the information propagation in the coupled GCM used here is realistic is difficult to answer and is linked to the question whether such models have skill in decadal predictions. The question whether the ocean state at the beginning of each assimilation decade is close enough to reality to be useful for bringing the ensemble members during the decadal assimilation cycle closer to reality can also not be answered here. The reasons why the ocean state might be unrealistic include a too-small ensemble size, errors in the assimilated, proxy-based temperature reconstructions, and lack of control over the ocean states by assimilation of atmospheric variables.

Due to the specific choices of the approach and due to the wide range of alternative choices, the study is only a first step in the characterization of the interest of the on-line versus off-line approach. The differences between the two approaches may be specific to the target selected for the evaluation of the performance, the period investigated, the variable assimilated, the number of members in the ensemble, the frequency of assimilation, the assimilation method, and many other factors. A different setup could produce different conclusions that could prove the on-line DA scheme more skilful than the off-line one. There could be various reasons why the on-line DA is not better than the off-line DA in following the proxy-based reconstructions in our setup, but it could be more skilful in a different setup. Firstly, the insufficient control of the ocean state could be due to the small ensemble size. If the ensemble size is too small to find a member that is close to the true climatic state, there will be no added skill by propagating this misleading information forward in time. A second reason for the initial state of the ocean not being accurately enough determined throughout the on-line assimilation could be that the selection of the best member was based on the atmospheric temperature state. A correct atmospheric state cannot guarantee that the ocean state is also determined correctly. A differently defined cost function, considering for example the global or direct average of the PAGES 2K continental reconstructions or different timescales, could also change the performance of the two schemes. Another aspect that could have influenced our approaches is the proxy data sets. The use of proxies with the minimum possible noise would give a better chance for the on-line approach to capture the true climatic state, as they would represent the true climate better and the correct information would be propagated when applying the on-line approach, whereas the off-line one would not be benefited to the same extent, as it is an a posteriori selection. Finally, the use of a full particle filter rather than a degenerate one might produce a bigger ensemble spread for the ocean, giving again a better possibility to the on-line DA scheme to capture the true ocean state more closely.

# 4   Conclusions

Two main approaches have so far been employed to reconstruct the past climate: empirical and dynamical methods. Direct assimilation of proxy-based reconstructions into climate model simulations addresses some of the weaknesses of the two methods. Here, we have compared two ensemble-based DA schemes, an off-line and an on-line one, with the test case corresponding to the climate of the period leading into the Maunder Minimum, i.e. AD 1600–1700.

The two DA schemes outperform the simulations without DA. The correlations between simulations and proxy-based reconstructions for the analyses of the DA schemes were higher than the correlations of the individual members, whilst the rms errors were lower. The rms errors of the ensemble means were lower than the errors of most of the individual members either due to the influence of forcings, or simply due to the lower variance of the ensemble mean compared to the individual members, but the DA analyses perform better, implying that some of the internal variability has been successfully captured by the DA. No big difference was found between the two approaches. The majority of the cost functions for the best member and the ensemble mean of the on-line DA method were found to be slightly lower than the ones of the off-line DA method, but the correlations and the rms errors, at both the continental and the hemispheric level were very close to each other. The results suggest that there is either no skilful information propagation on the decadal timescales, i.e. no substantial predictability that could give the on-line DA an advantage over the off-line DA, or that the ocean states that are used at the beginning of each decade for generating the on-line ensembles are not sufficiently close to reality, and thus even if there were skilful predictability in the real and in the model world, the on-line DA could not benefit from it.

These results raise the question of which approach should be preferred in the future. In some cases, since the reconstruction skill of the on-line approach is not improved compared to the off-line equivalent, it would appear natural to use the less complicated off-line approach to DA, especially when computationally less expensive alternatives of off-line DA schemes can be used, for example when employing simulations that already exist. The temporal consistency of the simulation is eliminated in these cases though, which does not happen in the on-line approach. In the majority of the cases, and especially in the cases where the computational cost of the two methods is equal, the on-line approach should be preferred, as a result of the temporally consistent states that it provides.

However, we cannot be sure through these experiments whether a different setup could produce a better agreement for the on-line DA. Validation is only done with respect to the proximity to the proxy-based reconstructions, which is only a first step. We do not validate against the unknown true climate, as this would require pseudo-proxy studies, which are beyond the scope of this paper. A differently defined cost function or different performance measures could also alter the comparison. Special care must be taken to make sure that the initial state of the ocean is being captured correctly throughout the on-line assimilation. A future direction for our work would be to test different setups, by employing the full rather than the degenerate particle filter, or by defining the cost function based on 1- or 30-year means instead of decadal means, in order to check whether ocean memory on those timescales leads to different results and maybe improvements in the on-line approach. More tests could be carried out by enhancing the ensemble size for both approaches or by using different proxy data sets.

*Acknowledgements.* A. Matsikaris is supported by a NERC studentship, the University of Birmingham and the Max Planck Institute for Meteorology in Hamburg. We would like to thank Helmuth Haak and Davide Zanchettin from MPI Hamburg for their support and guidelines on running the model and for useful discussions during the implementation of this work.

The service charges for this open access publication have been covered by the Max Planck Society.

Edited by: V. Rath

# References

Annan, J. D. and Hargreaves, J. C.: Identification of climatic state with limited proxy data, Clim. Past, 8, 1141–1151, doi:10.5194/cp-8-1141-2012, 2012.

Annan, J. D., Crucifix, M., Edwards, T. L., and Paul, A.: Parameter estimation using paleodata assimilation, PAGES news, 21, 78–79, 2013.

Bhend, J., Franke, J., Folini, D., Wild, M., and Brönnimann, S.: An ensemble-based approach to climate reconstructions, Clim. Past, 8, 963–976, doi:10.5194/cp-8-963-2012, 2012.

Branstator, G., Teng, H. Y., Meehl, G. A., Kimoto, M., Knight, J. R., Latif, M., and Rosati, A.: Systematic Estimates of Initial-Value Decadal Predictability for Six AOGCMs, J. Climate, 25, 1827–1846, doi:10.1175/jcli-d-11-00227.1, 2012.

Bretherton, C. S., Widmann, M., Dymnikov, V. P., Wallace, J. M., and Blade, I.: The effective number of spatial degrees of freedom of a time-varying field, J. Climate, 12, 1990–2009, 1999.

Briffa, K. R., Jones, P. D., and Schweingruber, F. H.: Summer temperatures across Northern North-America – Regional reconstructions from 1760 using tree-ring densities, J. Geophys. Res.-Atmos., 99, 25835–25844, doi:10.1029/94jd02007, 1994.

Bronnimann, S., Franke, J., Breitenmoser, P., Hakim, G., Goosse, H., Widmann, M., Crucifix, M., Gebbie, G., Annan, J., and van der Schrier, G.: Transient state estimation in paleoclimatology using data assimilation, PAGES news, 21, 74–75, 2013.

Crespin, E., Goosse, H., Fichefet, T., and Mann, M. E.: The 15th century Arctic warming in coupled model simulations with data assimilation, Clim. Past, 5, 389–401, doi:10.5194/cp-5-389-2009, 2009.

Crowley, T. J. and Lowery, T. S.: How warm was the medieval warm period?, Ambio, 29, 51–54, doi:10.1579/0044-7447-29.1.51, 2000.

Crowley, T. J. and Unterman, M. B.: Technical details concerning development of a 1200-yr proxy index for global volcanism, Earth Syst. Sci. Data Discuss., 5, 1–28, doi:10.5194/essdd-5-1-2012, 2012.

Dirren, S. and Hakim, G. J.: Toward the assimilation of time-averaged observations, Geophys. Res. Lett., 32, L04804, doi:10.1029/2004GL021444, 2005.

Goosse, H., Renssen, H., Timmermann, A., Bradley, R. S., and Mann, M. E.: Using paleoclimate proxy-data to select optimal realisations in an ensemble of simulations of the climate of the past millennium, Clim. Dynam., 27, 165–184, 2006.

Goosse, H., Crespin, E., de Montety, A., Mann, M. E., Renssen, H., and Timmermann, A.: Reconstructing surface temperature changes over the past 600 years using climate model simulations with data assimilation, J. Geophys. Res.-Atmos., 115, D09108, doi:10.1029/2009jd012737, 2010.

Goosse, H., Crespin, E., Dubinkina, S., Loutre, M. F., Mann, M. E., Renssen, H., Sallaz-Damaz, Y., and Shindell, D.: The role of forcing and internal dynamics in explaining the "Medieval Climate Anomaly", Clim Dynam, 39, 2847–2866, 2012.

Hakim, G. J., Annan, J., Brönnimann, S., Crucifix, M., Edwards, T., Goosse, H., Paul, A., van der Schrier, G., and Widmann, M.: Overview of data assimilation methods, PAGES news, 21, 72–73, 2013.

Hawkins, E. and Sutton, R.: Decadal Predictability of the Atlantic Ocean in a Coupled GCM: Forecast Skill and Optimal Perturbations Using Linear Inverse Modeling, J. Climate, 22, 3960–3978, doi:10.1175/2009jcli2720.1, 2009a.

Hawkins, E. and Sutton, R.: The potential to narrow uncertainty in regional climate predictions, B. Am. Meteorol. Soc., 90, 1095–1107, doi:10.1175/2009bams2607.1, 2009b.

Huntley, H. S. and Hakim, G. J.: Assimilation of time-averaged observations in a quasi-geostrophic atmospheric jet model, Clim. Dynam., 35, 995–1009, 2010.

Jansen, E., Overpeck, J., Briffa, K., Duplessy, J.-C., Joos, F., Masson-Delmotte, V., Olago, D., Otto-Bliesner, B., Peltier, W., Rahmstorf, S. A. R. R., Raynaud, D., Rind, D., Solomina, O., Villalba, R., and Zhang, D.: Palaeoclimate, in: Climate change 2007: the physical science basis. Contribution of working group 1 to the Fourth Assessment Report of the Intergovernmental Panel on Climate Change, edited by: Solomon, S., Qin, D., Manning, M., Chen, Z., Marquis, M., Averyt, K., Tignor, M., and Miller, H., Cambridge University Press, Cambridge, pp. 433–497, 2007.

Jones, P. D. and Mann, M. E.: Climate over past millennia, Rev. Geophys., 42, RG2002, doi:10.1029/2003RG000143, 2004.

Jungclaus, J. H., Lorenz, S. J., Timmreck, C., Reick, C. H., Brovkin, V., Six, K., Segschneider, J., Giorgetta, M. A., Crowley, T. J., Pongratz, J., Krivova, N. A., Vieira, L. E., Solanki, S. K., Klocke, D., Botzet, M., Esch, M., Gayler, V., Haak, H., Raddatz, T. J., Roeckner, E., Schnur, R., Widmann, H., Claussen, M., Stevens, B., and Marotzke, J.: Climate and carbon-cycle variability over the last millennium, Clim. Past, 6, 723–737, doi:10.5194/cp-6-723-2010, 2010.

Keenlyside, N. S. and Ba, J.: Prospects for decadal climate prediction, Wiley Interdisciplinary Reviews-Climate Change, 1, 627–635, doi:10.1002/wcc.69, 2010.

Ljungqvist, F. C.: A new reconstruction of temperature variability in the extra-tropical Northern hemisphere during the last two millennia, Geogr. Ann. A 92, 339–351, 2010.

Luterbacher, J., Dietrich, D., Xoplaki, E., Grosjean, M., and Wanner, H.: European seasonal and annual temperature variability, trends, and extremes since 1500, Science, 303, 1499–1503, doi:10.1126/science.1093877, 2004.

Mairesse, A., Goosse, H., Mathiot, P., Wanner, H., and Dubink-ina, S.: Investigating the consistency between proxy-based reconstructions and climate models using data assimilation: a mid-Holocene case study, Clim. Past, 9, 2741–2757, doi:10.5194/cp-9-2741-2013, 2013.

Mann, M. E., Zhang, Z., Hughes, M. K., Bradley, R. S., Miller, S. K., Rutherford, S., and Ni, F.: Proxy-based reconstructions of hemispheric and global surface temperature variations over the past two millennia, P. Natl. Acad. Sci. USA, 105, 13252–13257, 2008.

Mann, M. E., Zhang, Z. H., Rutherford, S., Bradley, R. S., Hughes, M. K., Shindell, D., Ammann, C., Faluvegi, G., and Ni, F. B.: Global Signatures and Dynamical Origins of the Little Ice Age and Medieval Climate Anomaly, Science, 326, 1256–1260, 2009.

Marsland, S. J., Haak, H., Jungclaus, J. H., Latif, M., and Roske, F.: The Max-Planck-Institute global ocean/sea ice model with orthogonal curvilinear coordinates, Ocean Model., 5, 91–127, doi:10.1016/s1463-5003(02)00015-x, 2003.

Moberg, A., Sonechkin, D. M., Holmgren, K., Datsenko, N. M., Karlen, W., and Lauritzen, S. E.: Highly variable Northern Hemisphere temperatures reconstructed from low- and high-resolution proxy data, Nature, 439, 1014–1014, doi:10.1038/nature04575, 2005.

PAGES 2K Consortium: Continental-scale temperature variability during the past two millennia, Nature Geoscience, 6, 339–346, 2013.

Pendergrass, A. G., Hakim, G. J., Battisti, D. S., and Roe, G.: Coupled Air-Mixed Layer Temperature Predictability for Climate Reconstruction, J. Climate, 25, 459–472, 2012.

Pongratz, J., Reick, C., Raddatz, T., and Claussen, M.: A reconstruction of global agricultural areas and land cover for the last millennium, Glob. Biogeochem. Cycl., 22, Gb3018, doi:10.1029/2007gb003153, 2008.

Raddatz, T. J., Reick, C. H., Knorr, W., Kattge, J., Roeckner, E., Schnur, R., Schnitzler, K. G., Wetzel, P., and Jungclaus, J.: Will the tropical land biosphere dominate the climate-carbon cycle feedback during the twenty-first century?, Clim. Dynam., 29, 565–574, doi:10.1007/s00382-007-0247-8, 2007.

Schmidt, G. A., Jungclaus, J. H., Ammann, C. M., Bard, E., Braconnot, P., Crowley, T. J., Delaygue, G., Joos, F., Krivova, N. A., Muscheler, R., Otto-Bliesner, B. L., Pongratz, J., Shindell, D. T., Solanki, S. K., Steinhilber, F., and Vieira, L. E. A.: Climate forcing reconstructions for use in PMIP simulations of the last millennium (v1.0), Geosci. Model Development, 4, 33–45, doi:10.5194/gmd-4-33-2011, 2011.

Steiger, N. J., Hakim, G. J., Steig, E. J., Battisti, D. S., and Roe, G. H.: Assimilation of Time-Averaged Pseudoproxies for Climate Reconstruction, J. Clim., 27, 426–441, doi:10.1175/jcli-d-12-00693.1, 2014.

Stevens, B., Giorgetta, M., Esch, M., Mauritsen, T., Crueger, T., Rast, S., Salzmann, M., Schmidt, H., Bader, J., Block, K., Brokopf, R., Fast, I., Kinne, S., Kornblueh, L., Lohmann, U., Pincus, R., Reichler, T., and Roeckner, E.: Atmospheric component of the Earth System Model: ECHAM6, J. Adv. Model. Earth Syst., 5, 146–172, doi:10.1002/jame.20015, 2013.

Van Den Dool, H. M.: Searching for analogues, how long must we wait, Tellus A, 46, 314–324, 1994.

Vieira, L. E. A., Solanki, S. K., Krivova, N. A., and Usoskin, I.: Evolution of the solar irradiance during the Holocene, Astron. Astrophys., 531, A6, doi:10.1051/0004-6361/201015843, 2011.

Widmann, M., Goosse, H., van der Schrier, G., Schnur, R., and Barkmeijer, J.: Using data assimilation to study extratropical Northern Hemisphere climate over the last millennium, Clim. Past, 6, 627–644, doi:10.5194/cp-6-627-2010, 2010.

Xoplaki, E., Luterbacher, J., Paeth, H., Dietrich, D., Steiner, N., Grosjean, M., and Wanner, H.: European spring and autumn temperature variability and change of extremes over the last half millennium, Geophys. Res. Lett., 32, L15713, doi:10.1029/2005gl023424, 2005.

# 10

# Photic zone changes in the north-west Pacific Ocean from MIS 4–5e

**G. E. A. Swann[1] and A. M. Snelling[2]**

[1]School of Geography, University of Nottingham, University Park, Nottingham, NG7 2RD, UK
[2]NERC Isotope Geosciences Facilities, British Geological Survey, Keyworth, Nottingham, NG12 5GG, UK

*Correspondence to:* G. E. A. Swann (george.swann@nottingham.ac.uk)

**Abstract.** In comparison to other sectors of the marine system, the palaeoceanography of the subarctic North Pacific Ocean is poorly constrained. New diatom isotope records of $\delta^{13}C$, $\delta^{18}O$, $\delta^{30}Si$ ($\delta^{13}C_{diatom}$, $\delta^{18}O_{diatom}$, and $\delta^{30}Si_{diatom}$) are presented alongside existing geochemical and isotope records to document changes in photic zone conditions, including nutrient supply and the efficiency of the soft-tissue biological pump, between Marine Isotope Stage (MIS) 4 and MIS 5e. Peaks in opal productivity in MIS 5b/c and MIS 5e are both associated with the breakdown of the regional halocline stratification and increased nutrient supply to the photic zone. Whereas the MIS 5e peak is associated with low rates of nutrient utilisation, the MIS 5b/c peak is associated with significantly higher rates of nutrient utilisation. Both peaks, together with other smaller increases in productivity in MIS 4 and 5a, culminate with a significant increase in freshwater input which strengthens/re-establishes the halocline and limits further upwelling of sub-surface waters to the photic zone. Whilst $\delta^{30}Si_{diatom}$ and previously published records of diatom $\delta^{15}N$ ($\delta^{15}N_{diatom}$) (Brunelle et al., 2007, 2010) show similar trends until the latter half of MIS 5a, the records become anti-correlated after this juncture and into MIS 4, suggesting a possible change in photic zone state such as may occur with a shift to iron or silicon limitation.

## 1 Introduction

The modern-day subarctic north-west Pacific Ocean represents a major component of the global oceanic system acting as the one of the terminuses of the deep water thermohaline circulation. Today high precipitation and low evaporation in the region maintain a year-round halocline in the water column (water depth = 100–150 m), reinforced in the summer/early autumn months by the presence of a seasonal thermocline (water depth = 50 m) (Emile-Geay et al., 2003; Antonov et al., 2010; Locarnini et al., 2010). This stratification exerts a major impact on the regional ocean by limiting the mixing of surface waters with underlying nutrient- and carbon-rich deep water and by preventing convection and formation of North Pacific Deep Water (Emile-Geay et al., 2003; Menviel et al., 2012).

The initial development of the halocline and stratified water column has been attributed to the onset of major Northern Hemisphere glaciation (NHG) at 2.73 Ma, which increased the flux of freshwater to the region, via increased monsoonal rainfall and/or glacial meltwater, and sea surface temperatures (SSTs) (Sigman et al., 2004; Haug et al., 2005; Swann et al., 2006; Nie et al., 2008). The decrease of abyssal water upwelling associated with this may have contributed to the establishment of globally cooler conditions and the expansion of glaciers across the Northern Hemisphere from 2.73 Ma (Haug et al., 2005). Whilst the halocline appears to have prevailed through the late Pliocene and early Quaternary glacial–interglacial cycles (Swann, 2010), other studies have shown that the stratification boundary may have broken down in the late Quaternary at glacial terminations and during the early part of interglacials (Sarnthein et al., 2004; Jaccard et al., 2005, 2009, 2010; Galbraith et al., 2007, 2008; Gebhardt et al., 2008; Brunelle et al., 2010; Kohfeld and Chase, 2011).

Developing a complete understanding of the nature of regional stratification in the subarctic North Pacific Ocean is important for a number of reasons. Firstly, the palaeoceanographic history of the region remains poorly constrained relative to other sectors of the global ocean. Secondly, with evidence of a pervasive link between the subarctic Pacific and Southern oceans (Haug et al., 2005; Jaccard et al., 2005,

2010) records from the former can be used to further investigate teleconnections between these regions (Haug and Sigman, 2009; Sigman et al., 2010). Thirdly, with subsurface waters in the ocean interior rich in carbon and nutrients (Galbraith et al., 2007; Gebhardt et al., 2008; Menviel et al., 2012), any weakening/removal of the halocline has potential implications for the regional soft-tissue biological pump and ocean–atmospheric exchanges of $CO_2$.

To further understand the subarctic north-west Pacific Ocean, diatom isotope measurements of $\delta^{13}C$, $\delta^{18}O$, $\delta^{30}Si$ ($\delta^{13}C_{diatom}$, $\delta^{18}O_{diatom}$, and $\delta^{30}Si_{diatom}$) are presented here from the open waters of ODP Site 882 between Marine Isotope Stage (MIS) 4 and MIS 5e (Fig. 1). Existing research from the region has revealed two periods of elevated opal concentration in this interval alongside large changes in proxies relating to nutrient supply and utilisation (Jaccard et al., 2005, 2009; Brunelle et al., 2007, 2010). The new diatom isotope data presented here will allow the changes in photic zone conditions and the response of the soft-tissue biological pump to be further constrained. Diatoms, unicellular siliceous algae, are ideally suited for this purpose as they (1) occupy the uppermost sections of the water column above the halocline, (2) dominate export production in high-latitude and upwelling zones (Nelson et al., 1995), and (3) represent a key component of the soft-tissue biological pump in transferring carbon into the ocean interior by incorporating ca. 23.5 % of all carbon produced by net primary production into their cellular organic matter (Mann, 1999).

## 2  Methods

ODP Site 882 is located on the western section of the Detroit Seamounts at a water depth of 3244 m (50°22′ N, 167°36′ E) (Fig. 1). The age model used in this study is derived from the astronomical calibration of high-resolution GRAPE density and magnetic susceptibility measurements with linear interpolation between selected tie points (Jaccard et al., 2009). Ages are constrained by two radiocarbon dates and verified by correlating magnetic susceptibility and benthic foraminifera $\delta^{18}O$ records from ODP Sites 882 and 883. Samples were prepared for diatom isotope analysis using techniques previously employed at this site (Swann et al., 2006, 2008) with the 75–150 μm fraction analysed. Diatom biovolumes, calculated following Hillebrand et al. (1999) and Swann et al. (2008), show that samples in this fraction are dominated by a single taxa *Coscinodiscus radiatus* (Ehrenb.) (Fig. 2) which blooms throughout the year with elevated fluxes often occurring in autumn/early winter (Takahashi, 1986; Takahashi et al., 1996; Onodera et al., 2005). Consequently, the diatom isotope measurements obtained here are interpreted as primarily reflecting annually averaged conditions with a slight bias towards autumn/early winter months. Smaller size fractions which contain a greater diversity of taxa were not analysed due to the potential for

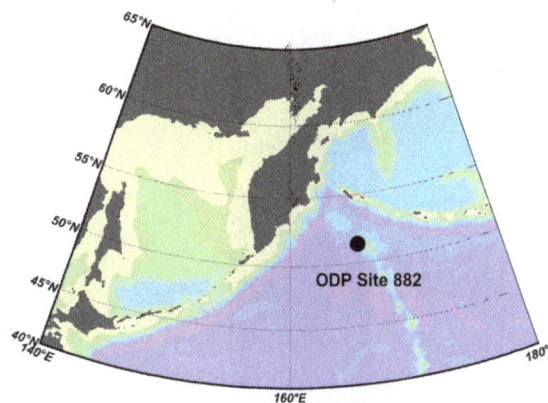

**Figure 1.** Location of ODP Site 882 in the subarctic north-west Pacific Ocean.

vital effects in $\delta^{18}O_{diatom}$ (Swann et al., 2008). Sample purity was assessed for all samples using light microscopy and SEM with unclean samples disregarded for isotope analysis. Both techniques show the excellent preservation of diatoms in the sediment record and suggest that issues of dissolution/diagenesis are not relevant to this study.

$\delta^{18}O_{diatom}$ and $\delta^{30}Si_{diatom}$ were analysed following a combined step-wise fluorination procedure at the NERC Isotope Geoscience Laboratory (UK) (Leng and Sloane, 2008) with measurements made on a Finnigan MAT 253 and values converted to the VSMOW and NBS28 scale respectively using the NIGL within-run laboratory diatom standard $BFC_{mod}$ which has been calibrated against NBS28. A small subset of the $\delta^{18}O_{diatom}$ data was previously published as part of an investigation in $\delta^{18}O_{diatom}$ vital effects in Swann et al. (2008) (see Table S1 in the Supplement). Where sufficient material remained following $\delta^{18}O_{diatom}$ and $\delta^{30}Si_{diatom}$ analysis, samples were analysed for $\delta^{13}C_{diatom}$ using a Costech elemental analyser linked to an Optima mass spectrometer via cold trapping (Hurrell et al., 2011). Replicate analyses of sample material across the analysed interval indicate an analytical reproducibility (1σ) of 0.4‰, 0.06‰ and 0.3‰ for $\delta^{18}O_{diatom}$, $\delta^{30}Si_{diatom}$ and $\delta^{13}C_{diatom}$ respectively.

## 3  Results

Through the analysed interval, $\delta^{13}C_{diatom}$ largely follows previously published siliceous productivity (opal) records from the region in indicating two intervals of higher productivity from 130 to 114 ka BP (MIS 5e) and from 101 to 86 ka BP (MIS 5b/c) (Jaccard et al., 2005, 2009) (Fig. 3). Before/after each of these intervals $\delta^{13}C_{diatom}$ is lower at < −18‰. These trends are also largely mirrored by the $\delta^{30}Si_{diatom}$ and $\delta^{15}N$ records of diatom-bound nitrogen ($\delta^{15}N_{diatom}$) (Brunelle et al., 2007, 2010) records of nutrient supply/utilisation, except during MIS 5e when values for both remain low and comparable to those in MIS 5d (Fig. 3).

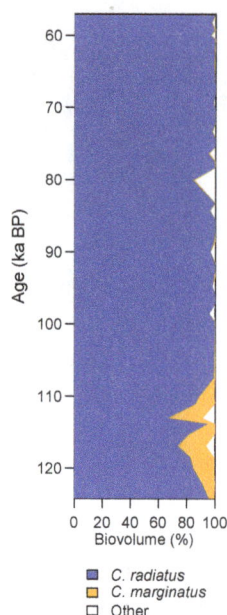

**Figure 2.** Relative diatom species biovolumes in samples analysed for $\delta^{18}O_{diatom}$.

Following a return to lower values in MIS 5a, all productivity/nutrient proxies show a series of abrupt oscillations that continue into MIS 4 with values in this interval equivalent to the peaks and minima documented in MIS 5b–e.

Measurements of $\delta^{18}O_{diatom}$ can be classified into three stages: (1) periods of relative stability in MIS 5e and MIS 5b/c (124–114 and 102–87 kaBP), (2) periods of significant decreases ($\geq 4‰$) in MIS 5d and MIS 5a (113–100 and 85–76 kaBP), and (3) periods of increase variability in MIS 5a-4 (75–57 kaBP) (Fig. 3). Intervals of high and stable $\delta^{18}O_{diatom}$ values in MIS 5e and MIS 5b/c coincide with peaks in $\delta^{13}C_{diatom}$ and opal concentrations. The termination of both productivity phases, as indicated by changes in $\delta^{13}C_{diatom}$, $\delta^{15}N_{diatom}$, $\delta^{30}Si_{diatom}$ and opal concentrations, are then concordant with the large reductions in $\delta^{18}O_{diatom}$ during MIS 5d and MIS 5a, suggesting a link between the processes controlling $\delta^{18}O_{diatom}$ and photic zone productivity/nutrient utilisation. This is reinforced by the often synchronous changes between $\delta^{18}O_{diatom}$, $\delta^{30}Si_{diatom}$ and opal concentrations during MIS 5a and into MIS 4.

## 4 Discussion

### 4.1 Environmental controls on diatom isotopes

Given the limited number of published diatom isotope records in palaeoceanography, the section below summarises the main controls on $\delta^{18}O_{diatom}$, $\delta^{30}Si_{diatom}$ and $\delta^{13}C_{diatom}$. Diatom isotopes act as an alternative proxy to records from planktonic foraminifera at sites, such as ODP Site 882, depleted in carbonates. Measurements of $\delta^{18}O_{diatom}$ can be interpreted in the same way as those of planktonic foraminifera ($\delta^{18}O_{foram}$) (Swann and Leng, 2009) with variations linked to changes in temperature ($-0.2‰\,°C^{-1}$) (Brandriss et al., 1998; Moschen et al., 2005; Dodd and Sharp, 2010; Crespin et al., 2010) and surface water $\delta^{18}O$ ($\delta^{18}O_{water}$). During biomineralisation diatoms uptake silicon, in the form of silicic acid ($H_4SiO_4$), with the lighter $^{28}Si$ preferentially used over $^{29}Si$ and $^{30}Si$. With an enrichment factor independent of temperature, the concentrations of $CO_2$ in the water ($pCO_{2\,(aq)}$) and other vital effects (De La Rocha et al., 1997; Milligan et al., 2004), $\delta^{30}Si_{diatom}$ reflects changes in photic zone silicic acid utilisation which is regulated by the biological demand for silicic acid, the rate at which nutrients are supplied to the photic zone and the $\delta^{30}Si$ composition of the silicic acid substrate ($\delta^{30}Si_{DSi}$) (De La Rocha, 2006; Reynolds et al., 2006).

A number of studies have examined the controls on $\delta^{13}C_{diatom}$ on carbon from bulk cellular diatom organic material including the cytoplasm. Whilst palaeoenvironmental reconstructions solely analyse the cell wall, which is preserved in the sediment and protected from dissolution by the diatom frustule (Abramson et al., 2009), it is assumed that the controls on cell wall $\delta^{13}C_{diatom}$ are similar to those for bulk $\delta^{13}C_{diatom}$ as the cell-wall organic matter forms a key template for diatom biomineralisation (Hecky et al., 1973; Swift and Wheeler, 1992; Kröger et al., 1999; Sumper et al., 2004). During photosynthesis, organic carbon matter is formed from both $HCO_3^-$ and $CO_{2\,(aq)}$ (Tortell et al., 1997) using both active and indirect transportation mechanisms (Sültemeyer et al., 1993) and $C_3$ and $C_4$ photosynthetic pathways (Reinfelder et al., 2000). Marine studies including those from the Bering Sea and North Pacific Ocean have demonstrated that the majority of diatom carbon originates from $HCO_3^-$ via direct transportation (Tortell and Morel, 2002; Cassar et al., 2004; Martin and Tortell, 2006; Tortell et al., 2006, 2008). Although $HCO_3^- : CO_{2\,(aq)}$ uptake ratios may alter with inter-species variations in cell morphologies (Martin and Tortell, 2008), no link exists with changes in $pCO_{2\,(aq)}$, Fe availability, growth rates, primary productivity or frustule area : volume ratios (Cassar et al., 2004; Martin and Tortell, 2006; Tortell et al., 2006, 2008).

With $^{12}C$ preferentially fractionated over $^{13}C$ (Laws et al., 1995), $\delta^{13}C_{diatom}$ predominantly reflects changes in photosynthetic carbon demand driven by variations in biological productivity or carbon cellular concentrations. Smaller magnitude variations in $\delta^{13}C_{diatom}$ may then arise with changes in the composition of the dissolved inorganic carbon substrate ($\delta^{13}C_{DIC}$) and through the intracellular and extra-cellular balance of $CO_2$ with an increase in photic zone $pCO_{2\,(aq)}$ reducing $\delta^{13}C_{diatom}$ (Laws et al., 1995; Rau et al., 1996, 1997). Whilst questions remain over the potential for $\delta^{13}C_{diatom}$ to be impacted by changes in $HCO_3^- : CO_2$ uptake, growth rates, amino acid composition, cell morphology as well as the diffusion of carbon into the cell by the enzyme RuBisCO (Laws et al., 1995, 1997, 2002; Rau et al., 1996,

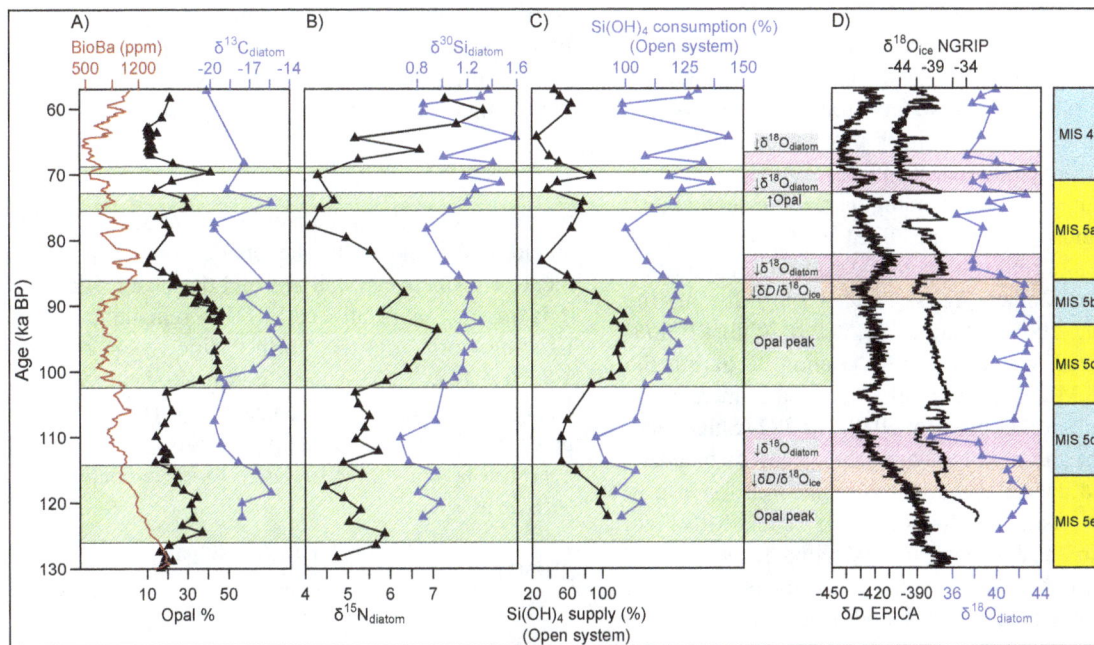

**Figure 3.** Data from ODP Site 882 showing changes in (**a**) productivity ($\delta^{13}C_{diatom}$, BioBa; Jaccard et al., 2005, and opal concentrations; Jaccard et al., 2009); (**b**) nutrient dynamics ($\delta^{30}Si_{diatom}$ and $\delta^{15}N_{diatom}$; Brunelle et al., 2010); (**c**) modelled $Si(OH)_4$ supply/consumption in an open system model; and (**d**) freshwater input ($\delta^{18}O_{diatom}$) together with EPICA Antarctic $\delta D$ (Jouzel et al., 2007) and NGRIP Greenland $\delta^{18}O_{ice}$ (NGRIP, 2004). Changes in the supply/consumption of $Si(OH)_4$ are relative to mean conditions in MIS 5e. Green/red shading indicates the increases in productivity and decreases in $\delta^{18}O_{diatom}$ respectively, which are discussed in the text.

1997, 2001; Popp et al., 1998; Cassar et al., 2006), many of these physiological processes as well as the impact of inter-species vital effects (Jacot des Combes et al., 2008) can be partially circumvented by analysing samples comprised of a single taxa. Consequently, with samples in this studies overwhelmingly dominated by *C. radiatus*, changes in $\delta^{13}C_{diatom}$ are primarily interpreted as reflecting changes in photic zone productivity (Fig. 2). We argue that the impact of a changes in $\delta^{13}C_{DIC}$ is negligible due to the aforementioned evidence that $\delta^{13}C_{DIC}$ exerts only a minimal impact on $\delta^{13}C_{diatom}$, likely within analytical error, although the lack of carbonates in the sediments prevents an independent $\delta^{13}C$ record being established to prove this beyond doubt. Similarly we argue that higher $\delta^{13}C_{diatom}$ values in MIS 5e, when higher $pCO_{2\,(aq)}$ should have acted to reduce $\delta^{13}C_{diatom}$, point towards $pCO_{2\,(aq)}$ not exerting a significant control on $\delta^{13}C_{diatom}$, although we are aware of the circular reasoning with this argument.

## 4.2 Changes in the regional biological pump (MIS 5e to MIS 5b)

Previously published opal concentration data (Jaccard et al., 2009) together with $\delta^{13}C_{diatom}$ data from this study indicates two intervals of high siliceous productivity at ODP Site 882 through the analysed interval: the first from 130 to 114 ka BP corresponding to the last interglacial (MIS 5e:

130–116 ka BP), the second from 101 to 86 ka BP covering the latter half of MIS 5c (105–93 ka BP) and most of MIS 5b (93–86 ka BP) (Jaccard et al., 2009) (Fig. 3, green shading). Whilst records of biogenic barium (BioBa) capture the MIS 5e peak (Jaccard et al., 2009), they fail to do so with the second flux event. Modern day calibrations have noted the lack of a relationship between BioBa and export production in the region (Serno et al., 2014) and speculated that the mismatch can be attributed to early diagenetic remobilisation of barium following a change in redox state (Gebhardt et al., 2008). On the other hand, all evidence points against an actual preservation/dissolution issue in this BioBa record (see Jaccard et al., 2009) and so, in line with Jaccard et al. (2009), we interpret BioBa as a measure of organic carbon export rather than siliceous productivity. With the isotope records reported here derived from diatoms and the siliceous fraction of the sediment record, we focus our discussion on the opal siliceous productivity record and only used BioBa as a proxy of organic carbon export.

Similar to the Southern Ocean, the modern-day subarctic north-west Pacific Ocean photic zone is largely limited by iron availability (Harrison et al., 2009; Tsuda et al., 2003). Accordingly, increases in bioavailable iron represent a plausible mechanism for explaining the two main (opal inferred) productivity peaks during MIS 5. Today iron supply is thought to primarily occur via aeolian dust deposition originating from East Asia and the Badain Juran Desert

(Yuan and Zhang, 2006) and other global regions (Hsu et al., 2012). Additional iron is then derived from volcanic activity (Banse and English, 1999), continental margins (Lam and Bishop, 2008), advection of waters from the Okhotsk Sea (Nishioka et al., 2007) and winter mixing of surface/sub-surface water (Shigemitsu et al., 2012). Both productivity peaks occur without a corresponding increase in aeolian dust at "Station 3" (close to ODP Site 882 at 50°00′ N, 164°59′ E) (Shigemitsu et al., 2007) or in East Asian winter monsoon records from the Chinese Loess Plateau and other marine sites (Sun et al., 2006; Zhang et al., 2009). Whilst a doubling in aeolian dust does occur at "Station 3" during the early stages of MIS 5c, this ceases before any increase in $\delta^{13}C_{diatom}$, opal or other proxy at ODP Site 882 (Shigemitsu et al., 2007). The absence of a significant increase in bioavailable iron would appear to rule out a major role for iron in driving the two productivity peaks in MIS 5e and MIS 5b/c. This would be in line with evidence indicating that productivity peaks during the last deglaciation across the North Pacific Ocean also occur without a corresponding increase in aeolian dust or other iron source input (Kohfeld and Chase, 2011). Others have also argued that iron only exerts a secondary or minor control on regional water column productivity in the palaeo-record (Kienast et al., 2004; Lam et al., 2013) whilst we are unable to account for possible changes in the flux of bioavailable iron from the Okhotsk Sea, winter mixing and other sources identified above.

### 4.2.1 Nutrient utilisation and supply

The deep and intermediate waters of the subarctic North Pacific Ocean contain some of the highest nutrient levels in the world (Whitney et al., 2013). Accordingly productivity peaks over glacial–interglacial cycles, including those covered in this study, have been linked to changes in the regional halocline and water column stratification which would alter the advection of nutrient- and carbon-rich sub-surface waters into the photic zone (Jaccard et al., 2005; Gebhardt et al., 2008). A key difference between the two productivity events in MIS 5e and MIS 5b/c is the response of the biological community to raised photic zone nutrient availability. Although productivity is high during MIS 5e, values are low for $\delta^{30}Si_{diatom}$ at ODP Site 882 ($< 1.0\permil$, $n = 3$) and for $\delta^{15}N_{diatom}$ ($< 6\permil$) at a nearby site (49°72′ N, 168°30′ E) (Brunelle et al., 2010) (Fig. 3). In contrast during MIS 5b/c the productivity peak is concordant with an increase in $\delta^{30}Si_{diatom}$ and $\delta^{15}N_{diatom}$ to ca. 1.2–1.3‰ and $> 6\permil$ respectively (Fig. 3).

Changes in $\delta^{30}Si_{diatom}$ may reflect either increased biological uptake of silicic acid (consumption) and/or changes in the supply of silicic acid to the photic zone. The modern-day regional stratified water column is best represented by a closed system model in which a finite amount of silicic acid exists for biomineralisation (Reynolds et al., 2006). In contrast an unstratified water column would be reflected by

an open system model with continual supply of silicic acid. By assuming that the two productivity peaks reflect a weakening in the stratification, an open system model can be used to investigate the controls on $\delta^{30}Si_{diatom}$:

$$\delta^{30}Si_{diatom} = \delta^{30}Si(OH)_4 + \epsilon \cdot f, \tag{1}$$

where $\delta^{30}Si(OH)_4$ is the isotopic composition of dissolved silicic acid supplied to the photic zone, $\epsilon$ is the enrichment factor between diatoms and dissolved silicic acid and $f$ is the fraction of utilised $Si(OH)_4$ remaining in the water. Existing work from the North Pacific Ocean has estimated $\delta^{30}Si(OH)_4$ at 1.23‰ and $\epsilon$ as 1.0 (Reynolds et al., 2006). Using changes in $Si(OH)_4$ consumption (Eq. 1) and siliceous productivity (opal), the supply of $Si(OH)_4$ into the photic zone can be constrained relative to mean conditions during MIS 5e as

$$Si(OH)_{4\,supply} = \frac{Opal_{sample}/Opal_{MIS\,5e}}{\left(1 - f_{consumed}^{sample}\right)/\left(1 - f_{consumed}^{MIS\,5e}\right)}. \tag{2}$$

Estimates of $Si(OH)_4$ consumption and supply from Eqs. (1) and (2) are only applicable for intervals when the water column represents an open system (e.g. the productivity peaks in MIS 5e and MIS 5b/c) and are dependant on modern-day estimates of $\delta^{30}Si(OH)_4$ being representative of past conditions. This assumption is based on evidence that $\delta^{30}Si(OH)_4$ is relatively resilient to change, outside of seasonal biological fluxes, over timescales similar to this study except in extreme circumstances linked to major reductions in the flux of riverine silicon into the ocean (Rocha and Bickle, 2005). The results show that the productivity peaks in MIS 5e and MIS 5b/c are both closely correlated with elevated levels of $Si(OH)_4$ being supplied to the photic zone (Fig. 3), supporting the suggestion that these intervals are linked to a reduction in water column stratification and an increase in the vertical flux of nutrients bearing sub-surface waters into the photic zone (Jaccard et al., 2005; Gebhardt et al., 2008). However, whilst the increase in $Si(OH)_4$ supply in MIS 5b/c is matched by a corresponding increase in biological consumption of $Si(OH)_4$, increasing the ratio of regenerated to preformed nutrients in the ocean interior, the opposite occurs during MIS 5e when the rates of $Si(OH)_4$ consumption are at their lowest over the analysed interval. (Fig. 3). Whilst reduced $Si(OH)_4$ consumption during MIS 5e could be linked to iron limitation, records indicate that aeolian dust deposition was equally low during both the MIS 5e and the MIS 5b/c productivity peaks. However, as before we are unable to account for changes in iron supply from non-aeolian sources.

### 4.2.2 Implications for $p$CO$_2$

Understanding the mechanisms that regulate changes in atmospheric concentrations of CO$_2$ ($p$CO$_2$) remains a key objective in palaeoclimatology. Previous research has demonstrated that the Southern Ocean and low-latitude oceans act as the dominant source/sink of atmospheric $CO_2$ over glacial–interglacial cycles (Pichevin et al., 2009; Fischer et al., 2010; Sigman et al., 2010). Whilst the North Pacific Ocean does not need to be invoked to explain the full amplitude of glacial–interglacial changes, recent work has advocated a potential role for the region in regulating atmospheric $pCO_2$ over the last termination (Rae et al., 2014).

Today the net annual ocean–atmosphere exchange of $CO_2$ in the subarctic north-west Pacific Ocean is close to zero, but alters from being a sink of atmosphere $CO_2$ in spring to a source in winter (Takahashi et al., 2006; Ayers and Lozier, 2012) (Fig. 4a). This seasonal variability can be attributed to changes in the biological pump and in SST which affects the solubility of $CO_2$ (Honda et al., 2002; Chierici et al., 2006; Ayers and Lozier, 2012). A weakening of the halocline stratification in MIS 5e and MIS 5b/c would have increased the advection of nutrient- and carbon-rich waters from the ocean interior, raising photic zone $pCO_{2\,(aq)}$ and the potential for $CO_2$ to be ventilated into the atmosphere due to an air–sea disequilibrium in $CO_2$. This, however, is dependant on the response and relative efficiency of the biological pump in taking advantage of the increased nutrient supply and altering the ratio of regenerated : performed nutrients to re-export carbon into the deep ocean (Sigman et al., 2004, 2010; Marinov et al., 2008). Whilst changes in the temperature/thermocline may also have been important, the only SST record for the region (Martínez-Garcia et al., 2010) does not contain the temporal resolution to investigate this further and does not provide a surface–subsurface depth temperature transect.

During MIS 5e a scenario of both higher $pCO_{2\,(aq)}$ and incomplete/low rates of nutrient utilisation suggests the regional ocean could have ventilated $CO_2$ into the atmosphere faster than the soft-tissue biological pump reabsorbed and sequestered $CO_2$ into the deep ocean (Fig. 4b) despite evidence for higher organic carbon export (BioBa) in this period (Fig. 3). The culmination of this interval at the end of MIS 5e would have resulted in the system returning to a stratified state, perhaps similar to the modern-day water column (Fig. 4a) with minimal air–sea fluxes of $CO_2$. In contrast the opal productivity peak in MIS 5b/c is marked by similar levels of photic zone $Si(OH)_4$ supply as in MIS 5e but with ca. 20 % higher rates of $Si(OH)_4$ consumption (Figs. 3 and 4c). The combination of high siliceous productivity (opal/$\delta^{13}C_{diatom}$) and a highly efficient biological pump ($\delta^{30}Si_{diatom}/\delta^{15}N_{diatom}$) during this interval suggests that the net flux of ocean–atmosphere $CO_2$ exchanges arising from the sea–air disequilibrium could have remained close to zero if photosynthetic carbon demand were similar to the rate of sub-surface carbon flux to the photic zone. Whilst a highly efficient soft-tissue biological pump raises the possibility for the region to have acted as a net sink of atmospheric $CO_2$, the potential and significance for this is limited by the relatively low proportion of surface waters which reach the deep ocean interior (Gebbie and Huybers, 2011) and low BioBa in this interval (Fig. 3).

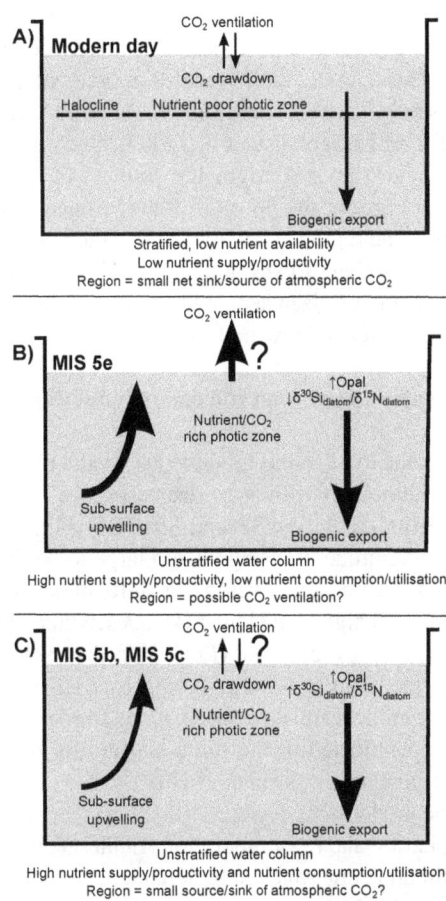

**Figure 4.** Schematic models showing subarctic north-west Pacific Ocean conditions for **(a)** modern day: halocline water column with nutrient poor surface waters limiting biological export; **(b)** MIS 5e: no halocline and enhanced upwelling of nutrient- and carbon-rich sub-surface waters leading to increased productivity. Low rates of nutrient utilisation suggest a possible increase in $pCO_2$ and release of $CO_2$ to the atmosphere; **(c)** MIS 5b/c: conditions similar to MIS 5e but with higher rates of nutrient consumption and a more efficient soft-tissue biological pump limiting/preventing ventilation of $CO_2$.

Although the data suggest that changes in the regional photic zone may have contributed to variations in atmospheric $pCO_2$ during MIS 5, both via the soft-tissue biological pump and associated changes in ocean alkalinity, it is not possible to quantify the magnitude of any fluxes or access whether they were accompanied by a change in diatom silicification and cellular Si : C ratios. Firstly, insufficient purified diatom material remains to measure diatom silicon concentrations. Secondly, although diatom elemental carbon measurements obtained during the analysis of $\delta^{13}C_{diatom}$ increase from < 0.3 wt% in MIS 5e to ca. 0.4 wt% in MIS 5b–d (see Supplement Table S1), the analytical reproducibility for $C_{diatom}$ is relatively high at 0.1 % ($1\sigma$) (Hurrell et al., 2011) and measurements are derived from the cell wall material and not the bulk cellular matter formed duringphotosynthesis. Furthermore, whilst other cores from the region show a similar double peak in opal productivity during MIS 5 (Narita et al., 2002) records at other sites suggest that the second peak is restricted to MIS 5b with no increase in MIS 5c (Shigemitsu et al., 2007). Such discrepancies either suggest poor stratigraphic controls on the age model for either core, or the potential for significant spatial variability across the region and reiterates that the magnitude of any ocean–atmosphere fluxes of $CO_2$ would be low compared to those occurring elsewhere in the marine system such as the Southern Ocean and low-latitude oceans.

### 4.2.3   Freshwater controls on siliceous productivity

Records show that the decline in siliceous productivity for both intervals culminates with large decreases in $\delta^{18}O_{diatom}$ of ca. 3–5‰ from ca. 113 ka BP and 85 ka BP (Fig. 3, red shading). The magnitude of change is too large to be driven by reductions in deep water upwelling or shifts in ocean water masses from both higher and lower latitudes, which would only alter $\delta^{18}O_{water}$ by ca. 1‰ (LeGrande and Schmidt, 2006). Instead the drop in $\delta^{18}O_{diatom}$ suggests an input of isotopically depleted freshwater that may be similar in origin to events documented at the same site during the late Pliocene/early Quaternary (Swann, 2010).

Although the modern-day regional halocline is maintained by high precipitation and low evaporation (Emile-Geay et al., 2003), it is difficult to envisage a sufficient increase in precipitation to initiate a 3–5‰ decrease in $\delta^{18}O_{diatom}$. This is reiterated by evidence that monsoonal activity was largely stable during MIS 5b–e (Sun et al., 2006; Zhang et al., 2009). At the same time the potential for a glacial source is questioned by evidence indicating a restricted glaciation in north-east Russia, closest to ODP Site 882, at the Last Glacial Maximum (LGM) (Barr and Clark, 2011, 2012), although other work suggests these ice sheets may have been considerably larger prior to the LGM (Bigg et al., 2008; Barr and Solomina, 2014). Recent work has shown that both of the major decreases in $\delta^{18}O_{diatom}$ coincide with increases in IRD accumulation in some, but not all, cores from the Okhotsk Sea (Nürnberg et al., 2011). It has also been argued that the regional water column was regulated by significant inputs of meltwater from the North American ice sheets during the last deglaciation (Lam et al., 2013). In either case, the decrease in $\delta^{18}O_{diatom}$ at the end of each siliceous productivity peak suggests that inputs of freshwater helped re-establish/strengthen the halocline, limiting the upwelling of nutrient-/$CO_2$-rich sub-surface waters and biological activity. However, with the decrease in $\delta^{18}O_{diatom}$ only occurring after the initial decline in productivity, freshwater can only be acting as a secondary control in re-establishing the halocline.

Previous work has suggested a link between changes in the subarctic north-west Pacific Ocean and the Southern Ocean (Jaccard et al., 2005, 2010; Brunelle et al., 2007; Shigemitsu et al., 2007; Galbraith et al., 2008; Sigman et al., 2010).

The most viable mechanisms for synchronous changes between polar regions are temperature- and salinity-driven variations in water column density (Brunelle et al., 2007). For example a cooling of polar SST would reduce the rate of sub-surface upwelling into the photic zone (de Boer et al., 2007), lowering nutrient availability and potentially triggering the initial decline in siliceous productivity. At the same time, a decrease in SST would increase the sensitivity of the water column to subsequent changes in salinity, making the region highly vulnerable to inputs of freshwater which would strengthen the water column and inhibit productivity (Sigman et al., 2004). Additional reductions in siliceous productivity may then arise from lower North Atlantic overturning and associated deep-water incursions and upwelling in the North Pacific (Schmittner, 2005). Support for a series of events similar to this at ODP Site 882 lies with the concordant decreases at ODP Site 882 between supplied $Si(OH)_4/\delta^{30}Si_{diatom}$/opal and Antarctic ($\delta D$)/NGRIP ($\delta^{18}O_{ice}$) ice-core records (NGRIP, 2004; Jouzel et al., 2007) at the start of each productivity decline from ca. 118 and 89 ka BP respectively (Fig. 3). The final switch to a low productivity system then coincides with the later decreases in $\delta^{18}O_{diatom}$ at 113 and 85 ka BP, suggesting that the climatic deterioration associated with lower $\delta D/\delta^{18}O_{ice}$ may have fuelled the increase in precipitation and/or an advancement of regional glaciers around the North Pacific Basin that triggered the increase in freshwater input. Whilst it remains unclear what initiated either siliceous productivity peak, it can be speculated that reductions in freshwater after 100 ka BP could have weakened the halocline and created the conditions for the second productivity bloom to eventually develop later in MIS 5b/c.

### 4.3   Photic zone changes from MIS 4-5a

Previous research has documented reduced levels of productivity in the north-west Pacific Ocean during the last glacial in response to surface water stratification (Narita et al., 2002; Jaccard et al., 2005, 2010; Brunelle et al., 2007, 2010; Shigemitsu et al., 2007; Galbraith et al., 2008; Gebhardt et al., 2008). From the latter half of MIS 5a onwards records of $\delta^{15}N_{diatom}$ and $\delta^{30}Si_{diatom}/Si(OH)_4$ consumption become anti-correlated (Fig. 3). Combined with a long-term shift to lower rates of $Si(OH)_4$ supply and higher rates of $Si(OH)_4$ consumption, this supports suggestions that changes in dust/iron inputs in the last glacial may have helped regulate the biological pump by altering the biological demand for individual nutrients (Brunelle et al., 2007, 2010; Galbraith et al., 2008; Shigemitsu et al., 2008), in this case by increasing biological uptake of silicon over nitrogen to the extent that $Si(OH)_4$ consumption in MIS 4 was up to 40 % higher than during MIS 5e. Elevated $Si(OH)_4$ consumption may also indicate that the availability of $Si(OH)_4$ rather than iron may have ultimately limited siliceous productivity over this interval, in line with a previous suggestion by Kienast et al. (2004).

Superimposed on a trend of low siliceous productivity during MIS 5a and MIS 4 are two small–moderate increases in opal at ca. 76–74 ka BP and ca. 70 ka BP (Fig. 3, green shading). The increase at 70 ka BP does not coincide with any samples analysed in this study, but the increase at 76–74 ka BP coincides with higher $\delta^{13}C_{diatom}$, $\delta^{30}Si_{diatom}$ and $Si(OH)_4$ supply/consumption (open model). Similar to before, both opal peaks culminate with a 2–3‰ reduction in $\delta^{18}O_{diatom}$ (Fig. 3, red shading), reiterating the role of freshwater in controlling photic zone dynamics in an era that coincides with increased monsoonal and thus precipitation variability (Sun et al., 2006; Shigemitsu et al., 2007; Zhang et al., 2009). However, whereas the earlier declines in siliceous productivity during MIS 5e and MIS 5b/c are accompanied by reductions in both $Si(OH)_4$ supply and consumption, here the declines initially occur with reduced $Si(OH)_4$ supply and higher rates of $Si(OH)_4$ consumption. This advocates the aforementioned suggestion that the photic zone shifted to a new state from the end of MIS 5a, highlighted by further large changes in $Si(OH)_4$ consumption in MIS 4 that do not coincide with a changes in siliceous productivity or $\delta^{18}O_{diatom}$ (Fig. 3).

## 5   Conclusions

Results here provide evidence for significant temporal changes in the strength and efficiency of the regional soft-tissue biological pump from MIS 4–5e, altering the ratio of regenerated to preformed nutrients in the water column. In particular the results show evidence of an inefficient soft-tissue biological pump from 124 to 114 ka BP, creating the potential for the region to have played a role in maintaining the warm climate of the last interglacial through the ventilation of oceanic $CO_2$ to the atmosphere. In addition to highlighting temporal changes in the biological pump, the data also reveal that the end of both these and other siliceous productivity fluxes over the analysed interval are linked to significant increases in freshwater input to the region, reestablishing/strengthening the halocline and limiting the subsurface supply of nutrient- and carbon-rich waters to the photic zone. However, further work is needed to resolve the source of these freshwater inputs and the mechanisms responsible for initiating the increase in siliceous productivity and $Si(OH)_4$ supply to the photic zone. Finally, whilst these findings reiterate earlier work in indicating a highly dynamic and changing water column in the subarctic North Pacific Ocean during the last glacial–interglacial cycle, further work is needed to assess the spatial representativeness of these results in other sectors of the subarctic North Pacific Ocean.

*Acknowledgements.* Thanks are owed to Eric Galbraith and Sam Jaccard for providing information on the ODP Site 882 age model and previously published BioBa and Opal data. Funding for GEAS was provided in part by a Natural Environment Research Council (NERC) postdoctoral fellowship award (NE/F012969/1). Finally we thank two anonymous reviewers whose comments helped to improve this manuscript.

Edited by: E. McClymont

## References

Abramson, L., Wirick, S., Lee, C., Jacobsen, C., and Brandes, J. A.: The use of soft $x$ ray spectromicroscopy to investigate the distribution and composition of organic matter in a diatom frustule and a biomimetic analog, Deep-Sea Res. Pt.-II, 56, 1369–1380, 2009.

Antonov, J. I., Seidov, D., Boyer, T. P., Locarnini, R. A., Mishonov, A. V., Garcia, H. E., Baranova, O. K., Zweng, M. M., and Johnson, D. R.: World Ocean Atlas 2009, Volume 2: Salinity, edited by: Levitus, S., in: NOAA Atlas NESDIS 69, US Government Printing Office, Washington DC, 184 pp., 2010.

Ayers, J. M. and Lozier, M. S.: Unraveling dynamical controls on the North Pacific carbon sink, J. Geophys. Res., 117, C01017, doi:10.1029/2011JC007368, 2012.

Banse, K. and English, D. C.: Comparing phytoplankton seasonality in the eastern and western subarctic Pacific and the western Bering sea, Prog. Oceanogr., 43, 235–288, 1999.

Barr, I. D. and Clark, C. D.: Glaciers and climate in Pacific Far NE Russia during the Last Glacial Maximum, J. Quaternary Sci., 26, 227–237, 2011.

Barr, I. D. and Clark, C. D.: Late Quaternary glaciations in Far NE Russia; combining moraines, topography and chronology to assess regional and global glaciation synchrony, Quaternary Sci. Rev., 53, 72–87, 2012.

Barr, I. D. and Solomina, O.: Pleistocene and Holocene glacier fluctuations upon the Kamchatka Peninsula, Global Planet. Change, 113, 110–120, 2014.

Bigg, G. R., Clark, C. D., and Hughes, A. L. C.: A last glacial ice sheet on the Pacific Russian coast and catastrophic change arising from coupled ice–volcanic interaction, Earth Planet. Sc. Lett., 265, 559–570, 2008.

Brandriss, M. E., O'Neil, J. R., Edlund, M. B., and Stoermer, E. F.: Oxygen isotope fractionation between diatomaceous silica and water, Geochim. Cosmochim. Ac., 62, 1119–1125, 1998.

Brunelle, B. G., Sigman, D. M., Cook, M. S., Keigwin, L. D., Haug, G. H., Plessen, B., Schettler, G., and Jaccard, S. L.: Evidence from diatom-bound nitrogen isotopes for subarctic Pacific stratification during the last ice age and a link to North Pacific denitrification changes, Paleoceanography, 22, PA1215, doi:10.1029/2005PA001205, 2007.

Brunelle, B. G., Sigman, D. M., Jaccard, S. L., Keigwin, L. D., Plessen, B., Schettler, G., Cook, M. S., and Haug, G. H.; Glacial/interglacial changes in nutrient supply and stratification in the western subarctic North Pacific since the penultimate glacial maximum, Quaternary Sci. Rev., 29, 2579–2590, 2010.

Cassar, N., Laws, E. A., Bidigare, R. R., and Popp, B. N.: Bicarbonate uptake by Southern Ocean phytoplankton, Global Biogeochem. Cy., 18, GB2003, doi:10.1029/2003GB002116, 2004.

Cassar, N., Laws, E. A., and Popp, B. N.: Carbon isotopic fractionation by the marine diatom Phaeodactylum tricornutum under nutrient- and light-limited growth condition, Geochim. Cosmochim. Ac., 70, 5323–5335, 2006.

Chierici, M., Fransson, A., and Nojiri, Y.: Biogeochemical processes as drivers of surface $fCO_2$ in contrasting provinces in the subarctic North Pacific Ocean, Global Biogeochem. Cy., 20, GB1009, doi:10.1029/2004GB002356, 2006.

Crespin, J., Sylvestre, F., Alexandre, A., Sonzogni, C., Pailles, C., and Perga, M.-E.: Re-examination of the temperature-dependent relationship between $\delta^{18}O_{diatom}$ and $\delta^{18}O_{lake\ water}$ and implications for paleoclimate inferences, J. Paleolimnol., 44, 547–557, 2010.

de Boer, A. M., Sigman, D. M., Toggweiler, J. R., and Russell, J. L.: Effect of global ocean temperature change on deep ocean ventilation, Paleoceanography, 22, PA2210, doi:10.1029/2005pa001242, 2007.

De La Rocha, C. L.: Opal-based isotopic proxies of paleoenvironmental conditions, Global Biogeochem. Cy., 20, GB4S09, doi:10.1029/2005GB002664, 2006.

De La Rocha, C. L and Bickle, M. J.: Sensitivity of silicon isotopes to whole-ocean changes in the silica cycle, Mar. Geol., 271, 267–282, 2005.

De La Rocha, C. L., Brzezinski, M. A., and DeNiro, M. J.: Fractionation of silicon isotopes by marine diatoms during biogenic silica formation, Geochim. Cosmochim. Ac., 61, 5051–5056, 1997.

Dodd, J. P. and Sharp, Z. D.: A laser fluorination method for oxygen isotope analysis of biogenic silica and a new oxygen isotope calibration of modern diatoms in freshwater environments, Geochim. Cosmochim. Ac., 74, 1381–1390, 2010.

Emile-Geay, J., Cane, M. A., Naik, N., Seager, R., Clement, A. C., and van Green, A.: Warren revisited: atmospheric freshwater fluxes and "Why is no deep water formed in the North Pacific", J. Geophys. Res., 108, 3178, doi:10.1029/2001JC001058, 2003.

Fischer, H., Schmitt, J., Lüthi, D., Stocker, T. F., Tschumi, T., Parekh, P., Joos, F., Köhler, P., Völker, C., Gersonde, R., Barbante, C., Le Floch, M., Raynaud, D., and Wolff, E.: The role of Southern Ocean processes in orbital and millennial $CO_2$ variations – a synthesis, Quaternary Sci. Rev., 29, 193–205, 2010.

Galbraith, E. D., Jaccard, S. L., Pedersen, T. F., Sigman, D. M., Haug, G. H., Cook, M., Southon, J. R., and Francois, R.: Carbon dioxide release from the North Pacific abyss during the last deglaciation, Nature, 449, 890–894, 2007.

Galbraith, E. D., Kienast, M., Jaccard, S. L., Pedersen, T. F., Brunelle, B. G., Sigman, D. M., and Kiefer, T.: Consistent relationship between global climate and surface nitrate utilization in the western subarctic Pacific throughout the last 500 ka, Paleoceanography, 23, PA2212, doi:10.1029/2007PA001518, 2008.

Gebhardt, H., Sarnthein, M., Grootes, P. M., Kiefer, T., Kuehn, H., Schmieder, F., and Rohl, U.: Paleonutrient and productivity records from the subarctic North Pacific for Pleistocene glacial terminations I to V, Paleoceanography, 23, PA4212, doi:10.1029/2007PA001513, 2008.

Gebbie, G. and Huybers, P.: How is the ocean filled?, Geophys. Res. Lett., 38, L06604, doi:10.1029/2011GL046769, 2011.

Harrison, P. J., Boyd, P. W., Varela, D. E., Takeda, S., Shiomoto, A., and Odate, T.: Comparison of factors controlling phytoplankton productivity in the NE and NW subarctic Pacific gyres, Prog. Oceanogr., 43, 205–234, 1999.

Haug, G. H. and Sigman, D. M.: Polar twins, Nat. Geosci., 2, 91–92, 2009.

Haug, G. H., Ganopolski, A., Sigman, D. M., Rosell-Mele, A., Swann, G. E. A., Tiedemann, R., Jaccard, S, Bollmann, J., Maslin, M. A., Leng, M. J., and Eglinton, G.: North Pacific seasonality and the glaciation of North America 2.7 million years ago, Nature, 433, 821–825, 2005.

Hecky, R. E., Mopper, K., Kilham, P., and Degens, E. T.: The amino acid and sugar composition of diatom cell-walls, Mar. Biol., 19, 323–331, 1973.

Hillebrand, H., Dürselen, C.-D., Kirschtel, D., Pollingher, U., and Zohary, T.: Biovolume calculation for pelagic and benthic microalgae, J. Phycol., 35, 403–424, 1999.

Honda, M. C., Imai, K., Nojiri, Y., Hoshi, F., Sugawarad, T., and Kusakabe, M.: The biological pump in the northwestern North Pacific based on fluxes and major components of particulate matter obtained by sediment-trap experiments (1997–2000), Deep-Sea Res. Pt.-II, 49, 5595–5625, 2002.

Hsu, S.-C., Huh, C.-A., Lin, C.-Y., Chen, W.-N., Mahowald, N. M., Liu, S.-C., Chou, C. C. K., Liang, M.-C., Tsai, C.-J., Lin, F.-J., Chen, J.-P., and Huang, Y.-T.: Dust transport from non-East Asian sources to the North Pacific, Geophys. Res. Lett., 39, L12804, doi:10.1029/2012GL051962, 2012.

Hurrell, E. R., Barker, P. A., Leng, M. J, Vane, C. H., Wynn, P., Kendrick, C. P., Verschuren, D., and Street-Perrott, F.: Developing a methodology for carbon isotope analysis of lacustrine diatoms, Rapid Commun. Mass Sp., 25, 1567–1574, 2011.

Jaccard, S. L., Haug, G. H., Sigman, D. M., Pedersen, T. F., Thierstein, H. R., and Röhl, U.: Glacial/interglacial changes in subarctic North Pacific stratification, Science, 308, 1003–1006, 2005.

Jaccard, S. L., Galbraith, E. D., Sigman, D. M., Haug, G. H., Francois, R., Pedersen, T. F., Dulski, P., and Thierstein, H. R.: Subarctic Pacific evidence for a glacial deepening of the oceanic respired carbon pool, Earth Planet. Sc. Lett., 277, 156–165, 2009.

Jaccard, S. L., Galbraith, E. D., Sigman, D. M., and Haug, G. H.: A pervasive link between Antarctic ice core and subarctic Pacific sediment records over the past 800 kyrs, Quaternary Sci. Rev., 29, 206–212, 2010.

Jacot des Combes, H., Esper, O., De La Rocha, C. L., Abelmann, A., Gersonde, R., Yam, R., and Shemesh, A.: Diatom $\delta^{13}C$, $\delta^{15}N$, and C/N since the Last Glacial Maximum in the Southern Ocean: potential impact of species composition, Paleoceanography, 23, PA4209, doi:10.1029/2008PA001589, 2008.

Jouzel, J., Masson-Delmotte, V., Cattani, O., Dreyfus, G., Falourd, S., Hoffmann, G., Minster, B., Nouet, J., Barnola, J. M., Chappellaz, J., Fischer, H., Gallet, J. C., Johnsen, S., Leuenberger, M., Loulergue, L., Luethi, D., Oerter, H., Parrenin, F., Raisbeck, G., Raynaud, D., Schilt, A., Schwander, J., Selmo, E., Souchez, R., Spahni, R., Stauffer, B., Steffensen, J. P., Stenni, B., Stocker, T. F., Tison, J. L., Werner, M., and Wolff, E. W.: Orbital and millennial Antarctic climate variability over the past 800000 years, Science, 371, 793–796, 2007.

Kienast, S. S., Hendy, I. L., Crusius, J., Pedersen, T. F., and Calvert, S. E.: Export production in the subarctic North Pacific over the last 800 kyrs: no evidence for iron fertilization?, J. Oceanogr., 60, 189–203, 2004.

Kohfeld, K. E. and Chase, Z.: Controls on deglacial changes in biogenic fluxes in the North Pacific Ocean, Quaternary Sci. Rev., 30, 3350–3363, 2011.

Kröger, N., Deutzmann, R., and Sumper, M.: Polycationic peptides from diatom biosilica that direct silica nanosphere formation, Science, 286, 1129–1132, 1999.

Lam, P. J. and Bishop, J. K. B.: The paleoclimatic record provided by eolian deposition in the deep-sea: the geologic history of wind, Geophys. Res. Lett., 35, L07608, doi:10.1029/2008GL033294, 2008.

Lam, P. J., Robinson, L. F., Blusztajn, J., Li, C., Cook, M. S., McManus, J. F., and Keigwin, L. D.: Transient stratification as the cause of the North Pacific productivity spike during deglaciation, Nat. Geosci., 6, 622–626, 2013.

Laws, E. A., Popp, B. N., Bidigare, R. R., Kennicutt, M. C., and Macko, S. A.: Dependence of phytoplankton carbon isotopic composition on growth rate and $(CO_2)_{aq}$: theoretical considerations and experimental results, Geochim. Cosmochim. Ac., 59, 1131–1138, 1995.

Laws, E. A., Bidigare, R. R., and Popp, N. B.: Effect of growth rate and $CO_2$ concentration on carbon isotope fractionation by the marine diatom Phaeodactylum tricornutum, Limnol. Ocanogr., 42, 1552–1560, 1997.

Laws, E. A., Popp, B. N., Cassar, N., and Tanimoto, J.: $^{13}C$ discrimination patterns in oceanic phytoplankton: likely influence of $CO_2$ concentrating mechanisms, and implications for palaeoreconstructions, Funct. Plant Biol., 29, 323–333, 2002.

LeGrande, A. N. and Schmidt, G. A.: Global gridded data set of the oxygen isotopic composition in seawater, Geophys. Res. Lett., 33, L12604, doi:10.1029/2006GL026011, 2006.

Leng, M. J. and Sloane, H. J.: Combined oxygen and silicon isotope analysis of biogenic silica, J. Quaternary Sci., 23, 313–319, 2008.

Locarnini, R. A., Mishonov, A. V., Antonov, J. I., Boyer, T. P., Garcia, H. E., Baranova, O. K., Zweng, M. M., and Johnson, D. R.: World Ocean Atlas 2009, Volume 1: Temperature, in: NOAA Atlas NESDIS 68, edited by: Levitus, S., US Government Printing Office, Washington DC, 184 pp., 2010.

Mann, D. G.: The species concept in diatoms, Phycologia, 38, 437–495, 1999.

Marinov, I., Follows, M., Gnanadesikan, A., Sarmiento, J. L., and Slater, R. D.: How does ocean biology affect atmospheric $pCO_2$? Theory and models, J. Geophys. Res., 113, C07032, doi:10.1029/2007JC004598, 2008.

Martin, C. L. and Tortell, P. D.: Bicarbonate transport and extracellular carbonic anhydrase activity in Bering Sea phytoplankton assemblages: results from isotope disequilibrium experiments, Limnol. Oceanogr., 51, 2111–2121, 2006.

Martin, C. L. and Tortell, P. D.: Bicarbonate transport and extracellular carbonic anhydrase in marine diatoms, Physiol. Plantarum, 133, 106–116, 2008.

Martínez-Garcia, A., Rosell-Melé, A., McClymont, E. L., Gersonde, R., and Haug, G. H.: Subpolar link to the emergence of the modern equatorial Pacific cold tongue, Science, 328, 1550–1553, 2010.

Menviel, L., Timmermann, A., Timm, O., Mouchet, A., Abe-Ouchi, A., Chikamoto, M. O., Harada, N., Ohgaito, R., and Okazaki, Y.: Removing the North Pacific halocline: effects on global climate, ocean circulation and the carbon cycle, Deep-Sea Res. Pt.-II, 61–64, 106–113, 2012.

Milligan, A. J., Varela, D. E., Brzezinski, M. A., and Morel, F. M. M.; Dynamics of silicon metabolism and silicon isotopic discrimination in a marine diatom as a function of $pCO_2$, Limnol. Oceanogr., 49, 322–329, 2004.

Moschen, R., Lücke, A., and Schleser, G.: Sensitivity of biogenic silica oxygen isotopes to changes in surface water temperature and palaeoclimatology, Geophys. Res. Lett., 32, L07708, doi:10.1029/2004GL022167, 2005.

Narita, H., Sato, M., Tsunogai, S., Murayama, M., Ikehara, M., Nakatsuka, T., Wakatsuchi, M., Harada, N., and Ujiié, Y.: Biogenic opal indicating less productive northwestern North Pacific during the glacial ages, Geophys. Res. Lett., 29, 1732, doi:10.1029/2001GL014320, 2002.

Nelson, D. M., Tréguer, P., Brzezinski, M. A., Leynaert, A., and Quéguiner, B.: Production and dissolution of biogenic silica in the ocean: revised global estimates, comparison with regional data and relationship to biogenic sedimentation, Global Biogeochem. Cy., 9, 359–372, 1995.

North Greenland Ice Core Project members: High-resolution record of Northern Hemisphere climate extending into the last interglacial period, Nature, 431, 147–151, 2004.

Nie, J., King, J., Liu, Z., Clemens, S., Prell, W., and Fang, X.: Surface-water freshening: a cause for the onset of North Pacific stratification from 2.75 Ma onward?, Global Planet. Change, 64, 49–52, 2008.

Nishioka, J., Ono, T., Saito, H., Nakatsuka, T., Takeda, S., Yoshimura, T., Suzuki, K., Kuma, K., Nakabayashi, S., Tsumune, D., Mitsudera, H., Johnson, W. K., and Tsuda, A.: Iron supply to the western subarctic Pacific: importance of iron export from the Sea of Okhotsk, J. Geophys. Res., 112, C10012, doi:10.1029/2006JC004055, 2007.

Nürnberg, D., Dethleff, D., Tiedemann, R., Kaiser, A., and Gorbarenko, S. A.: Okhotsk Sea ice coverage and Kamchatka glaciation over the last 350 ka – evidence from ice-rafted debris and planktonic $\delta^{18}O$, Palaeogeogr. Palaeocl., 310, 191–205, 2011.

Onodera, J., Takahashi, K., and Honda, M. C.: Pelagic and coastal diatom fluxes and the environmental changes in the northwestern North Pacific during 1997–2000, Deep-Sea Res. Pt.-II, 52, 2218–2239, 2005.

Pichevin, L. E., Reynolds, B. C., Ganeshram, R. S., Cacho, I., Pena, L., Keefe, K., and Ellam, R. M.: Enhanced carbon pump inferred from relaxation of nutrient limitation in the glacial ocean, Nature, 459, 1114–1118, 2009.

Popp, B. N., Laws, E. A., Bidigare, R. R., Dore, J. E., Hanson, K. L., and Wakeham, S. G.: Effect of phytoplankton cell geometry on carbon isotopic fractionation, Geochim. Cosmochim. Ac., 62, 69–77, 1998.

Rae, J. W. B., Sarnthein, M., Foster, G. L., Ridgwell, A., Grootes, P. M., and Elliott, T.: Deep water formation in the North Pacific and deglacial $CO_2$ rise, Paleoceanography, 29, 645–667, 2014.

Rau, G. H., Riebeseell, U., and Wolf-Gladrow, D.: A model of photosynthetic $^{13}C$ fractionation by marine phytoplankton based on diffusive molecular $CO_2$ uptake, Mar. Ecol.-Prog. Ser., 133, 275–285, 1996.

Rau, G. H., Riebesell, U., and Wolf-Gladrow, D.: $CO_{2\,aq}$-dependent photosynthetic $^{13}C$ fractionation in the ocean: a model versus measurements, Global Biogeochem. Cy., 11, 267–278, 1997.

Rau, G. H., Chavez, F. P., and Friederich, G. E.: Plankton $^{13}C/^{12}C$ variations in Monterey Bay, California: evidence of non-diffusive inorganic carbon uptake by phytoplankton in an upwelling environment, Deep-Sea Res. Pt.-I, 48, 79–94, 2001.

Reinfelder, J. R., Kraepiel, A. M. L., and Morel, F. M. M.: Unicellular C4 photosynthesis in a marine diatom, Nature, 407, 996–999, 2000.

Reynolds, B. C., Frank, M., and Halliday, A. N.: Silicon isotope fractionation during nutrient utilization in the North Pacific, Earth Planet. Sc. Lett., 244, 431–443, 2006.

Sarnthein, M., Gebhardt, H., Kiefer, T., Kucera, M., Cook, M., and Erlenkeuser, H.: Mid Holocene origin of the sea-surface salinity low in the subarctic North Pacific, Quaternary Sci. Rev., 23, 2089–2099, 2004.

Schmittner, A.: Decline of the marine ecosystem caused by a reduction in the Atlantic overturning circulation, Nature, 434, 628–633, 2005.

Serno, S., Winckler, G., Anderson, R. F., Hayes, C. T., Ren, H., Gersonde, R., and Haug, G. H.: Using the natural spatial pattern of marine productivity in the Subarctic North Pacific to evaluate paleoproductivity proxies, Paleoceanography, 29, 438–453, 2014.

Shigemitsu, M., Narita, H., Watanabe, Y. W., Harada, N., and Tsunogai, S.: Ba, Si, U, Al, Sc, La, Th, C and $^{13}C/^{12}C$ in a sediment core in the western subarctic Pacific as proxies of past biological production, Mar. Chem., 106, 442–455, 2007.

Shigemitsu, M., Watanabe, Y. W., and Narita, H.: Time variations of $\delta^{15}N$ of organic nitrogen in deep western subarctic Pacific sediment over the last 145 ka, Geochem. Geophy. Geosy., 9, Q10012, doi:10.1029/2008GC001999, 2008.

Shigemitsu, M., Okunishi, T., Nishioka, J., Sumata, H., Hashioka, T., Aita, M. N., Smith, S. L., Yoshie, N., Okada, N., and Yamanaka, Y.: Development of a one-dimensional ecosystem model including the iron cycle applied to the Oyashio region, western subarctic Pacific, J. Geophys. Res., 117, C06021, doi:10.1029/2011JC007689, 2012.

Sigman, D. M., Jaccard, S. L., and Haug, G. H.: Polar ocean stratification in a cold climate, Nature, 428, 59–63, 2004.

Sigman, D. M., Hain, M. P., and Haug, G. H.: The polar ocean and glacial cycles in atmospheric $CO_2$ concentration, Nature, 47–55, 2010.

Sültemeyer, D., Schmidt, C., and Fock, H. P.: Carbonic anhydrases in higher plants and aquatic microorganisms, Physiol. Plantarum, 88, 179–190, 1993.

Sumper, M. and Kröger, N.: Silica formation in diatoms: the function of long-chain polyamines and silaffins, J. Mater. Chem., 14, 2059–2065, 2004.

Sun, Y., Clemens, S. C., An, Z., and Yu, Z.: Astronomical timescale and palaeoclimatic implication of stacked 3.6-Myr monsoon records from the Chinese Loess Plateau, Quaternary Sci. Rev., 25, 33–48, 2006.

Swann, G. E. A.: Salinity changes in the North West Pacific Ocean during the late Pliocene/early Quaternary from 2.73 Ma to 2.53 Ma, Earth Planet. Sc. Lett., 297, 332–338, 2010.

Swann, G. E. A. and Leng, M. J.: A review of diatom $\delta^{18}O$ in palaeoceanography, Quaternary Sci. Rev., 28, 384–398, 2009.

Swann, G. E. A., Maslin, M. A., Leng, M. J., Sloane, H. J., and Haug, G. H.: Diatom $\delta^{18}O$ evidence for the development of the modern halocline system in the subarctic northwest Pacific at the onset of major Northern Hemisphere glaciation, Paleoceanography, 21, PA1009, doi:10.1029/2005PA001147, 2006.

Swann, G. E. A., Leng, M. J., Sloane, H. J., and Maslin, M. A.: Isotope offsets in marine diatom $\delta^{18}O$ over the last 200 ka, J. Quaternary Sci., 23, 389–400, 2008.

Swift, D. M. and Wheeler, A. P.: Evidence of an organic matrix from diatom biosilica, J. Phycol., 28, 202–290, 1992.

Takahashi, K.: Seasonal fluxes of pelagic diatoms in the subarctic Pacific, 1982–1983, Deep-Sea Res., 33, 1225–1251, 1986.

Takahashi, K., Hisamichi, K., Yanada, M., and Maita, Y.: Seasonal changes of marine phytoplankton productivity: a sediment trap study, Kaiyo Monthly, 10, 109–115, 1996.

Takahashi, T., Sutherland, S. C., Feely, R. A., and Wanninkhof, R.: Decadal change of the surface water $pCO_2$ in the North Pacific: a synthesis of 35 years of observations, J. Geophys. Res., 111, C07S05, doi:10.1029/2005JC003074, 2006.

Tortell, P. D. and Morel, F. M. M.: Sources of inorganic carbon for phytoplankton in the eastern Subtropical and equatorial Pacific Ocean, Limnol. Oceanogr., 47, 1012–1022, 2002.

Tortell, P. D., Reinfelder, J. R., and Morel, F. M. M.: Active uptake of bicarbonate by diatoms, Nature, 390, 243–244, 1997.

Tortell, P. D., Martin, C. L., and Corkum, M. E.: Inorganic carbon uptake and intracellular assimilation by subarctic Pacific phytoplankton assemblages, Limnol. Oceanogr., 51, 2102–2110, 2006.

Tortell, P. D., Payne, C., Gueguen, C., Strzepek, R. F., Boyd, P. W., and Rost, B.: Inorganic carbon uptake by Southern Ocean phytoplankton, Limnol. Oceanogr., 45, 1485–1500, 2008.

Tsuda, A., Takeda, S., Saito, H., Nishioka, J., Nojiri, Y., Kudo, I., Kiyosawa, H., Shiomoto, A., Imai, K., Ono, T., Shimamoto, A., Tsumune, D., Yoshimura, T., Aono, T., Hinuma, A., Kinugasa, M., Suzuki, K., Sohrin, Y., Noiri, Y., Tani, H., Deguchi, Y., Tsurushima, N., Ogawa, H., Fukami, K., Kuma, K., and Saino, T.: A mesoscale iron enrichment in the Western Subarctic Pacific induces a large centric diatom bloom, Science, 300, 958–961, 2003.

Whitney, F. A., Bograd, S. J., and Ono, T.: Nutrient enrichment of the subarctic Pacific Ocean pycnocline, Geophys. Res. Lett., 40, 2200–2205, 2013.

Yuan, W. and Zhang, J.: High correlations between Asian dust events and biological productivity in the western North Pacific, Geophys. Res. Lett., 33, L07603, doi:10.1029/2005GL025174, 2006.

Zhang, Y. G., Ji, J., Balsam, W., Liu, L., and Chen, J.: Mid-Pliocene Asian monsoon intensification and the onset of Northern Hemisphere glaciation, Geology, 37, 599–602, 2009.

# 11

# Using palaeo-climate comparisons to constrain future projections in CMIP5

G. A. Schmidt[1], J. D. Annan[2], P. J. Bartlein[3], B. I. Cook[1], E. Guilyardi[4,5], J. C. Hargreaves[2], S. P. Harrison[6,7], M. Kageyama[8], A. N. LeGrande[1], B. Konecky[9], S. Lovejoy[10], M. E. Mann[11], V. Masson-Delmotte[8], C. Risi[12], D. Thompson[13], A. Timmermann[14], L.-B. Tremblay[10], and P. Yiou[8]

[1]NASA Goddard Institute for Space Studies, 2880 Broadway, New York, NY 10025, USA

[2]Research Institute for Global Change, JAMSTEC, Yokohama Institute for Earth Sciences, Yokohama, Japan

[3]University of Oregon, Eugene, OR 97403, USA

[4]NCAS-Climate, University of Reading, Whiteknights, P.O. Box 217, Reading, Berkshire, RG6 6AH, UK

[5]Laboratoire d'Océanographie et du Climat: Expérimentation et Approches Numériques/Institut Pierre Simon Laplace, CNRS-IRD-UPMC – UMR7617, 4 place Jussieu, 75252 Paris Cedex 05, France

[6]Centre for Past Climate Change and School of Archaeology, Geography and Environmental Sciences (SAGES), University of Reading, Whiteknights, P.O. Box 217, Reading, Berkshire, RG6 6AH, UK

[7]Macquarie University, Sydney, NSW 2109, Australia

[8]Laboratoire des Sciences du Climat et de l'Environnement, Institut Pierre Simon Laplace, CEA-CNRS-UVSQ – UMR8212, CE Saclay l'Orme des Merisiers, 91191 Gif-sur-Yvette, France

[9]Brown University, 324 Brook St., P.O. Box 1846, Providence, RI 02912, USA

[10]McGill University, 805 Sherbrooke Street West Montreal, Quebec, H3A 0B9, Canada

[11]Pennsylvania State University, University Park, PA 16802, USA

[12]Laboratoire de Météorologie Dynamique/Institut Pierre Simon Laplace, 4, place Jussieu, 75252 Paris Cedex 05, France

[13]University of Arizona, Department of Geosciences, Gould-Simpson Building #77, 1040 E 4th St., Tucson, AZ 85721, USA

[14]University of Hawaii, 2525 Correa Road, Honolulu, HI 96822, USA

*Correspondence to:* G. A. Schmidt (gavin.a.schmidt@nasa.gov)

**Abstract.** We present a selection of methodologies for using the palaeo-climate model component of the Coupled Model Intercomparison Project (Phase 5) (CMIP5) to attempt to constrain future climate projections using the same models. The constraints arise from measures of skill in hindcasting palaeo-climate changes from the present over three periods: the Last Glacial Maximum (LGM) (21 000 yr before present, ka), the mid-Holocene (MH) (6 ka) and the Last Millennium (LM) (850–1850 CE). The skill measures may be used to validate robust patterns of climate change across scenarios or to distinguish between models that have differing outcomes in future scenarios. We find that the multi-model ensemble of palaeo-simulations is adequate for addressing at least some of these issues. For example, selected benchmarks for the LGM and MH are correlated to the rank of future projections of precipitation/temperature or sea ice extent to indicate that models that produce the best agreement with palaeo-climate information give demonstrably different future results than the rest of the models. We also explore cases where comparisons are strongly dependent on uncertain forcing time series or show important non-stationarity, making direct inferences for the future problematic. Overall, we demonstrate that there is a strong potential for the palaeo-climate simulations to help inform the future projections and urge all the modelling groups to complete this subset of the CMIP5 runs.

# 1  Introduction

The Coupled Model Intercomparison Project (Phase 5) (CMIP5) is an ongoing coordinated project instigated by the Working Group on Coupled Modelling (WGCM) at the World Climate Research Programme (WCRP) and consisting of contributions from over 25 climate modelling groups (and over 30 climate models) from around the world (Taylor et al., 2012). Multiple experiments are being coordinated, including historical simulations (1850–2005), future simulations following multiple representative concentration pathways (RCPs) and crucially, for the first time in CMIP, three sets of palaeo-climate simulations for the Last Glacial Maximum (LGM) (21 ka BP – Before Present), the mid-Holocene (MH) (6 ka BP) and the Last Millennium (850–1850 CE). The palaeo-climate simulations are also part of the Paleoclimate Model Intercomparison Project (Phase 3) (PMIP3) initiative.

The CMIP5/PMIP3 palaeo-simulations are true "out-of-sample" tests in that none of the models have been "tuned" to produce better palaeo-climates. Such tuning is not necessarily unwise (see Schneider von Deimling et al., 2006 for an example), but would complicate some of the potential analyses. Because the same models are being used for both past and future simulations, this archive of model output is a unique resource for research into the connections between model skill and model predictions, and has the potential to greatly improve assessments of future climate change.

There were many uncertainties in regional aspects of future climate projections highlighted in the Intergovernmental Panel on Climate Change (IPCC) 4th Assessment Report (AR4) (Meehl et al., 2007). These affected, for example, the future of sub-tropical rainfall, El Niño–Southern Oscillation (ENSO) changes, potential declines in the North Atlantic meridional circulation, and the fate of Arctic sea ice. Reducing the uncertainties in the projections could therefore have significant real world consequences for both adaptation and mitigation strategies.

There are three main classes of prediction uncertainty which relate to (a) the choice of scenario, (b) internal variability (sometimes described as initial condition uncertainty), and (c) the imperfections in the model (or structural uncertainty) (Hawkins and Sutton, 2009). Scenario uncertainties inevitably grow in importance with time, particularly after about 30 yr due to the timescales associated with economic change, $CO_2$ residence time and ocean thermal inertia. Initial condition uncertainty is globally important on scales of a few years (and longer at smaller spatial scales) but predictability is fundamentally limited by the chaotic dynamics of the atmosphere and upper ocean. Thus at the multi-decadal time horizon, reducing and/or better characterising structural uncertainty is the only way to potentially reduce overall uncertainty. These structural uncertainties (given a specific scenario of future emissions and other drivers) arise from a combination of model divergence – i.e. a large spread in model predictions given the same future scenario, and model inadequacy – i.e. models that are collectively either incomplete, inaccurate or are missing processes or feedbacks. The first effect is explicit (though not completely explored) in a multi-model ensemble, while the second is implicit and needs to be assessed independently.

Observations provide the means to test the models and reduce these uncertainties but instrumental records of useful data targets are few (essentially limited to in situ networks of temperature and rainfall prior to the satellite era). Additionally, and perhaps more crucially, changes in the recent past are relatively small compared to projections for the future. Furthermore, the majority of skill metrics in historical (20th century) simulations do not provide much guidance for future projections: models that are either good or bad at simulating some aspect of modern climate – the climatology, seasonal cycle, or interannual variability – often give essentially the same spread for the future (Santer et al., 2009; Knutti et al., 2010b). The reasons for this can range from the tuning procedure in the models, disconnects between the important physics at different timescales or in response to different drivers, or the very different magnitudes of change. Palaeo-climate changes offer a substantially larger signal that is commensurate with projected future changes and although palaeo-climate records are often affected by substantial noise and difficulties in interpretation (Schmidt, 2010), the most robust reconstructions can provide a crucial test of model performance over a wider range than is possible with the 20th century climate alone.

There have been many previous evaluations of palaeo-climate simulations via earlier incarnations of PMIP, as well as in many individual studies (see the review by Braconnot et al., 2012). However, there has been a lack of analyses that quantitatively link future simulations or forecasts with skill or sensitivity in the palaeo-climate simulations (though see Hargreaves et al., 2013 for an example). This is partly because (prior to CMIP5) palaeo-simulations were not done with exactly the same versions of the models being used for future projections and partly due to a lack of suitable reconstructions for model evaluation. This paper is therefore specifically focused on making the connections between palaeo-climate changes and the future rather than on understanding palaeo-climate change for its own sake.

We break this task into three main areas: (1) examples of metrics that are robust across palaeo- and future simulations, where skill in palaeo-climate evaluations builds credibility for the projections going forward; (2) examples of metrics that discriminate between different models in the past and in the future, and thus may be used to weight model projections; and (3) examples where important caveats come into play that prevent constraints from being useful. We specifically include examples in the third section where important caveats currently limit the palaeo-climate constraints to provide guidance to others on pitfalls that can occur.

The scope of the paper is as follows: Sect. 2 discusses some background on dealing with the multi-model ensemble, issues arising from the use of palaeo-climate proxy data and the use of data-synthesis products; Sect. 3 discusses specific examples of skill metrics that may have predictive power in future simulations by showing robust behaviour across palaeo and future experiments; Sect. 4 gives examples that discriminate between future projections; Sect. 5 presents some exploratory analysis of additional potentially useful metrics that are problematic for various reasons; Sect. 6 concludes and discusses the potential for further work in this area.

## 2 Methodologies

### 2.1 Palaeo-climate reconstructions

Many of the problems in dealing with reconstructing climate from palaeo-data are specific to the type of record, the time period and resolution concerned – for instance, annually resolved tree rings have issues distinct from lower resolution ocean sediment or pollen records (e.g. Kohfeld and Harrison, 2000; Ramstein et al., 2007; Jones et al., 2009; Harrison and Bartlein, 2012). There are however a number of general issues that affect the use of such data for model evaluation, including the potential for multiple climate controls on a given record, the scale over which they are representative, the need to quantify (and take into account) reconstruction uncertainties, and the sparse and uneven site coverage.

Records used for palaeo-climate reconstructions are in general influenced by several different aspects of climate as well as, potentially, non-climatic factors. For instance, oxygen or hydrogen isotopes from ice cores, carbonates or organic matter, are climatically meaningful variables, but do not necessarily have a one-to-one, stationary relationship with temperature or precipitation (e.g. Werner et al., 2000; Schmidt et al., 2007; Masson-Delmotte et al., 2011). Vegetation, in addition to being influenced by several aspects of seasonal climate, is directly influenced by the atmospheric $CO_2$ concentration (Prentice and Harrison, 2009). There are several approaches that have been adopted to overcome this type of problem: the use of multi-proxy reconstruction techniques, forward modelling of the system within a climate model or using climate model output (see an example related to coral carbonate isotopes in Sect. 5.1) or other climate prior, and model inversion or data assimilation. Multi-proxy reconstructions rely on the idea that different types of record will be sensitive to different aspects of climate, and that pooling the information from each of these records therefore provides a more robust reconstruction of any specific climate variable. In the sense that forward modelling (and by extension model inversion techniques) are based on physical and or physiological knowledge of the given system, the use of these approaches may be a more robust way of dealing with

the non-stationarity issue – however, as with climate models, the results are constrained by the quality of the models and the degree to which the system is well-understood (see for example the discussion of $CO_2$ fertilisation in Denman et al., 2007).

The scale over which a record is representative can be a major issue in comparing palaeo-data and model output. All types of records are responding to basically local conditions, though the scale over which the record is representative will depend greatly on the variable and the resolved timescale. Many records, such as tropical ice core $\delta^{18}O$, may have strong correlations with climate further afield (e.g. Schmidt et al., 2007). Comparisons at local or regional scales often require some form of dynamical or statistical downscaling of model output, though there are many associated issues with this (Wilby and Wigley, 1997). Alternatively, upscaling reconstructions (for instance, through the use of gridding) can often reveal large-scale patterns that models could be expected to resolve, although this requires a sufficiently dense network of sites (see Sect. 3 for examples). Other approaches include the use of cluster analysis to classify types of model behaviour and to determine cohesive regions for comparison with the large-scale patterns in the observations (e.g. Bonfils et al., 2004; Brewer et al., 2007).

Palaeo-climate reconstructions are usually accompanied by estimates of measurement or structural uncertainty. However, in practice these uncertainties have rarely been propagated into large-scale synthetic products (except in terms of non-quantitative quality control measures, see e.g. COHMAP Members, 1988) and even more rarely taken into account when the reconstructions were used for model evaluation. However, quantitative measures of uncertainty have been included in more recent palaeo-climate syntheses (e.g. MARGO Project Members, 2009; Bartlein et al., 2011) and the use of fuzzy-distance measures (Guiot et al., 1999; Harrison et al., 2013) provides an explicit way to take account of data uncertainties if these cannot be expressed with Euclidean distance. It is worth noting that model-data differences cannot be expected to be smaller than the data uncertainties themselves.

### 2.2 Modelling issues

There are two particular issues that are more problematic in palaeo-climate simulations than, for instance, simulations of the 20th century: model drift and forcing uncertainty. The issue of coupled climate model drift arises because of the long ($\sim$ thousands of years) time required to bring the deep ocean into equilibrium in coupled ocean-atmosphere models. In some cases, insufficient spin-up time may have been allowed before specific experiments are started. While drift also affects transient historical simulations, the magnitude of the forcings in the 20th century means that residual drift is usually a small component of the transient response. For simulations of the last millennium though, the forcings are much smaller, and drift in the early centuries of the simulation will be a larger fraction of the modelled change (Osborn et al., 2006; Fernández-Donado et al., 2013). One proposal to deal with this is via a correction using the drift in the control simulation (i.e. calculating a smooth trend and removing it from the perturbed simulation prior to analysis). While this works well for temperature, it is not very good for variables that exhibit threshold behaviour such as sea ice extent or precipitation. In practice, this issue needs to be assessed for each proposed comparison.

Second, there are important uncertainties in the forcings used for the palaeo-climate experiments. This is also true for aerosols in the historical simulations but such issues are more prevalent in palaeo-simulations. For example, the magnitudes of solar and volcanic forcing over the last millennium, and the size and height of ice sheets at the LGM are sources of major uncertainty. In the last millennium experiments, multiple forcing choices were proposed (Schmidt et al., 2011, 2012), but few groups have attempted (as yet) to comprehensively explore all the options, and this is also true for uncertainties associated with other time periods. If an insufficient range of different forcings is tested, it is plausible that mismatches between observations and simulations may be wrongly attributed to the model (or observations), when in fact they were related to a misspecified forcing (e.g. Kageyama et al., 2001).

Third, there are many aspects of past climate changes that are (currently) outside the scope of the available modelling within CMIP5 (and more widely). Variability in the last glacial period that involves complex ocean/ice sheet dynamics (such as Dansgaard–Oeschger events) are beyond what can be analysed directly since the CMIP5-class of models does not have sufficiently interactive dynamic ice sheets. There are also common biases across different models that have more to do with the state of computational technology than physics (for instance, poor or non-existent resolution of ocean eddies). Other examples can easily be found.

For clarity in the rest of the text, we define the term "ensemble" to denote the full multi-model database of results across all CMIP5 scenarios (which encompasses all palaeo-climate, historical, idealised and future projection simulations). The future projections used here consist of the four RCP scenarios (rcp26, rcp45, rcp6, rcp85) (future possibilities that correspond roughly to greenhouse gas radiative forcing at the year 2100, relative to the pre-industrial, of 2.6, 4.5, 6.0, and 8.5 W m$^{-2}$, respectively) along with idealised simulations that have been included to provide clean comparisons across models. The idealised simulations include a 1 % increasing $CO_2$ simulation, the response to an abrupt increase to $4xCO_2$, atmosphere-only simulations such as amip, amip4xCO2 and amipFuture (where all models are forced by the same pattern of ocean temperatures from the historical period, with $4xCO_2$, and with a warm anomaly imposed respectively), or sstClim and sstClim4xCO2 simulations (where ocean temperatures are held constant under pre-industrial or $4xCO_2$ conditions). We use CMIP5 to refer to the entire database, including the PMIP3 simulations. Specific model simulations are referred to by their name in the CMIP5 database (i.e. rcp85, past1000, piControl etc.), while the scenarios or periods are referred to more generally using a standard abbreviation or name (e.g. the LGM, MH, RCP 4.5). We list the models that we have used in analyses in this paper, along with the specific experiments and simulation IDs, in Table 1. While the multi-model ensemble is a useful source for addressing structural uncertainty, it should be noted that the ensemble is not a controlled sample from a well-defined distribution of plausible simulations.

## 2.3 Approaches to comparing reconstructions and simulations

There has been a gradual evolution in the approaches for comparing reconstructed changes and climate model simulations from essentially qualitative graphical comparisons of output and reconstructions of the corresponding climatic variables (e.g. Braconnot et al., 2007) to more quantitative approaches that measure model-data mismatch via some "metric" or distance function (e.g. Sundberg et al., 2012; Izumi et al., 2013). Metrics based on correlations or rms differences between fields of data and model output have been commonly used in model evaluation for current climate (e.g. Taylor, 2001; Schmidt et al., 2006; Gleckler et al., 2008). These methods provide opportunities for both inter- and intra-generational model comparisons (Reichler and Kim, 2008; Harrison et al., 2013). The concept of "skill" as adopted in the numerical weather prediction community is useful as a quantitative test of model performance: that is, does a model produce a more accurate prediction (match to the palaeo-climate record), than that which would be achieved by a simple null hypothesis (Hargreaves et al., 2013)? Most studies and metrics have focused on time slice or time series comparisons, though it is worth pointing out that nothing precludes comparing the simulations and palaeo-record in the frequency domain (e.g. Lovejoy and Schertzer, 2012b).

While most standard comparisons focus on evaluating individual model simulations against the reconstructions, a different approach is to focus on the collective performance of the ensemble as a whole. For instance, Hargreaves et al. (2011) tested the ability of the PMIP2 ensemble to represent the Last Glacial Maximum in terms of its "reliability", defined as the adequacy of the ensemble, considered in probabilistic terms, in predicting the changes documented in the palaeo-climate archives during that interval. Multi-model ensemble means can be informative and will generally outperform individual models (Annan and Hargreaves, 2011), but care must be taken to assess the suitability of each included model and (any) weighting of individual models needs to be well justified (Knutti et al., 2010a).

**Table 1.** List of models, institutions and experiments used in the analyses in this paper. Experiment names use the CMIP5 database shorthand, and run numbers are the "rip" coding for each experiment.

| Model name | Model institution | Experiments | Run numbers |
|---|---|---|---|
| ACCESS-1.0 | CSIRO (Commonwealth Scientific and Industrial Research Organisation, Australia), and BOM (Bureau of Meteorology, Australia) | historical | r1i1p1 |
| BCC-CSM1 | Beijing Climate Center, China Meteorological Administration, China | piControl<br>midHolocene<br>rcp85 | r1i1p1<br>r1i1p1<br>r1i1p1 |
| BNU-ESM | College of Global Change and Earth System Science, Beijing Normal University | piControl<br>amip<br>sstClim<br>sstClim4xCO2<br>rcp85 | r1i1p1<br>r1i1p1<br>r1i1p1<br>r1i1p1<br>r1i1p1 |
| CanESM2 | Canadian Centre for Climate Modelling and Analysis, Canada | historical<br>rcp45 | r[1-5]i1p1<br>r[1-5]i1p1 |
| CNRM-CM5 | Centre National de Recherches Météorologiques/Centre Européen de Recherche et Formation Avancée en Calcul Scientifique, France | piControl<br>historical<br>midHolocene<br>lgm<br>1pctCO2<br>abrupt4xCO2<br>rcp45<br>rcp85 | r1i1p1<br>r[1-10]i1p1<br>r1i1p1<br>r1i1p1<br>r1i1p1<br>r1i1p1<br>r1i1p1<br>r1i1p1 |
| CSIRO-Mk3-6-0 | Commonwealth Scientific and Industrial Research Organisation in collaboration with the Queensland Climate Change Centre of Excellence, Australia | piControl<br>historical<br>midHolocene<br>rcp45<br>rcp85 | r1i1p1<br>r[1–10]i1p1<br>r1i1p1<br>r[1–10]i1p1<br>r1i1p1 |
| EC-EARTH | EC-Earth consortium | piControl<br>historical<br>midHolocene<br>rcp45<br>rcp85 | r1i1p1<br>r7i1p1<br>r1i1p1<br>r[1, 2, 6–9, 11, 12, 14]i1p1<br>r1i1p1 |
| FGOALS-g2 | LASG, Institute of Atmospheric Physics, Chinese Academy of Sciences; and CESS, Tsinghua University, China | piControl<br>midHolocene<br>rcp85 | r1i1p1<br>r1i1p1<br>r1i1p1 |
| FGOALS-s2 | LASG, Institute of Atmospheric Physics, Chinese Academy of Sciences; and CESS, Tsinghua University, China | amip<br>midHolocene<br>piControl<br>rcp85<br>sstClim4xCO2<br>sstClim | r1i1p1<br>r1i1p1<br>r1i1p1<br>r1i1p1<br>r1i1p1<br>r1i1p1 |
| CMCC-CMS | Centro Euro-Mediterraneo per I Cambiamenti Climatici | piControl<br>rcp85 | r1i1p1<br>r1i1p1 |
| GFDL-CM2.1 | NOAA Geophysical Fluid Dynamics Laboratory, US | historical | r1i1p1 |
| GFDL-CM3 | NOAA Geophysical Fluid Dynamics Laboratory, US | piControl<br>amip<br>amip4xCO2<br>sstClim<br>rcp85 | r1i1p1<br>r1i1p1<br>r1i1p1<br>r1i1p1<br>r1i1p1 |

**Table 1.** Continued.

| Model name | Model institution | Experiments | Run numbers |
|---|---|---|---|
| GFDL-ESM2G | NOAA Geophysical Fluid Dynamics Laboratory, US | piControl<br>historical<br>midHolocene<br>rcp85 | r1i1p1<br>r1i1p1<br>r1i1p1<br>r1i1p1 |
| GFDL-ESM2M | NOAA Geophysical Fluid Dynamics Laboratory, US | piControl<br>historical<br>midHolocene<br>rcp85 | r1i1p1<br>r1i1p1<br>r1i1p1<br>r1i1p1 |
| GISS-E2-H | NASA Goddard Institute for Space Studies, US | piControl<br>historical | r1i1p1<br>r[1–5]i1p[12] |
| GISS-E2-R | NASA Goddard Institute for Space Studies, US | piControl<br>historical<br>past1000<br>midHolocene<br>lgm<br>1pctCO2<br>abrupt4xCO2<br>rcp45<br>rcp85 | r1i1p1, r1i1p141<br>r1i1p[12], r[45]i1p3<br>r1i1p12[1–8]<br>r1i1p1<br>r1i1p15[01]<br>r1i1p1<br>r1i1p1<br>r[1–5]i1p1<br>r1i1p1 |
| HadCM3 | Hadley Center, UK Met. Office, UK | historical | r[1–10]i1p1 |
| HadGEM2-CC | Hadley Center, UK Met. Office, UK | piControl<br>historical<br>midHolocene<br>rcp45<br>rcp85 | r1i1p1<br>r1i1p1<br>r1i1p1<br>r1i1p1<br>r1i1p1 |
| HadGEM2-ES | Hadley Center, UK Met. Office, UK | piControl<br>historical<br>midHolocene<br>rcp45<br>rcp85 | r1i1p1<br>r1i1p1<br>r1i1p1<br>r[1–3]i1p1<br>r1i1p1 |
| INM-CM4 | Institute for Numerical Mathematics, Russia | piControl<br>historical<br>midHolocene<br>rcp45<br>rcp85 | r1i1p1<br>r1i1p1<br>r1i1p1<br>r1i1p1<br>r1i1p1 |
| IPSL-CM5A-LR | Institut Pierre-Simon Laplace, France | piControl<br>historical<br>midHolocene<br>lgm<br>1pctCO2<br>abrupt4xCO2<br>rcp45<br>rcp85 | r1i1p1<br>r[1–4]i1p1<br>r1i1p1<br>r1i1p1<br>r1i1p1<br>r1i1p1<br>r[1–4]i1p1<br>r1i1p1 |
| IPSL-CM5A-MR | Institut Pierre-Simon Laplace, France | piControl<br>historical<br>midHolocene<br>rcp45<br>rcp85 | r1i1p1<br>r1i1p1<br>r1i1p1<br>r1i1p1<br>r1i1p1 |

**Table 1.** Continued.

| Model name | Model institution | Experiments | Run numbers |
|---|---|---|---|
| MIROC-ESM | Japan Agency for Marine-Earth Science and Technology, Atmosphere and Ocean Research Institute (The University of Tokyo), and National Institute for Environmental Studies, Japan | piControl<br>midHolocene<br>lgm<br>past1000<br>1pctCO2<br>abrupt4xCO2<br>rcp85 | r1i1p1<br>r1i1p1<br>r1i1p1<br>r1i1p1<br>r1i1p1<br>r1i1p1<br>r1i1p1 |
| MIROC5 | Atmosphere and Ocean Research Institute (The University of Tokyo), National Institute for Environmental Studies, and Japan Agency for Marine-Earth Science and Technology, Japan | piControl<br>historical<br>midHolocene<br>rcp45<br>rcp85 | r1i1p1<br>r1i1p1<br>r1i1p1<br>r[1–3]i1p1<br>r1i1p1 |
| MPI-ESM-P | Max Planck Institute for Meteorology, Hamburg, Germany | piControl<br>historical<br>past1000<br>lgm<br>midHolocene<br>1pctCO2<br>abrupt4xCO2<br>rcp85 | r1i1p1<br>r1i1p1<br>r1i1p1<br>r1i1p1<br>r1i1p1<br>r1i1p1<br>r1i1p1<br>r1i1p1 |
| MPI-ESM-LR | Max Planck Institute for Meteorology, Hamburg, Germany | piControl<br>historical<br>rcp85 | r1i1p1<br>r[1–3]i1p1<br>r1i1p1 |
| MPI-ESM-MR | Max Planck Institute for Meteorology, Hamburg, Germany | piControl<br>amip<br>amip4xCO2<br>amipFuture<br>sstClim<br>sstClim4xCO2<br>historical<br>rcp85 | r1i1p1<br>r1i1p1<br>r1i1p1<br>r1i1p1<br>r1i1p1<br>r1i1p1<br>r[1–3]i1p1<br>r1i1p1 |
| MRI-CGCM3 | Meteorological Research Institute, Tsukuba, Japan | piControl<br>midHolocene<br>lgm<br>1pctCO2<br>abrupt4xCO2<br>rcp85 | r1i1p1<br>r1i1p1<br>r1i1p1<br>r1i1p1<br>r1i1p1<br>r1i1p1 |
| NCAR-CCSM4 | National Center for Atmospheric Research, US/Dept. of Energy/NSF | piControl<br>amip<br>amip4xCO2<br>amipFuture<br>sstClim<br>sstClim4xCO2<br>midHolocene<br>lgm<br>1pctCO2<br>abrupt4xCO2<br>rcp45<br>rcp85 | r1i1p1<br>r1i1p1<br>r1i1p1<br>r1i1p1<br>r1i1p1<br>r1i1p1<br>r1i1p1<br>r1i1p1<br>r1i1p1<br>r1i1p1<br>r1i1p1<br>r1i1p1 |
| NCAR-CESM1 | National Center for Atmospheric Research, US/Dept. of Energy/NSF | historical | r[5–7]i1p1 |

**Table 1.** Continued.

| Model name | Model institution | Experiments | Run numbers |
|---|---|---|---|
| NCAR-CESM1-CAM5 | National Center for Atmospheric Research, US/Dept. of Energy/NSF | piControl<br>amip<br>amipFuture<br>rcp85 | r1i1p1<br>r1i1p1<br>r1i1p1<br>r1i1p1 |
| NCAR-CESM1-BGC | National Center for Atmospheric Research, US/Dept. of Energy/NSF | piControl<br>rcp85 | r1i1p1<br>r1i1p1 |
| NorESM1-M | Norwegian Climate Centre, Norway | piControl<br>historical<br>midHolocene<br>rcp45<br>rcp85 | r1i1p1<br>r[1–3]i1p1<br>r1i1p1<br>r1i1p1<br>r1i1p1 |
| NorESM1-ME | Norwegian Climate Centre, Norway | piControl<br>rcp85 | r1i1p1<br>r1i1p1 |

## 2.4 Linking past and future

The key task of this paper is to provide guidance and examples for deciding on whether the palaeo-climate simulations have a connection to the future projections, and if so, what the comparison to palaeo-reconstructions can imply for the future. We stress that robust links between past and future simulations can only be derived if the model configurations used are the same in the different experiments. A previously common practice of using a lower resolution or differently tuned or scoped model for past simulations than for future projections, while perhaps convenient for efficiency, is not appropriate because such variations often have lead to substantial differences in sensitivity. Thus, all the examples discussed below link models that were identical (excepting boundary conditions and forcings) in the past and future CMIP5 simulations.

We distinguish two ways in which palaeo-data-model comparisons can be used as a guide to the future: (1) as a validation of a robust relationship between diagnostics across models and scenarios, or (2) as a method to discriminate between differently skillful models. In the first case, one would search for properties or correlations that we expect to be features of all climates within the ensemble, determine whether that is the case, and use the palaeo-data to provide some independent support for that relationship. In the second case, there is a prerequisite that for the diagnostic chosen, the "skill" metric when it is compared to a reconstruction actually correlates to future outcomes within the ensemble. If this is not the case, then the skill in that diagnostic is orthogonal to the spread in the projections and cannot be used to constrain them. Even when such a relationship is found, we need to consider whether it is physically meaningful to be confident that it has not arisen either though chance due to a small sample size or as an artifact of the model or the experimental design. To gain confidence in such palaeo-constraint, we also need to understand the physical processes that explain the connections between past and future.

While connections may in principle be highly complex, it is natural as a first step to consider whether a correlation exists between past and future behaviour in the same diagnostic. The search for useful metrics (in this sense) using modern data has generally been disappointing (Knutti et al., 2010b), although there have been a small number of cases where apparently meaningful relationships have been found (Boé et al., 2009; Hall and Qu, 2006; Brient and Bony, 2012; Fasullo and Trenberth, 2012). It is notable that the first three examples relate future climate changes to externally forced changes in the modern climate (decadal or seasonal variations), rather than using metrics based on the climatological mean state alone. This lends support to our working hypothesis that past variations seen in palaeo-climate simulations will be informative about the future.

Where a credible relationship between past and future is found, there is a range of methods that can be applied to use observations to constrain future predictions (Collins et al., 2012). One method, applied by both Boé et al. (2009) and Hall and Qu (2006), is to take the observational estimate, and use the relationship (often linear) embodied in the correlation between past and future model output to project this value into the future. An attractive feature of this approach, beyond its simplicity, is that it readily allows extrapolation of the observed relationship in the case where the true value is suspected of lying outside the model range. An alternative approach, which has been widely applied to perturbed physics ensembles, is more explicitly Bayesian and considers the ensemble as a probabilistic sample. For the prior, equal weight is typically assigned to each ensemble member. Probabilistic weights are then calculated for each member of the ensemble, according to their performance in reproducing the observations. This weighted ensemble now represents the posterior estimate of future change. This method uses the model spread as a prior constraint which, depending on one's viewpoint and the specific case in question, may be considered either a strength or weakness (Collins et al., 2012).

# 3 Robust relationships in past and future simulations

In this section we highlight examples of physically based correlations between key diagnostics that show similar relationships in the palaeo-climate simulations and in future projections (or the more idealised warming scenarios) and whose fidelity can be assessed using the palaeo-climate record. If these conditions are realised, the observations can be used to support the model results, and thus help provide contingent future predictions of one diagnostic given a potential change in the other.

An important issue for assessing future climate impacts is to what extent the large-scale mean temperature response can be used as an index for more regional changes. We consider the relationships between global mean temperature and temperature changes in the tropics and other regions in Sect. 3.1, and relationships between land and ocean temperatures in Sect. 3.2.

## 3.1 Relationships between regional and global temperature change

A common feature in future and palaeo-simulations is that some parts of the world warm or cool at different rates. In future climate simulations, the high latitudes warm more than the low latitudes, as is also observed during the recent instrumental era. This "polar amplification" is also present in LGM simulations and data, with a stronger cooling in the high latitudes than in the tropics (Masson-Delmotte et al., 2006a, b). Izumi et al. (2013) investigate high vs. low latitude temperature changes in lgm, midHolocene, historical, 1pctCO2 and abrupt4xCO2 PMIP3/CMIP5 simulations and find broadly consistent relationships for lgm, historical and increased GHG forcings, between mean annual SST changes, w.r.t to piControl, over the northern extratropics and the northern tropics. However, the relationship is not consistent for mean annual air temperatures in the lgm simulations compared to the others because of the particular impact of the northern ice sheets. Here we examine the relationships between changes in global mean temperature and change over large-scale regions. The uneven distribution of the palaeo-climatic reconstructions indeed suggests a focus on specific regions, rather than the globe.

The main climate forcings for the LGM are the lower concentrations in atmospheric greenhouse gases and the presence of Laurentide and Fennoscandian ice sheets in the northern extratropics. The ice sheets have a strong local albedo effect (e.g. Braconnot et al., 2012) but also affect the mid-latitude large-scale atmospheric circulation due to the associated change in topography (e.g. Pausata et al., 2011;

Rivière et al., 2009; Laîné et al., 2009). However, away from the direct ice sheet perturbations, we expect that the greenhouse gas forcing would be the main forcing for the LGM climate change and thus patterns of response may be similar to future warmer climates (Hewitt and Mitchell, 1997).

We analyse the comparison between the mean annual surface air temperature change over a region compared to the global mean change for the abrupt4xCO2, 1pctCO2 and lgm CMIP5 simulations from the 8 models for which the results were available at the time of the analysis. We have considered the tropics (land + oceans) and the tropical oceans as targets, because they have been used previously in perturbed physics ensemble studies (Schneider von Deimling et al., 2006; Hargreaves et al., 2007), East Antarctica, for which the temperature change has been shown to scale with global temperature change for the LGM and the CMIP3 2xCO2 and 4xCO2 changes (Masson-Delmotte et al., 2006a, b) and the mid-latitude region of the North Atlantic and Europe.

Figure 1 shows a clear relationship between the tropical and global temperature change for the 1pctCO2 and abrupt4xCO2 anomalies, both for the combined land and ocean grid cells (top-left panel) and for ocean grid cells alone (bottom-left panel), and this relationship is consistent across these two experiments. The relationship for the LGM is ambiguous because the results for 7 out of the 8 models cluster around the same values. These appear to fall outside the relationship which can be derived from the 1pctCO2 and abrupt4xCO2 simulations, with a smaller LGM tropical temperature change for a given global temperature change. This may be because of an outsize influence of the LGM northern hemisphere ice sheets on the global mean for this particular climate. Furthermore, the models which simulate the smallest (largest) warming for increased $CO_2$ are not those which simulate the smallest (largest) cooling for LGM. This implies that either the impact from the lower GHG concentrations are not symmetric compared to those for increased GHG concentrations, or that the ice sheet remote impact extends to the tropics (as inferred by Laîné et al., 2009). The relationship appears more consistent across experiments for East Antarctica (Fig. 1, bottom-right panel) and, surprisingly given the proximity of the ice sheets, over the North Atlantic/Europe region (Fig. 1, top right panel).

In the second row of Fig. 1, we indicate the range of the reconstructed LGM regional response. In the case of the tropical oceans (bottom-left plot), this range is computed from the MARGO (2009) data. Uncertainties are derived using a bootstrap method, randomly drawing 1000 samples (of random size, and with replacement) from the initial MARGO data set. For each drawn site, we assume a Gaussian probability function centered on the mean reconstruction and with a standard deviation equal to the uncertainty given in the data set and we draw a possible value considering this probability distribution function. We obtain 1000 estimates of the mean value and compute its mean ±2 standard deviations, which defines the shaded blue band on the Fig. 1.

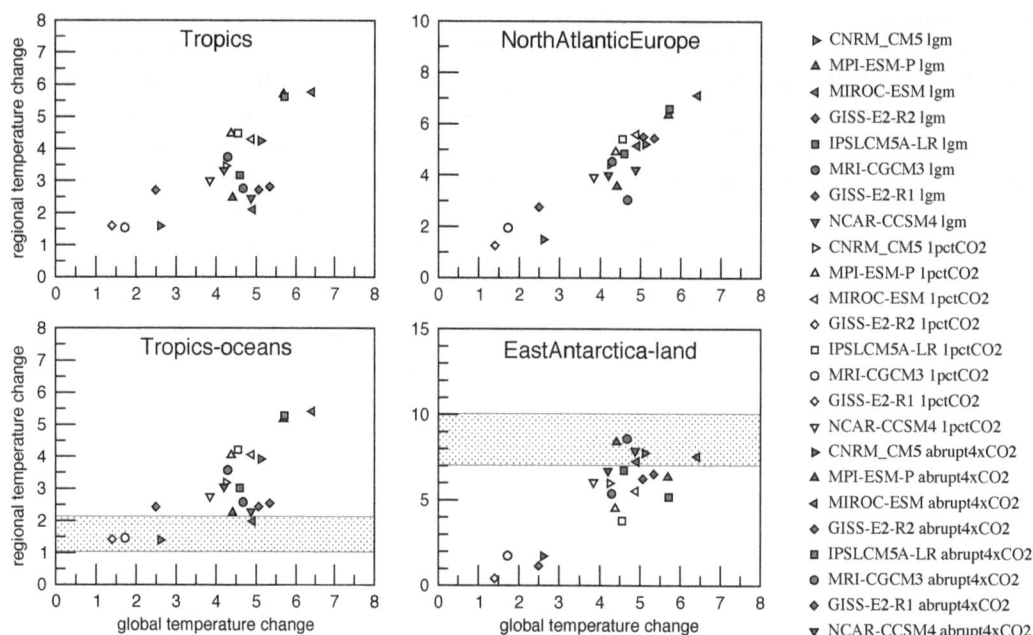

**Fig. 1.** Model average regional vs. global temperature changes for the glacial (in blue, the pre-industrial – LGM difference is shown), years 120 to 140 of the 1pctCO2 simulations (in yellow, in comparison to piControl) and years 100 to 150 of the abrupt4xCO2 simulations (in red, in comparison to piControl). For the bottom plots, the regional model average is taken only over grid boxes that correspond to proxy data sites within the defined region (reconstruction range shown in blue shading). For the tropical oceans (bottom-left panel), the blue shaded band shows the average $\pm 2 \times$ standard deviations of the present – LGM warming as evaluated by the bootstrap method from the MARGO (2009) reconstructions, taking into account the uncertainty on the reconstructions (see main text). For East Antarctica, the blue shaded band corresponds to the range of available reconstructions (5 sites) $\pm 1\,°C$ (Braconnot et al., 2012). Definition of the regions: Tropics: $23°\,S–23°\,N$; North Atlantic Europe: $45°\,W–90°\,E$; $35–45°\,N$; East Antarctica: $5°\,W–165°\,E$; $70–80°\,S$. The results have been computed for all models in the database on 23 July 2012 for which there were results for the lgm, piControl, 1pctCO2 and abrupt4xCO2 simulations.

For East Antarctica, we "only" have 5 points, so we simply consider the uncertainty of $\pm 1\,°C$ on the reconstruction (Masson-Delmotte et al., 2006) and the range of available reconstructions. In both cases, the available data discriminate between the models, with 2 models out of 8 falling in the range of the reconstructions for the tropical oceans and of 4 models out of 8 in the case of East Antarctica.

In summary, the range of model results for increased $CO_2$ scenarios shows that there is a relationship between regional and global temperature changes for all regions considered here. The range of simulated LGM regional/global average temperature change is smaller than in the increased $CO_2$ runs. The results are consistent with the relationship derived in future scenarios for East Antarctica and the North Atlantic/Europe region. For the tropics, the LGM ratio is smaller than that seen in future scenarios, which could be due to the impact of ice sheets on the global mean temperature change. Both data and models suggest an amplification of changes from the tropics to Antarctica and the data can help constrain the global LGM temperature change to 4.2 to $5\,°C$, but only weakly constrain the expected sensitivity to abrupt4xCO2 forcing (from 4.2 to $6.5\,°C$). Additional sensitivity experiments will be needed to test the individual impacts of $CO_2$ and ice sheets and better understand the full

LGM response and the inter-model differences. These results are based on only 8 models and will need to be revisited when a larger number of simulations are available.

## 3.2   Land–ocean contrasts

Even though models show biases in the LGM when directly compared to reconstructions (Fig. 1) there are large-scale relationships which appear to be consistent for different climates. For instance, model results have consistently shown that for the LGM, the continents cooled more than the ocean (e.g. Braconnot et al., 2007, 2012; Laîné et al., 2009), while, in a symmetric manner, predictions for future climate show a stronger warming over land than over the oceans (e.g. Sutton et al., 2007; Drost et al., 2012). The ratio between cooling over non-glaciated land and cooling over the ocean for the LGM tropics was $\sim 1.3$ in the PMIP1 computed sea surface temperature (SST) simulations (Pinot et al., 1999), a result close to the ratio of $\sim 1.5$ found in both the PMIP2 fully coupled LGM experiments (Braconnot et al., 2012) and CMIP3 future projections (Sutton et al., 2007). Izumi et al. (2013) evaluated this land-sea ratio from the CMIP5 lgm, piControl, historical, 1pctCO2 and abrupt4xCO2 simulations and found consistent land-sea ratios for global changes and for

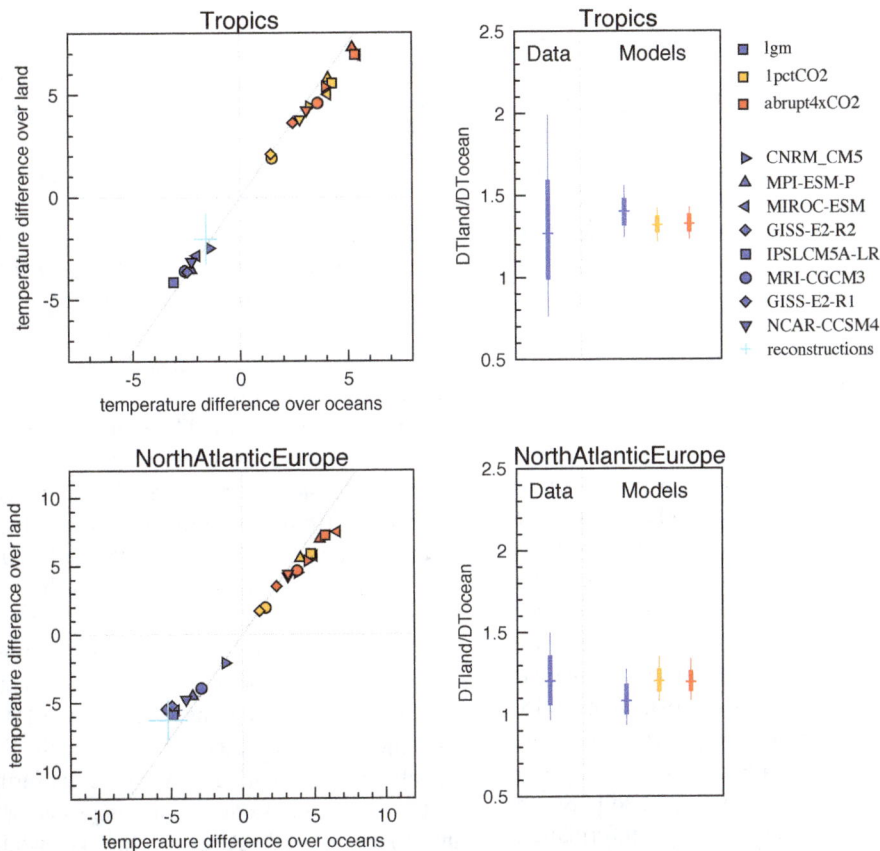

**Fig. 2.** Left panels: average surface air temperature change, compared to piControl, over land compared to over the oceans for the tropics (23° S–23° N) and the North Atlantic and Europe region (45° W–90° E, 35–45° N). LGM – piControl in blue, 1pctCO2 – piControl in orange, abrupt4xCO2 – piControl in red. For the latter 2 periods, the averages have been computed over the same years as Fig. 1. The results have been computed for all models in the database on 23 July 2012 for which there were results for the lgm, piControl, 1pctCO2 and abrupt4xCO2. The grey lines indicate the 1 : 1.5 ratio in both plots. The results from the reconstructions are based on the MARGO (2009) data for the oceans and on the Bartlein et al. (2011) data for the continents. The error bars show twice the standard deviation of the distribution of the mean temperature changes over the region, as estimated by the same bootstrap method as described for the bottom-left plot of Fig. 1 and in Sect. 3.1. Right panels: plots show the land-sea ratio computed from the data, again with the same bootstrap method. The horizontal bar shows the median ratio, the thick vertical line shows the 25–75th percentiles and the thin lines the 10–90th percentiles. The model results are computed by a bootstrap method applied on all model results, as explained in the text. The definition of the thin/thick vertical lines and horizontal bar are the same as for the data.

the northern and southern extratropics as well as for the tropics, though the ratio varies with latitude and is smallest in the tropics.

Figure 2 shows temperature changes over land vs. oceans for the tropics and North Atlantic/Europe. As also shown by Izumi et al. (2013), the relationship between temperature changes over land and over the oceans appears to be broadly consistent for the lgm, 1pctCO2 and abrupt4xCO2 results, though the degree of agreement in the land-sea contrast varies across regions. Furthermore, even though models appear to overestimate temperature changes both over land and oceans in the tropics, and to underestimate them in the North Atlantic/Europe region, the land-ocean ratio appears to be consistent with the data. The right-hand panels in Fig. 2 were built by estimating the distribution of the ratios between

mean temperature changes over land and over the oceans using a bootstrap method on the data from Bartlein et al. (2011) for continental data and MARGO (2009) for the ocean data and taking their uncertainties into account (as for Fig. 1). A similar approach is used for the models, selecting only from points where target reconstructions exists. Coefficients are taken from a linear regression constrained to pass through the origin from 1000 trials. The results show that for the tropics, the land-sea ratio is consistent for the different periods, and model-derived ratios are themselves consistent with the reconstructions. This is also true for the North Atlantic/Europe region, although in this case, the LGM results are more offset from the increased $CO_2$ results.

We conclude that these relationships are robust, although the reasons for this appear to be imperfectly understood (Lambert et al., 2011) and will require, as for the results from Sect. 3.1, additional sensitivity and process-based analyses. Incidentally, it is worthwhile to note that the land–ocean relationship was previously used to highlight the inconsistency between an earlier compilation of tropical LGM sea surface temperatures and adjacent continental reconstructions (Rind and Peteet, 1985).

## 4  Palaeo-derived measures of skill that discriminate between models

In this section we highlight diagnostics for which we have commensurate palaeo-climate information and for which the skill metrics across the ensemble serve to discriminate between models that show different behaviours in future projections. This requires that we demonstrate that differences in future sensitivity are correlated to past sensitivities, and that palaeo-reconstructions exist that can effectively weight the projections from models with more realistic sensitivity in the past more highly in an ensemble projection. We illustrate this with three examples: in Sect. 4.1, we look at a simple binary grouping of model behaviour related to South American rainfall that can be evaluated using information from the mid-Holocene. Section 4.2 revisits attempts to constrain overall climate sensitivity using information from the LGM, and Sect. 4.3 looks at the potential to estimate sea ice sensitivity to Arctic warming through results from the mid-Holocene.

### 4.1  Rainfall change in South America

Projections of precipitation change in South America have a large spread in the CMIP3 (Meehl et al., 2007) and CMIP5 (Knutti and Sedláček, 2012) archives. In future projections, most models simulate a dipole of precipitation change in northern South America. However, the sign and magnitude of this dipole depends on the model: some models simulate drier conditions in Guyana, Venezuela and Colombia and wetter conditions in Nordeste and eastern Brazil, while some model simulate the opposite changes (Fig. 3).

We define the precipitation dipole as the annual-mean precipitation averaged over 0–8° N; 50–60° W (hereafter "Guyana") minus the annual-mean precipitation averaged over 5–15° S, 35–45° W (hereafter "Nordeste"). We divide 28 different models from the CMIP5 archive into two equal groups. Models where the dipole is weak or negative in the changes in precipitation between rcp85 and piControl are placed in group 1; models which have a strong positive dipole are in group 2. All of the models simulate similar patterns of present-day precipitation, although models in group 2 tend to have a more pronounced double ITCZ. Among the models,

midHolocene output was available for 7 models in group 1 and for 5 models in group 2.

Figure 3 shows a link between precipitation change in the future and in the MH. Models in group 1 simulate wetter conditions in Guyana and drier conditions in Nordeste, associated with a northward shift of the ITCZ in the rcp85 and a broadening of the ITCZ in the MH simulations. Conversely, group 2 models simulate drier conditions in "Guyana" and wetter conditions in "Nordeste", associated with a southward shift of the Intertropical Convergence Zone (ITCZ). They show a similar dipole in the MH, with a strong southward shift of the ITCZ. Thus the models from a particular group show essentially the same change in the dipole pattern and the same shift in the ITCZ in both future and MH simulations. These patterns are robust relative to the numbers of groups or the number of models included in any group. Palaeo-data from South America show drying everywhere except northeastern Brazil (Prado et at., 2013), a response which is more consistent with group 1 than group 2.

The processes underlying these patterns can be investigated using a variety of other CMIP5 simulations. Table 2 shows correlations between precipitation changes and other features of the simulations. Shifts in the ITCZ in the future projections are associated with shifts in the SST dipole in the Atlantic: models that shift the ITCZ the furthest southwards are those with the strongest warming south of the Equator relative to the rest of the Atlantic. However, while ITCZ shifts in response to SST dipoles are expected (e.g. Kang et al., 2008), this is not the dominant pattern for the MH to PI change. Some of the model behaviours are seen in the amipFuture and sstClim4xCO2 simulations, indicating that the intrinsic response of the atmosphere to a given SST change plays a key role in the formation of the dipole. This is consistent with the fast atmospheric response to $CO_2$ being an important component of the total precipitation response in global warming (e.g. Bala et al., 2010; Bony et al., 2013). Models that have reduced precipitation over northern South America in the MH simulations also have reduced precipitation in the projections and under 4xCO2. These models have the strongest land surface warming in response to both 4xCO2 and MH forcing. Although the precipitation response of the different groups of models to a change in forcing differs, within each model group the response to different forcing (SST changes, orbital forcing, 4xCO2) is similar. This suggests that common mechanisms are involved in the precipitation response to all forcings, and that we can expect future changes to resemble those predicted by the group 1 models. A more quantitative assessment of these changes still remains to be finalised.

### 4.2  LGM constraints on climate sensitivity

The LGM has been a prime target for assessments of climate sensitivity since it is a quasi-stable period with significant climate differences from today, with reasonably

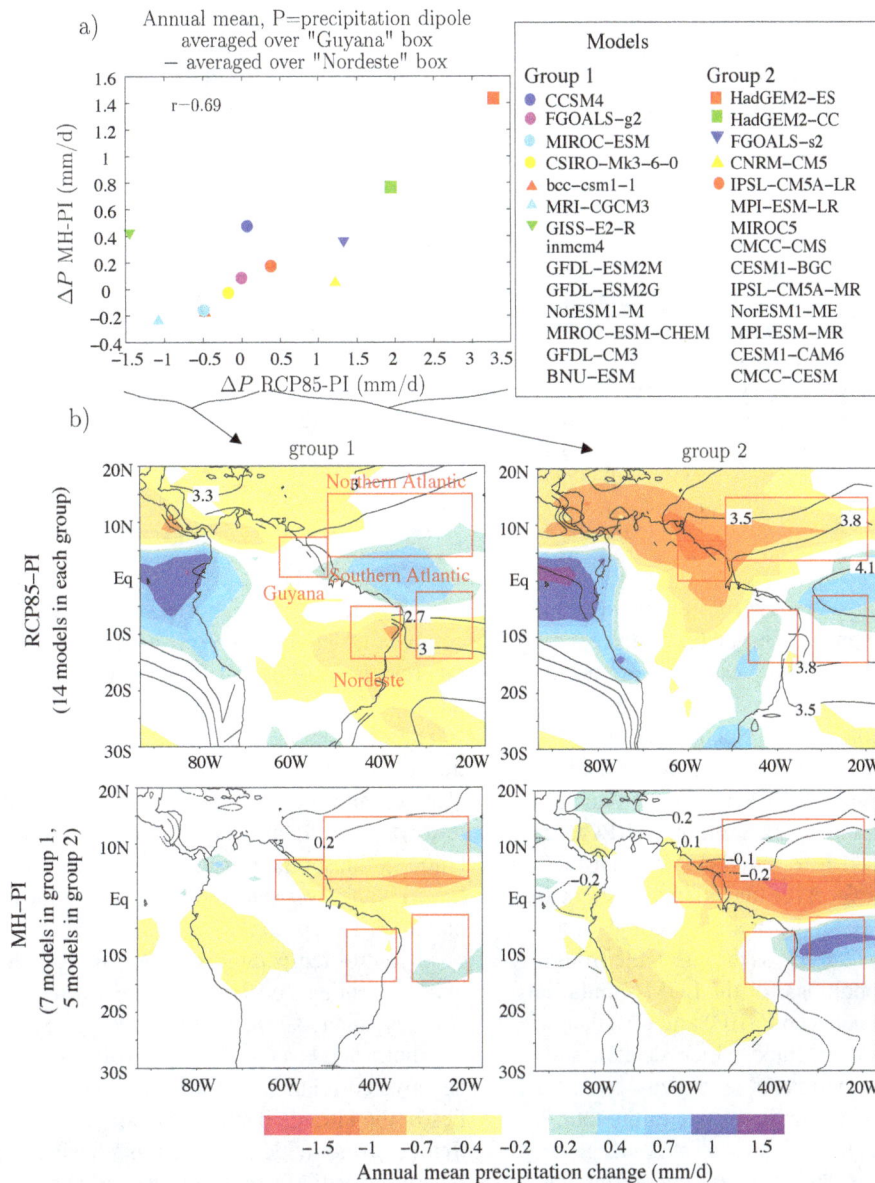

**Fig. 3. (a)** Relationship between the precipitation dipole change from pre-industrial to future climate under RCP 8.5 for the 2080–2100 and the precipitation dipole change from pre-industrial to mid-Holocene. The precipitation dipole is defined as the difference of precipitation change in RCP 8.5 between the "Guyana" region and the "Nordeste" region. Only those models within each group that had both rcp85 and midHolocene data available at the time of the analysis are plotted. Other models that provided only rcp85 data are listed for completeness, but without any markers. **(b)** Maps of precipitation changes from piControl to rcp85 (top panels) and from piControl to midHolocene (bottom panels) in average over all available models in group 1 (left panels) and in group 2 (right panels). Contours show corresponding SST changes. The boxes over land and ocean show the areas used in the dipole definitions.

well-known boundary conditions and sufficient data to reconstruct large-scale climate shifts (e.g. Lorius et al., 1990; Edwards et al., 2007; Köhler et al., 2010; Schmittner et al., 2011; PALAEOSENS Project Members, 2012). This provides a good opportunity to apply the methods described in Sect. 2 as a proof-of-concept estimate of the equilibrium climate sensitivity based on the CMIP5 LGM simulations.

We use an ensemble of opportunity consisting of 7 models which participated in the PMIP2 experiment, together with 7 CMIP5 models for which sufficient data were available (at time of writing). Estimates of the climate sensitivities of these models were obtained from a variety of sources and were derived using a range of methods: For the PMIP2/CMIP3 models, sensitivity was generally calculated using a slab ocean coupled to the atmospheric component

**Table 2.** Correlation of different variables with future precipitation change in the RCP 8.5 scenario. Precipitation changes are defined as in Fig. 3: annual-mean precipitation averaged over 0–8° N; 60–50° W minus annual-mean precipitation averaged over 5–15° S; 45–35° W; SST dipole changes are defined as the annual-mean change in SST averaged over 3–15° N; 50–20° W minus the annual-mean change in SST averaged over 3–15° S; 20–30° W (see boxes on Fig. 3b). Land surface warming is the annual-mean warming averaged over 0–15° S; 50–70° W. The double ITCZ index is defined as the annual-mean precipitation averaged over the southern branch (3–7° S; 20–35° W) minus the annual-mean precipitation averaged over the northern branch (3–7° N; 20–35° W).

| Variable | Correlation (r) | No. of models | p value |
|---|---|---|---|
| midHolocene-piControl: Δ precip | 0.69 | 12 | 0.013 |
| rcp85-piControl: Δ SST dipole | 0.47 | 28 | 0.012 |
| midHolocene-piControl: Δ SST dipole | 0.22 | 12 | Not significant |
| amipFuture-amip: Δ precip | 0.55 | 9 | 0.125 |
| sstClim4xCO2-sstClim: Δ precip | 0.67 | 12 | 0.017 |
| sstClim4xCO2-sstClim: Δ SAT (land) | 0.62 | 12 | 0.032 |
| midHolocene-piControl: Δ SAT (land) | 0.55 | 12 | 0.063 |
| double ITCZ index in piControl | −0.48 | 28 | 0.010 |

(Meehl et al., 2007), whereas in CMIP5, the most readily available estimates use a regression based on a transient simulation (Andrews et al., 2012). These estimates are not perfectly commensurate, with some models reporting a 10 % difference in the two methods (e.g. Schmidt et al., 2014). Unfortunately, some of the PMIP2 models used for the LGM simulations differ from the CMIP3 versions for which the sensitivity estimates were made (for example, MIROC3.2). Thus, while the values used here may be somewhat inconsistent and imprecise, we expect the uncertainty arising from these sources (around 0.5 °C) to be modest in comparison to the range of values represented across the ensemble (roughly 2–5 °C). The boundary conditions for the LGM simulations are essentially unchanged between PMIP2 and CMIP5 (save for changes in the shape of the imposed ice sheets), allowing us to consider these experiments as broadly equivalent though there are some systematic biases due to the total ice volume and resulting changes in land/sea mask (Kageyama et al., 2013). Limitations in the boundary conditions (such as the exclusion of dust and vegetation effects) which we do not attempt to account for here, could introduce additional bias and uncertainty into our result. For these and other reasons discussed below, these results should be considered as a proof of concept rather than conclusive.

The LGM was associated with a large negative radiative forcing with respect to the pre-industrial including substantially lower concentrations of greenhouse gases (e.g. Köhler et al., 2010). However, the ensemble does not show the expected negative correlation between climate sensitivities and their globally averaged LGM temperature anomalies (over the full 100 yr of simulation output) (Fig. 4a, see also Crucifix, 2006). Hargreaves et al. (2012) analysed the PMIP2 ensemble on a regional basis and found their LGM temperature changes in the tropics to exhibit a negative correlation with climate sensitivity, most strongly in the latitude band 20° S–30° N. Results from the PMIP3 models are consistent

with this relationship, but do not strengthen it. When we combine all models into one ensemble, the correlation over this region weakens to −0.54 but it is significant at the 95 % level.

The correlation is generally insignificant at higher latitudes where the feedbacks in response to large cryospheric changes may be very different to those exhibited in a future warmer climate. There is also a strong positive correlation in the southern ocean (i.e. colder LGM anomalies are linked with *lower* sensitivity), possibly due to the large range of biases in the control climate (Fig. 4c). The correlation of piControl temperatures to sensitivity points to the Arctic and the southern oceans as regions where base climatology strongly impacts sensitivity, probably via cloud effects (see Trenberth and Fasullo, 2010 for a discussion). The significant negative correlation between the LGM temperature anomalies in the latitude band 20° S–30° N, and the climate sensitivities of the models (Fig. 5), is physically plausible, since this region is far from the cryospheric and sea ice changes of the LGM, and the forcing here is dominated by the reduction in greenhouse gas concentrations. Assuming that the correlation with tropical temperatures provides a valid constraint on the real climate system, we can use this correlation to project a reconstruction of past change onto the future, as in Boé et al. (2009).

Annan and Hargreaves (2013) generated a new estimate of LGM temperature changes, based on a combination of several multiproxy data sets, and the ensemble of PMIP2 models. The method does not depend on the magnitude of changes estimated by the models, but only their spatial patterns. Due to the suspicion that the tropical temperatures at the LGM from the MARGO synthesis are too warm (e.g. Telford et al., 2013), we focus here on the results from the sensitivity test in Annan and Hargreaves (2013), where reconstructed tropical temperatures were decreased uniformly by 1 °C. Using the resulting estimate of LGM

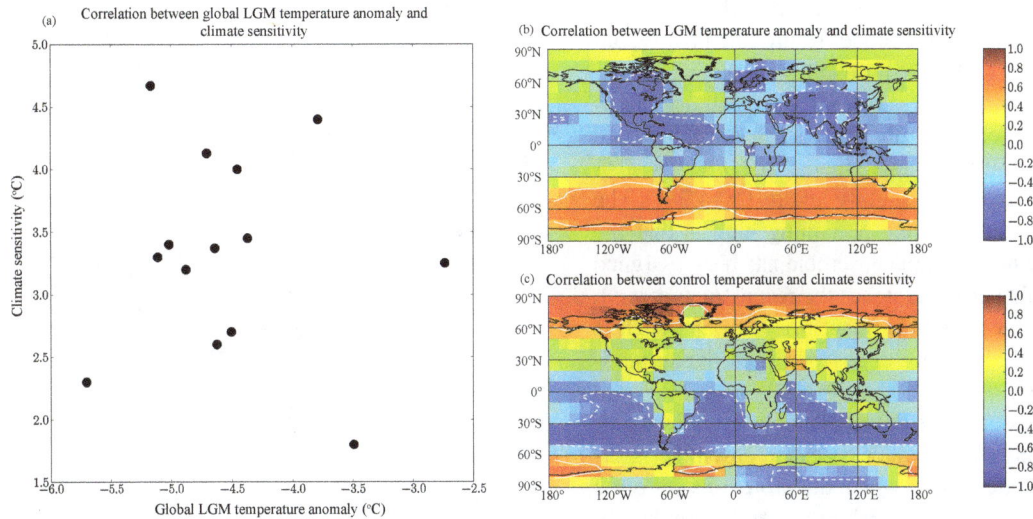

Fig. 4. (a) Global mean LGM temperature change versus overall climate sensitivity to 2xCO$_2$. (b) correlation between local LGM air temperature anomaly and climate sensitivity across the model ensemble. (c) correlation across the model ensemble between control run temperatures and climate sensitivity.

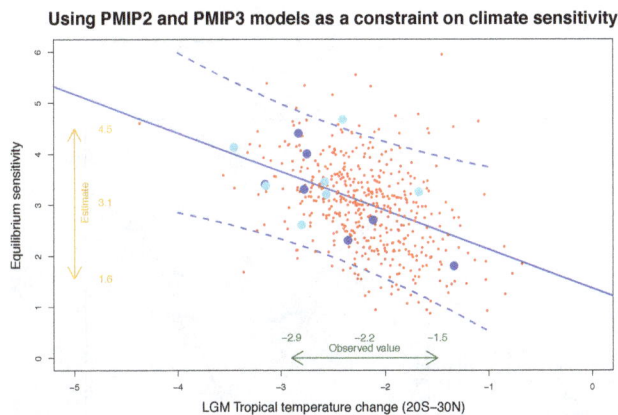

Fig. 5. Using LGM tropical temperature as a constraint on climate sensitivity. Cyan and blue dots represent PMIP2 and CMIP5 simulations respectively. Linear regression and predictive uncertainty range are plotted as solid and dashed blue lines respectively. Small red dots represent a Monte Carlo sample from the estimated proxy-derived reconstruction, mapped onto the climate sensitivity.

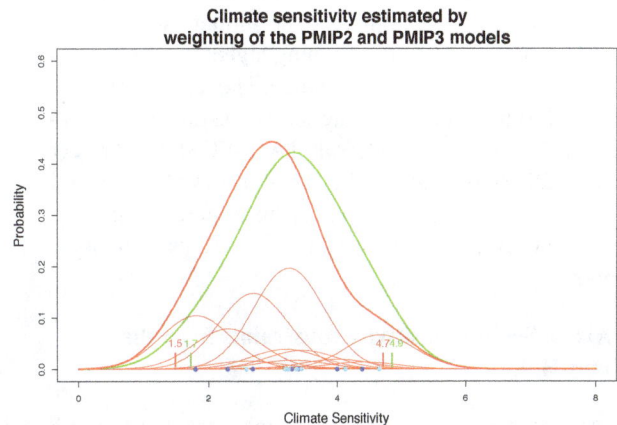

Fig. 6. Climate sensitivity estimated through weighting of the PMIP models. Blue and cyan dots represent PMIP2 and CMIP5 simulations respectively. Green curve shows prior distribution of climate sensitivity (based on equal weighting of the models). Thick red curve shows posterior distribution, after weighting according to match to the LGM tropical temperature. Thin red curves show the individual models' contributions to the posterior after weighting. Vertical bars indicate 5, 50 and 95 percentiles.

temperature change in the 20° S–30° N latitude band of $-2.2 \pm 0.7$ °C (at 90 % confidence), the predicted value for climate sensitivity arising from the correlation is 3.1 °C, with a 90 % interval of 1.6–4.5 °C calculated by Monte Carlo sampling, but this range is sensitive (by up to 0.4 °C) to the reconstruction uncertainties (Annan and Hargreaves, 2013).

In a more explicitly Bayesian approach, we can initially assign equal probability to each model in the ensemble. This choice can be questioned, given both the range of model complexities, and also the possible inter- or intra-generational similarities between models of related origins (Masson and Knutti, 2011). However, quantifying these

issues is far from straightforward, so we make our choice for reasons of practicality and in order to demonstrate the utility of the overall method. A standard kernel density estimation based on the ensemble leads to the prior distribution presented as the green curve in Fig. 6, which has a 90 % range of 1.7–4.9 °C and a median of 3.3 °C. The observationally derived estimate of tropical temperature gives rise to the natural Gaussian likelihood function $L(M|\overline{O}) = P(O|\overline{M})$ from which the weights are calculated (where $O$ represents the reconstructed observations, $M$ the model simulation, $L(M|\overline{O})$ the likelihood of the model result given the reconstructed data, and $P(O|\overline{M})$ the probability of the reconstructions assuming the models are correct). The posterior distribution is shown in red, the bulk of which has been shifted to lower values with the median reducing to 3.0 °C. Its 90 % probability range has moved slightly less to 1.5–4.7 °C. The reason for the upper limit here remaining high is that the highest sensitivity model in the ensemble has been assigned a fairly large weight since it matches the reconstruction well. The small size of the ensemble means that this approach is rather sensitive to the presence or absence of particular models in the ensemble.

These two approaches differ considerably in their use of the model ensemble. In the latter case, the ensemble is directly used as a prior estimate, which therefore imposes quite a strong constraint on climate sensitivity even before the observational constraints are used. The former method may be considered as roughly equivalent to using a prior that is uniform in the observed variable (here tropical temperature), although this approach is rarely presented in explicitly Bayesian terms. Despite the different assumptions and approaches, these methods both generate similar estimates for the climate sensitivity – both assigning higher probability towards the lower end of the model range. The ranges are comparable with other palaeo-climate-derived estimates of 2.3–4.8 °C (68 % confidence interval, PALAEOSENS Project Members, 2012) but, given the small ensemble size and possible naïvety of the assumptions made here, these estimates may not be robust and need to be tested using a larger ensemble.

### 4.3   Arctic Sea ice sensitivity constraints from the mid-Holocene

The rate and pattern of Arctic sea ice change in future decades is of interest due both to the surprisingly rapid changes currently occurring and the large spread in model estimates in, for instance, the onset of summertime "ice-free" conditions (Stroeve et al., 2012; Massonet et al., 2012). Recent studies (Mahlstein and Knutti, 2012; Abe et al., 2011) have demonstrated that biases in sea ice volume have a strong impact on the simulated responses to radiative perturbations, and that there may be a possibility to discriminate among models based on interannual modes of sea ice variability. The mid-Holocene simulations (driven mainly by changes in orbital forcing) provide an orthogonal test of Arctic sea ice sensitivity since MH insolation changes imply that NH summers were warmer than summers today (see Kutzbach, 1981 and many subsequent papers). Palaeo-data from the circum-Arctic region indicates that this warmth was accompanied by reductions in sea ice extent at least during some months of the year (Dyke and Savelle, 2001; de Vernal et al., 2005; McKay et al., 2008; Funder et al., 2011; Polyak et al., 2010; Moros et al., 2006).

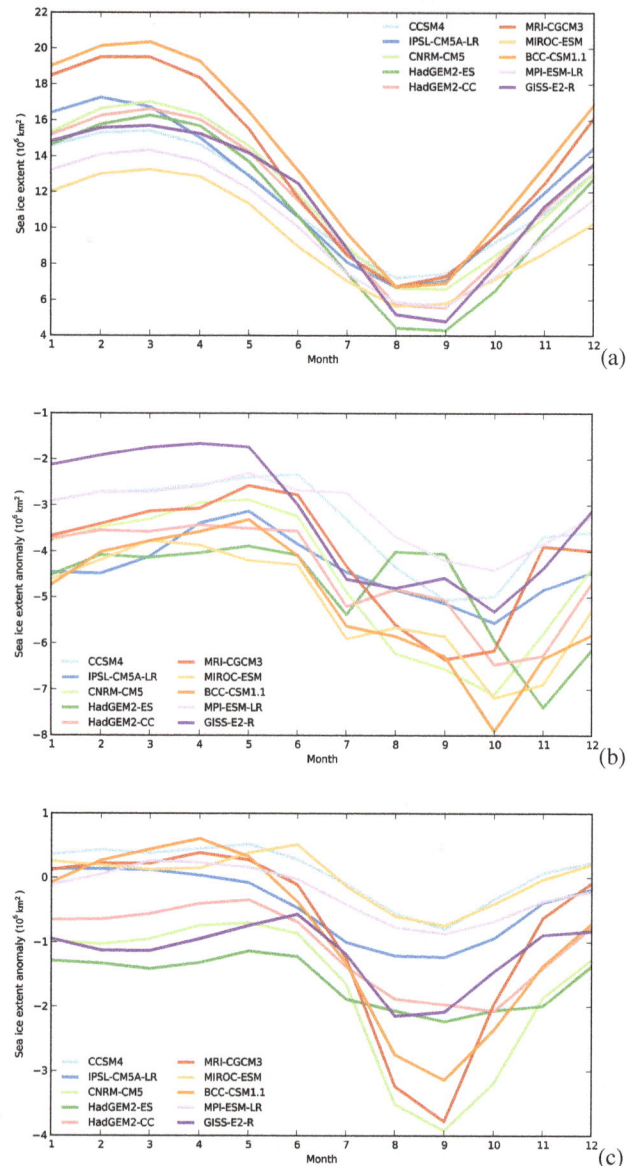

**Fig. 7.** Sea-ice extent in CMIP5 models in $10^6$ km$^2$. (**a**) 30-yr mean seasonal cycle for the period 1870–1900, (**b**) the anomaly in sea ice extent for the period 2036–2065 in RCP 8.5, and (**c**) the anomaly at the mid-Holocene.

The CMIP5 MH simulations (Fig. 7c) consistently show decreases in sea ice extent from July/August through to November relative to the pre-industrial. Changes in winter months (December–February) do not agree in sign across the models, though these changes are not well characterised in the palaeo-data either. There is a relationship (Fig. 8) between the size of the anomaly at the MH and in future projections (using 2036–2065 in rcp85), presumably reflecting the underlying sensitivity of the sea ice model and Arctic climate in general (see also O'ishi and Abe-Ouchi, 2011). We focus on the 30 yr period centered on 2050 since that is when there

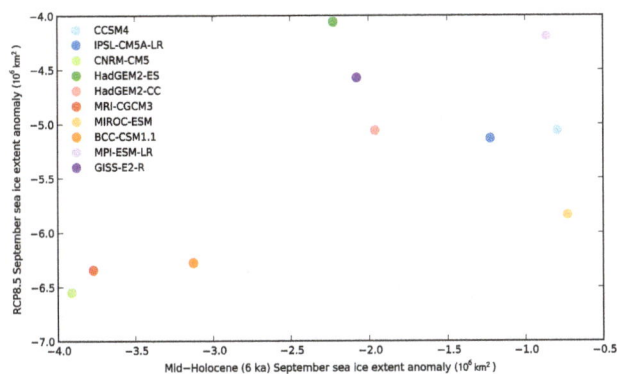

**Fig. 8.** Relationship between the September RCP 8.8 anomaly (derived from Fig. 7b) and the September MH anomaly (from Fig. 7c) across the CMIP5 models.

is a very large spread in individual model projections in the RCP 8.5 simulations (Fig. 7).

This correlation exists despite the variations in the cause of the ice loss (summer insolation versus greenhouse-gas-related forcing). Although the small size of the ensemble raises questions about the robustness of the relationships, the MH ice extent anomaly can be used to estimate the likely loss in future projections. Given the qualitative nature of the palaeo-data, we are not yet able to make a quantitative projection, but there is some support given to the models with a larger projected change. A more detailed approach using more specific and local diagnostics in comparison to a wider proxy network will likely give a more quantitative result (Tremblay et al., 2014).

## 5   Exploratory metrics: potential and limitations

While the examples given above show direct connections between past and future in ways that can be used relatively straightforwardly, there are a number of reasons why other diagnostics may not be as useful. In this section we provide examples of where the palaeo-climate information has yet to be explored, is ambiguous, or where connections seen in palaeo-climate changes do not translate easily into the future for some reason. This may be related to forcing ambiguities, climate-change-related non-stationarity in climate/proxy relationships, or potentially, a poor understanding or representation of the dominant processes. While these examples are not directly informative about the future, they illustrate how the palaeo-simulations can be explored in ways that illuminate key uncertainties and, potentially, provide more opportunities in the future.

Section 5.1 focuses on the potential for shorter-term extremes in temperature and precipitation at the regional scale to be predicted by large-scale seasonal anomalies. Section 5.2 deals with the issue of the evaluation of models over the historical period in the tropical Pacific using forward

models of coral-based proxies. Section 5.3 addresses diagnostics in the frequency domain that are strongly affected by uncertainties in the forcing fields, rather than intrinsic properties of the models. Finally, Sect. 5.4 provides an example of how connections between past and future hydroclimate diagnostics may be non-stationary.

### 5.1   Regional extremes

Extreme climate events such as heat waves and cold spells can have long-lasting impacts on society or ecosystems (IPCC SREX, 2012) and there have been analyses of the impact of heat waves during recent centuries in Europe (Le Roy Ladurie, 2004, 2006; Barriendos and Rodrigo, 2006; Camuffo et al., 2010). The development of such events spans days to a few weeks, so that they are largely intra-seasonal by nature (Seneviratne et al., 2012). In such a context, the generally linear relationship between palaeo-climate reconstructions and actual climate can be strongly distorted. Since extreme events are by definition rare, large numbers of examples are required to get good statistics. Simulations of the past millennium offer a promising tool to investigate modelled extremes since they sample a longer time series and bigger range of possible cases than in most other simulations. The strongest limitation for an application of this method to palaeoclimatic data has been the necessity of dealing with daily data in order to resolve extreme value distributions (which may be non-Gaussian) and the need for palaeo-archives that record extreme variables (e.g. Donnelly and Woodruff, 2007). However, if we can demonstrate the robustness of the relationships between short and longer-term statistics over long periods of time, and/or their dependence on external forcings, we can potentially constrain the behaviour of temperature extremes in the future.

The statistical analyses of (daily) temperature hot extremes of the 20th century have shown that temperature is generally a bounded variable, for which the upper bound can be computed from the statistical parameters of extremes (Parey et al., 2010a, b). Diagnostic studies focusing on the probability distribution of temperature and precipitation extremes are often based on the application of Extreme Value Theory (EVT), though simpler metrics have also been used (e.g. Hansen et al., 2010). EVT describes the behaviour of the probability distribution near the tails, and allows one to estimate return periods for extremes that are longer than the period of observation (Coles, 2001). It has been applied to meteorological observations (Parey et al., 2010a), reanalysis data (Nogaj et al., 2006) and model simulations (Kharin et al., 2005, 2007) in order to estimate trends in extremes.

Extremes of hot and cold temperatures are correlated with mean temperatures over the northern extra-tropics (Yiou et al., 2009). Since very few models have archived daily outputs of temperature or precipitation on multi-century timescales, there has been no assessment of whether this is true over the longer term (Jansen et al., 2007). However, daily resolution

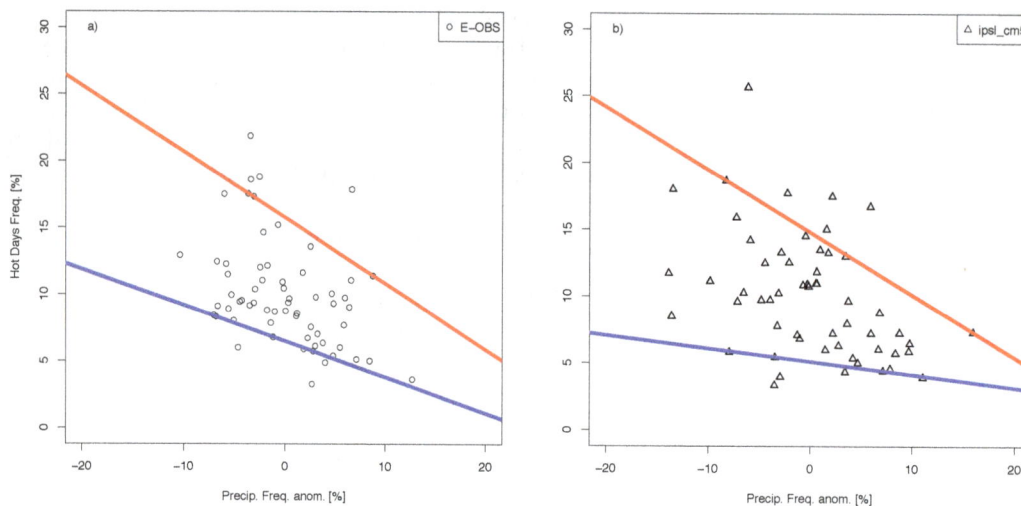

**Fig. 9.** Illustration of quantile regressions between the percentage of summer hot days (i.e. exceeding the 90th quantile of daily mean temperature in June-July-August) in western Europe (10° W–30° E; 36–61° N) and the precipitation frequency anomaly with respect to the 1948–2011 mean in winter-spring (January to May). The precipitation frequency is computed over southern Europe (10° W–30° E; 36–46° N) and is defined as the percentage of days with precipitation exceeding 0.5 mm. The quantile regressions are computed for the 10th and 90th quantiles of the hot day frequency, following Quesada et al. (2012). The red lines are for the 90th quantile regressions and the blue lines are for the 10th quantile regression. (**a**) shows the quantile regression for western Europe from the updated EOBS data set (Haylock et al., 2008) between 1950 and 2012 where each point represents a year. (**b**) is for the "historical" simulation (1960–2008) of the IPSL-CM5A-MR model (Dufresne et al., 2013). Both panels show a widening of the quantile regression for low values of precipitation frequency, indicating a consistency of the model simulation with observations.

data was requested for simulations in the CMIP5 archive (Yiou et al., 2012).

Summer heat waves are generally preceded by droughts in the winter and spring in extratropical regions (Fischer et al., 2007; Vautard et al., 2007). The mechanism involves a positive feedback between sensible heat fluxes, evapotranspiration and temperature (Schär et al., 1999), and this has also been found in global and regional simulations of the future (Seneviratne et al., 2006, 2010; Quesada et al., 2012). Quantile regression provides a useful statistic metric to investigate the linkage between precipitation in winter and spring and summer temperature. Ordinary least-squares regression focuses on the mean values of related variables, but by setting a threshold based on the upper/lower quantiles of the variable to be predicted, regression coefficients related to the high (or low) values of this variable are obtained (Koenker, 2005). The purpose of quantile regression is to investigate the conditional dependence between variables: for instance, the dependence structure could be different for small and large predictors. Hence, differences of slopes for small and high quantiles show that the relation between the predictand and predictor depends on the value of the predictor. An interesting feature is that quantile regression is not very sensitive to outliers, because the regression is performed on the ranks rather than the values themselves. We illustrate this diagnostic in Fig. 9, by computing the quantile regression for 90th and 10th deciles of the summer hot day frequency and winter-spring precipitation frequency anomaly in the IPSL-CM5A-MR historical

simulation and the E-OBS gridded data set (Haylock et al., 2008).

As in Quesada et al. (2012), the frequency of hot days is defined by the percentage of days in western Europe (10° W–30° E; 36–61° N) between June and August whose temperature anomaly exceeds the 90th quantile over a reference period (1948–2011). The frequency of rainy days is the percentage of days in southwestern Europe (10° W–30° E; 36–46° N) between January and May whose precipitation exceeds 0.5 mm. This is a simplistic index for soil moisture (or drought) but it does have a significant predictive skill to European summer temperature variations (Vautard et al., 2007). More sophisticated indices of drought or soil moisture marginally improve the predictive skill (Seneviratne and Koster, 2012). In Fig. 9 the quantile regression slopes illustrate the asymmetry of the precipitation or temperature dependence for hot or cool summers in western Europe (Hirschi et al., 2011; Mueller and Seneviratne, 2012; Quesada et al., 2012; Seneviratne and Koster, 2012).

The 90th and 10th quantile regression lines are not parallel. Thus while the general picture is that a dry winter/spring tends to favor a hot summer and wet winter-spring conditions are generally followed by cool summers, dry winter-spring conditions can be followed by cool summers as well as heat waves (large spread between low and high quantiles). This is due to the fact that the genesis of heat waves can be broken in just a few days, due to fast variations of the synoptic atmospheric circulation (Hirschi et al., 2011; Quesada et al., 2012). This feature has been tested on CMIP3 and some CMIP5 simulations for the present and future scenarios and shows that the seasonal predictability of large European hot summers decreases under drier conditions in southern Europe, although their frequency increases (Quesada et al., 2012). By looking at the last millennium simulations we will be able to examine the stability in time of these patterns, and hence potentially constrain future changes.

## 5.2 20th-century changes in tropical Pacific climate

The response of the tropical Pacific Ocean to anthropogenic climate change is uncertain, partly because we do not have a good understanding of how the region has responded to drivers in the past. Instrumentally based estimates of SST over the 20th century are not internally consistent (Deser et al., 2010), and model simulations have a wide spread of 20th-century trends (Thompson et al., 2011). Trends in the tropical Pacific are particularly challenging because the instrumental record is sparse even for the early 20th century and long-term in situ measurements of SST are uncommon. High-resolution palaeo-climate records, particularly the large network of tropical Pacific coral $\delta^{18}O_{calcite}$ records, can be used to extend the observational record and assess long-term trends. These $\delta^{18}O_{calcite}$ records respond to the combined effects of SST and the isotopic composition of seawater ($\delta^{18}O_{sw}$) (which is strongly correlated to sea surface salinity, SSS) and can reveal changes on longer timescales.

To address the limitations of historical observations, model simulations and coral proxy records in the tropical Pacific, Thompson et al. (2011) used a forward-modelling approach to generate synthetic coral records (i.e. pseudo-corals) from observational and climate model output and test whether these pseudo-corals are in agreement with the network of coral $\delta^{18}O_c$ observations. The forward model for $\delta^{18}O_c$ calculates isotopic variations as a function of SST and SSS anomalies, with an SST-$\delta^{18}O_c$ slope of $-0.22\permil\,°C^{-1}$ and the SSS-$\delta^{18}O_{sw}$ slope varying by region (LeGrande and Schmidt, 2006). When driven with historical SST and SSS data, the simple model of $\delta^{18}O_c$ is able to capture the spatial and temporal pattern of ENSO and the linear trend observed in 23 Indo-Pacific coral records between 1958 and 1990 (Thompson et al., 2011). The observed trends were driven primarily by warming at the coral sites, though SSS was responsible for approximately 40 % of the shared $\delta^{18}O_c$ trend. However, pseudo-coral records calculated from CMIP3 and CMIP5 historical simulations could not reproduce the magnitude of the secular trend (Fig. 10, upper panel), the change in mean state, or the change in ENSO-related variance observed in the coral network from 1890 to 1990. While the observational coral network suggests a reduction in ENSO-related variance and an El Niño-like trend over the 20th century, CMIP3 and CMIP5 simulations vary greatly on both points.

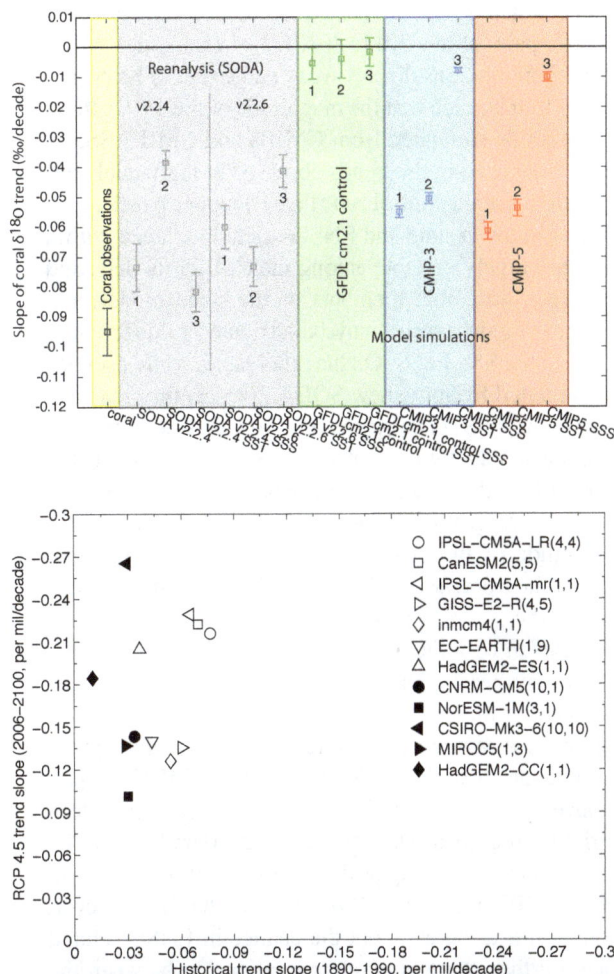

**Fig. 10.** Upper panel: magnitude of the trend in $\delta^{18}O_c$ (‰/decade, computed from a simple linear regression through the trend PC) in corals (far left), Simple Ocean Data Assimilation (SODA) 20th-century reanalysis (Carton and Giese, 2008; Giese and Ray, 2011; Compo et al., 2011), a 500 yr control run from GFDL CM2.1 (Wittenberg, 2009), and the CMIP3 and CMIP5 multi-model ensembles. In each case, $\delta^{18}O_c$ was modelled from SST and SSS (1), SST only (2), and SSS (3). Lower Panel: magnitude of the $\delta^{18}O_c$ trend (‰/decade, computed from a simple linear regression through the trend PC) over 1890–1990 in pseudocorals modelled from CMIP5 historical simulations and over 2006–2100 in the RCP 4.5 projections where numbers in parenthesis indicate the number of runs in the historical and RCP 4.5 ensemble, respectively.

The differences between observed and GCM-derived $\delta^{18}O_c$ trends may stem from the simplicity of the forward model for $\delta^{18}O_c$, bias in the coral records, and/or errors in the GCM SST and SSS responses, or indicate an important role for unforced variability. Isotope-enabled coupled control simulations highlight uncertainties in the SSS-$\delta^{18}O_{sw}$ relationship and suggest that short-term isotope variability may play a minor role (Russon et al., 2013; Thompson et al., 2013). Previous work has also highlighted potential biases in simulated salinity fields as a source of the discrepancy (Thompson et al., 2011; 2013). For example, CMIP3 and CMIP5 simulations display weak and spatially heterogeneous SSS trends, such that the magnitude of the $\delta^{18}O_c$ trend in pseudo-corals simulated from CMIP3 and CMIP5 SSS is indistinguishable from the trends observed in individual centuries of an unforced control run (Fig. 10, upper panel). Further, trends in mean state and ENSO-related variance within the basin are highly variable among the CMIP5 models, and even between ensemble members of the same model, and much of this model spread may be attributed to differences in the simulated SSS fields. On the other hand, while pseudo-corals, modelled from the new SODA 20th-century reanalysis of SST and SSS, display greater agreement with the observed coral trends, two recent versions of this product disagree regarding the relative contribution of SST and SSS. These results suggest that more work is needed to constrain the magnitude of the observed 20th-century salinity trend throughout the tropical Pacific Ocean. This work provides an example of the utility of forward models in investigating potential biases in both the models and proxy data, which may be used for further model development and exploration and improvement of model metrics.

Despite the disagreement among models and runs regarding the change over the 20th century, the CMIP5 projections converge upon a more El Niño-like (e.g. warmer eastern equatorial Pacific) mean state change by 2100 under RCP 4.5 (with only one model suggesting the opposite), consistent with the CMIP3 projections (Meehl et al., 2007). However, the models still disagree about the change in ENSO-related variance. Further, there is no clear relationship between the magnitude of the simulated 20th-century $\delta^{18}O_c$ trend and the projected future $\delta^{18}O_c$ trend in the CMIP5 ensemble (Fig. 10, lower panel). This suggests that an agreement of the simulated 20th-century change in the tropical Pacific with that of the observational coral network would not be a reliable indicator of future trends. Nonetheless, this work highlights key uncertainties in the observed and simulated salinity trends within the basin and thus provides a basis for further development of the models and this potential metric. More generally, it shows the utility of a forward modelling approach in palaeo-model/data comparisons to highlight key functional dependencies in specific proxies and investigate potential biases in both models and reconstructions.

## 5.3   Decadal to multi-decadal variability

In contrast to the spatial domain used in other examples here, this section highlights two analyses in the frequency domain that illustrate the important role of relatively uncertain forcings in assessing skill in model simulations of decadal to multi-decadal variability. Given the short instrumental period, it might be hoped that longer time series from proxy reconstructions for the last millennium could be used to

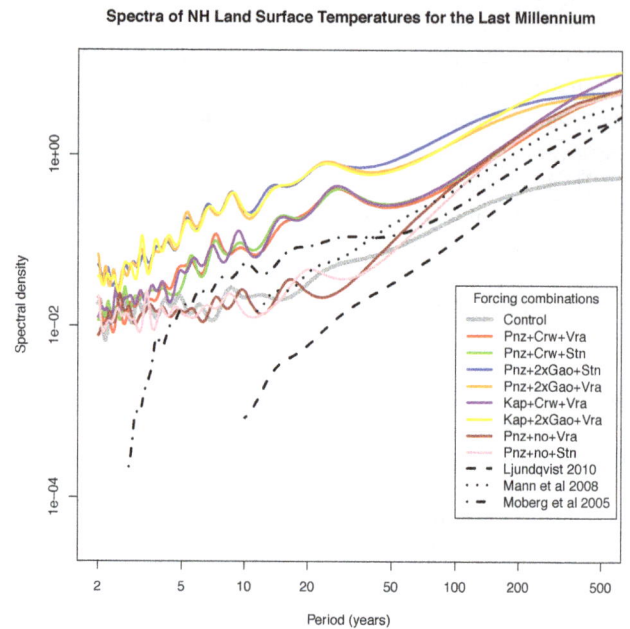

**Spectra of NH Land Surface Temperatures for the Last Millennium**

**Fig. 11.** Spectra from an ensemble of LM simulations using the same model but driven with different sets of forcings compared with Ljundqvist (2010), Mann et al. (2008) and Moberg et al. (2005) reconstructions. The clustering of simulations is driven entirely by changes in the volcanic forcing data set used, with the simulations with the most decadal and multi-decadal variability using the Gao et al. (2008) reconstruction. Only in the examples where no volcanic forcing is used at all is the impact of different solar forcing reconstructions detectable. Spectra derived using MEM with 30 poles, from 850 to 2005, after correction for control run drift using a loess low-frequency estimate derived from the control run. Key abbreviations: Land use: Pnz (Pongratz et al., 2008), Kap (Kaplan et al., 2011); Solar: Vra (Vieira et al., 2011), Stn (Steinhilber et al., 2009); Volcanic: 2xGao (twice the forcing from Gao et al., 2008), Crw (Crowley and Unterman, 2013).

constrain internal variability, and hence the unforced spread in projections over the next few decades.

In Fig. 11, we show the maximum-entropy method (MEM) spectra (using 30 poles) for the NH mean land surface temperature over 8 last millennium simulations (850–2005) with the GISS-E2-R model that were run with different combinations of plausible solar, volcanic and land use forcings (Schmidt et al., 2011, 2012). The spectra are similar for models that have the same volcanic forcing, and significantly different when the volcanic forcing is derived from a different data set or where no volcanic forcing was imposed at all. Specifically, interannual to multi-decadal variability is much larger when volcanoes are imposed, and the larger the volcanic forcing, the greater the variability, with the largest response in simulations using the Gao et al. (2008) reconstruction (Gao), compared to the Crowley and Unterman (2013) reconstruction (Crw). Note that the implementation of the Gao et al. volcanic forcing in these simulations was misspecified and gave roughly twice the expected radiative forcing. However, given the uncertainties in specifying volcanic forcing (for instance, associated with the effective radius of the particles), the exercise is nonetheless useful in highlighting the role of forcings in determining variance. In contrast, the difference between two different solar forcings (Vieira et al., 2011; Steinhilber et al., 2009) is not detectable in this metric.

The no-volcano simulations underestimate the decadal/multi-decadal variance seen in two of the three reconstructions, while the with-volcano simulations overestimate it. The lowest-frequency bands in the models (primarily driven by orbital forcing, and the 20th century anthropogenic trend) have slightly larger variance than in the reconstructions.

Another analysis of variability as a function of timescale is one focused on power law scaling (Lovejoy and Schertzer, 1986). Several scaling studies of GCMs demonstrate that they generally simulate the statistics (including spectral scaling exponents) reasonably well up to $\approx 10\,\mathrm{yr}$ scales (e.g. Fraedrich and Blender, 2003; Zhu et al., 2006; Rybski et al., 2008; Lovejoy and Schertzer, 2013; Vyushin et al., 2012). However, tests at lower frequencies are strongly affected by the solar and volcanic forcings as well as the possible impacts of slow processes such as deep ocean or land-ice dynamics which are perhaps poorly represented or missing.

Following Lovejoy and Schertzer (2012a), we calculate the root mean square (rms) fluctuation as a function of timescale, from months to centuries, for the NH land temperatures using the same eight runs of the GISS-E2-R model used above, for the period 1500–1900 CE (Fig. 12). Since simulations are strongly clustered according to changes in the volcanic forcing used (Fig. 11), for simplicity we averaged over the three Gao and three Crw volcanic and the two no-volcanic runs. For comparison, we show the mean of the same metric from three multiproxy reconstructions (Huang et al., 2000; Moberg et al., 2005; Ljundqvist, 2010). The multiproxy average is processed with and without the 20th century to indicate the importance of that period for the scaling behaviour – in all cases the variance in the multi-decadal to century scale is greatly enhanced by the recent anthropogenic trend. These curves show fluctuations stable with scale over the low frequency weather regime (years to decades) but increasing in the climate regime (decades to centuries) (Fig. 12).

The comparison with the GISS-E2-R simulations is illuminating. First, we note that the slopes for the simulations show decreasing variance from annual to centennial scale, in contrast to the reconstructions. Only the volcano-free runs (bottom) clear have increasing variance with scale in the centennial and longer periods, though with a magnitude of variance at all scales that is too low. Volcanic forcings add variance at all scales, but producing larger magnitudes that inferred from the reconstructions.

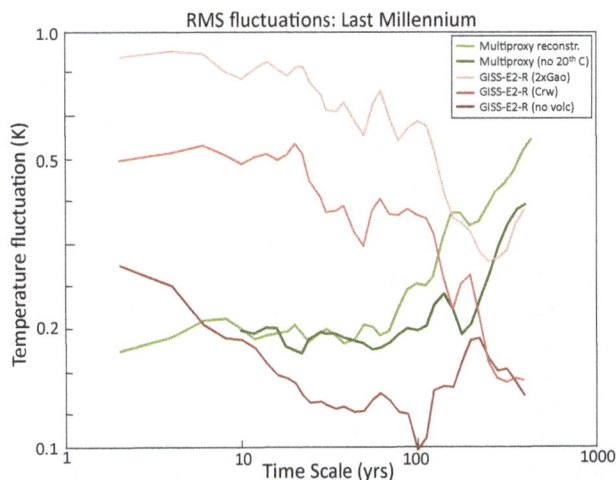

**Fig. 12.** rms fluctuations of instrumental and palaeo-climate reconstructions compared to drift-corrected simulations of the Northern Hemisphere land temperature for the period 1500–1900. 2xGao and Crw refer to GISS-E2-R simulations using the (2x) Gao et al. (2008) and Crowley and Unterman (2013) reconstructions of volcanic forcing. The multiproxy reconstruction used is an average of three NH estimates, and the rms fluctuations are separately shown for the periods 1000–1900 and 1000–1980.

Both these results demonstrate clear mismatches in behaviour between the models' simulated variance at different scales and the inferred variability from multi-proxy reconstructions. However, there are strong sensitivities to the (uncertain) external forcing functions, precluding a straightforward attribution of the mismatch to potentially misspecified forcings, missing mechanisms, insufficient "slow" variability or problems in the reconstructions. Specifically, reconstructions may have frequency-dependent biases that vary depending on the methodology and source data. For instance, boreholes used in Huang et al. (2000) do not have high frequency variability, while low frequency variability in tree-ring-based reconstructions is hard to capture. In models, the importance of decadal and multi-decadal variance in the Pacific or Atlantic sectors vary widely and are poorly constrained from observations, and there may be significant issues with the forcing functions themselves. Other analyses (i.e. Schurer et al., 2013) have examined the coherence of last millennium simulations and the proxy reconstructions and found that while signatures of multiple forcings can be determined, there is a mismatch in the magnitude of the response to volcanoes coherent with the conclusions drawn above. It therefore remains unclear what implications these tests have for future projections, but improvements in the forcing data sets or a focus on more specific comparisons may prove fruitful in future analyses.

## 5.4 Non-stationarity in hydroclimate diagnostics

Hydroclimate variability can be quantified using a range of variables, including precipitation, soil moisture, lake levels, or other synthetic indices (e.g. Nigam and Ruiz-Barradas, 2006). Most models provide output for these diagnostics, but often these variables are not directly derivable from palaeo-climate archives, creating a challenge when conducting model-data comparisons. However, calibrations of networks of precipitation-sensitive tree ring widths have been used to reconstruct the Palmer Drought Severity Index (PDSI) in North America and Asia over the Common Era (Cook et al., 2004, 2010). PDSI is calculated using temperature-derived estimates of the evapo-transpiration and precipitation, and nominally represents a normalised index of soil moisture, with negative values indicating drought and positive values indicate wetter than normal conditions. There are many outstanding issues with using variations of the index globally to assess drought, in definition and availability and quality of inputs and sensitivity (e.g. contrast Sheffield et al., 2012 and Dai, 2013). However, we focus here on the question of how well this index, if derived from GCM output, reflects simulated soil moisture and whether this relationship changes over time.

From two GCMs (GISS-E2-R and MIROC-ESM), we calculated PDSI using simulated temperature and precipitation from the GISS-E2-R and MIROC-ESM models using the Thornthwaite method. We compared this index against the standardised (zero mean, unit standard deviation, over the 1850–1950 period, 10 yr smoothing) total column soil moisture model output for the Central Plains of North America (105–90° W; 32–48° W) (Fig. 13). Prior to the start of the industrial period in 1850, PDSI and soil moisture track each other closely in both models (GISS-E2-R: $r = 0.82$; MIROC-ESM: $r = 0.50$). Beginning near the middle of the twentieth century, however, the two indices diverge dramatically. In one model (GISS-E2-R) the correlation weakens considerably ($r = 0.33$), while in the other model (MIROC-ESM) the sign of the correlation reverses ($r = -0.29$). The PDSI changes over the 21st century would suggest severe and unprecedented drought. In contrast, the simulated soil moisture trends indicate a more modest shift towards drying (GISS-E2-R) or even wetter conditions over the coming decades (MIROC-ESM). The divergence in projections is related to the treatment of evapo-transpiration (ET) in the model versus in the PDSI (Thornthwaite) calculation. In this PDSI calculation, temperature is used as a proxy for the energy available while in the GCMs the soil energy and moisture budgets are calculated directly using explicit physical models. In reality, ET becomes increasingly decoupled from temperature as the temperature increases, a factor reflected in the model soil moisture but not in the PDSI index. For time periods with strong transient changes in temperature (e.g. the late 20th century and into the future), our analysis suggests that the

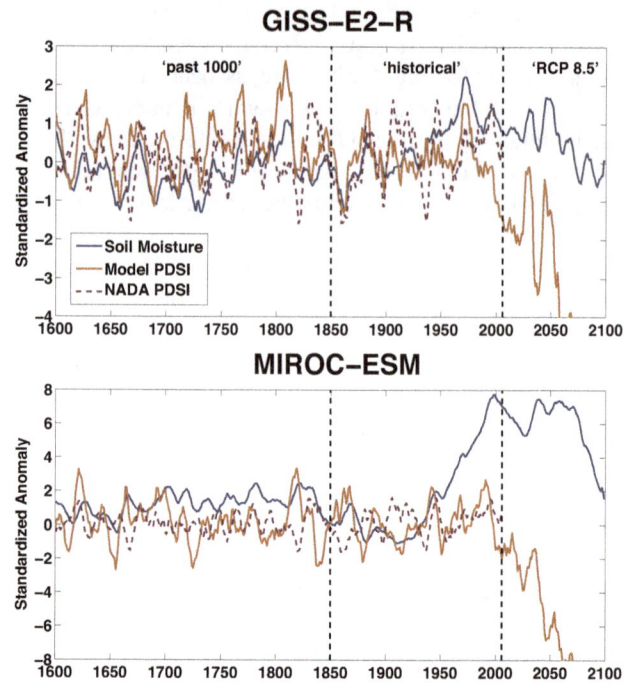

**Fig. 13.** Standardised anomalies for PDSI and soil moisture in two models (GISS-E2-R and MIROC-ESM) using a past1000 simulation, and a historical+rcp85 continuation. For reference, the tree-ring-based reconstruction is plotted (dashed line) (Cook et al., 2010), though this would not be expected to line up exactly with the model simulations. All data smoothed with a 10 yr running mean.

usefulness of PDSI for projecting drought and hydroclimate trends is limited.

While this example of a diagnostic divergence is specific to the PDSI and soil moisture, there are wider implications that might need to be explored in other metrics. It is not unusual for a proxy to have a non-stationary response to a climate variable of interest (see Sect. 5.2 for another example), and it is incumbent on the investigator to ensure that any consequences of this are fully explored.

## 6 Conclusions and recommendations

In this paper, we have focused on the opportunities provided by "out-of-sample" palaeo-climate experiments within the CMIP5 framework, and specifically how measures of skill in modelling palaeo-climate change might inform future projections of climate change.

We have given examples that show that some relationships are robust across the multi-model ensemble, over multiple simulations and in the palaeo-data (Sect. 3) and examples of skill measures that are well correlated with the simulated magnitude of future change, thus allowing the likely magnitude of future changes to be constrained (Sect. 4). We also give examples of cases (Sect. 5) where there is a need for caution because of the limitations with models, the experimental setup used in CMIP5, or with the palaeo-climate data itself.

Our examples illustrate the general requirements for attempts to use the palaeo-climate simulations to quantitatively constrain future projections. Each example makes use of a specific target (or targets) from a palaeo-climate reconstruction of change that is within the scope of the modelled system, defines a metric of skill that quantifies the accuracy of the modelled changes and assesses the connection to a future prediction. The successes and problems discussed above lead naturally to a set of guidelines that could profitably guide future research for both modellers and the palaeo-data community:

1. Palaeo-simulations need to be performed with models that are also being used for future projections and produce model diagnostics that are commensurate in all experiments (as in CMIP5).

2. The more extensive the structural uncertainty examined (across models, boundary conditions etc.) the more robust any resulting constraints will be. Some of our analyses (i.e. Sects. 4.1 and 4.2) are limited by the small number of palaeo-simulations currently available in the CMIP5 database, and we hope that the demonstration of their potential to address questions relevant to the future will encourage other modelling groups to complete and archive these simulations.

3. Palaeo-data targets should be spatially representative synthesis products with well-characterised uncertainties. Our analyses rely heavily on the use of synthesis data products, for instance the MARGO data set for the LGM (MARGO, 2009), pollen-based reconstructions for the mid-Holocene (Bartlein et al., 2011), multi-proxy reconstructions of hemispheric temperature (e.g. Moberg et al., 2005), or gridded tree-ring-based reconstructions of PDSI for the last millennium (Cook et al., 2010). Such products are invaluable, but there is a need for increased transparency of included uncertainties and continued expansion of targets (e.g. see Müller et al., 2011 for sea ice extent). Increasing model complexity and scope, for instance by including a carbon cycle, fire models or online tracers such as water isotopes, necessitates the creation of new synthesis products (e.g. charcoal records: Daniau et al., 2012; or sea surface carbonate isotopes: Oppo et al., 2007) if useful comparisons are to be made. Examples in Sects. 4.3 and 5.1 illustrate the need for more efforts in this direction.

4. Skill metrics may be impacted by uncertainties in external forcing (and thus not solely characterise the realism of modelled processes; as in the spectra generated in Sect. 5.3), or may have non-stationary relationships with impacts of interest (Sect. 5.4). Improved forward modelling of palaeo-data (as in Sect. 5.2) will be increasingly important.

5. Relationships between targets in the past and the future predictions should be examined and not assumed. Not all mismatches of palaeo-models and reconstructions are related to factors important for future sensitivities, and not all divergences in future projections are correlated to differences in palaeo-climate skill.

The periods and hypotheses tested using palaeo-climate simulations are far more limited than the number of interesting features in the palaeo-climate record. The three periods selected for CMIP5 were chosen on the basis of their relative maturity (the existence of prior sets of experiments, already tested issues, existing data syntheses), but additional periods are also potentially useful – the mid-Pliocene (2.5 million years ago), the transient 8.2 ka event, the last interglacial, the peak Eocene, etc. (see Schmidt, 2010 for justifications). Some of these periods are already being examined in a coordinated fashion (e.g. Haywood et al., 2013 and Dolan et al., 2012 for the Pliocene), and we hope that more coordinated experiments will be started. Further expansions of the model experiments will lead to increases in the production of higher frequency diagnostics (daily and sub-daily variations), and include perturbed physics ensembles to better characterise the model structural uncertainty. On the data side, much greater efforts to create palaeo-data synthesis products with robust uncertainty estimates are possible. All of these expansions will create possibilities for more and better tests of model performance and hence potentially lead to better constraints on future projections. In the meantime, there is still a huge untapped scope for more informative palaeo-model comparisons that can be made using the existing databases.

*Acknowledgements.* This paper arose from a workshop at the Bishop Museum, Honolulu in March 2012 organised by the PAGES-CLIVAR Intersection Panel. Funding from the NASA Modeling and Analysis Program, NOAA, NSF, AORI, PAGES and CLIVAR is gratefully acknowledged. We acknowledge the WCRP WGCM, which is responsible for CMIP, and we thank the climate modelling groups (listed in Table 1 of this paper) for producing and making available their model output. For CMIP, the US Department of Energy's Program for Climate Model Diagnosis and Intercomparison provides coordinating support and led development of software infrastructure in partnership with the Global Organization for Earth System Science Portals. Individual researchers were funded from the French ANR ISOTROPIC and CHEDAR projects, NSF grant ATM-0902133, the Japanese Ministry of Environment S-5 fund, NOAA grant NA10OAR4310115, and the Canadian Sea Ice and Snow Evolution (CanSISE) Network. We thank Rumi Ohgaito and Tetsuo Sueyoshi (JAMSTEC) for help in some of the analysis, Richard Healy for model support, and Françoise Vimeux and Sandrine Bony for discussions. Comments from Tamsin Edwards and an anonymous reviewer greatly improved the clarity of the paper.

All authors contributed to the conception and scope of the paper. Sections 1, 2 and 6 were jointly written by all authors. Principle contributions for each technical section are as follows: Sect. 3.1: M. Kageyama; Sect. 3.2: M. Kageyama, S. P. Harrison; Sect. 4.1: C. Risi; Sect. 4.2: J. D. Annan, J. C. Hargreaves; Sect. 4.3: L.-B. Tremblay; Sect. 5.1: P. Yiou; Sect. 5.2: D. Thompson; Sect. 5.3: A. N. LeGrande, G. A. Schmidt, S. Lovejoy; Sect. 5.4: B. I. Cook.

Edited by: M. Crucifix

# References

Abe, M., Shiogama, H., Nozawa, T., and Emori, S.: Estimation of future surface temperature changes constrained using the future-present correlated modes in inter-model variability of CMIP3 multimodel simulations, J. Geophys. Res., 116, D18104, doi:10.1029/2010JD015111, 2011.

Andrews, T., Gregory, J. M., Webb, M. J., and Taylor, K. E.: Forcing, feedbacks and climate sensitivity in CMIP5 coupled atmosphere–ocean climate models, Geophys. Res. Lett., 39, L09712, doi:10.1029/2012GL051607, 2012.

Annan, J. D. and Hargreaves, J. C., Understanding the CMIP3 multimodel ensemble, J. Climate, 24, 4529–4538, doi:10.1175/2011JCLI3873.1, 2011.

Annan, J. D. and Hargreaves, J. C.: A new global reconstruction of temperature changes at the Last Glacial Maximum, Clim. Past, 9, 367–376, doi:10.5194/cp-9-367-2013, 2013.

Bala, G., Caldeira, K., and Nemani, R.: Fast versus slow response in climate change: implications for the global hydrological cycle, Clim. Dynam., 35, 423–434, doi:10.1007/s00382-009-0583-y, 2010.

Barriendos, M. and Rodrigo, F.: Study of historical flood events on Spanish rivers using documentary data, Hydrolog. Sci. J., 51, 765–783, 2006.

Bartlein, P. J., Harrison, S. P., Brewer, S., Connor, S., Davis, B. A. S., Gajewski, K., Guiot, J., Harrison-Prentice, T. I., Henderson, A., Peyron, O., Prentice, I. C., Scholze, M., Seppä, H., Shuman, B., Sugita, S., Thompson, R. S., Viau, A. E., Williams, J., and Wu, H. Pollen-based continental climate reconstructions at 6 and 21 ka: a global synthesis, Clim. Dynam., 37, 775–802, 2011.

Boé, J. L., Hall, A., and Qu, X.: September sea-ice cover in the Arctic Ocean projected to vanish by 2100, Nat. Geosci., 2, 1–3, 2009.

Bonfils, C., de Noblet-Ducoudré, N., Guiot, J., and Bartlein, P.: Some mechanisms of mid-Holocene climate change in Europe, inferred from comparing PMIP models to data, Clim. Dynam., 23, 79–98, 2004.

Bony, S., Bellon, G., Klocke, D., Sherwood, S., Fermepin, S., and Denvil, S.: Robust direct effect of carbon dioxide on tropical circulation and regional precipitation, Nat. Geosci., 6, 447–451, 2013.

Braconnot, P., Otto-Bliesner, B., Harrison, S., Joussaume, S., Peterchmitt, J.-Y., Abe-Ouchi, A., Crucifix, M., Driesschaert, E., Fichefet, Th., Hewitt, C. D., Kageyama, M., Kitoh, A., Laîné, A., Loutre, M.-F., Marti, O., Merkel, U., Ramstein, G., Valdes, P., Weber, S. L., Yu, Y., and Zhao, Y.: Results of PMIP2 coupled simulations of the Mid-Holocene and Last Glacial Maximum – Part 1: experiments and large-scale features, Clim. Past, 3, 261–277, doi:10.5194/cp-3-261-2007, 2007.

Braconnot, P., Harrison, S. P., Kageyama, M., Bartlein, P. J., Masson-Delmotte, V., Abe-Ouchi, A., Otto-Bliesner, B., and Zhao, Y.: Evaluation of climate models using palaeoclimatic data, Nat. Clim. Change, 2, 417–424, doi:10.1038/nclimate1456, 2012.

Brewer, S., Guiot, J., and Torre, F.: Mid-Holocene climate change in Europe: a data-model comparison, Clim. Past, 3, 499–512, doi:10.5194/cp-3-499-2007, 2007.

Brient, F. and Bony, S.: How may low-cloud radiative properties simulated in the current climate influence low-cloud feedbacks under global warming?, Geophys. Res. Lett., 39, L20807, doi:10.1029/2012GL053265, 2012.

Camuffo, D., Bertolin, C., Diodato, N., Barriendos, M., Dominguez-Castro, F., Cocheo, C., della Valle, A., Garnier, E., and Alcoforado, M. J.: The western Mediterranean climate: how will it respond to global warming?, Climatic Change, 100, 137–142, 2010.

Carton, J. A. and Giese, B. S.: A reanalysis of ocean climate using Simple Ocean Data Assimilation (SODA), Mon. Weather Rev., 136, 2999–3017, 2008.

COHMAP Members: Climatic changes of the last 18,000 years: observations and model simulations, Science, 241, 1043–1052, 1988.

Coles, S.: An introduction to statistical modeling of extreme values, Springer, London, New York, 2001.

Collins, M., Chandler, R. E., Cox, P. M., Huthnance, J. M., Rougier, J., and Stephenson, D. B.: Quantifying future climate change, Nat. Clim. Change, 2, 403–409, 2012.

Compo, G. P., Whitaker, J. S., Sardeshmukh, P. D., Matsui, N., Allan, R. J., Yin, X., Gleason, B. E., Vose, R. S., Rutledge, G., Bessemoulin, P., Brönnimann, S., Brunet, M., Crouthamel, R. I., Grant, A. N., Groisman, P. Y., Jones, P. D., Kruk, M., Kruger, A. C., Marshall, G. J., Maugeri, M., Mok, H. Y., Nordli, Ø., Ross, T. F., Trigo, R. M., Wang, X. L., Woodruff, S. D., and Worley, S. J.: The Twentieth Century Reanalysis Project, Q. J. Roy. Meteorol. Soc., 137, 1–28, 2011.

Cook, E. R., Woodhouse, C. A., Eakin, C. M., Meko, D. M., and Stahle, D. W.: Long-Term Aridity Changes in the Western United States, Science, 306, 1015–1018, 2004.

Cook, E. R., Seager, R., Heim Jr., R. R., Vose, R. S., Herweijer, C., and Woodhouse, C.: Megadroughts in North America: placing IPCC projections of hydroclimatic change in a long-term palaeoclimate context, J. Quaternary Sci., 25, 48–61, 2010.

Crowley, T. J. and Unterman, M. B.: Technical details concerning development of a 1200 yr proxy index for global volcanism, Earth Syst. Sci. Data, 5, 187–197, doi:10.5194/essd-5-187-2013, 2013.

Crucifix, M.: Does the Last Glacial Maximum constrain climate sensitivity?, Geophys. Res. Lett., 33, L18701, doi:10.1029/2006GL027137, 2006.

Dai, A.: Increasing drought under global warming in observations and models, Nat. Clim. Change, 3, 52–58, doi:10.1038/nclimate1633, 2013.

Daniau, A.-L., Bartlein, P. J., Harrison, S. P., Prentice, I. C., Brewer, S., Friedlingstein, P., Harrison-Prentice, T. I., Inoue, J., Izumi, K., Marlon, J. R., Mooney, S., Power, M. J., Stevenson, J., Tinner, W., Andrič, M., Atanassova, J., Behling, H., Black, M., Blarquez, O., Brown, K. J., Carcaillet, C., Colhoun, E. A., Colombaroli, D., Davis, B. A. S., D'Costa, D., Dodson, J., Dupont, L.,

Eshetu, Z., Gavin, D. G., Genries, A., Haberle, S., Hallett, D. J., Hope, G., Horn, S. P., Kassa, T. G., Katamura, F., Kennedy, L. M., Kershaw, P., Krivonogov, S., Long, C., Magri, D., Marinova, E., McKenzie, G. M., Moreno, P. I., Moss, P., Neumann, F. H., Norström, E., Paitre, C., Rius, D., Roberts, N., Robinson, G. S., Sasaki, N., Scott, L., Takahara, H., Terwilliger, V., Thevenon, F., Turner, R., Valsecchi, V. G., Vannière, B., Walsh, M., Williams, N., and Zhang, Y.: Predictability of biomass burning in response to climate changes, Global Biogeochem. Cy., 26, GB4007, doi:10.1029/2011GB004249, 2012.

Denman, K. L., Brasseur, G., Chidthaisong, A., Ciais, P., Cox, P. M., Dickinson, R. E., Hauglustaine, D., Heinze, C., Holland, E., Jacob, D., Lohmann, U., Ramachandran, S., da Silva Dias, P. L., Wofsy, S. C., and Zhang, X.: Couplings between changes in the climate system and biogeochemistry, in: IPCC, Climate Change 2007: The Physical Science Basis, Contribution of Working Group I to the Fourth Assessment Report of the Intergovernmental Panel on Climate Change, edited by: Solomon, S., Qin, D., Manning, M., Chen, Z., Marquis, M., Averyt, K., Tignor, M., and Miller, H., Cambridge University Press, Cambridge, UK and USA, 499–587, 2007.

Deser, C., Phillips, A. S., and Alexander, M. A.: Twentieth century tropical sea surface temperature trends revisited, Geophys. Res. Lett., 37, L10701, doi:10.1029/2010GL043321, 2010.

de Vernal, A., Hillaire-Marcel, C., and Darby, D. A.: Variability of sea ice cover in the Chukchi Sea (western Arctic Ocean) during the Holocene, Paleoceanography, 20, PA4018, doi:10.1029/2005PA001157, 2005.

de Vernal, A., Hillaire-Marcel, C., Solignac, S., Radi, T., and Rochon, A.: Reconstructing seaice conditions in the Arctic and Subarctic prior to human observations, in: Arctic Sea Ice Decline: Observations, Projections, Mechanisms and Implications, edited by: de Weaver, E., AGU Monograph Series, 180, 27–45, 2008.

Dolan, A. M., Koenig, S. J., Hill, D. J., Haywood, A. M., and DeConto, R. M.: Pliocene Ice Sheet Modelling Intercomparison Project (PLISMIP) – experimental design, Geosci. Model Dev., 5, 963–974, doi:10.5194/gmd-5-963-2012, 2012.

Donnelly, J. P. and Woodruff, J. D.: Intense hurricane activity over the past 5,000 years controlled by El Niño and the West African monsoon, Nature, 447, 465–468, 2007.

Drost, F., Karoly, D., and Braganza, K.: Communicating global climate change using simple indices: an update, Clim. Dynam., 39, 989–999, doi::10.1007/s00382-011-1227-6, 2012.

Dufresne, J.-L., Foujols, M.-A., Denvil, S., Caubel, A., Marti, O., Aumont, O., Balkanski, Y., Bekki, S., Bellenger, H., Benshila, R., Bony, S., Bopp, L., Braconnot, P., Brockmann, P., Cadule, P., Cheruy, F., Codron, F. F., Cozic, A., Cugnet, D., de Noblet, N., Duvel, J.-P., Ethé, C., Fairhead, L., Fichefet, T., Flavoni, S., Friedlingstein, P., Grandpeix, J.-Y., Guez, L., Guilyardi, E., Hauglustaine, D., Hourdin, F., Idelkadi, A., Ghattas, J., Joussaume, S., Kageyama, M., Krinner, G., Labetoulle, S., Lahellec, A., Lefèbvre, M.-P., Lefèvre, F., Lévy, C., Li, Z. X., Lloyd, J., Lott, F., Madec, G., Mancip, M., Marchand, M., Masson, S., Meurdesoif, Y., Mignot, J., Musat, I., Parouty, S., Polcher, J., Rio, C., Schulz, M., Swingedouw, D., Szopa, S., Talandier, C., Terray, P., and Viovy, N.: Climate change projections using the IPSL-CM5 Earth System Model: from CMIP3 to CMIP5, Clim. Dynam., 40, 2123–2165, doi:10.1007/s00382-012-1636-1, 2013.

Dyke, A. S. and Savelle, J. M.: Holocene history of the Bering Sea bowhead whale (Balaena mysticetus) in its Beaufort Sea summer grounds off southwestern Victoria Island, Western Canadian Arctic, Quatern. Res., 55, 371–379, 2001.

Edwards, T. L., Crucifix, M., and Harrison, S. P.: Using the past to constrain the future: how the palaeorecord can improve estimates of global warming, Prog. Phys. Geogr., 31, 481–500, doi:10.1177/0309133307083295, 2007.

Fasullo, J. T. and Trenberth, K. E.: A less cloudy future: The role of subtropical subsidence in climate sensitivity, Science, 338, 792–794, 2012.

Fernández-Donado, L., González-Rouco, J. F., Raible, C. C., Ammann, C. M., Barriopedro, D., García-Bustamante, E., Jungclaus, J. H., Lorenz, S. J., Luterbacher, J., Phipps, S. J., Servonnat, J., Swingedouw, D., Tett, S. F. B., Wagner, S., Yiou, P., and Zorita, E.: Large-scale temperature response to external forcing in simulations and reconstructions of the last millennium, Clim. Past, 9, 393–421, doi:10.5194/cp-9-393-2013, 2013.

Fischer, E. M., Seneviratne, S. I., Lüthi, D., and Schär, C.: Contribution of land-atmosphere coupling to recent European summer heat waves, Geophys. Res. Lett., 34, L06707, doi:10.1029/2006GL029068, 2007.

Fraedrich, K. and Blender, K.: Scaling of atmosphere and ocean temperature correlations in observations and climate models, Phys. Rev. Lett., 90, 108501, doi:10.1103/PhysRevLett.90.108501, 2003.

Funder, S., Goosse, H., Jepsen, H., Kaas, E., Kjær, K. H., Korsgaard, N. J., Larsen, N. K., Linderson, H., Lyså, A., Möller, P., Olsen, J., and Willerslev, E.: A 10,000-year record of Arctic Ocean sea-ice variability: View from the beach, Science, 333, 747–750, doi:10.1126/science.1202760, 2011.

Gao, C., Robock, A., and Ammann, C.: Volcanic forcing of climate over the last 1500 years: An improved ice-core based index for climate models, J. Geophys. Res., 113, D2311, doi:10.1029/2008JD010239, 2008.

Giese, B. S. and Ray, S.: El Niño variability in simple ocean data assimilation (SODA), 1871–2008, J. Geophys. Res., 116, C02024, doi:10.1029/2010JC006695, 2011.

Gleckler, P. J., Taylor, K. E., and Doutriaux, C.: Performance metrics for climate models, J. Geophys. Res., 113, D06104, doi:10.1029/2007JD008972, 2008.

Guiot, J., Boreux, J. J., Braconnot, P., and Torre, F.: Data-model comparison using fuzzy logic in paleoclimatology, Clim. Dynam., 15, 569–581, 1999.

Hall, A. and Qu, X.: Using the current seasonal cycle to constrain snow albedo feedback in future climate change, Geophys. Res. Lett., 33, L03502, doi:10.1029/2005GL025127, 2006.

Hansen, J., Sato, M., Ruedy, R., Kharecha, P., Lacis, A., Miller, R. L., Nazarenko, L., Lo, K., Schmidt, G. A., Russell, G., Aleinov, I., Bauer, S., Baum, E., Cairns, B., Canuto, V., Chandler, M., Cheng, Y., Cohen, A., Del Genio, A., Faluvegi, G., Fleming, E., Friend, A., Hall, T., Jackman, C., Jonas, J., Kelley, M., Kiang, N. Y., Koch, D., Labow, G., Lerner, J., Menon, S., Novakov, T., Oinas, V., Perlwitz, Ja., Perlwitz, Ju., Rind, D., Romanou, A., Schmunk, R., Shindell, D., Stone, P., Sun, S., Streets, D., Tausnev, N., Thresher, D., Unger, N., Yao, M., and Zhang, S.: Climate simulations for 1880–2003 with GISS modelE, Clim. Dynam., 29, 661–696, doi:10.1007/s00382-007-0255-8, 2007.

Hansen, J. E., Sato, M., and Ruedy, R.: Perception of climate change, P. Natl. Acad. Sci., 109, 14726–14727, doi:10.1073/pnas.1205276109, 2012.

Hargreaves, J. C., Abe-Ouchi, A., and Annan, J. D.: Linking glacial and future climates through an ensemble of GCM simulations, Clim. Past, 3, 77–87, doi:10.5194/cp-3-77-2007, 2007.

Hargreaves, J. C., Paul, A., Ohgaito, R., Abe-Ouchi, A., and Annan, J. D.: Are paleoclimate model ensembles consistent with the MARGO data synthesis?, Clim. Past, 7, 917–933, doi:10.5194/cp-7-917-2011, 2011.

Hargreaves, J. C., Annan, J. D., Yoshimori, M., and Abe-Ouchi, A.: Can the last glacial maximum constrain climate sensitivity?, Geophys. Res. Lett., 39, L24702, doi:10.1029/2012GL053872, 2012.

Hargreaves, J. C., Annan, J. D., Ohgaito, R., Paul, A., and Abe-Ouchi, A.: Skill and reliability of climate model ensembles at the Last Glacial Maximum and mid-Holocene, Clim. Past, 9, 811–823, doi:10.5194/cp-9-811-2013, 2013.

Harrison, S. P. and Bartlein, P. J.: Records from the past, lessons for the future: what the palaeo-record implies about mechanisms of global change, in: The Future of the World's Climates, edited by: Henderson-Sellers, A. and McGuffie, K., Elsevier, Amsterdam, the Netherlands, 403–436, 2012.

Harrison, S. P., Bartlein, P. J., Brewer, S., Prentice, I. C., Boyd, M., Hessler, I., Holmgren, K., Izumi, K., and Willis, K.: Model benchmarking with glacial and mid-Holocene climates, Clim. Dynam., doi:10.1007/s00382-013-1922-6, in press, 2013.

Hawkins, E. and Sutton, R.: The potential to narrow uncertainty in regional climate predictions, B. Am. Meteorol. Soc., 90, 1095–1107, doi:10.1175/2009BAMS2607.1, 2009.

Haylock, M. R., Hofstra, N., Tank, A. M. G. K., Klok, E. J., Jones, P. D., and New, M.: A European daily high-resolution gridded data set of surface temperature and precipitation for 1950–2006, J. Geophys. Res., 113, D20119, doi:10.1029/2008JD010201, 2008.

Haywood, A. M., Hill, D. J., Dolan, A. M., Otto-Bliesner, B. L., Bragg, F., Chan, W.-L., Chandler, M. A., Contoux, C., Dowsett, H. J., Jost, A., Kamae, Y., Lohmann, G., Lunt, D. J., Abe-Ouchi, A., Pickering, S. J., Ramstein, G., Rosenbloom, N. A., Salzmann, U., Sohl, L., Stepanek, C., Ueda, H., Yan, Q., and Zhang, Z.: Large-scale features of Pliocene climate: results from the Pliocene Model Intercomparison Project, Clim. Past, 9, 191–209, doi:10.5194/cp-9-191-2013, 2013.

Hewitt, C. D. and Mitchell, J. F. B.: Radiative forcing and response of a GCM to ice age boundary conditions: Cloud feedback and climate sensitivity, Clim. Dynam., 13, 821–834, 1997.

Hirschi, M., Seneviratne, S. I., Alexandrov, V., Boberg, F., Boroneant, C., Christensen, O. B., Formayer, H., Orlowsky, B., and Stepanek, P.: Observational evidence for soil-moisture impact on hot extremes in southeastern Europe, Nat. Geosci., 4, 17–21, 2011.

Huang, S. P., Pollack, H. N., and Shen, P.-Y.: Temperature trends over the past five centuries reconstructed from borehole temperatures, Nature, 403, 756–758, 2000.

IPCC SREX: Managing the Risks of Extreme Events and Disasters to Advance Climate Change Adaptation, in: A Special Report of Working Groups I and II of the Intergovernmental Panel on Climate Change, edited by: Field, C. B., Barros, V., Stocker, T. F., Qin, D., Dokken, D. J., Ebi, K. L., Mastrandrea, M. D., Mach, K. J., Plattner, G.-K., Allen, S. K., Tignor, M., and Midgley, P. M., Cambridge University Press, Cambridge, UK, and New York, NY, USA, 582 pp., 2012.

Izumi, K., Bartlein, P. J., and Harrison, S. P.: Consistent behaviour of the climate system in response to past and future forcing. Geophys. Res. Lett., 40, 1817–1823, doi:10.1002/grl.50350, 2013.

Jansen, E., Overpeck, J., Briffa, K., Duplessy, J.-C., Joos, F., Masson-Delmotte, V., Olago, D., Otto-Bliesner, B., Peltier, W., Rahmstorf, S., Ramesh, R., Raynaud, D., Rind, D., Solomina, O., Villalba, R., and Zhang, D.: Palaeoclimate, in: Climate Change 2007: The Physical Science Basis, Contribution of Working Group I to the Fourth Assessment Report of the Intergovernmental Panel on Climate Change, edited by: Solomon, S., Qin, D., Manning, M., Chen, Z., Marquis, M., Averyt, K., Tignor, M., and Miller, H., Cambridge University Press, Cambridge, 2007.

Jones, P. D., Briffa, K. R., Osborn, T. J., Lough, J. M., van Ommen, T. D., Vinther, B. M., Luterbacher, J., Wahl, E. R., Zwiers, F. W., Mann, M. E., Schmidt, G. A., Ammann, C. M., Buckley, B. M., Cobb, K. M., Esper, J., Goosse, H., Graham, N., Jansen, E., Kiefer, T., Kull, C., Küttel, M., Mosley-Thompson, E., Overpeck, J. T., Riedwyl, N., Schulz, M., Tudhope, A. W., Villalba, R., Wanner, H., Wolff, E., and Xoplaki, E.: High-resolution palaeoclimatology of the last millennium: A review of current status and future prospects, Holocene, 19, 3–49, doi:10.1177/0959683608098952, 2009.

Kageyama, M., Peyron, O., Pinot, S., Tarasov, P., Guiot, J., Joussaume, S., and Ramstein, G.: The Last Glacial Maximum climate over Europe and Western Siberia: a PMIP comparison between models and data, Clim. Dynam., 17, 23–43, 2001.

Kageyama, M., Braconnot, P., Bopp, L., Caubel, A., Foujols, M.-A., Guilyardi, E., Khodri, M., Lloyd, J., Lombard, F., Mariotti, V., Marti, O., Roy, T., and Woillez, M.-N.: Mid-Holocene and Last Glacial Maximum climate simulations with the IPSL model, Part I: comparing IPSL_CM5A to IPSL_CM4, Clim. Dynam., 40, 2447–2468, doi:10.1007/s00382-012-1488-8, 2013.

Kang, S. M., Held, I. M., Frierson, D. M. W., and Zhao, M.: The response of the ITCZ to extratropical thermal forcing: Idealized slab-ocean experiments with a GCM, J. Climate, 21, 3521–3532, doi:10.1175/2007JCLI2146.1, 2008.

Kaplan, J. O., Krumhardt, K. M., Ellis, E. C., Ruddiman, W. F., Lemmen, C., and Klein Goldewijk, K.: Holocene carbon emissions as a result of anthropogenic land cover change, Holocene, 21, 775–791, doi:10.1177/0959683610386983, 2011.

Kharin, V. V., Zwiers, F. W., and Zhang, X. B.: Intercomparison of near-surface temperature and precipitation extremes in AMIP-2 simulations, reanalyses, and observations, J. Climate, 18, 5201–5223, 2005.

Kharin, V. V., Zwiers, F. W., Zhang, X. B., and Hegerl, G. C.: Changes in temperature and precipitation extremes in the IPCC ensemble of global coupled model simulations, J. Climate, 20, 1419–1444, 2007.

Kohfeld, K. E. and Harrison, S. P.: How well can we simulate past climates? Evaluating earth system models using global palaeoenvironmental datasets, Quaternary Sci. Rev., 19, 321–346, 2000.

Köhler, P., Bintanja, R., Fischer, H., Joos, F., Knutti, R., Lohmann, G., and Masson-Delmotte, V.: What caused Earth's temperature variations during the last 800,000 years? Data-based evidence on radiative forcing and constraints on climate sensitivity, Quaternary Sci. Rev., 29, 129–145, doi:10.1016/j.quascirev.2009.09.026, 2010.

Knutti, R., Abramowitz, G., Collins, M., Eyring, V., Gleckler, P. J., Hewitson, B., and Mearns, L.: Good Practice Guidance Paper on Assessing and Combining Multi Model Climate Projections, in: Meeting Report of the Intergovernmental Panel on Climate Change Expert Meeting on Assessing and Combining Multi Model Climate Projections, edited by: Stocker, T. F., Qin, D., Plattner, G.-K., Tignor, M., and Midgley, P. M., IPCC Working Group I Technical Support Unit, University of Bern, Bern, Switzerland, 2010a.

Knutti, R., Furrer, R., Tebaldi, C., Cermak, J., and Meehl, G. A.: Challenges in combining projections from multiple climate models, J. Climate, 23, 2739–2758, 2010b.

Knutti, R. and Sedláček, J.: Robustness and uncertainties in the new CMIP5 climate model projections, Nat. Clim. Change, 3, 369–373, 2012.

Koenker, R.: Quantile regression, Cambridge University Press, Cambridge, 2005.

Kutzbach, J. E.: Monsoon climate of the early Holocene: Climate experiment with the Earth's orbital parameters for 9000 years ago, Science, 214, 59–61, 1981.

Laîné, A., Kageyama, M., Braconnot, P., and Alkama, R.: Impact of greenhouse gas concentration changes on the surface energetics in the IPSL-CM4 model: Regional warming patterns, land/sea warming ratio, glacial/interglacial differences, J. Climate, 22, 4621–4635, 2009.

Lambert, F. H., Webb, M. J., and Joshi, M. M.: The relationship between land–ocean surface temperature contrast and radiative forcing, J. Climate, 24, 3239–3256, doi:10.1175/2011JCLI3893.1, 2011.

LeGrande, A. N. and Schmidt, G. A.: Global gridded data set of the oxygen isotopic composition in seawater, Geophys. Res. Lett., 33, L12604, doi:10.1029/2006GL026011, 2006.

Le Roy Ladurie, E.: Histoire humaine et comparée du climat, Canicules et glacier, XIIIè–XVIIIè siècle, Fayard, Paris, 2004.

Le Roy Ladurie, E.: Histoire humaine et comparée du climat, Disettes et révolutions 1740–1860, Fayard, Paris, 2006.

Ljundqvist, F. C.: A new reconstruction of temperature variability in the extra-tropical Northern Hemisphere during the last two millennia, Geograf. Ann. A, 92, 339–351, 2010.

Lorius, C., Jouzel, J., Raynaud, D., Hansen, J., and Treut, H. L.: The ice-core record: Climate sensitivity and future greenhouse warming, Nature, 347, 139–145, 1990.

Lovejoy, S. and Schertzer, D.: Scale invariance in climatological temperatures and the spectral plateau, Ann. Geophys., 4B, 401–410, 1986.

Lovejoy, S. and Schertzer, D.: Haar wavelets, fluctuations and structure functions: convenient choices for geophysics, Nonlin. Processes Geophys., 19, 513–527, doi:10.5194/npg-19-513-2012, 2012a.

Lovejoy, S. and Schertzer, D.: Stochastic and scaling climate sensitivities: solar, volcanic and orbital forcings, Geophys. Res. Lett., 39, L11702, doi:10.1029/2012GL051871, 2012b.

Lovejoy, S. and Schertzer, D.: The Weather and Climate: Emergent Laws and Multifractal Cascades, Cambridge University Press, Cambridge, 496 pp., 2013.

Mahlstein, I. and Knutti, R.: September Arctic sea ice predicted to disappear near $2\,°C$ global warming above present, J. Geophys. Res., 117, D06104, doi:10.1029/2011JD016709, 2012.

Mann, M. E., Zhang, Z., Hughes, M. K., Bradley, R. S., Miller, S. K., Rutherford, S., and Ni, F.: Proxy-based reconstructions of hemispheric and global surface temperature variations over the past two millennia; P. Natl. Acad. Sci., 105, 13252–13257, doi:10.1073/pnas.0805721105, 2008.

MARGO Project Members: Constraints on the magnitude and patterns of ocean cooling at the Last Glacial Maximum, Nat. Geosci., 2, 127–132, doi:10.1038/NGEO411, 2009.

Masson, D. and Knutti, R.: Climate model genealogy, Geophys. Res. Letts., 38, L08703, doi:10.1029/2011GL046864, 2011.

Masson-Delmotte, V., Kageyama, M., Braconnot, P., Charbit, S., Krinner, G., Ritz, C., Guilyardi, E., Jouzel, J., Abe-Ouchi, A., Crucifix, M., Gladstone, R., Hewitt, C., Kitoh, A., LeGrande, A., Marti, O., Merkel, U., Motoi, T., Ohgaito, R., Otto-Bliesner, B., Peltier, W., Ross, I., Valdes, P., Vettoretti, G., Weber, S., Wolk, F., and Yu, Y.: Past and future polar amplification of climate change: climate model intercomparisons and ice-core constraints, Clim. Dynam., 26, 513–529, doi:10.1007/s00382-005-0081-9, 2006a.

Masson-Delmotte, V., Kageyama, M., Braconnot, P., Charbit, S., Krinner, G., Ritz, C., Guilyardi, E., Jouzel, J., Abe-Ouchi, A., Crucifix, M., Gladstone, R., Hewitt, C., Kitoh, A., LeGrande, A., Marti, O., Merkel, U., Motoi, T., Ohgaito, R., Otto-Bliesner, B., Peltier, W., Ross, I., Valdes, P., Vettoretti, G., Weber, S., Wolk, F., and Yu, Y.: Past and future polar amplification of climate change: climate model intercomparisons and ice-core constraints, Clim. Dynam., 27, 437–440, doi:10.1007/s00382-006-0149-1, 2006b.

Masson-Delmotte, V., Braconnot, P., Hoffmann, G., Jouzel, J., Kageyama, M., Landais, A., Lejeune, Q., Risi, C., Sime, L., Sjolte, J., Swingedouw, D., and Vinther, B.: Sensitivity of interglacial Greenland temperature and $\delta^{18}$O: ice core data, orbital and increased $CO_2$ climate simulations, Clim. Past, 7, 1041–1059, doi:10.5194/cp-7-1041-2011, 2011.

Massonnet, F., Fichefet, T., Goosse, H., Bitz, C. M., Philippon-Berthier, G., Holland, M. M., and Barriat, P.-Y.: Constraining projections of summer Arctic sea ice, The Cryosphere, 6, 1383–1394, doi:10.5194/tc-6-1383-2012, 2012.

McKay, J. L., de Vernal, A., Hillaire-Marcel, C., Not, C., Polyak, L., and Darby, D.: Holocene fluctuations in Arctic sea-ice cover: dinocyst-based reconstructions for the Eastern Chukchi Sea, Can. J. Earth Sci., 45, 1377–1397, 2008.

Meehl, G. A., Stocker, T. F., Collins, W. D., Friedlingstein, P., Gaye, A. T., Gregory, J. M., Kitoh, A., Knutti, R., Murphy, J. M., Noda, A., Raper, S. C., Watterson, I. G., Weaver, A. J., and Zhao, Z.-C.: Global climate projections, in: Climate Change 2007: The Physical Science Basis, Contribution of Working Group I to the Fourth Assessment Report of the Intergovernmental Panel on Climate Change, edited by: Solomon, S., Qin, D., Manning, M., Chen, Z., Marquis, M., Averyt, K. B., Tignor, M., and Miller, H. L., Cambridge University Press, Cambridge, UK and New York, NY, USA, 2007.

Moberg, A., Sonnechkin, D. M., Holmgren, K., and Datsenko, N. M.: Highly variable Northern Hemisphere temperatures reconstructed from low- and high-resolution proxy data, Nature, 433, 613–617, 2005.

Moros, M., Andrews, J. T., Eberl, D. D., and Jansen, E.: The Holocene history of drift ice in the northern North Atlantic – evidence for different spatial and temporal modes, Palaeoceanography, 21, PA2017, doi:10.1029/2005PA001214, 2006.

Mueller, B. and Seneviratne, S. I.: Hot days induced by precipitation deficits at the global scale, P. Natl. Acad. Sci., 109, 12398–12403, doi:10.1073/pnas.1204330109, 2012.

Müller, J., Wagner, A., Fahl, K., Stein, R., Prange, M., and Lohmann, G.: Towards quantitative sea ice reconstructions in the northern North Atlantic: A combined biomarker and numerical modelling approach, Earth Planet. Sc. Lett., 306, 137–148, 2011.

Nigam, S. and Ruiz-Barradas, A.: Seasonal hydroclimate variability over North America in global and regional reanalyses and AMIP Simulations: Varied representation, J. Climate, 19, 815–837, doi:10.1175/JCLI3635.1, 2006.

Nogaj, M., Yiou, P., Parey, S., Malek, F., and Naveau, P.: Amplitude and frequency of temperature extremes over the North Atlantic region, Geophys. Res. Lett., 33, L10801, doi:10.1029/2005GL024251, 2006.

O'ishi, R. and Abe-Ouchi, A.: Polar amplification in the mid-Holocene derived from dynamical vegetation change with a GCM, Geophys. Res. Lett., 38, L14702, doi:10.1029/2011GL048001, 2011.

Oppo, D. W., Schmidt, G. A., and LeGrande, A. N.: Seawater isotope constraints on tropical hydrology during the Holocene, Geophys. Res. Lett., 34, L13701, doi:10.1029/2007GL030017. 2007.

Osborn, T., Raper, S., and Briffa, K.: Simulated climate change during the last 1,000 years: comparing the ECHO-G general circulation model with the MAGICC simple climate model, Clim. Dynam., 27, 185–197, 2006.

PALAEOSENS Project Members: Making sense of palaeoclimate sensitivity, Nature, 491, 683–691, doi:10.1038/nature11574, 2012.

Parey, S., Dacunha-Castelle, D., and Hoang, T. T. H.: Mean and variance evolutions of the hot and cold temperatures in Europe, Clim. Dynam., 34, 345–359, 2010a.

Parey, S., Hoang, T. T. H., and Dacunha-Castelle, D.: Different ways to compute temperature return levels in the climate change context, Environmetrics, 21, 698–718, 2010b.

Pausata, F. S. R., Li, C., Wettstein, J. J., Kageyama, M., and Nisancioglu, K. H.: The key role of topography in altering North Atlantic atmospheric circulation during the last glacial period, Clim. Past, 7, 1089–1101, doi:10.5194/cp-7-1089-2011, 2011.

Polyak, L., Alley, R. B., Andrews, J. T., Brigham-Grette, J., Cronin, T. M., Darby, D. A., Dyke, A. S., Fitzpatrick, J. J., Funder, S., Holland, M., Jennings, A. E., Miller, G. H., O'Regan, M., Savelle, J., Serreze, M., St. John, K., White, J. W. C., and Wolff, E.: History of sea ice in the Arctic, Quaternary Sci. Rev., 29, 1757–1778, doi:10.1016/j.quascirev.2010.02.010, 2010.

Pongratz, J., Reick, C. H., Raddatz, T., and Claussen, M.: A reconstruction of global agricultural areas and land cover for the last millennium, Global Biogeochem. Cy., 22, GB3018, doi:10.1029/2007GB003153, 2008.

Prado, L. F., Wainer, I., Chiessi, C. M., Ledru, M.-P., and Turcq, B.: A mid-Holocene climate reconstruction for eastern South America, Clim. Past, 9, 2117–2133, doi:10.5194/cp-9-2117-2013, 2013.

Prentice, I. C. and Harrison, S. P.: Ecosystem effects of $CO_2$ concentration: evidence from past climates, Clim. Past, 5, 297–307, doi:10.5194/cp-5-297-2009, 2009.

Quesada, B., Vautard, R., Yiou, P., Hirschi, M., and Seneviratne, S. I.: Asymmetric European summer heat predictability from wet and dry southern winters and springs, Nat. Clim. Change, 2, 736–741, doi:10.1038/nclimate1536, 2012.

Ramstein, G., Kageyama, M., Guiot, J., Wu, H., Hély, C., Krinner, G., and Brewer, S.: How cold was Europe at the Last Glacial Maximum? A synthesis of the progress achieved since the first PMIP model-data comparison, Clim. Past, 3, 331–339, doi:10.5194/cp-3-331-2007, 2007.

Reichler, T. and Kim, J.: How well do coupled models simulate today's climate?, B. Am. Meteorol. Soc., 89, 303–311, 2008.

Rind, D. and Peteet, D.: Terrestrial conditions at the last glacial maximum and CLIMAP sea-surface temperature estimates: are they consistent?, J. Geophys. Res., 94, 12851–12871, 1985.

Russon, T., Tudhope, A. W., Hegerl, G. C., Collins, M., and Tindall, J.: Inter-annual tropical Pacific climate variability in an isotope-enabled CGCM: implications for interpreting coral stable oxygen isotope records of ENSO, Clim. Past, 9, 1543–1557, doi:10.5194/cp-9-1543-2013, 2013.

Rybski, D., Bunde, A., and von Storch, H.: Long-term memory in 1000-year simulated temperature records, J. Geophys. Res., 113, D02106, doi:10.1029/2007JD008568, 2008.

Santer, B. D., Taylor, K. E., Gleckler, P. J., Bonfils, C., Barnett, T. P., Pierce, D. W., Wigley, T. M. L., Mears, C., Wentz, F. J., Brüggemann, W., Gillett, N. P., Klein, S. A., Solomon, S., Stott, P. A., and Wehner, M. F.: Incorporating model quality information in climate change detection and attribution studies, P. Natl. Acad. Sci., 106, 14778–14783, 2009.

Schär, C., Lüthi, D., Beyerle, U., and Heise, E.: The soil-precipitation feedback: A process study with a regional climate model, J. Climate, 12, 722–741, 1999.

Schmidt, G. A.: Enhancing the relevance of palaeoclimate model/data comparisons for assessments of future climate change, J. Quaternary Sci., 25, 79–87, doi:10.1002/jqs.1314, 2010.

Schmidt, G. A., Ruedy, R., Hansen, J. E., Aleinov, I., Bell, N., Bauer, M., Bauer, S., Cairns, B., Canuto, V., Cheng, Y., Del Genio, A., Faluvegi, G., Friend, A. D., Hall, T. M., Hu, Y., Kelley, M., Kiang, N. Y., Koch, D., Lacis, A. A., Lerner, J., Lo, K. K., Miller, R. L., Nazarenko, L., Oinas, V., Perlwitz, Ja., Perlwitz, Ju., Rind, D., Romanou, A., Russell, G. L., Sato, M., Shindell, D. T., Stone, P. H., Sun, S., Tausnev, N., Thresher, D., and Yao, M.-S.: Present day atmospheric simulations using GISS ModelE: Comparison to in-situ, satellite and reanalysis data, J. Climate, 19, 153–192, doi:10.1175/JCLI3612.1, 2006.

Schmidt, G. A., LeGrande, A., and Hoffmann, G.: Water isotope expressions of intrinsic and forced variability in a coupled ocean–atmosphere model, J. Geophys. Res., 112, D10103, doi:10.1029/2006JD007781, 2007.

Schmidt, G. A., Jungclaus, J. H., Ammann, C. M., Bard, E., Braconnot, P., Crowley, T. J., Delaygue, G., Joos, F., Krivova, N. A., Muscheler, R., Otto-Bliesner, B. L., Pongratz, J., Shindell, D. T., Solanki, S. K., Steinhilber, F., and Vieira, L. E. A.: Climate forcing reconstructions for use in PMIP simulations of the last millennium (v1.0), Geosci. Model Dev., 4, 33–45, doi:10.5194/gmd-4-33-2011, 2011.

Schmidt, G. A., Jungclaus, J. H., Ammann, C. M., Bard, E., Braconnot, P., Crowley, T. J., Delaygue, G., Joos, F., Krivova, N. A., Muscheler, R., Otto-Bliesner, B. L., Pongratz, J., Shindell, D. T., Solanki, S. K., Steinhilber, F., and Vieira, L. E. A.: Climate forcing reconstructions for use in PMIP simulations of the Last Millennium (v1.1), Geosci. Model Dev., 5, 185–191, doi:10.5194/gmd-5-185-2012, 2012.

Schmidt, G. A., Kelley, M., Nazarenko, L., Ruedy, R., Russell, G. L., Aleinov, I., Bauer, M., Bauer, S., Bhat, M. K., Bleck, R., Canuto, V., Chen, Y., Cheng, Y., Clune, T. L., DelGenio, A., de Fainchtein, R., Faluvegi, G., Hansen, J. E., Healy, R. J., Kiang, N. Y., Koch, D., Lacis, A. A., LeGrande, A. N., Lerner, J., Lo, K. K., Matthews, E. E., Menon, S., Miller, R. L., Oinas, V., Oloso, A., Perlwitz, J., Puma, M. J., Putman, W. M., Rind, D., Romanou, A., Sato, M., Shindell, D. T., Sun, S., Syed, R., Tausnev, N., Tsigaridis, K., Unger, N., Voulgarakis, A., Yao, M.-S., and Zhang, J.: Configuration and assessment of the GISS ModelE2 contributions to the CMIP5 archive, J. Adv. Model. Earth Syst., doi:10.1002/2013MS000265, in press, 2014.

Schmittner, A., Urban, N. M., Shakun, J. D., Mahowald, N. M., Clark, P. U., Bartlein, P. J., Mix, A. C., and Rosell-Melé, A.: Climate Sensitivity Estimated from Temperature Reconstructions of the Last Glacial Maximum, Science, 334, 1385–1388, doi:10.1126/science.1203513, 2011.

Schneider von Deimling, T., Ganopolski, A., Held, H., and Rahmstorf, S.: How cold was the Last Glacial Maximum?, Geophys. Res. Lett., 33, L14709, doi:10.1029/2006GL026484, 2006.

Schurer, A. P., Hegerl, G. C., Mann, M. E., Tett, S. F. B., and Phipps, S. J.: Separating forced from chaotic climate variability over the past millennium, J. Climate, 26, 6954-6973, doi:10.1175/JCLI-D-12-00826.1, 2013.

Seneviratne, S. I. and Koster, R. D.: A revised framework for analyzing soil moisture memory in climate data: Derivation and interpretation, J. Hydrometeorol., 13, 404–412, 2012.

Seneviratne, S. I., Lüthi, D., Litschi, M., and Schär, C.: Land-atmosphere coupling and climate change in Europe, Nature, 443, 205–209, 2006.

Seneviratne, S. I., Corti, T., Davin, E. L., Hirschi, M., Jaeger, E. B., Lehner, I., Orlowsky, B., and Teuling, A. J.: Investigating soil moisture-climate interactions in a changing climate: A review, Earth Sci. Rev., 99, 125–161, 2010.

Seneviratne, S. I., Nicholls, N., Easterling, D., Goodess, C. M., Kanae, S., Kossin, J., Luo, Y., Marengo, J., McInnes, K., Rahimi, M., Reichstein, M., Sorteberg, A., Vera, C., and Zhang, X.: Changes in climate extremes and their impacts on the natural physical environment, in: A Special Report of Working Groups I and II of the Intergovernmental Panel on Climate Change (IPCC SREX Report), edited by: Field, C. B., Barros, V., Stocker, T. F., Qin, D., Dokken, D. J., Ebi, K. L., Mastrandrea, M. D., Mach, K. J., Plattner, G.-K., Allen, S. K., Tignor, M., and Midgley, P. M., Cambridge University Press, Cambridge, 2012.

Sheffield, J., Wood, E. F., and Roderick, M. L.: Little change in global drought over the past 60 years, Nature, 491, 435–438, doi:10.1038/nature11575, 2012.

Steinhilber, F., Beer, J., and Fröhlich, C.: Total solar irradiance during the Holocene, Geophys. Res. Lett., 36, L19704, doi:10.1029/2009GL040142, 2009.

Stroeve, J. C., Kattsov, V., Barrett, A., Serreze, M., Pavlova, T., Holland, M., and Meier, W. N.: Trends in Arctic sea ice extent from CMIP5, CMIP3 and observations, Geophys. Res. Lett., 39, L16502, doi:10.1029/2012GL052676, 2012.

Sundberg, R., Moberg, A., and Hind, A.: Statistical framework for evaluation of climate model simulations by use of climate proxy data from the last millennium – Part 1: Theory, Clim. Past, 8, 1339–1353, doi:10.5194/cp-8-1339-2012, 2012.

Sutton, R. T., Dong, B., and Gregory, J. M.: Land/sea warming ratio in response to climate change: IPCC AR4 model results and comparison with observations, Geophys. Res. Lett., 34, L02701, doi:10.1029/2006GL028164, 2007.

Taylor, K. E.: Summarizing multiple aspects of model performance in a single diagram, J. Geophys. Res., 106, 7183–7192, 2001.

Taylor, K. E., Stouffer, R. J., and Meehl, G. A.: An Overview of CMIP5 and the experiment design, B. Am. Meteorol. Soc., 93, 485–498, doi:10.1175/BAMS-D-11-00094.1, 2012.

Telford, R. J., Li, C., and Kucera, M.: Mismatch between the depth habitat of planktonic foraminifera and the calibration depth of SST transfer functions may bias reconstructions, Clim. Past, 9, 859–870, doi:10.5194/cp-9-859-2013, 2013.

Thompson, D. M., Ault, T. R., Evans, M. N., Cole, J. E., and Emile-Geay, J.: Comparison of observed and simulated tropical climate trends using a forward model of coral $\delta^{18}O$, Geophys. Res. Lett., 38, L14706, doi:10.1029/2011GL048224, 2011.

Thompson, D. M., Ault, T. R., Evans, M. N., Cole, J. E., Emile-Geay, J., and LeGrande, A. N.: Coral-CGCM comparison highlights role of salinity in long-term trends, in: El Niño Southern Oscillation: observation and modeling, edited by: Braconnot, P., Brierley, C., Harrison, S. P., and von Gunten, L., PAGES News, 21, 60–61, 2013.

Tremblay, L.-B., Huard, D., Schmidt, G. A., and de Vernal, A.: Mid Holocene constraints on future Arctic climate change, in preparation, 2014.

Trenberth, K. E. and Fasullo, J. T.: Simulation of present-day and twenty-first-Century energy budgets of the Southern Oceans, J. Climate, 23, 440–454, doi:10.1175/2009JCLI3152.1, 2010.

Vautard R., Yiou, P., D'Andrea, F., de Noblet, N., Viovy, N., Cassou, C., Polcher, J., Ciais, P., Kageyama, M., and Fan, Y.: Summertime European heat and drought waves induced by wintertime Mediterranean rainfall deficit, Geophys. Res. Lett., 34, L07711, doi:10.1029/2006GL028001, 2007.

Vieira, L. E. A., Solanki, S. K., Krivova, N. A., and Usoskin, I.: Evolution of the solar irradiance during the Holocene, Astron. Astrophys., 531, A6, doi:10.1051/0004-6361/201015843, 2011.

Vyushin, D. I., Kushner, P. J., and Zwiers, F.: Modeling and understanding persistence of climate variability, J. Geophys. Res., 117, D21106, doi:10.1029/2012JD018240, 2012.

Werner, M., Mikolajewicz, U., Heimann, M., and Hoffmann, G.: Borehole Versus Isotope Temperatures on Greenland: Seasonality Does Matter, Geophys. Res. Lett., 27, 723–726, 2000.

Wilby, R. L. and Wigley, T. M. L.: Downscaling general circulation model output: a review of methods and limitations, Prog. Phys. Geogr., 21, 530–548, 1997.

Wittenberg, A. T.: Are historical records sufficient to constrain ENSO simulations?, Geophys. Res. Lett., 36, L12702, doi:10.1029/2009GL038710, 2009.

Yiou, P., Dacunha-Castelle, D., Parey, S., and Hoang, T. T. H.: Statistical representation of temperature mean and variability in Europe, Geophys. Res. Lett., 36, L04710, doi:10.1029/2008GL036836, 2009.

Yiou, P., Servonnat, J., Yoshimori, M., Swingedouw, D., Khodri, M., and Abe-Ouchi, A.: Stability of weather regimes during the last millennium from climate simulations, Geophys. Res. Lett., 39, L08703, doi:10.1029/2012GL051310, 2012.

Zhu, X., Fraederich, L., and Blender, R.: Variability regimes of simulated Atlantic MOC, Geophys. Res. Lett., 33, L21603, doi:10.1029/2006GL027291, 2006.

# 12

# Nutrient utilisation and weathering inputs in the Peruvian upwelling region since the Little Ice Age

**C. Ehlert**[1,*]**, P. Grasse**[1]**, D. Gutiérrez**[2]**, R. Salvatteci**[2,3]**, and M. Frank**[1]

[1]GEOMAR Helmholtz Centre for Ocean Research Kiel, Kiel, Germany
[2]Instituto del Mar del Perú (IMARPE), Dirección de Investigaciones Oceanográficas, Callao, Peru
[3]Institute of Geoscience, Department of Geology, Kiel University, Ludewig-Meyn-Str. 10, 24118 Kiel, Germany
[*]now at: Max Planck Research Group for Marine Isotope Geochemistry, Institute for Chemistry and Biology of the Marine Environment (ICBM), University of Oldenburg, Oldenburg, Germany

*Correspondence to:* C. Ehlert (cehlert@mpi-bremen.de)

**Abstract.** For this study two sediment cores from the Peruvian shelf covering the time period between the Little Ice Age (LIA) and present were examined for changes in productivity (biogenic opal concentrations (bSi)), nutrient utilisation (stable isotope compositions of silicon ($\delta^{30}Si_{opal}$) and nitrogen ($\delta^{15}N_{sed}$)), as well as in ocean circulation and material transport (authigenic and detrital radiogenic neodymium ($\varepsilon_{Nd}$) and strontium ($^{87}Sr/^{86}Sr$) isotopes).

For the LIA the proxies recorded weak primary productivity and nutrient utilisation reflected by low average bSi concentrations of $\sim 10\%$, $\delta^{15}N_{sed}$ values of $\sim 5\%$ and intermediate $\delta^{30}Si_{opal}$ values of $\sim 0.9\%$. At the same time, the radiogenic isotope composition of the detrital sediment fraction indicates dominant local riverine input of lithogenic material due to higher rainfall in the Andean hinterland. These patterns were most likely caused by permanent El Niño-like conditions characterised by a deeper nutricline, weak upwelling and low nutrient supply. At the end of the LIA, $\delta^{30}Si_{opal}$ dropped to low values of $+0.6\%$ and opal productivity reached its minimum of the past 650 years. During the following transitional period of time the intensity of upwelling, nutrient supply and productivity increased abruptly as marked by the highest bSi contents of up to 38 %, by $\delta^{15}N_{sed}$ of up to $\sim 7\%$, and by the highest degree of silicate utilisation with $\delta^{30}Si_{opal}$ reaching values of $+1.1\%$. At the same time, detrital $\varepsilon_{Nd}$ and $^{87}Sr/^{86}Sr$ signatures documented increased wind strength and supply of dust to the shelf due to drier conditions. Since about 1870, productivity has been high but nutrient utilisation has remained at levels similar to the LIA, indicating significantly increased nutrient availability.

Comparison between the $\delta^{30}Si_{opal}$ and $\delta^{15}N_{sed}$ signatures suggests that during the past 650 years the $\delta^{15}N_{sed}$ signature in the Peruvian upwelling area has to a large extent been controlled by surface water utilisation and not, as previously assumed, by subsurface nitrogen loss processes in the water column, which only had a significant influence during modern times (i.e. since $\sim$ AD 1870).

## 1 Introduction

Global climate of the late Holocene has been disrupted by major anomalies, the most recent of which being the Little Ice Age (LIA) between ca. AD 1400 and 1850 (Lamb, 1965; Grove, 2001). During that time a weakening of the Walker circulation (Conroy et al., 2008), a reduced influence of the South Pacific subtropical high (SPSH) along the Peruvian margin (Sifeddine et al., 2008; Gutiérrez et al., 2009; Salvatteci et al., 2014a), and a southward shift of the mean position of the Intertropical Convergence Zone (ITCZ) and the associated precipitation belt compared to today (Sachs et al., 2009) caused pronounced changes in rainfall patterns in the tropics. El Niño-like warmer conditions in the eastern South Pacific were accompanied by an intensified South American summer monsoon (Bird et al., 2011), resulting in $\sim 10\%$ higher precipitation in northeastern Peru ($\sim 5°$ S; Rabatel et al., 2008) and up to 20–30 % higher precipitation in the Bolivian Andes

($\sim 16°$ S; Reuter et al., 2009). On the one hand this caused growth and extension of the Andean glaciers (Vuille et al., 2008) and on the other it enabled human settlements in the presently hyperarid southern Peruvian Andes (Unkel et al., 2007). In the upwelling areas off Peru and the western South American shelf regions, the main consequence of these climatic conditions during the LIA was a deepening of the nutricline and a strongly diminished biological productivity (Vargas et al., 2007; Sifeddine et al., 2008; Valdés et al., 2008; Gutiérrez et al., 2009).

Sediment cores from the Peruvian shelf covering the period of time from the LIA until present indicate that the marine realm was characterised by an abrupt biogeochemical regime shift towards modern conditions at the end of the LIA due to the northward movement of the ITCZ and an expansion of the SPSH. While low productivity and a more oxygenated water column prevailed during the LIA, markedly increased biological productivity and pronounced oxygen depletion over wide areas of the shelf have characterised the system since the end of the LIA (Vargas et al., 2007; Sifeddine et al., 2008; Gutiérrez et al., 2009; Salvatteci et al., 2014a).

In this study the stable silicon isotope composition of sedimentary diatoms ($\delta^{30}Si_{opal}$) covering the period of time from the LIA to the present is analysed. The main goal is the reconstruction of the factors controlling the dynamics of nutrient cycling together with oxygen in the Peruvian upwelling, in particular a comparison between the $\delta^{30}Si_{opal}$ and the stable nitrogen isotope composition ($\delta^{15}N_{sed}$) of sedimentary organic matter. Both $\delta^{30}Si_{opal}$ and $\delta^{15}N_{sed}$ provide information about utilisation of silicic acid ($Si(OH)_4$) and nitrate ($NO_3^-$) during primary productivity, e.g. during the formation of diatom frustules and associated organic matter, respectively (Altabet and Francois, 1994; De La Rocha et al., 1997). Diatoms preferentially incorporate the lighter isotopes from the dissolved $Si(OH)_4$ and $NO_3^-$ pools, leaving the residual dissolved nutrients enriched in the heavier isotopes (Wada and Hattori, 1978; Altabet et al., 1991; De La Rocha et al., 1997). Si isotope fractionation is mainly controlled by the utilisation of $Si(OH)_4$ in surface waters by biota (diatoms) (e.g. De La Rocha et al., 1998; Brzezinski et al., 2002; Egan et al., 2012). The $\delta^{15}N$ of $NO_3^-$ is partly controlled by $NO_3^-$ utilisation of marine organisms but is also affected by N-loss processes in the water column (denitrification, anammox) (Codispoti et al., 2001; Dalsgaard et al., 2003), resulting in a marked enrichment of the upwelling source waters in the heavier $^{15}NO_3^-$ (Liu and Kaplan, 1989; Lam et al., 2009; given that it is currently not possible to distinguish between different N-loss processes from the sediments, we will use the term denitrification for simplicity). Consequently, sedimentary $\delta^{15}N_{sed}$ records from areas dominated by oxygen-depleted waters such as the shelf region off Peru are usually interpreted to directly reflect changes in the intensity of subsurface $NO_3^-$ loss and the extent and strength of oxygen depletion (e.g. De Pol-Holz et al., 2007, 2009; Agnihotri et

al., 2008; Gutiérrez et al., 2009), whereas the effect of $NO_3^-$ utilisation on the preserved $\delta^{15}N_{sed}$ is often neglected. Comparison of both isotope systems can therefore provide information about the degree of utilisation of $NO_3^-$ and $Si(OH)_4$ versus the influence of $NO_3^-$ loss processes. Increasing nutrient utilisation should result in a consistent increase in both $\delta^{30}Si_{opal}$ and $\delta^{15}N_{sed}$. In contrast, a change in $NO_3^-$ reduction due to varying oxygen depletion in the water column would affect only the $\delta^{15}N_{sed}$, leaving the $\delta^{30}Si_{opal}$ unaffected.

The main forces driving surface productivity and subsurface oxygenation off Peru at centennial timescales during the past two millennia have been changes in the strength of the Walker circulation and in the expansion/contraction of the SPSH (Gutierrez et al., 2009; Salvatteci et al., 2014a). Therefore, the radiogenic isotope compositions of neodymium ($\varepsilon_{Nd}$) and strontium ($^{87}Sr / ^{86}Sr$) of the authigenic ferromanganese (Fe-Mn) oxyhydroxide coatings of the sedimentary particles, which are expected to record the radiogenic isotope compositions of past bottom waters, as well as of the detrital fraction of the sediment were examined. These proxy data provide information about changes in (surface ocean) circulation and transport processes, provenance of the sediments, and input mechanisms of terrigenous material as a function of changes in precipitation on land during the transition from wetter LIA conditions to drier modern conditions. Weathering of continental source rocks delivers lithogenic particles of different origin and age to the shelf, which have distinct radiogenic isotope signatures ($\varepsilon_{Nd\ detritus}$, $^{87}Sr / ^{86}Sr_{detritus}$) that can be used to trace their source areas (Goldstein et al., 1984). Central Peruvian Andean rocks have more radiogenic $\varepsilon_{Nd}$ signatures whereas southern Peruvian rocks are characterised by less radiogenic $\varepsilon_{Nd}$ signatures (Sarbas and Nohl, 2009), which is also reflected in the sediments along the shelf (Ehlert et al., 2013). Changes in detrital material input and transport pathways are generally closely related to climatic changes causing variations in the supply from the respective source areas (e.g. Grousset et al., 1988). It should therefore be possible to detect the transition from wetter LIA conditions with higher local input from central Peru via rivers due to higher precipitation rates towards the drier presently prevailing conditions with an increased influence of aeolian material transport from further south in the Atacama Desert (Molina-Cruz, 1977) and deposition along the shelf after the LIA.

## 2  Material and methods

### 2.1  Core locations and age models

For the reconstruction of surface water $Si(OH)_4$ utilisation and terrestrial material input and transport for the period of time between the LIA and present, two sediment cores with high sedimentation rates were analysed. Box core B0405-6 was recovered from the upper continental slope off Pisco at

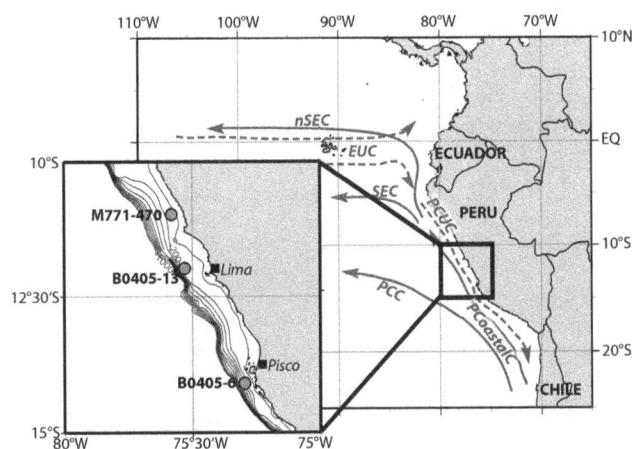

**Figure 1.** Schematic circulation patterns in the eastern equatorial Pacific. Surface currents (solid lines): (n)SEC, (northern) South Equatorial Current; PCC, Peru–Chile Current; PCoastalC, Peru Coastal Current. Subsurface currents (dashed lines): EUC, Equatorial Undercurrent; PCUC, Peru–Chile Undercurrent (after Brink, 1983; Kessler, 2006). The inset shows the location of cores M771-470, B0405-6 and B0405-13 (grey dots) in greater detail. The bathymetry is given for 0 to 1000 m water depth in 100 m increments.

14°07.9′ S, 76°30.1′ W at a water depth of 299 m with the Peruvian RV *José Olaya Balandra* in 2004 (Fig. 1) (Gutiérrez et al., 2006). The age model was previously published by Gutiérrez et al. (2009) and is based on downcore profiling of the activities of $^{241}$Am, excess $^{210}$Pb, and on radiocarbon ages obtained from bulk sedimentary organic carbon, which document that the core covers the past $\sim 650$ years. The second core, multicorer M771-470, was taken at 11° S, 77°56.6′ W at 145 m water depth during cruise M77/1 with the German RV *Meteor* in 2008 (Fig. 1). The age model was obtained by measuring excess $^{210}$Pb activities and modelling of the resulting profiles as described by Meysman et al. (2005) (for details see Supplement). Ages prior to $\sim$ AD 1850 were inferred using sedimentation rates from nearby core B0405-13 (Gutiérrez et al., 2009; Salvatteci et al., 2014b).

## 2.2 Methods

### 2.2.1 Biogenic opal and silicon isotope analyses

The biogenic opal (bSi) contents in both cores were measured following the sequential leaching techniques described by DeMaster (1981) and Müller and Schneider (1993). Si isotope analyses were performed on the 11–32 μm diatom fraction that was extracted from the sediment applying the procedures described by Morley et al. (2004).

Approximately 300 mg of sediment was treated with 30 % $H_2O_2$ and 35 % HCl to remove organic matter and carbonate. Afterwards the sediment was wet-sieved to separate the 11–32 μm fraction. In a third step a heavy-liquid solution (sodium polytungstate, 2.1–2.2 g mL$^{-1}$) was applied in several steps to separate diatoms from the detrital lithogenic silicate material. All samples were screened under the microscope to verify their purity with respect to the detrital (clay) fraction.

The diatom samples were then transferred into Teflon vials and dissolved in 1 mL of 0.1 M NaOH and diluted with Milli-Q water according to Reynolds et al. (2008). More details are provided in Ehlert et al. (2012). Si concentrations of the dissolved diatom samples were measured colorimetrically using a photospectrometer (Hansen and Koroleff, 1999). Chromatographic separation and purification of the Si was achieved with 1mL pre-cleaned AG50W-X8 cation exchange resin (mesh 200–400) (Georg et al., 2006; as modified by de Souza et al., 2012). Si isotope ratios were measured on a Nu Plasma HR MC-ICPMS (Nu Instruments) at GEOMAR equipped with an adjustable source-defining slit, which can be set to medium resolution to ensure separation of the $^{30}$Si peak from molecular interferences. The measurements were carried out applying standard-sample bracketing (Albarède et al., 2004). All solutions were measured at a Si concentration of 14–21 μmol kg$^{-1}$ of samples and standards depending on the performance of the instrument on the respective measurement day and were introduced into the plasma via a Cetac Aridus II desolvating nebuliser system equipped with a PFA nebuliser operated at a 60 to 80 μL min$^{-1}$ uptake rate. Si isotope compositions are reported in the $\delta^{30}$Si notation as deviations of the measured $^{30}$Si / $^{28}$Si from the NIST standard NBS28 in parts per thousand (‰). Repeated measurements of the reference materials IRMM018 and Big Batch gave average $\delta^{30}$Si values of $-1.52 \pm 0.18$ ‰ ($2\sigma_{(sd)}$) and $-10.84 \pm 0.18$ ‰ ($2\sigma_{(sd)}$), respectively, which are in good agreement with values obtained by other laboratories (Reynolds et al., 2007). Samples were measured three to five times within a 1-day session and measurements were repeated on at least 2 separate days. The resulting uncertainties ranged between 0.04 and 0.23 ‰ ($2\sigma_{(sd)}$) (Tables 1, 2). Replicate measurements of an in-house diatom matrix standard over longer periods of time gave an external reproducibility of 0.11 ‰ ($2\sigma_{(sd)}$). Error bars provided in the figures correspond to that external reproducibility unless the uncertainties of the repeated sample measurements were higher.

### 2.2.2 Neodymium and strontium isotope analyses

To obtain the radiogenic isotope composition of past bottom seawater at the sites of the sediment cores from the early diagenetic Fe-Mn coatings of the sediment particles, previously published methods were applied (Gutjahr et al., 2007; see Supplement for details). The residual detrital material was leached repeatedly to remove remaining coatings and was then treated with a mixture of concentrated HF-HNO$_3$-HCl for total dissolution. The separation and purification of Nd

**Table 1.** Downcore records of core M771-470 for $\delta^{30}Si_{opal}$ (‰); bSi content (wt %); and $^{143}Nd/^{144}Nd$, $\varepsilon_{Nd}$ and $^{87}Sr/^{86}Sr$ of detrital material. $2\sigma_{(sd)}$ represents the external reproducibilities of repeated sample (Si) and standard (Nd, Sr) measurements.

| Depth (cm) | $\delta^{30}Si_{opal}$ (‰) | $2\sigma_{(sd)}$ | bSi (wt %) | $^{143}Nd/^{144}Nd_{detritus}$ | $\varepsilon_{Nd}$ detritus | $2\sigma_{(sd)}$ | $^{87}Sr/^{86}Sr_{detritus}$ | $2\sigma_{(sd)}$ |
|---|---|---|---|---|---|---|---|---|
| 0.5 | 1.03 | 0.15 | 18.8 | – | – | – | – | – |
| 1.5 | – | – | 18.6 | – | – | – | – | – |
| 2.5 | – | – | 22.2 | – | – | – | – | – |
| 3.5 | 0.93 | 0.08 | 16.9 | 0.512369 | −5.2 | 0.3 | 0.709315 | 1.5e-05 |
| 4.5 | – | – | 16.3 | – | – | – | – | – |
| 5.5 | – | – | 17.2 | 0.512381 | −5.0 | 0.3 | 0.709356 | 1.5e-05 |
| 7 | – | – | 19.5 | – | – | – | – | – |
| 9 | 0.96 | 0.09 | 19.8 | 0.512398 | −4.7 | 0.3 | 0.708822 | 1.5e-05 |
| 11 | – | – | 18.8 | – | – | – | – | – |
| 13 | – | – | 15.9 | – | – | – | – | – |
| 15 | – | – | – | 0.512383 | −5.0 | 0.3 | 0.708737 | 1.5e-05 |
| 16 | 0.96 | 0.07 | 19.3 | – | – | – | – | – |
| 19 | – | – | – | 0.512386 | −4.9 | 0.3 | 0.708552 | 1.5e-05 |
| 20 | 1.05 | 0.10 | 18.9 | 0.512410 | −4.5 | 0.3 | 0.708412 | 8.0e-06 |
| 23 | – | – | – | 0.512393 | −4.8 | 0.3 | 0.708720 | 1.5e-05 |
| 24 | 1.15 | 0.13 | 26.9 | – | – | – | – | – |
| 26 | – | – | – | 0.512387 | −4.9 | 0.3 | 0.707482 | 8.0e-06 |
| 27 | – | – | – | 0.512397 | −4.7 | 0.3 | 0.707555 | 1.5e-05 |
| 28 | 1.00 | 0.14 | 14.0 | – | – | – | – | – |
| 29 | – | – | – | 0.512452 | −3.6 | 0.3 | 0.706549 | 1.5e-05 |
| 32 | 0.55 | 0.17 | 10.1 | 0.512442 | −3.8 | 0.3 | 0.706763 | 1.5e-05 |
| 32 | – | – | – | 0.512445 | −3.8 | 0.3 | 0.706469 | 8.0e-06 |
| 36 | 1.10 | 0.15 | 14.4 | 0.512419 | −4.3 | 0.3 | 0.706767 | 8.0e-06 |
| 40 | 0.79 | 0.11 | 12.3 | 0.512408 | −4.5 | 0.3 | 0.706964 | 8.0e-06 |
| 44 | 0.91 | 0.18 | 15.0 | 0.512421 | −4.2 | 0.3 | 0.707057 | 8.0e-06 |
| 48 | 0.75 | 0.05 | – | 0.512395 | −4.7 | 0.3 | 0.707816 | 8.0e-06 |

**Table 2.** Downcore records of core B0405-6 for $\delta^{30}Si_{opal}$ (‰); bSi content (wt %); and $^{143}Nd/^{144}Nd$, $\varepsilon_{Nd}$ and $^{87}Sr/^{86}Sr$ of detrital material. $2\sigma_{(sd)}$ represents the external reproducibilities of repeated sample (Si) and standard (Nd, Sr) measurements.

| Year AD | $\delta^{30}Si_{opal}$ (‰) | $2\sigma_{(sd)}$ | bSi (wt %) | $^{143}Nd/^{144}Nd_{detritus}$ | $\varepsilon_{Nd}$ detritus | $2\sigma_{(sd)}$ | $^{87}Sr/^{86}Sr_{detritus}$ | $2\sigma_{(sd)}$ |
|---|---|---|---|---|---|---|---|---|
| 1950 | 0.91 | 0.15 | 21.7 | 0.512507 | −2.6 | 0.1 | 0.708372 | 8.0e-06 |
| 1925 | 0.83 | 0.15 | 21.0 | 0.512460 | −3.5 | 0.3 | 0.707923 | 8.0e-06 |
| 1903 | 0.62 | 0.10 | 18.9 | 0.512487 | −2.9 | 0.3 | 0.707715 | 8.0e-06 |
| 1857 | 1.02 | 0.16 | 34.4 | 0.512471 | −3.3 | 0.3 | 0.707829 | 8.0e-06 |
| 1857 | 1.22 | 0.14 | 37.7 | 0.512481 | −3.1 | 0.1 | 0.707736 | 8.0e-06 |
| 1818 | 0.56 | 0.15 | 12.6 | 0.512468 | −3.3 | 0.3 | 0.707702 | 8.0e-06 |
| 1793 | 0.82 | 0.14 | 15.8 | 0.512446 | −3.7 | 0.3 | 0.707265 | 8.0e-06 |
| 1761 | 0.71 | 0.16 | 13.5 | 0.512627 | −0.2 | 0.3 | 0.707296 | 8.0e-06 |
| 1698 | 0.73 | 0.09 | 17.3 | 0.512462 | −3.4 | 0.3 | 0.707278 | 8.0e-06 |
| 1564 | 0.81 | 0.12 | 20.8 | 0.512467 | −3.3 | 0.3 | 0.707281 | 8.0e-06 |
| 1475 | 0.77 | 0.04 | 17.1 | 0.512427 | −4.1 | 0.3 | 0.707959 | 8.0e-06 |
| 1370 | 0.80 | 0.23 | 34.2 | 0.512509 | −2.5 | 0.3 | 0.707111 | 8.0e-06 |

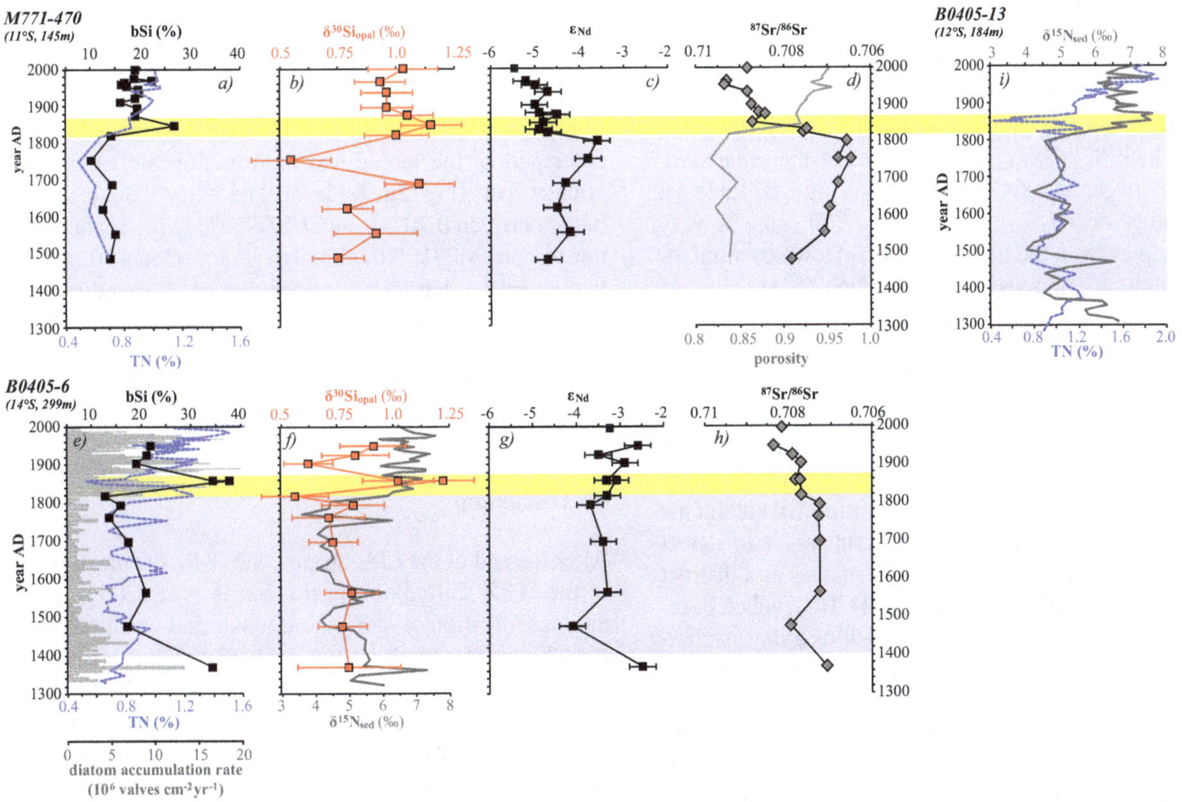

**Figure 2.** Downcore records for core M771-470 (upper panel) and core B0405-6 (lower panel). The blue and yellow shadings indicate the age range of the LIA and the transitional period, respectively. (**a, e**) bSi concentration (black squares), (**a, e, i**) total N concentration (dashed blue curve), (**e**) diatom accumulation rate (grey bars) (Gutiérrez et al., 2009), (**b, f**) $\delta^{30}Si_{opal}$ (red squares), (**f**) bulk $\delta^{15}N_{sed}$ (grey curve) (Gutiérrez et al., 2009), (**c, g**) $\varepsilon_{Nd}$ detritus (black squares), (**d, h**) $^{87}Sr/^{86}Sr_{detritus}$ (grey diamonds; $x$ axis is inverted), (**d**) sediment porosity (grey curve). Error bars represent $2\sigma_{(sd)}$ external reproducibilities of repeated standard or sample measurements. For comparison, (**i**) shows the total N content and $\delta^{15}N_{sed}$ of core B0405-13 (Gutiérrez et al., 2009).

and Sr in the leachates and in the completely dissolved detrital sediment fraction followed previously published procedures for Nd (Cohen et al., 1988) and Sr (Horwitz et al., 1992) applying ion exchange chromatography for separation of Rb/Sr from the rare earth elements (REEs) (0.8 mL AG50W-X12 resin, mesh 200–400) followed by separation of Sr from Rb (50 µL Sr-Spec resin, mesh 50–100), and separation of Nd from the other REEs (2 mL Eichrom Ln-Spec resin, mesh 50–100). All radiogenic isotope measurements were performed on the Nu Plasma HR MC-ICPMS (Nu Instruments) at GEOMAR. Measured Nd isotope compositions were corrected for instrumental mass bias using a $^{146}Nd/^{144}Nd$ ratio of 0.7219 and were normalised to the accepted $^{143}Nd/^{144}Nd$ literature value of 0.512115 of the JNdi-1 standard (Tanaka et al., 2000). All values are given as $\varepsilon_{Nd}$, which corresponds to the measured $^{143}Nd/^{144}Nd$, normalised to the chondritic uniform reservoir (CHUR) (0.512638), multiplied by 10 000. The external reproducibility was estimated via repeated measurements of the JNdi-1 standard and was always better than 20 ppm ($2\sigma_{(sd)}$, Tables 1, 2). Measured $^{87}Sr/^{86}Sr$ ratios were corrected for instrumen-

tal mass bias using $^{88}Sr/^{86}Sr = 8.3752$ and were normalised to the accepted value for NIST SRM987 of 0.710245. The $2\sigma_{(sd)}$ external reproducibility of repeated standard measurements was always better than 36 ppm ($2\sigma_{(sd)}$, Tables 1, 2). Procedural Nd and Sr blanks for leachates and total dissolutions of the detrital material were $\leq 83$ pg and 2.1 ng, respectively, and thus negligible compared to the concentrations of the samples.

## 3 Results

### 3.1 Core M771-470 (Callao)

Sediment core M771-470, from a location at 11° S 145 m water depth, is characterised by bSi concentrations between 10.1 and 26.9 % and total N contents between 0.5 and 1.1 % (Fig. 2a, Table 1), whereby the lowest values occurred just prior to the end of the LIA. The maximum bSi concentrations were found during the transition period. In contrast, the highest nitrogen (N) content occurred later in the youngest part of the record. The $\delta^{30}Si_{opal}$ varied between +0.6 and

$+1.1$‰ (Fig. 2b) and followed bSi concentrations with the maximum and minimum isotope values corresponding to the same respective depths for both parameters.

The $\varepsilon_{\text{Nd detritus}}$ is characterised by values between $-3.6$ and $-5.2$, with a mean value of $-4.5 \pm 1.0$ ($2\sigma_{\text{(sd)}}$) (Fig. 2c, Table 1). The $^{87}\text{Sr}/^{86}\text{Sr}_{\text{detritus}}$ signatures of the same samples range between 0.70647 and 0.70936 (Fig. 2d, Table 1). The variability of $\varepsilon_{\text{Nd detritus}}$ and $^{87}\text{Sr}/^{86}\text{Sr}_{\text{detritus}}$ is very similar. Samples from the LIA show a trend towards more radiogenic $\varepsilon_{\text{Nd detritus}}$ and less radiogenic $^{87}\text{Sr}/^{86}\text{Sr}_{\text{detritus}}$. At the beginning of the transition period, both records indicate a marked change to less radiogenic $\varepsilon_{\text{Nd detritus}}$ and more radiogenic $^{87}\text{Sr}/^{86}\text{Sr}_{\text{detritus}}$ values, which was more pronounced in the Sr than in the Nd isotope data, resulting in the youngest samples having the least radiogenic $\varepsilon_{\text{Nd detritus}}$ and the most radiogenic $^{87}\text{Sr}/^{86}\text{Sr}_{\text{detritus}}$ signatures.

In theory, the radiogenic isotope composition of authigenic Fe-Mn oxyhydroxide coatings is a useful tracer to detect changes on the prevailing bottom water masses at a distinct location. The Peru–Chile Undercurrent (PCUC), which dominates the bottom waters at the core locations today, is characterised by radiogenic $\varepsilon_{\text{Nd}}$ signatures of $-1.8$ (Lacan and Jeandel, 2001; Grasse et al., 2012). A deepening of the nutricline and a vertical expansion of surface water masses during the LIA could change that value towards less radiogenic signatures typical for water masses originating from the South Pacific (Piepgras and Wasserburg, 1982; Grasse et al., 2012). However, as has been shown before (Ehlert et al., 2013), the authigenic coating fraction from sediments along the Peruvian shelf does not necessarily represent changes in water mass advection and is therefore not a reliable proxy (see also Supplement for details).

### 3.2 Core B0405-6 (Pisco)

In core B0405-6, from a location near $14°$ S off Pisco at 299 m water depth, the range of bSi concentrations and its maximum value are higher than in core M771-470 and varied between 12.6 and 37.7 % (Fig. 2e, Table 2). The trends are very similar to those of core M771-470, and bSi content correlates closely with the diatom accumulation rate (Fig. 2e). The lowest values of both parameters occurred at the end of the LIA and highest values were found right after the end of the LIA at the beginning of the transition period. The N content ranges from 0.5 % around AD 1860 to 1.8 % in the youngest sample of the record (Fig. 2e) (Gutiérrez et al., 2009) with maximum N content in core B0405-6 also being higher than in core M771-470. The $\delta^{30}\text{Si}_{\text{opal}}$ record shows the same range from $+0.6$ to $+1.1$‰ as core M771-470 and a very similar trend with the lowest values near the end of the LIA and the highest values immediately thereafter during the transition period (Fig. 2f, Table 3). The $\delta^{15}\text{N}_{\text{sed}}$ ranges between 3.6 and 7.6‰ and shows a trend from lower mean values around 4 to 5‰ during the LIA to higher values between 6 and 7‰ in the modern sediments (Fig. 2f).

The $\varepsilon_{\text{Nd detritus}}$ signatures are characterised by overall somewhat more radiogenic values than of core M771-470 ranging from $-4.1$ to $-2.5$ (mean value $-3.2 \pm 0.9$, $2\sigma_{\text{(sd)}}$ excluding the value of $-0.2$ $\varepsilon_{\text{Nd}}$ at AD 1761, which is considered an outlier), with slightly less radiogenic values in the older part of the record and more radiogenic values in the younger part (Fig. 2g, Table 2). The $^{87}\text{Sr}/^{86}\text{Sr}_{\text{detritus}}$ values range between 0.70711 and 0.70796 (Fig. 2h, Table 2). Similar to core M771-470, although less pronounced, the main feature in the detrital Sr isotope record observed is a trend from less radiogenic $^{87}\text{Sr}/^{86}\text{Sr}_{\text{detritus}}$ values in the older part of the record towards more radiogenic values in the youngest part, with a shift at the end of the LIA and during the early transition period.

## 4   Discussion

After the end of the LIA, around AD 1820, the mean position of the ITCZ shifted northward (Sachs et al., 2009), causing an intensification of alongshore winds and enhanced coastal upwelling off the Peruvian coast (Sifeddine et al., 2008; Gutiérrez et al., 2009), diminished coastal sea surface temperatures (Vargas et al., 2007), and a decrease in precipitation on land (Rabatel et al., 2008; Bird et al., 2011). Records of productivity and redox conditions based on $\delta^{15}\text{N}_{\text{sed}}$ and the Mo and Cd content of the sediments indicate a rapid change in the biogeochemical composition of the source waters to higher nutrient concentrations, causing higher biological productivity and lower subsurface oxygen, which have persisted until the present day (Sifeddine et al., 2008; Gutiérrez et al., 2009; Salvatteci et al., 2014a). The shift after the end of the LIA constitutes a major anomaly of late Holocene climate in the eastern Pacific, which was of the same order of magnitude as the changes in conditions off Chile during the Younger Dryas (De Pol-Holz et al., 2006). This study focuses on the reconstruction of the regime shift from the LIA and a transitional period towards modern conditions and its controlling factors, including the evolution of nutrient utilisation and changes in the advection of water masses and material transport.

### 4.1   Changes in biological productivity and nutrient consumption

#### 4.1.1   Evolution of surface water productivity and nutrient utilisation

The pronounced change in the biogeochemical regime from low productivity during the LIA to higher productivity during the transitional and modern period thereafter is documented by several sedimentary records from the eastern equatorial Pacific (EEP) region and has been dated at $\sim$ AD 1820 (Sifeddine et al., 2008; Gutierrez et al., 2009; Díaz-Ochoa et al., 2009, 2011; Salvatteci et al., 2014a). Similarly, both cores M771-470 from $11°$ S and B0405-6 from $14°$ S off

Pisco show the characteristic coeval pronounced increase in bSi and total N content (Fig. 2a, e) and $C_{org}$ concentration (not shown here) after the end of the LIA and during the transition period. Therefore, three time periods that show distinct differences in productivity and nutrient utilisation have been identified from our records and will be discussed in the following: the LIA, the transition period from the LIA to modern conditions between $\sim 1820$ and $\sim$ AD 1870, and modern conditions after $\sim$ AD 1870.

Both cores recorded a 2- to 3-fold increase in bSi content from 10 to 12 % prior to the end of the LIA to values of up to 27 % in M771-470 and up to 38 % in B0405-6 during the transition period (Fig. 2a, e). Afterwards the bSi contents decreased again but since then have remained at a level of $\sim 20$ % since then and are thus significantly higher than prior to the end of the LIA. The increase in bSi content is also reflected by a marked increase in diatom accumulation rate in core B0405-6 (Fig. 2e) (Gutiérrez et al., 2009). Analyses of the downcore diatom assemblages have shown that the high diatom accumulation rates and bSi content in core B0405-6 during the transition period were associated with diatom layers dominated by *Skeletonema costatum* (Gutiérrez et al., 2009), a species that is today more abundant when upwelling is more intense during austral winter/spring.

Both cores are characterised by a very high correlation between total N and $C_{org}$ content ($r^2 = 0.95$ for core M771-470 and 0.8 for core B0405-6) (Gutiérrez et al., 2009, 2015). In contrast, bSi and total N contents do not co-vary throughout the records (Fig. 3a). Surface sediments from the Peruvian shelf region between the Equator and $\sim 18°$ S show a relatively weak but positive correlation between bSi and N contents ($r^2 = 0.5$, Fig. 3a) (Ehlert et al., 2012; Mollier-Vogel et al., 2012). Similar to the surface sediments, bSi and total N concentrations in core M771-470 are positively correlated, whereas they essentially do not correlate in core B0405-6. This is because the bSi maximum at the end of the transition period was more pronounced in core B0405-6 and higher than surface sediment bSi contents anywhere along the shelf region off Peru. At the same time, only a rather gradual increase in total N content occurred, with some excursions to low values during the transition period (Fig. 2a, e). The total N concentration also did not always co-vary with $\delta^{15}N_{sed}$ (Fig. 3c). In particular, the samples from the late transition period show very low total N concentrations but high $\delta^{15}N_{sed}$ and high $\delta^{30}Si_{opal}$ and bSi content.

Sedimentary $\delta^{15}N_{sed}$ data, which are only available for core B0405-6, show a shift from lower values around $+4$ to $+5‰$ during the LIA to higher values around $+7‰$ after the end of the LIA and have remained at that level since then (Fig. 2f) (Gutiérrez et al., 2009). The values in the younger part of the record are in good agreement with surface sediment $\delta^{15}N_{sed}$ data measured in the main Peruvian upwelling region ranging from $+6$ to $+9‰$ (Mollier-Vogel et al., 2012). Bulk $\delta^{15}N_{sed}$ signatures measured in core B0405-13 from $12°$ S (184 m water depth) close to the location of

core M771-470 can be used for comparison and show very similar values, amplitude, and variability to core B0405-6 (Fig. 2i) (Gutiérrez et al., 2009). In contrast to $\delta^{15}N_{sed}$, the $\delta^{30}Si_{opal}$ signatures, which mainly reflect changes in surface water nutrient utilisation, are not only characterised by a simple increase at the end of the LIA. Instead, both $\delta^{30}Si_{opal}$ records closely follow the evolution of the bSi concentrations and show intermediate $\delta^{30}Si_{opal}$ signatures between $+0.8$ and $+0.9‰$ during the LIA; a pronounced short-term decrease to $+0.6‰$ at the end of the LIA, which was followed by a marked increase to values around $+1.1‰$ during the transition period; and finally a return to intermediate values between $+0.8$ and $+1.0‰$ in the modern part of the records (Fig. 2b, f). The correspondence between bSi content and $\delta^{30}Si_{opal}$ is more pronounced in core B0405-6 (Fig. 3b), which shows a higher variability and amplitude of bSi content. The difference in the $\delta^{30}Si_{opal}$, $\delta^{15}N_{sed}$, bSi and total N content records during the transition from LIA to modern conditions reflects the different environmental factors controlling the proxies, which will be discussed in the following sections.

### 4.1.2 Present-day surface water utilisation versus subsurface nitrate loss

Diatoms are the dominant phytoplankton group of the Peruvian upwelling region (Estrada and Blasco, 1985; Bruland et al., 2005). The $\delta^{30}Si_{opal}$ of these diatoms is primarily controlled by surface water diatom productivity and $Si(OH)_4$ utilisation (De La Rocha et al., 1998; Brzezinski et al., 2002; Egan et al., 2012). Off Peru the $\delta^{30}Si_{opal}$ has also been shown to be dependent on the isotopic signature of the advected surface and subsurface water masses (Ehlert et al., 2012; Grasse et al., 2013). Similarly, the $\delta^{15}N_{sed}$ of the organic matter is controlled by N isotope fractionation during $NO_3^-$ uptake by phytoplankton, mostly diatoms. Off Peru, however, the $NO_3^-$ supplied to the surface waters has previously been enriched in $^{15}NO_3^-$ due to upwelling of oxygen-depleted subsurface waters, which had undergone significant $NO_3^-$-loss processes (mostly denitrification, but also anammox processes, associated with a high fractionation of up to $20–30‰$) (Lam et al., 2009; Altabet et al., 2012). Bulk sediment $\delta^{15}N_{sed}$ in areas with oxygen-depleted waters is therefore usually interpreted to reflect changes in the intensity of subsurface $NO_3^-$ reduction and the extent and strength of the oxygen minimum zone (Altabet et al., 1999; De Pol-Holz et al., 2007; Agnihotri et al., 2008; Gutiérrez et al., 2009). The direct comparison of $\delta^{30}Si_{opal}$, reflecting mostly utilisation, and $\delta^{15}N_{sed}$, reflecting both utilisation and $NO_3^-$ reduction, from core B0405-6 off Pisco will therefore provide insights into the strength of $NO_3^-$ reduction in the past.

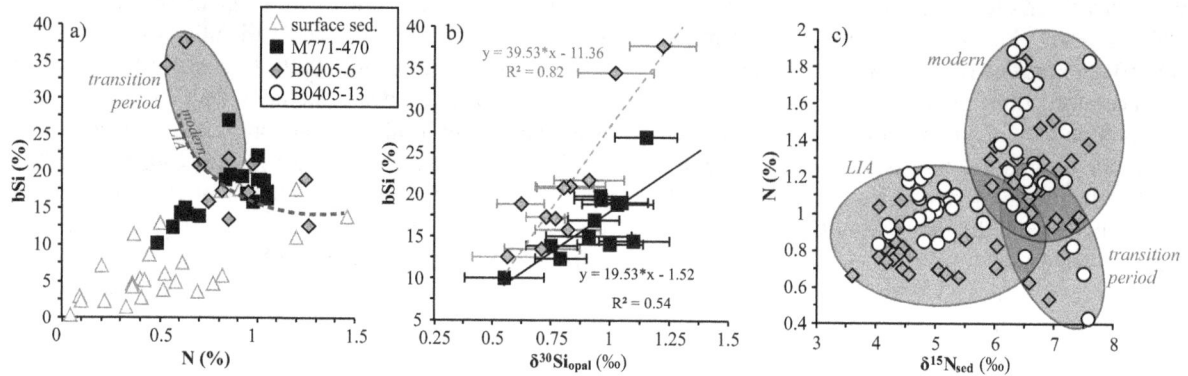

**Figure 3.** Surface sediment (white triangles) and downcore data (core M771-470: black squares; B0405-6: grey diamonds; B0405-13: white circles) for **(a)** total N versus bSi concentrations (the dashed line marks the end of the LIA), **(b)** $\delta^{30}Si_{opal}$ versus bSi concentration and **(c)** $\delta^{15}N_{sed}$ versus total N concentrations. Error bars represent $2\sigma_{(sd)}$ external reproducibilities.

Subsurface water column $\delta^{15}N_{NO_3^-}$ data from the present-day Peruvian shelf are isotopically very heavy, in particular along the southern shelf region between 10 and 17° S, where values of up to +25‰ are reached due to the increasing oxygen deficit and intensification of $NO_3^-$-loss processes (Mollier-Vogel et al., 2012; Altabet et al., 2012). These isotopically enriched waters are upwelled along the shelf and represent the source for organic matter production in the surface waters. Therefore, it is expected that the deposited sedimentary organic matter reflects these enriched subsurface water signatures. However, the latitudinal increase in surface sediment $\delta^{15}N_{sed}$ from the same shelf region to maximum mean values around +9‰ is much lower than that measured in the water column (Mollier-Vogel et al., 2012). The reason for this observation is that the $\delta^{15}N_{sed}$ signal in the southern shelf region (10–17° S) did not fully record the $^{15}NO_3^-$ enrichment in the water column but is a combination of the isotopic effects associated with subsurface $NO_3^-$ loss and incomplete surface water $NO_3^-$ utilisation and water mass mixing. Direct comparison of $\delta^{30}Si_{opal}$ and $\delta^{15}N_{sed}$ allows for the relative importance of these processes to be investigated and distinguished (Fig. 4a). Diatoms off Peru preferentially take up $Si(OH)_4$ and $NO_3^-$ at a ratio of ~ 1 : 1 or below (Brzezinski, 1985; Takeda, 1998; Hutchins et al., 2002). If utilisation were the only driving factor, the sedimentary $\delta^{30}Si_{opal}$ and $\delta^{15}N_{sed}$ should all plot close to a line that reflects the enrichment during increasing utilisation, i.e. 1.1‰ for $\delta^{30}Si_{opal}$ (De La Rocha et al., 1997) and ~ 5‰ for $\delta^{15}N$ (DeNiro and Epstein, 1981; Minagawa and Wada, 1984). Under the influence of denitrification with an enrichment of ~ 20‰ (Lam et al., 2009), however, the relationship between $\delta^{30}Si_{opal}$ and $\delta^{15}N_{sed}$ would be very different (Fig. 4a).

Most modern shelf samples plot either on or above the theoretical curve for utilisation implying, if at all, $Si(OH)_4$ limiting conditions (Fig. 4a). Very few samples are shifted towards the theoretical curve for denitrification, indicating

a weak influence of $NO_3^-$-loss processes on the preserved isotope signatures. Especially along the central shelf region (green curves in Fig. 4a), where the cores were sourced, surface sediment signatures closely reflect the utilisation in surface waters with only little influence of $NO_3^-$ loss in the water column and sediments.

### 4.1.3 Past surface water utilisation versus subsurface nitrate loss

Assuming that source water isotope composition (+1.5‰ $\delta^{30}Si_{Si(OH)_4}$, +9‰ $\delta^{15}N_{NO_3}$) and isotope enrichment during utilisation and denitrification (utilisation: −1.1‰ for $\delta^{15}Siopal$ and ~ −5‰ for $\delta^{15}NNO_3^-$, denitrification: ~ −20‰ for $\delta^{15}NNO_3^-$) in the past were similar to the conditions of the present-day shelf region (Ehlert et al., 2012; Mollier-Vogel et al., 2012), the samples of core B0405-6 indicate variable utilisation/$NO_3^-$-loss conditions (Fig. 4b). Samples from the LIA and the transition period generally plot on or above the utilisation curve, indicating stronger $Si(OH)_4$ than $NO_3^-$ utilisation. This implies that in the Peruvian upwelling system has rather been a $Si(OH)_4$-limited system during that time, similar to today (Fig. 4a). During the transition period, when strong upwelling conditions caused intense blooming of *Skeletonema costatum*, utilisation of $Si(OH)_4$ and $NO_3^-$ was very close to a 1 : 1 ratio. In contrast, the samples from the end of the LIA and especially the recent samples are shifted slightly towards the denitrification curve, indicating a higher influence of $NO_3^-$-loss processes. This is particularly the case for the samples from the end of the LIA, which have the lowest $\delta^{30}Si_{opal}$ but at the same time already show a strong increase in $\delta^{15}N_{sed}$ to values of near +6‰. The most likely explanation is that upwelling was strongly increased during those brief periods, resulting in high nutrient supply, high productivity, and either more complete $NO_3^-$ utilisation (Gutiérrez et al., 2009) or increased $NO_3^-$ loss caused by enhanced subsurface oxygen

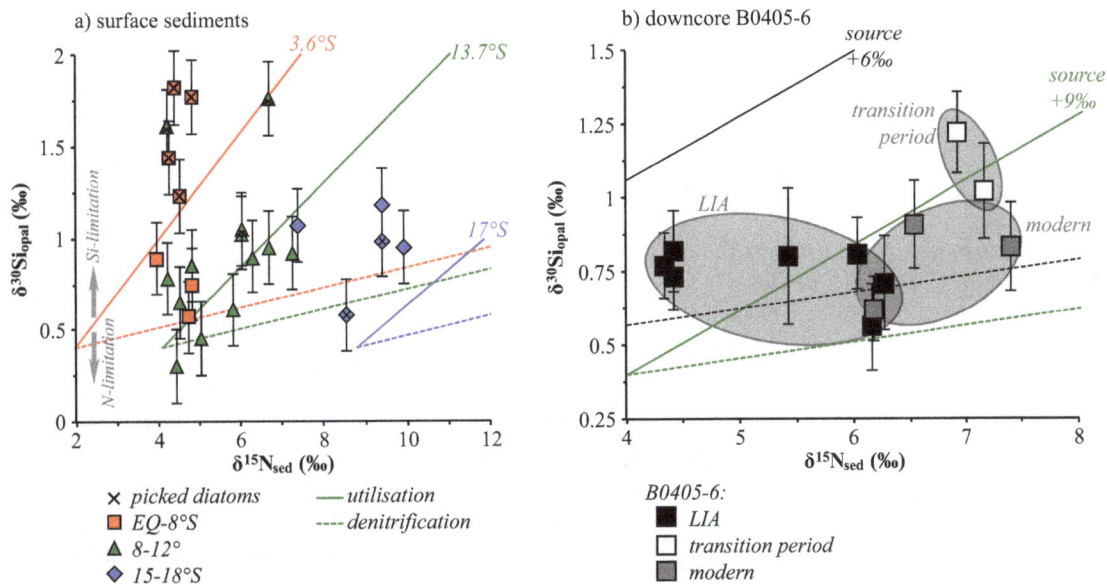

**Figure 4.** Relationship between $\delta^{15}N_{sed}$ versus $\delta^{30}Si_{opal}$ for (**a**) surface sediments and (**b**) downcore data from core B0405-6. The crosses in (**a**) indicate $\delta^{30}Si$ data obtained from hand-picked diatoms, which reflect a different growth season than bulk $\delta^{30}Si_{opal}$ and which are influenced by stronger $Si(OH)_4$ limitation (higher $\delta^{30}Si$) (Ehlert et al., 2012). The solid lines reflect theoretical utilisation (assuming 1 : 1 utilisation of $Si(OH)_4$ and $NO_3^-$ by the diatoms) and the dashed lines mark the theoretically expected line for denitrification, which represent the expected signal preserved in the sediments, based on present-day measurements: $\delta^{30}Si$ source signature and enrichment factor $\varepsilon_{diatom-Si(OH)_4}$ are always +1.5 ‰ (Ehlert et al., 2012) and −1.1 ‰ (De La Rocha et al., 1997), respectively. $\delta^{15}N_{sed}$ source signature and $\varepsilon_{organic-NO_3}$ vary with latitude (Mollier-Vogel et al., 2012); in the north at 3.6° S, source signature and $\varepsilon$ are +5.7 and −3.7 ‰ (red curves); along the central shelf at 13.7° S, source signature and $\varepsilon$ are +8.9 and −4.8 ‰ (green curves); and in the south at 17° S, source signature and $\varepsilon$ were measured to be +14.5 and −5.7 ‰ (blue curves), respectively. The samples are colour-coded according to their location on the shelf and relative to the $NO_3^-$ utilisation/$NO_3^-$ loss that they experienced. Data points that plot above the utilisation curves reflect predominant $Si(OH)_4$ limitation, whereas data points below record stronger $NO_3^-$ limitation. The isotopic enrichment during denitrification was always set to be +20 ‰. For the downcore data (**b**) two different assumed source signatures are displayed: +9 ‰ (green lines, corresponding to the modern conditions along the central shelf region in **a**) and +6 ‰ (grey lines). Data points are colour-coded according to the respective time periods (black: LIA; white: transition period; grey: modern). Error bars represent $2\sigma_{(sd)}$ external reproducibilities.

depletion. Overall, however, the utilisation signal appears to have dominated both the Si and N isotope records.

If, however, the $\delta^{15}N_{sed}$ is dominated by utilisation it is interesting that $\delta^{15}N_{sed}$ and proxies for sediment redox conditions (e.g. molybdenum concentrations) in the cores (both B0405-6 and B0405-13) are strongly coupled throughout the record (Sifeddine et al., 2008; Gutiérrez et al., 2009). One direct interpretation could be that the diatom blooms, and subsequently the degradation of the organic matter, strongly control the oxygen availability in the sediments after sedimentation and burial. Therefore, increased diatom productivity and higher $Si(OH)_4$ and $NO_3^-$ utilisation would result in an increase in $\delta^{15}N_{sed}$. At the same time, more oxygen is consumed during degradation of the organic matter in the sediments, causing more reducing conditions in the sediments. Consequently, a change in the subsurface water column structure, e.g. enhanced re-supply of oxygen via ocean currents, may not be reflected in the $\delta^{15}N_{sed}$ record.

### 4.1.4 Modelling the surface water utilisation

Following the above considerations we will try to quantify past utilisation based on our data. The theoretical relationship between the degree of surface water nutrient utilisation and the stable isotope composition of Si and N can be described assuming either Rayleigh-type (single input followed by no additional nutrients newly supplied to a particular parcel of water followed by fractional loss as a function of production and export) or steady-state (continuous supply and partial consumption of nutrients causing a dynamic equilibrium of the dissolved nutrient concentration and the product) fractionation behaviour (Fig. 5) (Mariotti et al., 1981). The lighter isotopes are preferentially incorporated into the diatom frustules and the organic matter, respectively, leaving the dissolved fraction enriched in the heavier isotopes (Wada and Hattori, 1978; Altabet et al., 1991; De La Rocha et al., 1997). The fractionation between $\delta^{30}Si$ in seawater and $\delta^{30}Si$ in the produced diatom opal has generally been assumed to be −1.1 ‰ (De La Rocha et al., 1997), whereas between

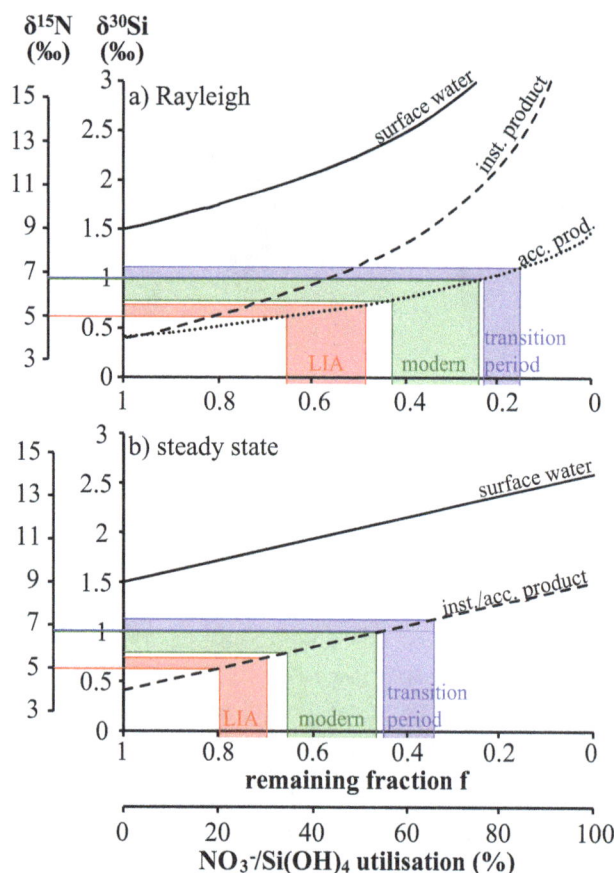

**Figure 5.** Theoretical changes in $\delta^{30}$Si and $\delta^{15}$N values of seawater and the instantaneous and accumulated product as a function of $f$ (remaining nutrients from the available pool = [nutrient$_{observed}$]/[nutrient$_{initial}$]) with an initial $\delta^{30}$Si$_{Si(OH)4}$ value of +1.5‰ and $\delta^{15}$N$_{NO_3^-}$ of +9‰. The formation of the product, e.g. diatom opal, follows either (a) Rayleigh-type fractionation or (b) steady-state-type fractionation behaviour, with enrichment factors $\varepsilon$ of −1.1‰ ($\delta^{30}$Si) and −5‰ ($\delta^{15}$N) (corresponding to conditions along the modern central Peruvian shelf; see Fig. 4). The colour shadings mark the range of measured mean $\delta^{30}$Si$_{opal}$ (both cores) and $\delta^{15}$N$_{sed}$ (B0405-6 only) in the cores for the LIA (red), the transition period (blue) and modern sediments (green), from which the respective nutrient utilisation (%) can be deduced.

$\delta^{15}$N$_{NO_3^-}$ of seawater and $\delta^{15}$N of the newly formed organic matter it is usually between −3 and −6‰ (DeNiro and Epstein, 1981; Minagawa and Wada, 1984). Here we adopted −5‰, which corresponds to present-day conditions along the central Peruvian shelf (Mollier-Vogel et al., 2012).

Along the Peruvian shelf region biological productivity in the euphotic zone is mainly driven by upwelling of nutrients from subsurface waters. For the calculation of the utilisation of these nutrients, an initial $\delta^{30}$Si$_{Si(OH)4}$ of +1.5‰ (Ehlert et al., 2012) as well as an initial $\delta^{15}$N$_{NO_3^-}$ of +9‰ (Mollier-Vogel et al., 2012) for the upwelled water masses

at 14° S is assumed. The lower mean $\delta^{15}$N$_{sed}$ of about +5‰ and $\delta^{30}$Si$_{opal}$ of +0.7‰ signatures during the LIA in the southerly core B0405-6 correspond to a dissolved $\delta^{15}$N$_{NO_3^-}$ and $\delta^{30}$Si$_{Si(OH)4}$ isotope signature of the surface waters of +10 and +1.8‰ and a calculated NO3_ and Si(OH)$_4$ utilisation of only 20–30% for steady-state-type fractionation (Fig. 5b) and 35–50% for Rayleigh-type fractionation (Fig. 5a) behaviour. The highest mean values of +1.1‰ for $\delta^{30}$Si$_{opal}$ and +6.8‰ for $\delta^{15}$N$_{sed}$ for the transition period correspond to a much higher utilisation of ∼60% for steady-state-type fractionation and ∼80% assuming Rayleigh-type fractionation. Consequently, the calculated utilisation of available Si(OH)$_4$ and NO$_3^-$ more than doubled, whereby bSi concentrations and diatom accumulation rates increased by about a factor of 3 (Fig. 2e).

The changes in Si(OH)$_4$ and NO$_3^-$ utilisation were of the same order of magnitude and reflect low nutrient utilisation during the LIA and a much higher degree of utilisation thereafter. The large increase in $\delta^{15}$N$_{sed}$ at the end of the LIA has been interpreted to reflect an expansion of nutrient-rich, oxygen-poor subsurface waters (Gutiérrez et al., 2009). However, comparison with $\delta^{30}$Si$_{opal}$ shows that the increase in $\delta^{15}$N$_{sed}$ may indeed have occurred as a consequence of the extension of the oxygen minimum zone and increasing subsurface NO$_3^-$ loss but can also be explained by higher surface water utilisation. As Mollier-Vogel et al. (2012) have shown, the subsurface enrichment of $\delta^{15}$N$_{NO_3^-}$ caused by NO$_3^-$-loss processes can only be reflected in the sediments under near-complete surface water NO$_3^-$ utilisation, which obviously did not occur at our studied sites.

In the modern samples the $\delta^{30}$Si$_{opal}$ are characterised by a slight decrease after the transition period from mean value of +1.12 to +0.82‰, whereas the $\delta^{15}$N$_{sed}$ values remain at the same level around +7‰ (Fig. 2). This corresponds to a ∼20% higher NO$_3^-$ than Si(OH)$_4$ utilisation (Fig. 5). However, when assuming that diatoms are the dominating primary producers with a NO$_3^-$ / Si(OH)$_4$ uptake ration of ∼1, these 20% could be interpreted to reflect the increase in subsurface NO$_3^-$ loss that is not observable during the LIA or the transition period; that is, during the LIA and the transition period utilisation was the dominating process influencing the $\delta^{15}$N$_{sed}$ signal, whereas during modern times NO$_3^-$ loss enhanced the signal. By combining the $\delta^{30}$Si$_{opal}$ and the $\delta^{15}$N$_{sed}$ records this "additional" signal can be quantified here: if only utilisation were to play a role, the expected $\delta^{15}$N$_{sed}$ signal would be ∼+6‰ (corresponding to the measured ∼+0.8‰ for $\delta^{30}$Si$_{opal}$, Fig. 5). The additional 1‰ $\delta^{15}$N$_{sed}$ must be due to NO$_3^-$ loss.

The overall relatively low $\delta^{30}$Si$_{opal}$ signatures between 0.8 and 1.0‰ during the LIA and in the modern part of the records (Figs. 2, 5) document that the utilisation of Si(OH)$_4$ only changed slightly during the investigated period of time although the accumulation rate of produced diatoms was much higher after the LIA (Fig. 2e) (Gutiérrez et al., 2009).

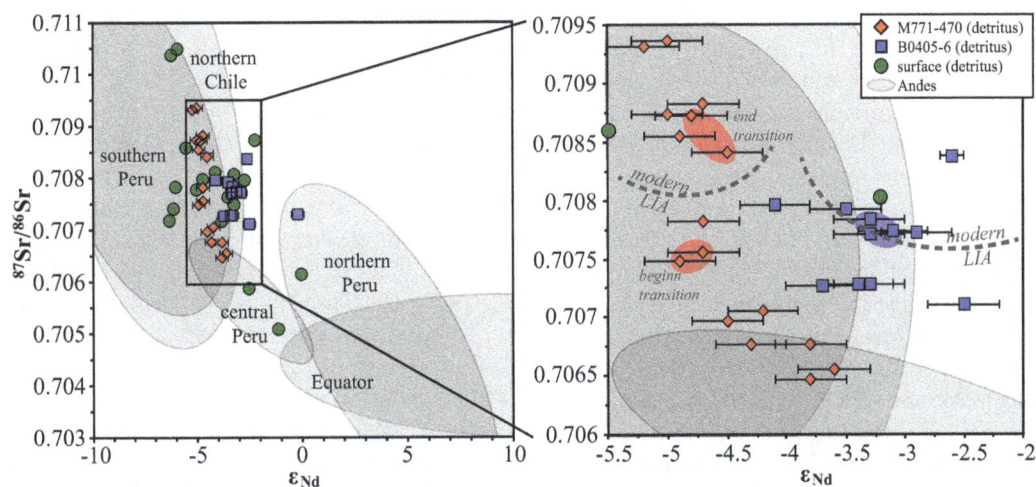

**Figure 6.** $\varepsilon_{Nd\,detritus}$ versus $^{87}Sr/^{86}Sr_{detritus}$ for core M771-470 (red diamonds) and B0405-6 (blue squares). Error bars represent $2\sigma_{(sd)}$ external reproducibilities. The green dots are data obtained from surface sediment samples at different sites on the Peruvian shelf. The grey shadings indicate potential sources and provenance end members of the detrital material.

This suggests that the nutrient concentrations in the upwelled subsurface source waters must have been lower during the LIA than they are today. During the LIA, large-scale circulation changes, i.e. a weak Walker circulation and a contraction of the SPSH (Conroy et al., 2008; Lamy et al., 2001), caused permanent El Niño-like conditions along the Peruvian upwelling system. During such conditions, the alongshore winds weakened and caused a deepening of the thermo-, oxy- and nutricline, and therefore a reduction of vertical pumping of nutrient-rich and oxygen-depleted subsurface waters off Peru. Such a reduced nutrient supply to the euphotic zone from subsurface waters resulted in an increase in nutrient deficit in surface waters and decreased biological productivity. Enhanced water column oxygenation and lower organic matter flux led to decreased organic matter preservation in the sediments.

### 4.1.5 Factors influencing the reconstruction of the utilisation signals

There are two main factors that can influence the reconstruction of nutrient utilisation in the past: (1) a change in the dominating diatom assemblages has to be considered, and (2) the interpretation strongly depends on the assumptions for the environmental conditions, e.g. source water signature and isotope enrichment during utilisation.

Varying upwelling and nutrient supply conditions also cause changes in the dominating diatom assemblages. Recent results from culturing experiments suggest species-dependent enrichment factors for diatom $\delta^{30}Si$ ($-0.5$ to $-2.1‰$; Sutton et al., 2013) and also diatom frustule-bound $\delta^{15}N$ ($-1.9$ to $-11.2‰$; Horn et al., 2011). This raises the question of whether a change in diatom assemblages may have been the cause of the observed downcore

$\delta^{30}Si_{opal}$ and, to a lesser extent, the bulk $\delta^{15}N_{sed}$ variations. The quasi-monospecific diatom layers from the transition period AD 1820–1870 consist mainly of *Skeletonema costatum* (Gutiérrez et al., 2009), for which an enrichment factor $\varepsilon$ of $-1.0‰$, similar to the applied $-1.1‰$, was determined (De la Rocha et al., 1997). The younger sediments also contain abundant upwelling-indicative species such as *Thalassionema nitzschioides* and *Chaetoceros* sp. (Abrantes et al., 2007), whereby *Chaetoceros brevis*, a species from the Southern Ocean, has been shown to have a much higher $\varepsilon$ of $-2.1‰$ (Sutton et al., 2013). That means, assuming the same surface water $\delta^{30}Si_{Si(OH)_4}$ signatures, a dominance of *Chaetoceros* sp. in the sediments should result in a lower $\delta^{30}Si_{opal}$, whereas assemblages dominated by *Skeletonema costatum* should be characterised by higher $\delta^{30}Si_{opal}$ signatures, which is exactly what core B0405-6 shows. Consequently, the difference in $\delta^{30}Si_{opal}$ over time could reflect the change in diatom assemblage and not a change in nutrient utilisation. On the other hand, *Chaetoceros brevis* is a polar species and it is not clear whether it undergoes the same high fractionation factor off Peru during frustule growth. It has been shown that the offset between modern surface water $\delta^{30}Si_{Si(OH)4}$ and surface sediment $\delta^{30}Si_{opal}$ along the central Peruvian shelf is between $-1.1$ and $-1.3‰$ (Ehlert et al., 2012), which indicates that either the enrichment factor for the dominating *Chaetoceros* species off Peru does not deviate significantly from $-1.1‰$ or that the mixing of different diatoms in the sediment samples overprints any isotopic excursions of single species caused by higher or lower fractionation factors. Given the paucity of data on the fractionation factors for the dominant diatom species off Peru, the importance of the role of downcore changes in the assemblage composition is hard to determine.

The assumed source water $\delta^{30}Si_{Si(OH)4}$ and $\delta^{15}N_{NO_3^-}$ values of +1.5 and +9‰ (Figs. 5, 6), respectively, were measured in the present-day subsurface waters under strong upwelling conditions during which high amounts of nutrients are supplied to the euphotic zone (Ehlert et al., 2012; Mollier-Vogel et al., 2012). Under strong upwelling conditions the bottom waters on the shallow shelf are today dominated by the southward-directed high-nutrient PCUC (Fig. 1) (Brink et al., 1983). Under LIA conditions (prevailing El Niño-like conditions), however, atmospheric and oceanic circulation was different; the nutricline was deeper as a consequence of a weak Walker circulation and the winds driving the upwelling were weaker as a consequence of the SPSH contraction (Salvatteci et al. 2014a). In fact, the pumped waters were likely nutrient-depleted, because the Ekman layer did not reach the subsurface nutrient-rich waters. Under these conditions, the subtropical and equatorial nutrient-depleted surface water masses may have occupied the entire surface layer in the coastal realm because they expanded both latitudinally and vertically in the water column (Montes et al., 2011). This may have changed the source water isotopic signatures and would therefore also change the calculated degrees of utilisation. If, for example, the assumed source water $\delta^{15}N_{NO_3^-}$ was +6‰ instead of +9‰ (e.g. due to weaker subsurface $NO_3^-$ loss and weaker $^{15}N$ enrichment during the LIA), the downcore $\delta^{15}N_{sed}$ data of core B0405-6 would all plot closer to the denitrification curve (Fig. 4b). Redox proxies from the records indeed indicate a weaker oxygen minimum zone (Gutiérrez et al., 2009), which would make a lower $\delta^{15}N_{NO_3^-}$ in source waters likely. However, to date there is no reliable information regarding if and how much the source water $\delta^{30}Si_{Si(OH)4}$ and $\delta^{15}N_{NO_3^-}$ signatures changed over time.

In the following we thus investigate variations in past water mass circulation, upwelling conditions, and material input and transport to reconstruct the source water conditions without considering the potential influence of changes in diatom assemblages.

### 4.2 Changes in water mass circulation, detrital material input and transport

The radiogenic isotope composition of the lithogenic particles ($\varepsilon_{Nd\ detritus}$ and $^{87}Sr/^{86}Sr_{detritus}$) of the sediments provides useful information about the source region of material and therefore about changes in material input and transport, either aeolian or via ocean currents (e.g. Grousset et al., 1988). Surface sediments along the Peruvian shelf show highly variable signatures, which have overall more radiogenic $\varepsilon_{Nd\ detritus}$ values in the north and much less radiogenic $\varepsilon_{Nd\ detritus}$ values in the south off southern Peru and northern Chile (Ehlert et al., 2013). This north–south trend is a consequence of the southward increasing contributions of material input from the adjacent Andean hinterland rocks. The Andean rocks along the northwestern South American region

display a wide range in $\varepsilon_{Nd}$ and $^{87}Sr/^{86}Sr$ signatures (Fig. 6) (Sarbas and Nohl, 2009), varying from highly radiogenic $\varepsilon_{Nd}$ around 0 and unradiogenic $^{87}Sr/^{86}Sr$ values around 0.704 in the equatorial region in northern Peru to much less radiogenic $\varepsilon_{Nd}$ mostly below −4 and more radiogenic $^{87}Sr/^{86}Sr$ mostly above 0.705 in southern Peru and northern Chile. The sedimentary $\varepsilon_{Nd\ detritus}$ and $^{87}Sr/^{86}Sr_{detritus}$ records of the two cores off Callao at 11° S and off Pisco at 14° S show broad similarities, but also some differences. Both cores recorded a significant change in $\varepsilon_{Nd\ detritus}$ and $^{87}Sr/^{86}Sr_{detritus}$, and therefore a change in provenance, at the end of the LIA and during the transition period. Core M771-470, although from a location further north, is overall characterised by less radiogenic $\varepsilon_{Nd\ detritus}$ values than core B0405-6 (Figs. 2, 6). The $\varepsilon_{Nd\ detritus}$ of core M771-470 recorded a trend from less radiogenic towards more radiogenic values prior to the end of the LIA, followed by a step of 1.5 $\varepsilon_{Nd}$ units towards less radiogenic values, which afterwards remained at that level. In contrast, the $\varepsilon_{Nd\ detritus}$ record of core B0405-6 remained at a level around −3.6 during the LIA and then slightly increased to maximum values of up to −2.5 in the younger part. The $^{87}Sr/^{86}Sr_{detritus}$ record in both cores is mainly characterised by a rapid shift towards more radiogenic values at the end of the LIA, whereby the change was much more pronounced in core M771-470 (Figs. 2d, h). The youngest samples of the cores are in good agreement with measurements of surface sediments from the same area (Figs. 2, 6) (Ehlert et al., 2013). The variability in core M771-470 displays the same magnitude as the complete glacial–interglacial variation in core SO147-106KL, from a location at 12° S off Lima (Ehlert et al., 2013). All data of both M771-470 and B0405-6 plot within the provenance fields of southern Peru and northern Chile (Fig. 6).

Today, material input along the Peruvian shelf occurs mostly via riverine and minor aeolian input (Molina-Cruz, 1977; Scheidegger and Krissek, 1982). The LIA, however, was characterised by wetter conditions (Haug et al., 2001; Gutiérrez et al., 2009). These changes in precipitation were associated with the position of the ITCZ, changes in Walker circulation, and expansion/contraction of the SPSH (Salvatteci et al., 2014a). Additionally, there has been a tight connection to northern hemispheric climate. Speleothem records from the central Peruvian Andes, for example, indicate a pronounced link to North Atlantic climate (Kanner et al., 2013). During cold periods like the LIA, the Peruvian upwelling region exhibited an El Niño-like mean state (Salvatteci et al., 2014a) due to the mean southward migration of the ITCZ and the associated precipitation belt, which also caused more intense rainfall in the central Andean hinterland (Rabatel et al., 2008; Reuter et al., 2009). Most terrigenous particles and weathering products such as clay minerals from the LIA show indications of increased riverine transport and discharge (Sifeddine et al., 2008; Salvatteci et al., 2014a). Consequently, material input during the LIA was dominated by local sources due to the higher river discharge.

After the end of the LIA the region experienced a northward displacement of the ITCZ and the northern rim of the SPSH to their modern position, coupled with an enhancement of the atmospheric Walker circulation (Gutiérrez et al., 2009). The climate in the EEP became drier and alongshore winds became stronger, riverine input diminished, and aeolian dust input increased. The wind-blown dust has mainly originated from the area of the Atacama Desert located in the southern Peruvian and northern Chilean Andes (Molina-Cruz, 1977). This material has less radiogenic $\varepsilon_{Nd}$ and much more radiogenic $^{87}Sr / ^{86}Sr$ values (Fig. 6) (Sarbas and Nohl, 2009). The record of core M771-470 is in agreement with this. The LIA sediments indicate a local origin, probably via riverine input, whereas the younger sediments display characteristics from a more southerly origin and therefore increased aeolian sources, possibly from the Atacama Desert. The signatures and overall small variations in core B0405-6 are much more difficult to interpret. There are fewer rivers in southern Peru around Pisco compared to the Callao region. Therefore, riverine-derived material from northern and central Peru, which is transported via the PCUC, can get dispersed further south and can be deposited in the Pisco region. On the other hand, the influence of aeolian deposition should be much higher at the southern core location. During the LIA, river input increased in southern Peru as well, whereas aeolian deposition was low. The invariant signature observed might be the result of mixing of sediment from the different sources. Also, in comparison to core M771-470, core B0405-6 is from a location much closer to the coast, which most likely diminished the differences in material input and transport between the LIA and modern conditions.

In summary, our combined proxy information coherently hints at the same controlling processes that we have already identified on glacial–interglacial timescales (Ehlert et al., 2013) and at different El Niño–Southern Oscillation patterns during the LIA (enhanced El Niño-like conditions) and in modern times (La Niña-like conditions). The locally sourced radiogenic isotope signatures show that precipitation and runoff from the hinterland was higher during the LIA, but this could not compensate for the lower nutrient supply via diminished upwelling. Eolian wind forcing was low and the source waters of the upwelling carried less nutrients. Consequently, diatom productivity and nutrient utilisation were low. In contrast, after the end of the LIA, radiogenic isotopes indicate diminished river runoff and increased dust transport, which is in agreement with an overall drier climate, probably driven by an expansion of the SPSH, and a shoaling of the thermocline/nutricline due to a stronger atmospheric Walker circulation. Especially in more recent times, the efficient remineralisation of nutrients from subsurface waters fuelled enhanced diatom productivity most likely responsible for higher nutrient utilisation in surface waters as well as enhanced oxygen demand and $NO_3^-$ loss in subsurface waters.

## 5 Conclusions

Proxies of productivity, nutrient utilisation and material provenance (bSi and N content, $\delta^{30}Si_{opal}$, $\delta^{15}N_{sed}$, $\varepsilon_{Nd\ detritus}$ and $^{87}Sr / ^{86}Sr_{detritus}$) from two cores from the Peruvian shelf recorded significant changes in surface water $Si(OH)_4$ and $NO_3^-$ concentration and utilisation due to changes in upwelling intensity and nutrient supply. During the LIA the overall nutrient content in the water column and in surface waters was low because the upwelling source waters contained less nutrients. Consequently, the Peruvian upwelling regime was characterised by persistently reduced primary productivity. The reasons for this were most likely a contraction of the South Pacific subtropical high and a weaker Walker circulation that resulted in a weakening of alongshore winds and a deepening of the nutricline.

The enhanced rainfall associated with higher moisture on land during prevailing El Niño-like conditions during the LIA was recorded by the radiogenic isotope composition of the detrital material along the shelf, which was mainly transported via rivers from the Andean hinterland. At the end of the LIA, in accordance with a northward shift of the ITCZ and an intensification of wind strength, a higher dust transport of particles associated with drier conditions and aeolian forcing is reflected by the radiogenic isotope composition of the detrital sediments. These conditions were also reflected in increasing upwelling strength, a rapid shoaling of the thermocline and nutricline, and enhanced nutrient supply and productivity to the surface waters. During a transition period a marked increase in diatom blooming events doubled the $Si(OH)_4$ and $NO_3^-$ utilisation compared to the LIA, and was also higher than present-day utilisation. After that transition period, more persistent non-El Niño conditions favoured a high productivity accompanied by moderate utilisation of nutrients. Utilisation was similar to the LIA but productivity was much higher, which reflects the much higher concentrations of nutrients in surface waters.

Most studies of past coastal upwelling regions have argued so far that the sedimentary $\delta^{15}N_{sed}$ records were dominated by the large N isotope fractionation signature occurring during $NO_3^-$-loss processes (denitrification or anammox) in oxygen-depleted subsurface waters upwelling. However, comparison between $\delta^{30}Si_{opal}$ and $\delta^{15}N_{sed}$ in the same sediment samples of our study, and assuming similar source water signatures to today, indicates that, except for the period of time since $\sim$ AD 1870, the $\delta^{15}N_{sed}$ signatures to a large extent reflect expected utilisation signals, which has important implications for the reconstruction of variations in the intensity of oxygen depletion, the N cycle of the past and its controlling factors.

*Acknowledgements*. This work is a contribution of Sonder-forschungsbereich 754 "Climate–Biogeochemistry Interactions in the Tropical Ocean" (www.sfb754.de), which is supported by the Deutsche Forschungsgemeinschaft. We acknowledge the help of Jutta Heinze in the laboratory of GEOMAR for the biogenic opal concentration measurements. We thank Ulrike Lomnitz and Klaus Wallmann for their help with the $^{210}$Pb dating and the establishment of the age model of core M771-470.

The service charges for this open access publication
have been covered by a Research Centre of the
Helmholtz Association.

Edited by: T. Kiefer

# References

Abrantes, F., Lopes, C., Mix, A. C., and Pisias, N. G.: Diatoms in Southeast Pacific surface sediments reflect environmental properties, Quaternary Sci. Rev., 26, 155–169, 2007.

Agnihotri, R., Altabet, M. A., Herbert, T. D., and Tierney, J. E.: Sub-decadally resolved paleoceanography of the Peru margin during the last two millennia, Geochem. Geophys. Geosys., 9, Q05013, doi:10.1029/2007GC001744, 2008.

Albarède, F., Telouk, P., Blichert-Toft, J., Boyet, M., Agranier, A., and Nelson, B.: Precise and accurate isotopic measurements using multiple-collector ICPMS, Geochim. Cosmochim. Ac., 68, 2725–2744, 2004.

Altabet, M. A. and Francois, R.: Sedimentary nitrogen isotopic ratio as a recorder for surface ocean nitrate utilisation, Global Biogeochem. Cy., 8, 103–116, 1994.

Altabet, M. A., Deuser, W. G., Honjo, S., and Stienen, C.: Seasonal and depth-related changes in the source of sinking particles in the North Atlantic, Nature, 354, 136–139, 1991.

Altabet, M. A., Pilskaln, C., Thunell, R. C., Pride, C. J., Sigman, D. M., Chavez, F. P., and Francois, R.: The nitrogen isotope biogeochemistry of sinking particles from the margin of the Eastern North Pacific, Deep-Sea Res. Pt. I, 46, 655–679, 1999.

Altabet, M. A., Ryabenko, E., Stramma, L., Wallace, D. W. R., Frank, M., Grasse, P., and Lavik, G.: An eddy-stimulated hotspot for fixed nitrogen-loss from the Peru oxygen minimum zone, Biogeosciences, 9, 4897–4908, doi:10.5194/bg-9-4897-2012, 2012.

Bird, B. W., Abbott, M. B., Vuille, M., Rodbell, D. T., Stansell, N. D., and Rosenmeier, M. F.: A 2,300-year-long annually resolved record of the South American summer monsoon from the Peruvian Andes, PNAS, 108, 8583–8588, 2011.

Brink, K. H., Halpern, D., Huyer, A., and Smith, R. L.: The Physical Environment of the Peruvian Upwelling System, Prog. Oceanogr., 12, 285–305, 1983.

Bruland, K. W., Rue, E. L., Smith, G. J., and DiTullio, G. R.: Iron, macronutrients and diatom blooms in the Peru upwelling regime: brown and blue waters of Peru, Mar. Chem., 93, 81–103, 2005.

Brzezinski, M. A.: The Si:C:N ratio of marine diatoms: interspecific variability and the effect of some environmental variables, Journal of Phycology, 21, 347–357, 1985.

Brzezinski, M. A., Pride, C. J., Franck, V. M., Sigman, D. M., Sarmiento, J. L., Matsumoto, K., Gruber, N., Rau, G. H., and

Coale, K. H.: A switch from Si(OH)4 to NO3- depletion in the glacial Southern Ocean, Geophys. Res. Lett., 29, 3–6, 2002.

Codispoti, L. A., Brandes, J. A., Christensen, J. P., Devol, A. H., Naqvi, S. W. A., Paerl, H. W., and Yoshinary, T.: The oceanic fixed nitrogen and nitrous oxide budgets: Moving targets as we enter the anthropocene?, Sci. Mar., 65, 85–105, 2001.

Cohen, A. S., O'Nions, R. K., Siegenthaler, R., and Griffin, W. L.: Chronology of the pressure-temperature history recorded by a granulite terrain, Contribut. Mineral. Petrol., 98, 303–311, 1988.

Conroy, J. L., Restrepo, A., Overpeck, J. T., Steinitz-Kannan, M., Cole, J. E., Bush, M. B., and Colinvaux, P. A.: Unprecedented recent warming of surface temperatures in the eastern tropical Pacific Ocean, Nature Geosci., 2, 46–50, 2008.

Dalsgaard, T., Canfield, D. E., Petersen, J., Thamdrup, B., and Acuna-González, J.: N2 production by the anammox reaction in the anoxic water column of Golfo Dulce, Costa Rica, Nature, 422, 606–608, 2003.

De La Rocha, C. L., Brzezinski, M. A., and DeNiro, M. J.: Fractionation of silicon isotopes by marine diatoms during biogenic silica formation, Geochim. Cosmochim. Ac., 61, 5051–5056, 1997.

De La Rocha, C. L., Brzezinski, M. A., DeNiro, M. J., and Shemesh, A.: Silicon-isotope composition of diatoms as an indicator of past oceanic changes, Nature, 395, 680–683, 1998.

DeMaster, D. J.: The supply and accumulation of silica in the marine environment, Geochim. Cosmochim. Ac., 45, 1715–1732, 1981.

DeNiro, M. J. and Epstein, S.: Influence of diet on the distribution of nitrogen isotopes in animals, Geochim. Cosmochim. Ac., 45, 341–351, 1981.

De Pol-Holz, R., Ulloa, O., Dezileau, L., Kaiser, J., Lamy, F., and Hebbeln, D.: Melting of the Patagonian Ice Sheet and deglacial perturbations of the nitrogen cycle in the eastern South Pacific, Geophys. Res. Lett., 33, L04704, doi:10.1029/2005GL024477, 2006.

De Pol-Holz, R., Ulloa, O., Lamy, F., Dezileau, L., Sabatier, P., and Hebbeln, D.: Late Quaternary variability of sedimentary nitrogen isotopes in the eastern South Pacific Ocean, Paleoceanography, 22, PA2207, doi:10.1029/2006PA001308, 2007.

De Pol-Holz, R., Robinson, R. S., Hebbeln, D., Sigman, D. M., and Ulloa, O.: Controls on sedimentary nitrogen isotopes along the Chile margin, Deep-Sea Res. Pt. II, 56, 1042–1054, 2009.

De Souza, G. F., Reynolds, B. C., Rickli, J., Frank, M., Saito, M. A., Gerringa, L. J. A., and Bourdon, B.: Southern Ocean control of silicon stable isotope distribution in the deep Atlantic Ocean, Global Biogeochem. Cy., 26, GB2035, doi:10.1029/2011GB004141, 2012.

Díaz-Ochoa, J. A., Lange, C. B., Pantoja, S., De Lange, G. J., Gutiérrez, D., Munoz, P., and Salamanca, M.: Fish scales in sediments from off Callao, central Peru, Deep-Sea Res. Pt. II, 56, 1113–1124, 2009.

Díaz-Ochoa, J. A., Pantoja, S., De Lange, G. J., Lange, C. B., Sánchez, G. E., Acuña, V. R., Muñoz, P., and Vargas, G.: Oxygenation variability in Mejillones Bay, off northern Chile, during the last two centuries, Biogeosciences, 8, 137–146, doi:10.5194/bg-8-137-2011, 2011.

Egan, K. E., Rickaby, R. E. M., Leng, M. J., Hendry, K. R., Sloane, H. J., Bostock, H. C., and Halliday, A. N.: Diatom silicon isotopes as a proxy for silicic acid utilisation: A Southern Ocean core top calibration, Geochim. Cosmochim. Ac., 96, 174–192, 2012.

Ehlert, C., Grasse, P., Mollier-Vogel, E., Böschen, T., Franz, J., De Souza, G. F., Reynolds, B. C., Stramma, L., and Frank, M.: Factors controlling the silicon isotope distribution in waters and surface sediments of the Peruvian coastal upwelling, Geochim. Cosmochim. Ac., 99, 128–145, 2012.

Ehlert, C., Grasse, P., and Frank, M.: Changes in silicate utilisation and upwelling intensity off Peru since the Last Glacial Maximum – insights from silicon and neodymium isotopes, Quaternary Sci. Rev., 72, 18–35, 2013.

Estrada, M. and Blasco, D.: Phytoplankton assemblages in coastal upwelling areas, in: Simposio Internacional Sobre Las Areas de Afloramiento Mas Importantes del Oeste Africano (Cabo Blanco y Benguela), edited by: Bas, C., Margalef, R., and Rubies, P., Barcelona, Instituto de Investigaciones Pesqueras, 379–402, 1985.

Georg, R. B., Reynolds, B. C., Frank, M., and Halliday, A. N.: New sample preparation techniques for the determination of Si isotopic compositions using MC-ICPMS, Chem. Geol., 235, 95–104, 2006.

Goldstein, S. L., O'Nions, R. K., and Hamilton, P. J.: A Sm–Nd isotopic study of atmospheric dusts and particulates from major river systems, Earth Planet. Sci. Lett., 70, 221–236, 1984.

Grasse, P., Stichel, T., Stumpf, R., Stramma, L., and Frank, M.: The distribution of neodymium isotopes and concentrations in the Eastern Equatorial Pacific: Water mass advection versus particle exchange, Earth Planet. Sci. Lett., 353/354, 198–207, 2012.

Grasse, P., Ehlert, C., and Frank, M.: The Influence of Water Mass Mixing on the Dissolved Si Isotope Composition in the Eastern Equatorial Pacific, Earth Planet. Sci. Lett., 380, 60–71, 2013.

Grousset, F. E., Biscaye, P. E., Zindler, A., Prospero, J., and Chester, R.: Neodymium isotopes as tracers in marine sediments and aerosols: North Atlantic, Earth Planet. Sci. Lett., 87, 367–378, 1988.

Grove, M. J.: The initiation of the "Little Ice Age" in regions round the North Atlantic, Clim. Change, 48, 53–82, 2001.

Gutiérrez, D., Sifeddine, A., Reyss, J.-L., Vargas, G., Velazco, F., Salvatteci, R., Ferreira-Bartrina, V., Ortlieb, L., Field, D. B., Baumgartner, T., Boussafir, M., Boucher, H., Valdés, J., Marinovic, L., Soler, P., and Tapia, P. M.: Anoxic sediments off Central Peru record interannual to multidecadal changes of climate and upwelling ecosystem during the last two centuries, Adv. Geosci., 6, 119–125, 2006, http://www.adv-geosci.net/6/119/2006/.

Gutiérrez, D., Sifeddine, A., Field, D. B., Ortlieb, L., Vargas, G., Chavez, F. P., Velazco, F., Ferreira-Bartrina, V., Tapia, P. M., Salvatteci, R., Boucher, H., Morales, M. C., Valdés, J., Reyss, J.-L., Campusano, A., Boussafir, M., Mandeng-Yogo, M., García, M., and Baumgartner, T.: Rapid reorganization in ocean biogeochemistry off Peru towards the end of the Little Ice Age, Biogeosciences, 6, 835–848, doi:10.5194/bg-6-835-2009, 2009.

Horn, M. G., Robinson, R. S., Rynearson, T. A., and Sigman, D. M.: Nitrogen isotopic relationship between diatom-bound and bulk organic matter of cultured polar diatoms, Paleoceanography, 26, PA3208, doi:10.1029/2010PA002080, 2011.

Horwitz, E. P., Chiarizia, R., and Dietz, M. L.: A Novel Strontium-Selective Extraction Chromatographic Resin, Solv. Extract. Ion Exch., 10, 313–336, 1992.

Hutchins, D. A., Hare, C. E., Weaver, R. S., Zhang, Y., Firme, G. F., DiTullio, G. R., Alm, M. B., Riseman, S. F., Maucher, J. M., Geesey, M. E., Trick, C. G., Smith, G. J., Rue, E. L., Conn, J., and Bruland, K. W.: Phytoplankton iron limitation in the Humboldt Current and Peru Upwelling, Limnol. Oceanogr., 47, 997–1011, 2002.

Kanner, L. C., Burns, S. J., Cheng, H., Edwards, R. L., and Vuille, M.: High-resolution variability of the South American summer monsoon over the last seven millennia: insights from a speleothem record from the central Peruvian Andes, Quaternary Sci. Rev., 75, 1–10, 2013.

Kessler, W. S.: The circulation of the eastern tropical Pacific: A review, Prog. Oceanogr., 69, 181–217, 2006.

Lacan, F. and Jeandel, C.: Tracing Papua New Guinea imprint on the central Equatorial Pacific Ocean using neodymium isotopic compositions and Rare Earth Element patterns, Earth Planet. Science Lett., 186, 497–512, 2001.

Lam, P., Lavik, G., Jensen, M. M., Van de Vossenberg, J., Schmid, M., Woebken, D., Gutiérrez, D., Amann, R., Jetten, M. S. M., and Kuypers, M. M. M.: Revising the nitrogen cycle in the Peruvian oxygen minimum zone, PNAS, 106, 4752–4757, 2009.

Lamb, H. H.: The early Medieval Warm Epoch and its sequel, Palaeogeography, Palaeoclimatology, Palaeoecology, 1, 13–37, 1965.

Lamy, F., Hebbeln, D., Röhl, U., and Wefer, G.: Holocene rainfall variability in southern Chile: a marine record of latitudinal shifts of the Southern Westerlies, Earth Planet. Science Lett., 185, 369–382, 2001.

Liu, K.-K. and Kaplan, I. R.: The eastern tropical Pacific as a source of 15N-enriched nitrate in seawater off southern Califomia, Limnol. Oceanogr., 34, 820–830, 1989.

Mariotti, A., Germon, J. C., Hubert, P., Kaiser, P., Letolle, R., Tardieux, A., and Tardieux, P.: Experimental determination of nitrogen kinetic isotope fractionation: some principles; illustration for the denitrification and nitrification processes, Plant Soil, 62, 413–430, 1981.

Meysman, F. J. R., Boudreau, B. P., and Middelburg, J. J.: Modeling reactive transport in sediments subject to bioturbation and compaction, Geochim. Cosmochim. Ac., 69, 3601–3617, 2005.

Minagawa, M. and Wada, E.: Stepwise enrichment of 15N along food chains: Further evidence and the relation between d15N and animal age, Geochim. Cosmochim. Ac., 48, 1135–1140, 1984.

Molina-Cruz, A.: The Relation of the Southern Trade Winds to Upwelling Processes during the Last 75,000 Years, Quaternary Res., 8, 324–338, 1977.

Mollier-Vogel, E., Ryabenko, E., Martinez, P., Wallace, D. W. R., Altabet, M. A., and Schneider, R. R.: Nitrogen isotope gradients off Peru and Ecuador related to upwelling, productivity, nutrient uptake and oxygen deficiency, Deep-Sea Res. Pt. I, 70, 14–25, 2012.

Montes, I., Schneider, W., Colas, F., Blanke, B., and Echevin, V.: Subsurface connections in the eastern tropical Pacific during La Niña 1999 – 2001 and El Niño 2002–2003, J. Geophys. Res., 116, C1202), doi:10.1029/2011JC007624, 2011.

Morley, D. W., Leng, M. J., Mackay, A. W., Sloane, H. J., Rioual, P., and Battarbee, R. W.: Cleaning of lake sediment samples for diatom oxygen isotope analysis, J. Paleilimno., 31, 391–401, 2004.

Müller, P. J. and Schneider, R. R.: An automated leaching method for the determination of opal in sediments and particulate matter, Deep-Sea Res. Pt. I, 40, 425–444, 1993.

Piepgras, D. J. and Wasserburg, G. J.: Isotopic Composition of Neodymium in Waters from the Drake Passage, Science, 217, 207–214, 1982.

Rabatel, A., Francou, B., Jomelli, V., Naveau, P., and Grancher, D.: A chronology of the Little Ice Age in the tropical Andes of Bolivia (16° S) and its implications for climate reconstruction, Quaternary Res., 70, 198–212, 2008.

Reuter, J., Stott, L., Khider, D., Sinha, A., Cheng, H., and Edwards, R. L.: A new perspective on the hydroclimate variability in northern South America during the Little Ice Age, Geophys. Res. Lett., 36, L21706, doi:10.1029/2009GL041051, 2009.

Reynolds, B. C., Aggarwal, J., André, L., Baxter, D. C., Beucher, C. P., Brzezinski, M. A., Engström, E., Georg, R. B., Land, M., Leng, M. J., Opfergelt, S., Rodushkin, I., Sloane, H. J., Van den Boorn, S. H. J. M., Vroon, P. Z., and Cardinal, D.: An interlaboratory comparison of Si isotope reference materials, J. Anal. Atom. Spectrom., 22, 561–568, doi:10.1039/b616755a, 2007.

Reynolds, B. C., Frank, M., and Halliday, A. N.: Evidence for a major change in silicon cycling in the subarctic North Pacific at 2.73 Ma, Paleoceanography, 23, PA4219, doi:10.1029/2007PA001563, 2008.

Sachs, J. P., Sachse, D., Smittenberg, R. H., Zhang, Z., Battisti, D. S., and Golubic, S.: Southward movement of the Pacific intertropical convergence zone AD 1400–1850, Nat. Geosci., 2, 519–525, 2009.

Salvatteci, R., Gutiérrez, D., Field, D. B., Sifeddine, A., Ortlieb, L., Bouloubassi, I., Boussafir, M., Boucher, H., and Cetin, F.: The response of the Peruvian Upwelling Ecosystem to centennial-scale global change during the last two millennia, Clim. Past, 10, 715–731, doi:10.5194/cp-10-715-2014, 2014a.

Salvatteci, R., Field, D. B., Sifeddine, A., Ortlieb, L., Ferreira-Bartrina, V., Baumgartner, T., Caquineau, S., Velazco, F., Reyss, J.-L., Sanchez-Cabeza, J. A., and Gutiérrez, D.: Cross-stratigraphies from a seismically active mud lens off Peru indicate horizontal extensions of laminae, missing sequences, and a need for multiple cores for high resolution records, Mar. Geol., 357, 72–89, 2014b.

Sarbas, B. and Nohl, U.: The GEOROC database – A decade of "online geochemistry", Geochim. Cosmochim. Ac., (Goldschmidt Abstracts), A1158, 2009.

Scheidegger, K. F. and Krissek, L. A.: Dispersal and deposition of eolian and fluvial sediments off Peru and northern Chile, Geol. Soc. Am. Bull., 93, 150–162, 1982.

Sifeddine, A., Gutiérrez, D., Ortlieb, L., Boucher, H., Velazco, F., Field, D. B., Vargas, G., Boussafir, M., Salvatteci, R., Ferreira-Bartrina, V., García, M., Valdés, J., Caquineau, S., Mandeng-Yogo, M., Cetin, F., Solis, J., Soler, P., and Baumgartner, T.: Laminated sediments from the central Peruvian continental slope: A 500 year record of upwelling system productivity, terrestrial runoff and redox conditions, Prog. Oceanogr., 79, 190–197, 2008.

Sutton, J. N., Varela, D. E., Brzezinski, M. A., and Beucher, C. P.: Species-dependent silicon isotope fractionation by marine diatoms, Geochim. Cosmochim. Ac., 104, 300–309, 2013.

Takeda, S.: Influence of iron availability on nutrient consumption ratio of diatoms in oceanic waters, Nature, 393, 774–777, 1998.

Tanaka, T., Togashi, S., Kamioka, H., Amakawa, H., Kagami, H., Hamamoto, T., Yuhara, M., Orihashi, Y., Yoneda, S., Shimizu, H., Kunimaru, T., Takahashi, K., Yanagi, T., Nakano, T., Fujimaki, H., Shinjo, R., Asahara, Y., Tanimizu, M., and Dragusanu, C.: JNdi-1: a neodymium isotopic reference in consistency with LaJolla neodymium, Chem. Geol., 168, 279–281, 2000.

Unkel, I., Kadereit, A., Mächtle, B., Eitel, B., Kromer, B., Wagner, G., and Wackler, L.: Dating methods and geomorphic evidence of palaeoenvironmental changes at the eastern margin of the South Peruvian coastal desert (14°30′ S) before and during the Little Ice Age, Quaternary Internat., 175, 3–28, 2007.

Valdés, J., Ortlieb, L., Gutiérrez, D., Marinovic, L., Vargas, G., and Sifeddine, A.: 250 years of sardine and anchovy scale deposition record in Mejillones Bay, northern Chile, Prog. Oceanogr., 79, 198–207, 2008.

Vargas, G., Pantoja, S., Rutllant, J. A., Lange, C. B., and Ortlieb, L.: Enhancement of coastal upwelling and interdecadal ENSO-like variability in the Peru-Chile Current since late 19th century, Geophys. Res. Lett., 34, L13607, doi:10.1029/2006GL028812, 2007.

Vuille, M., Francou, B., Wagnon, P., Juen, I., Kaser, G., Mark, B. G., and Bradley, R. S.: Climate change and tropical Andean glaciers: Past, present and future, Earth-Sci. Rev., 89, 79–96, 2008.

Wada, E. and Hattori, A.: Nitrogen isotope effects in the assimilation of inorganic nitrogenous compounds by marine diatoms, Geomicrobiol. J., 1, 85–101, 1978.

# 13

# The WAIS Divide deep ice core WD2014 chronology – Methane synchronization (68–31 ka BP) and the gas age–ice age difference

C. Buizert[1], K. M. Cuffey[2], J. P. Severinghaus[3], D. Baggenstos[3], T. J. Fudge[4], E. J. Steig[4], B. R. Markle[4], M. Winstrup[4], R. H. Rhodes[1], E. J. Brook[1], T. A. Sowers[5], G. D. Clow[6], H. Cheng[7,8], R. L. Edwards[8], M. Sigl[9], J. R. McConnell[9], and K. C. Taylor[9]

[1]College of Earth, Ocean, and Atmospheric Sciences, Oregon State University, Corvallis, OR 97331, USA
[2]Department of Geography, University of California, Berkeley, CA 94720, USA
[3]Scripps Institution of Oceanography, University of California, San Diego, La Jolla, CA 92093, USA
[4]Quaternary Research Center and Department of Earth and Space Sciences, University of Washington, Seattle, WA 98195, USA
[5]Department of Geosciences and Earth and Environmental Systems Institute, Pennsylvania State University, University Park, PA 16802, USA
[6]US Geological Survey, Boulder, CO 80309, USA
[7]Institute of Global Environmental Change, Xi'an Jiaotong University, Xi'an 710049, China
[8]Department of Geology and Geophysics, University of Minnesota, Minneapolis, MN 55455, USA
[9]Desert Research Institute, Nevada System of Higher Education, Reno, NV 89512, USA

*Correspondence to:* C. Buizert (buizertc@science.oregonstate.edu)

**Abstract.** The West Antarctic Ice Sheet Divide (WAIS Divide, WD) ice core is a newly drilled, high-accumulation deep ice core that provides Antarctic climate records of the past $\sim 68$ ka at unprecedented temporal resolution. The upper 2850 m (back to 31.2 ka BP) have been dated using annual-layer counting. Here we present a chronology for the deep part of the core (67.8–31.2 ka BP), which is based on stratigraphic matching to annual-layer-counted Greenland ice cores using globally well-mixed atmospheric methane. We calculate the WD gas age–ice age difference ($\Delta$age) using a combination of firn densification modeling, ice-flow modeling, and a data set of $\delta^{15}$N-N$_2$, a proxy for past firn column thickness. The largest $\Delta$age at WD occurs during the Last Glacial Maximum, and is $525 \pm 120$ years. Internally consistent solutions can be found only when assuming little to no influence of impurity content on densification rates, contrary to a recently proposed hypothesis. We synchronize the WD chronology to a linearly scaled version of the layer-counted Greenland Ice Core Chronology (GICC05), which brings the age of Dansgaard–Oeschger (DO) events into agreement with the U / Th absolutely dated Hulu Cave speleothem record. The small $\Delta$age at WD provides valuable opportunities to investigate the timing of atmospheric greenhouse gas variations relative to Antarctic climate, as well as the interhemispheric phasing of the "bipolar seesaw".

## 1 Introduction

Deep ice cores from the polar regions provide high-resolution climate records of past atmospheric composition, aerosol loading and polar temperatures (e.g., NGRIP community members, 2004; EPICA Community Members, 2006; Wolff et al., 2006; Ahn and Brook, 2008). Furthermore, the coring itself gives access to the ice sheet interior and bed, allowing investigation of glaciologically important processes such as ice deformation (Gundestrup et al., 1993), folding (NEEM community members, 2013), crystal fabric evolution

(Gow et al., 1997), and geothermal heat flow (Dahl-Jensen et al., 1998). Having a reliable ice core chronology (i.e., an age–depth relationship) is paramount for the interpretation of the climate records and comparison to marine and terrestrial paleoclimate archives.

The West Antarctic Ice Sheet Divide (WAIS Divide, WD) ice core (79.48° S, 112.11° W; 1766 m above sea level; −30 °C present-day mean annual temperature) was drilled and recovered to 3404 m depth (WAIS Divide Project Members, 2013). Drilling was stopped 50 m above the estimated bedrock depth to prevent contamination of the basal hydrology. Due to high accumulation rates of 22 cm ice a$^{-1}$ at present and $\sim$ 10 cm ice a$^{-1}$ during the Last Glacial Maximum (LGM), the WD core delivers climate records of unprecedented temporal resolution (Steig et al., 2013; Sigl et al., 2013) as well as gas records that are only minimally affected by diffusive smoothing in the firn column (Mischler et al., 2009; Mitchell et al., 2011, 2013; Marcott et al., 2014). The combination of high accumulation rates and basal melting at the WD site results in ice near the bed that is relatively young ($\sim$ 68 ka) compared to cores drilled in central East Antarctica.

In WD, annual layers can be identified reliably for the upper 2850 m of the core, reaching back to 31.2 ka BP (thousands of years before present, with present defined as 1950 CE). Below 2850 m depth an alternative dating strategy is needed. Several methods have been employed previously at other deep ice core sites. First, orbital tuning via $\delta O_2 / N_2$ has been applied successfully to several Antarctic cores (Bender, 2002; Kawamura et al., 2007). However, an age span of only $\sim$ 3 precessional cycles in WD, in combination with the low signal-to-noise ratio of $\delta O_2 / N_2$ data, makes this technique unsuitable for WD. The uncertainty in the orbital tuning is about one-fourth of a precessional cycle ($\sim$ 5 ka), making it a relatively low-resolution dating tool. Second, in Greenland, ice-flow modeling has been used to extend layer-counted chronologies (e.g., Johnsen et al., 2001; Wolff et al., 2010). This method requires assumptions about past accumulation rates, ice flow, and ice sheet elevation. Particularly for the oldest WD ice, the resulting uncertainty would be substantial. Third, several radiometric techniques have been proposed to date ancient ice. Radiocarbon ($^{14}$C) dating of atmospheric $CO_2$ trapped in the ice is unsuitable as it suffers from in situ cosmogenic production in the firn (Lal et al., 1990), and the oldest WD ice dates beyond the reach of $^{14}$C dating. Other absolute (radiometric) dating techniques, such as recoil $^{234}$U dating (Aciego et al., 2011), $^{81}$Kr dating (Buizert et al., 2014a), or atmospheric $^{40}$Ar buildup (Bender et al., 2008), currently suffer from uncertainties that are too large ($\geq$ 20 ka) to make them applicable at WD.

Instead, at WD we use stratigraphic matching to well-dated Greenland ice cores using globally well-mixed atmospheric methane ($CH_4$) mixing ratios (Blunier et al., 1998; Blunier and Brook, 2001; Blunier et al., 2007; Petrenko et al., 2006; EPICA Community Members, 2006; Capron

et al., 2010). This method is particularly suited to WD because of the small gas age–ice age difference ($\Delta$age, Sect. 3) and the high-resolution, high-precision $CH_4$ record available (Sect. 2.1). The method has three main sources of uncertainty: (i) the age uncertainty in the records one synchronizes to, (ii) $\Delta$age of the ice core being dated, and (iii) the interpolation scheme used in between the $CH_4$ tie points. We present several improvements over previous work that reduce and quantify these uncertainties: (i) we combine the layer-counted Greenland Ice Core Chronology (GICC05) and a recently refined version of the U / Th-dated Hulu speleothem record (Edwards et al., 2015; Reimer et al., 2013; Southon et al., 2012) to obtain a more accurate estimate of the (absolute) ages of abrupt Dansgaard–Oeschger (DO) events (Sect. 4.4); (ii) we combine firn densification modeling, ice-flow modeling, a new WD $\delta^{15}$N-$N_2$ data set that spans the entire core, and a Monte Carlo sensitivity study to obtain a reliable $\Delta$age estimate (Sect. 3); and (iii) we compare four different interpolation schemes to obtain an objective estimate of the interpolation uncertainty (Sect. 4.5).

This work is the first part in a series of two papers describing the WD2014 chronology for the WD core in detail. The second part describes the development of the annual layer count from both multi-parameter chemistry and electrical conductivity measurements. The WD2014 chronology is currently the recommended gas and ice timescale for the WD deep core, and as such it supersedes the previously published WDC06A-7 chronology (WAIS Divide Project Members, 2013).

## 2 Methods

### 2.1 Data description

*Measurements of water stable isotopes*. Water isotopic composition ($\delta^{18}$O and $\delta D = \delta^2$H) was measured at IsoLab, University of Washington. Procedures for the deep section of the core are identical to those used for the upper part of the core reported in WAIS Divide Project Members (2013) and Steig et al. (2013). Measurements were made at 0.25 to 0.5 m depth resolution using laser spectroscopy (Picarro L2120-*i* water isotope analyzer), and normalized to VSMOW-SLAP (Vienna Standard Mean Ocean Water – Standard Light Antarctic Precipitation). The precision of the measurements is better than 0.1 and 0.8 ‰ for $\delta^{18}$O and $\delta D$, respectively.

*Measurements of $CH_4$*. Two $CH_4$ data sets were used for WD. The first is from discrete ice samples, and was measured jointly at Pennsylvania State University (0–68 ka, 0.5–2 m resolution) and Oregon State University (11.4–24.8 ka, 1–2 m resolution). Air was extracted from $\sim$ 50 g ice samples using a melt–refreeze technique, and analyzed on a standard gas chromatograph equipped with a flame-ionization detector. Corrections for solubility, blank size and gravitational enrichment are applied (Mitchell et al., 2011; WAIS Divide

Project Members, 2013). The second data set is a continuous CH$_4$ record measured by coupling a laser spectrometer to a continuous flow analysis setup (Stowasser et al., 2012; Rhodes et al., 2013; Chappellaz et al., 2013), and was measured jointly by Oregon State University and the Desert Research Institute (Rhodes et al., 2015). The continuous data set is used to identify the abrupt DO transitions, as it provides better temporal resolution and analytical precision. Both records are reported on the NOAA04 scale (Dlugokencky et al., 2005). Analytical precision in the CH$_4$ data ($2\sigma$ pooled standard deviation) is around 3.2 and 14 ppb for the discrete data from Oregon State University and Pennsylvania State University, respectively, and 3 to 8 ppb for the continuous CH$_4$ data, depending on the analyzer used (Rhodes et al., 2015); the 14 ppb stated for the PSU discrete data may be an overestimation, as depth-adjacent (rather than true replicate) samples were used in the analysis.

*Measurements of $\delta^{15}N$.* Atmospheric N$_2$ isotopic composition ($\delta^{15}N$) was measured at Scripps Institution of Oceanography, University of California. Air was extracted from $\sim$ 12 gram ice samples using a melt–refreeze technique, and collected in stainless steel tubes at liquid-He temperature. $\delta^{15}N$ was analyzed using conventional dual-inlet isotope ratio mass spectrometry (IRMS) on a Thermo Finnigan Delta V mass spectrometer. Results are normalized to La Jolla (California, USA) air, and routine analytical corrections are applied (Sowers et al., 1989; Petrenko et al., 2006; Severinghaus et al., 2009). Duplicates were not run for most $\delta^{15}N$ data in this study, but the pooled standard deviations of Holocene WD $\delta^{15}N$ data sets with duplicate analyses are 0.003‰ (Orsi, 2013). We conservatively adopt an analytical uncertainty of 0.005‰ for this data set to allow for other sources of error.

*Measurements of [Ca].* Ca concentrations in the ice were measured at the Ultra Trace Chemistry Laboratory at the Desert Research Institute via continuous flow analysis. Longitudinal samples of ice (approximately 100 cm $\times$ 3.3 cm $\times$ 3.3 cm) were melted continuously on a melter head that divides the meltwater into three parallel streams. Elemental measurements were made on meltwater from the innermost part of the core with ultra-pure nitric acid added to the melt stream immediately after the melter head; potentially contaminated water from the outer part of the ice is discarded. Elemental analysis of the innermost meltwater stream is performed in parallel on two inductively coupled plasma mass spectrometers (ICPMS), each measuring a different set of elements; some elements were analyzed on both. The dual ICPMS setup allows for measurement of a broad range of 30 elements and data quality control (McConnell et al., 2002, 2007). Precision of the Ca measurements in WD glacial ice is estimated to be $\pm3$%, with a lower detection limit of 0.15 ng g$^{-1}$. Continuous Ca and CH$_4$ measurements are done on the same ice, and are exactly co-registered in depth.

## 2.2 Firn densification model description

Air exchange with the overlying atmosphere keeps the interstitial air in the porous firn layer younger than the surrounding ice matrix, resulting in an age difference between polar ice and the gas bubbles it contains, commonly referred to as $\Delta$age (Schwander and Stauffer, 1984). Here we use a coupled firn–densification–heat–diffusion model to calculate $\Delta$age back in time (Barnola et al., 1991; Goujon et al., 2003; Schwander et al., 1997; Rasmussen et al., 2013), constrained by measurements of $\delta^{15}N$ of N$_2$, a proxy for past firn column thickness (Sowers et al., 1992). The model is based on a dynamical description of the Herron–Langway model formulated in terms of overburden load (Herron and Langway, 1980), which is solved in a Lagrangian reference frame. This model has been applied previously to the Greenland NEEM, NGRIP, and GISP2 cores (Rasmussen et al., 2013; Seierstad et al., 2015; Buizert et al., 2014b), where it gives a good agreement to the Goujon densification model (Rasmussen et al., 2013; Goujon et al., 2003). The model allows for the inclusion of softening of firn in response to impurity loading (Horhold et al., 2012), following the mathematical description of Freitag et al. (2013a). The equations governing the model densification rates are given in Appendix A.

The model uses a 2-year time step and 0.5 m depth resolution down to 1000 m, the lower model boundary. A thick model domain is needed because of the long thermal memory of the ice sheet. At WD, downward advection of cold surface ice is strong due to the relatively high accumulation rates, and the geothermal gradient does not penetrate the firn column (Cuffey and Paterson, 2010). We further use a lock-in density that equals the mean close-off density (Martinerie et al., 1994) minus 17.5 kg m$^{-3}$ (as in Blunier and Schwander, 2000) and an empirical parameterization of lock-in gas age based on firn air measurements from 10 sites (Buizert et al., 2012, 2013).

We furthermore use the steady-state version of the Herron–Langway model (Herron and Langway, 1980) in performing sensitivity studies (Sect. 3.2) and the dynamical Arnaud model (Arnaud et al., 2000; Goujon et al., 2003) to validate our $\Delta$age solution.

## 2.3 Temperature reconstruction and ice-flow model

Our temperature reconstruction (Fig. 1a) is based on water $\delta$D, a proxy for local vapor condensation temperature, calibrated using a measured borehole temperature profile (following Cuffey et al., 1995; Cuffey and Clow, 1997) and, for the last 31.2 ka, adjusted iteratively to satisfy constraints on firn thickness provided by $\delta^{15}N$ and by the observed layer thickness $\lambda(z)$. Using $\delta^{18}O$ rather than $\delta$D in the temperature reconstruction leads to differences that are negligibly small. This borehole temperature calibration approach is possible at WD because the large ice thickness and relatively high accumulation rates help to preserve a memory of past

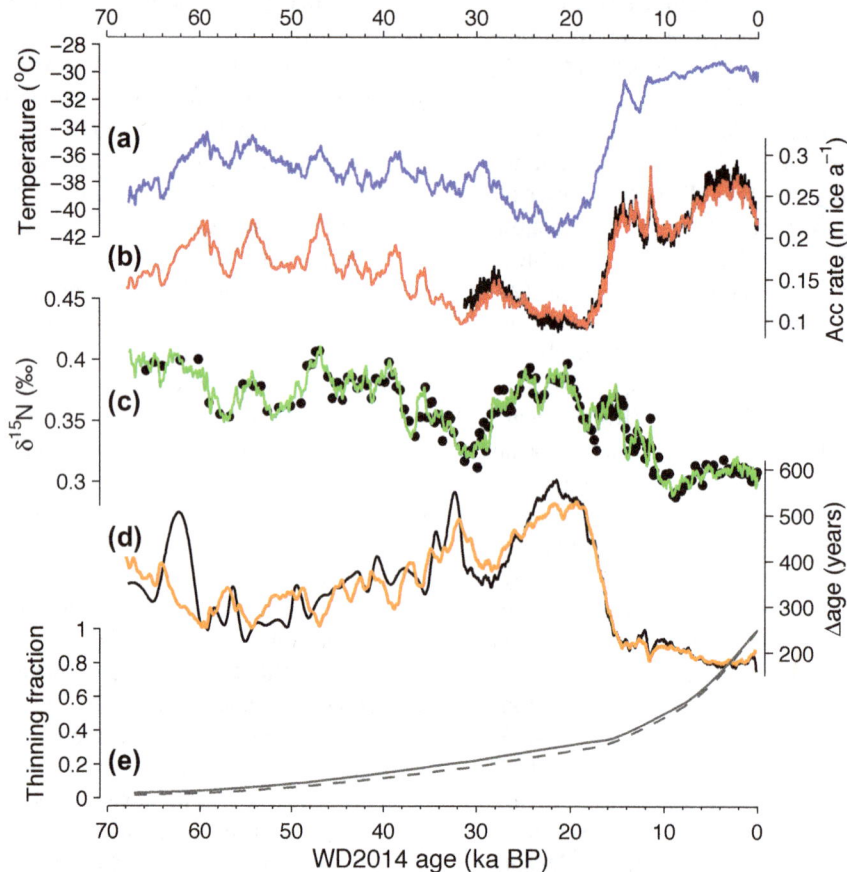

**Figure 1.** Modeling $\Delta$age for WAIS Divide. **(a)** Past temperatures reconstructed from water $\delta$D, calibrated to the borehole temperature profile. **(b)** Past accumulation rates as reconstructed by the firn densification inverse model (red), and from the annual-layer count (black). **(c)** $\delta^{15}$N data (black dots) with densification model output (green). **(d)** $\Delta$age calculated using the densification model (orange) and using the Parrenin $\Delta$depth method (black) with constant 4 m thick convective zone and no correction for thermal $\delta^{15}$N fractionation. **(e)** Modeled thinning function from ice-flow model (solid), and a simple Nye strain model for comparison (dashed); the Nye thinning function, which has a uniform strain rate as a function of depth, is given as $f_\lambda(z) = (H - z)/H$ with $H$ the ice sheet thickness (Cuffey and Paterson, 2010, p. 616).

temperatures in the ice sheet. A coupled 1-D ice-flow–heat-diffusion model converts surface $T(t)$ into a depth profile for comparison to measured borehole temperatures. The 1-D ice-flow model calculates the vertical ice motion, taking into account the surface snow accumulation, the variation in density with depth, and a prescribed history of ice thickness. Vertical motion is calculated by integrating a depth profile of strain rate and adding a rate of basal melt. As in the model of Dansgaard and Johnsen (1969), the strain rate maintains a uniform value between the surface and a depth equal to 80 % of the ice thickness, and then varies linearly to some value at the base of the ice. This basal value is defined by the "basal stretching parameter" $f_b$, the ratio of strain rate at the base to strain rate in the upper 80 % of the ice column. The basal ice is melting, so part of the ice motion likely occurs as sliding. The along-flow gradient in such sliding is unknown and thus so too is the parameter $f_b$. We overcome this problem by making both the current ice thickness and the basal melt

rate free parameters when optimizing models with respect to measured borehole temperatures. Because the basal melt rate and $f_b$ affect the vertical velocities in similar fashion, the optimization constrains a combination of melt rate and $f_b$ that is tightly constrained by the measured temperatures. Thus we find that varying $f_b$ through a large range, from 0.1 to 1.5, changes the reconstructed LGM temperature by less than 0.2 °C. Effects of the prescribed ice-thickness history are likewise minor; assuming a 150 m thickness increase from the LGM to 15 ka changes the reconstructed LGM temperature by less than 0.2 °C compared to a constant thickness. Note that the 1-D flow model used here is simpler than the one used by Cuffey and Clow (1997) in that it does not attempt to calculate changes in the shape of the strain rate profile; the unknown basal sliding motion at the WD site negates the usefulness of such an exercise.

One output of the 1-D flow model is the strain history of ice layers as a function of depth and time. The cumulative

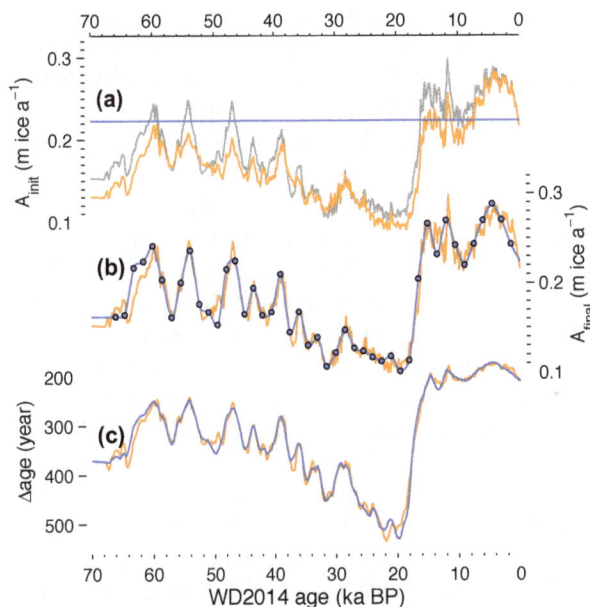

**Figure 2.** Reconstructing $A(t)$ and $\Delta$age from $\delta^{15}$N: the choice of accumulation template. **(a)** $A_{init}(t)$ based on $\lambda(z)$ from the annual-layer count (0–31.2 ka BP) and Clausius–Clapeyron scaling of water stable isotope data (34.2–68 ka) in orange; $A_{init} = 0.22$ m a$^{-1}$ in blue; and for comparison the final $A(t)$ solution (Fig. 1b) in grey. For the orange curve we have used the Nye thinning function (Fig. 1e); the final ice-flow model optimizes the agreement between $A(t)$ obtained from $\lambda(z)$ and from the inverse firn modeling approach. The WD2014 chronology uses the orange $A_{init}(t)$ scenario. **(b)** $A(t)$ found in the inverse firn modeling approach using both $A_{init}(t)$ scenarios; color coding as in panel **(a)**. The function $\xi(t)$ is found as follows. We use control points at 1500-year intervals (blue dots); the algorithm has the freedom to change the value of $\xi(t)$ at each of these points. In between the control points, $\xi(t)$ is found via linear interpolation. **(c)** Modeled $\Delta$age using both $A_{init}(t)$ scenarios; color coding as in panel **(a)**.

strain is represented by the thinning function $f_\lambda(z)$ (Cuffey and Paterson, 2010), the ratio of annual-layer thickness at depth in the ice sheet to its original ice-equivalent thickness at the surface when deposited. The modeled thinning function is shown in Fig. 1e (solid line). In the deep part of the ice sheet, $f_\lambda(z)$ becomes increasingly uncertain as the unknown basal melt rate and $f_b$ become the dominant controls. Here we optimize the model by comparing accumulation rates derived from $f_\lambda(z)$ with those implied by a firn densification model and the measured $\delta^{15}$N of $N_2$ (Sect. 3.1). While this has little effect on the temperature history reconstruction, it provides an important constraint on calculated basal melt rate, an interesting quantity for ice dynamics studies. Our analysis of basal melt rates and further details of the temperature optimization process and 1-D flow modeling will be presented elsewhere.

# 3   The gas age–ice age difference ($\Delta$age)

## 3.1   The WD2014 $\Delta$age reconstruction

The firn densification forward model uses past surface temperature $T(t)$ and accumulation $A(t)$ as model forcings, and provides $\Delta$age$(t)$ and $\delta^{15}$N$(t)$ as model output.

For the past 31.2 ka, WD has an annual-layer-counted chronology; for this period the annual-layer thickness $\lambda(z)$ provides a constraint on past accumulation rates via $\lambda(z) = A(z) \times f_\lambda(z)$. WD accumulation reconstructed from $\lambda(z)$ is plotted in black in Fig. 1b.

Prior to 31.2 ka we have no such constraint on $A(t)$, and an alternative approach is needed. We use the densification model as an inverse model, where we ask the model to find the $A(t)$ history that minimizes the root-mean-square (rms) deviation between measured and modeled $\delta^{15}$N, given the $T(t)$ forcing. The $\delta^{15}$N data and model fit are shown in Fig. 1c, the $A(t)$ history that optimizes the $\delta^{15}$N fit is shown in Fig. 1b (red), and the modeled $\Delta$age is shown in Fig. 1c (orange). The optimal $A(t)$ history is estimated in two steps. First, we make an initial estimate $A_{init}(t)$ for the past accumulation history. Second, we adjust the $A(t)$ forcing by applying a smooth perturbation $\xi(t)$ such that $A(t) = [1 + \xi(t)] \times A_{init}(t)$; an automated algorithm is used to find the curve $\xi(t)$ that optimizes the model fit to the $\delta^{15}$N data. For the last 31.2 ka we obtain a good agreement between $A$ obtained from $\lambda(z)$ and the modeled $f_\lambda(z)$ (Fig. 1b, black) and $A$ obtained from the inverse method (red). The solution we present here is therefore fully internally consistent, i.e., the $A$ and $T$ histories used in the firn densification modeling are the same as those used in the ice-flow modeling, and they provide a good fit to both the $\delta^{15}$N data and borehole temperature data. WD does not suffer from the $\delta^{15}$N model–data mismatch that is commonly observed for East Antarctic cores during the glacial period (Landais et al., 2006; Capron et al., 2013).

We base our $A_{init}$ values on $\lambda(z)$ for the past 31.2 ka; prior to that we use the common assumption that $A$ follows $\delta^{18}$O (i.e., Clausius–Clapeyron scaling); the fit to the $\delta^{15}$N data is optimized for $A = 24.2 \times \exp[0.1263 \times \delta^{18}$O]. To test the validity of the Clausius–Clapeyron assumption, we additionally run the scenario $A_{init}(t) = 0.22$ m a$^{-1}$ (i.e., constant accumulation at present-day level). The $A(t)$ and $\Delta$age reconstructed under both $A_{init}$ scenarios are similar at multi-millennial timescales (Fig. 2). In the layer-counted interval ($< 31.2$ ka BP), $A$ obtained from $\lambda(z)$ and $\delta^{18}$O is significantly coherent at all timescales longer than 3000 years, but not at higher frequencies. This is equivalent to the variability resolved in the $A_{init}(t) = 0.22$ m a$^{-1}$ scenario above. We conclude that the WD $\delta^{15}$N data support the idea that $A$ follows $\delta^{18}$O on multi-millennial timescales. However, there may not be a strong relationship at timescales less than a few thousand years, as is clear from the abrupt $A$ increase around 12 ka seen in $\lambda(z)$ that is not reflected in $\delta^{18}$O (Fig. 1a and b). For consistency between the upper and deeper part of the core we use the $\Delta$age values obtained with the inverse densification model for the entire core.

Recently, another $\delta^{15}$N-based approach has been suggested that uses $\Delta$depth, rather than $\Delta$age, in reconstructing gas chronologies (Parrenin et al., 2012). This method removes the dependence on $T(t)$ and replaces this with a dependence on the thinning function $f_\lambda(z)$. Note that this method is very successful in the upper part of an ice core, where $f_\lambda(z)$ is well constrained, but not very reliable near the base, where $f_\lambda(z)$ is highly uncertain. Therefore, the firn densification modeling approach should be considered to be more reliable at WD during marine isotope stages (MIS) 2 through 4. Results from the $\Delta$depth method are plotted in black in Fig. 1c, and generally show good agreement with the firn modeling approach. A notable exception is the 60–65 ka interval, where the $\Delta$depth method overestimates the $\Delta$age due to the fact that we have to compress $\lambda(z)$ strongly in order to fit age constraints derived from DO 18 (Sect. 4.5).

Last, we want to point out that the $\delta^{15}$N data support an early warming at WD, as reported recently (WAIS Divide Project Members, 2013). WD $\delta^{15}$N starts to decrease around 20.5 ka BP, suggesting a thinning of the firn column. The $\lambda(z)$ (as derived from the layer count) shows that accumulation did not change until 18 ka BP, at which point it started to increase (which would act to increase the firn thickness). The most plausible explanation for the $\delta^{15}$N decrease around 20.5 ka BP is therefore an early onset of West Antarctic deglacial warming, in agreement with increasing $\delta^{18}$O around that time. The warming enhances the densification rate of polar firn, thereby decreasing its thickness (e.g., Herron and Langway, 1980).

## 3.2 $\Delta$age sensitivity study

Besides $A$ and $T$ there are several model parameters that have the potential to influence the model outcome; these are the convective zone (CZ) thickness (Sowers et al., 1992; Kawamura et al., 2006), surface density ($\rho_0$), and sensitivity to ice impurity content. In this section we evaluate the sensitivity of the model output to all of these parameters. We performed 1000 model runs in which the model parameters were randomly perturbed. The spread in $\Delta$age model results is used to calculate the WD2014 age uncertainty.

**Convective zone thickness**. In the WD2014 model run (Sect. 3.1) we use a constant 3.5 m CZ, corresponding to the present-day situation (Battle et al., 2011). In the sensitivity study we vary the CZ by one of two methods: (1) we let the CZ be constant in time; its thickness is set by drawing from a Gaussian distribution with 3.5 m mean and 3.5 m $2\sigma$ width (i.e., 95 % probability of drawing a value in the 0–7 m range). (2) We let the CZ be a function of accumulation rate (Dreyfus et al., 2010), CZ $= 3.5 + k \times (A–0.22)$; we draw $k$ from a Gaussian distribution with mean of $-10$ and a $2\sigma$ width of 40 (at an LGM $A$ of 10 cm a$^{-1}$ this gives a CZ of 0–10 m

thickness). In both methods, whenever CZ values are selected that are smaller than 0 m, the CZ thickness is set to 0 m instead. For each of the 1000 model runs in the sensitivity study we randomly selected either of the two methods.

**Surface density**. In the WD2014 model run we use past surface densities ($\rho_0$) as given by the parameterization of Kaspers et al. (2004). In the sensitivity study we add a constant offset to the Kaspers values, the magnitude of which is drawn from a Gaussian distribution of zero mean and a $2\sigma$ width of 60 kg m$^{-3}$. This range corresponds to the full range of observed $\rho_0$ variability in Kaspers et al. (2004).

**Past temperatures**. Model temperature forcing is constrained by $\delta$D and measured borehole temperatures. There is, however, a range to the solutions allowed by the borehole temperature and ice-flow model; here we use the upper and lower extremes of this range, determined by Monte Carlo analysis using uncertainties of input variables. The scenarios were chosen to provide the maximum $T$ range for the glacial period rather than for the Holocene, because we are interested in the uncertainty in the methane synchronization (68–31.2 ka BP). In the sensitivity study we use $T(t) = T_{optimal}(t) + \kappa \times \Delta T(t)$, where $T_{optimal}$ is the forcing used in the WD2014 model run (Fig. 1a), $\Delta T(t)$ is half the difference between the maximum-$T$ and minimum-$T$ scenarios, and $\kappa$ is drawn from a Gaussian distribution of zero mean and unit $2\sigma$ width (giving 95 % probability that $T(t)$ is within the extreme range identified from the borehole, Fig. 3a).

**$\delta^{15}$N uncertainty**. We conservatively adopt an analytical uncertainty of 0.005 ‰ for this data set; in addition, the interpretation of $\delta^{15}$N in terms of firn thickness is subject to further uncertainty due to irregular firn layering and the stochastic nature of bubble trapping, as was observed for other atmospheric gases such as CH$_4$ (Etheridge et al., 1992; Rhodes et al., 2013). For each run of the sensitivity study, we therefore perturb each of the individual $\delta^{15}$N data points by adding an offset that is drawn from a Gaussian distribution of zero mean and a $2\sigma$ width of 0.015 ‰.

**Impurity-enhanced densification**. Following recent work we include the possibility that increased glacial impurity loading could have enhanced densification rates (Horhold et al., 2012; Freitag et al., 2013a). We use the mathematical formulation of Freitag et al. (2013a), in which the activation energy of the sintering process is a function of the Ca concentration in the firn. The value of $\beta$, the sensitivity to Ca, is drawn from a Gaussian distribution with 0.0015 mean and a $2\sigma$ width of 0.0015. The topic of impurity-enhanced densification is discussed in detail in Sect. 3.3.

The $A$ and $\Delta$age scenarios found in the sensitivity study are shown in Fig. 3b and c, respectively. The shaded areas in Fig. 3b and c give the total range of solutions, as well as the $\pm 2\sigma$ and $\pm 1\sigma$ confidence intervals. Note that the total range of solutions will depend on the number of model runs (here 1000) but that the position of the $\pm 2\sigma$ and $\pm 1\sigma$ envelopes will not. To investigate the distribution of values, we include

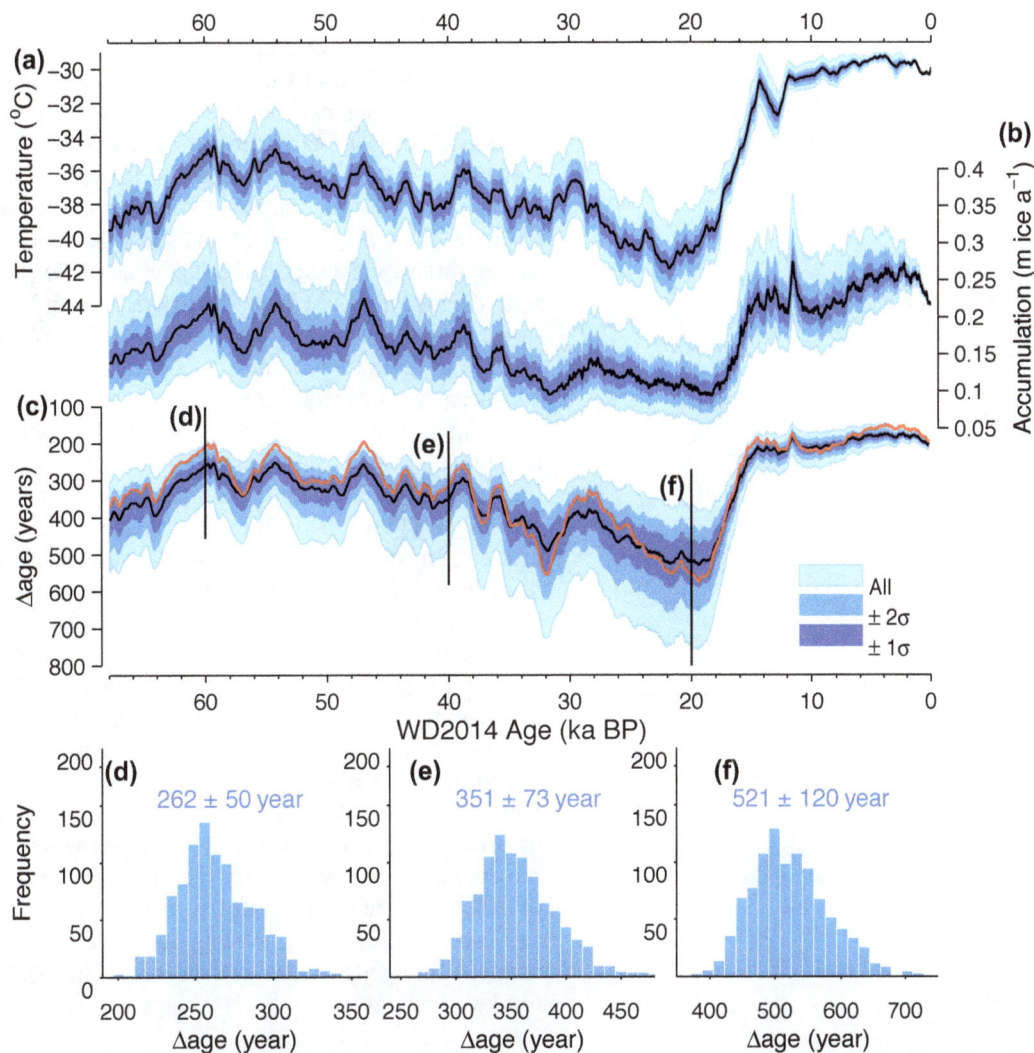

**Figure 3.** $\Delta$age sensitivity study. Shades of blue give the confidence intervals as marked; the black curves represent the values used in the WD2014 chronology; the red curve gives an alternative $\Delta$age solution using the Arnaud densification model. (**a**) Temperature forcing of the densification model. (**b**) Reconstructed accumulation rates. (**c**) Reconstructed $\Delta$age; note the reversed scale. Histograms of $\Delta$age distribution are shown for (**d**) 60 kaBP, (**e**) 40 kaBP, and (**f**) 20 kaBP. Distribution mean and $2\sigma$ uncertainty bound is stated in each panel.

histograms of $\Delta$age at 20 kyr intervals (Fig. 3d–f). Based on the sensitivity study, we estimate the WD $\Delta$age to be $521 \pm 120$ years ($2\sigma$ uncertainty) at the LGM ($\sim 20$ kaBP). The $\Delta$age value of $351 \pm 73$ years at 40 kaBP gives a representative $\Delta$age for MIS 3; $\Delta$age at 60 kaBP is $262 \pm 50$ years.

Additionally, we have repeated our $\Delta$age reconstruction using the firn densification physics described by Arnaud et al. (2000) rather than the Herron–Langway description used so far; the Arnaud model provides the physical basis for the commonly used firn densification model of Goujon et al. (2003). More details on the implementation of the Arnaud model are given in Appendix A. $\Delta$age found using the Arnaud model is plotted in red in Fig. 3c. Averaged over the entire core, $\Delta$age found with the Arnaud model is 19 years

(about 7 %) smaller than $\Delta$age from the Herron–Langway model. The root-mean-square (rms) difference between both solutions is 35 years, corresponding to 0.63 times the $2\sigma$ uncertainty found in the sensitivity study. Both solutions are thus found to be in good agreement. The Herron–Langway approach is preferred because the internally consistent solution of temperature, accumulation, and ice flow associated with it is in better agreement with borehole temperature data than the solutions associated with the Arnaud model. Furthermore, the Herron–Langway model is more successful in simulating the magnitude of the $\delta^{15}$N response to the accumulation anomaly at 12 ka (not shown), suggesting it has a more realistic sensitivity to accumulation variability.

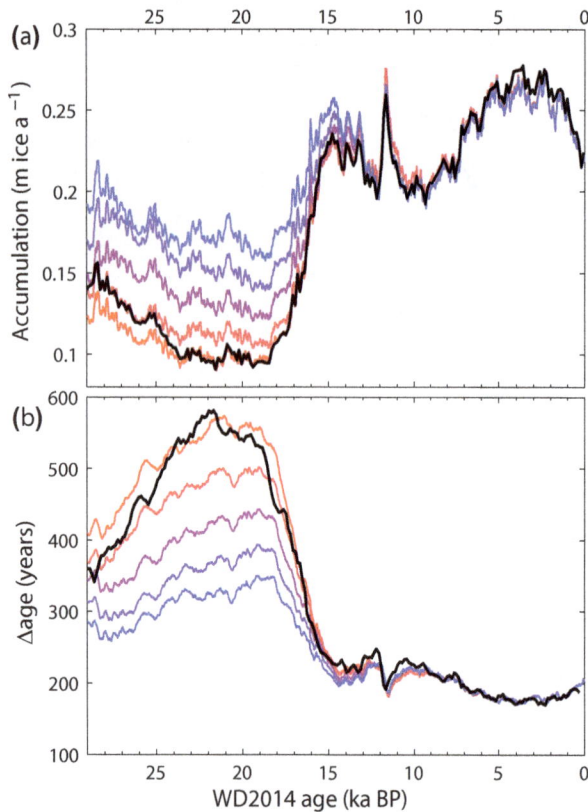

**Figure 4.** Impurity enhancement of densification rates at WD. Densification modeling results for **(a)** accumulation rates and **(b)** $\Delta$age. We use Ca sensitivities $\beta = 0$ (red) through $\beta = 1 \times 10^{-2}$ (blue), in steps of $2.5 \times 10^{-3}$ (shades of deep purple). Black curves give $A$ and $\Delta$age from ice-flow modeling and $\lambda(z)$.

## 3.3 Impurity softening of firn?

Recent work suggests a link between densification rates and impurity content (for which $[Ca^{2+}]$ is used as a proxy) in polar firn (Horhold et al., 2012; Freitag et al., 2013a). Here we measured total [Ca] by ICP-MS, but at WD nearly all Ca is in the form of $Ca^{2+}$. The influence of the impurity sensitivity $\beta$ (see Eq. A6 in the Appendix) on $\Delta$age at WD is shown in Fig. 4. The sensitivity recommended by Freitag et al. (2013a) from investigating present-day firn packs is $\beta = 1 \times 10^{-2}$. We reconstructed $A$ and $\Delta$age with the firn densification inverse model using five values of $\beta$ ranging from $\beta = 0$ (red) to $\beta = 1 \times 10^{-2}$ (blue) in steps of $2.5 \times 10^{-3}$. Average [Ca] is around $0.8$ ng g$^{-1}$ in the early Holocene and around $9$ ng g$^{-1}$ in the LGM; a change to about an order of magnitude. Following Freitag et al. (2013a) we use the total [Ca] rather than non-sea-salt Ca. If densification rates are sensitive to impurity loading (large $\beta$, blue curves), this results in increased firn compaction during the LGM. The densification model, which is trying to match the $\delta^{15}$N data, will compensate by increasing the $A$ forcing, which in turn results in a decreas-

ing $\Delta$age. Hence the model simulations with large $\beta$ (blue) give a higher $A$ and smaller $\Delta$age.

For the past 31.2 ka we have an independent $A$ estimate from $\lambda(z)$ that we can compare to the solutions from the firn model (Fig. 4, black curve). We also plotted $\Delta$age reconstructed via the $\Delta$depth method of Parrenin et al. (2012). Remarkably, we find consistent solutions only when using a Ca sensitivity $\beta \leq 2.5 \times 10^{-3}$, i.e., less than one-quarter of the sensitivity suggested by Freitag et al. (2013a). The best fit to the independent LGM (25–20 ka BP) $A$ and $\Delta$age estimates is obtained for $\beta = 0$. We conclude that WD does not provide any evidence for impurity (or, more specifically, Ca) enhancement of densification rates.

An important caveat is that our model uses 10-year-average [Ca] values, and therefore cannot resolve effects of interannual layering within the firn. Explicitly modeling the layering would require centimeter-scale resolution in the dynamical firn model, which is prohibitive from a computational point of view. Furthermore, [Ca] data at the required sub-annual resolution are difficult, if not impossible, to measure for the deepest part of the core, where $\lambda(z)$ is below $1$ cm a$^{-1}$. Increased firn layering and enhanced bulk densification affect the firn thickness in a similar manner; both lead to a shallower lock-in depth, and thereby a reduced $\delta^{15}$N. Therefore, in order to reconcile our WD results with the impurity hypothesis of Horhold et al. (2012), one would need to invoke a strong reduction in LGM firn layering relative to the present day to compensate for the impurity-driven increase in bulk densification rates. Recent work on the EDML core suggests that firn density layering may have been more pronounced during glacial times (Bendel et al., 2013); including firn layering is therefore likely to only exacerbate the problem.

Work on present-day firn has provided support for firn softening by impurity loading (Horhold et al., 2012; Freitag et al., 2013a, b). More work is needed to understand how densification rates are linked to impurity content in a mechanistic, rather than purely empirical, way. Perhaps such a microscopic description could provide an explanation why firn densification rates at WD, to first order, do not appear to be affected by order-of-magnitude variations in [Ca] loading. One possible explanation could be that densification rates are controlled by some parameter that co-varies with Ca in modern day firn yet does not change appreciably over glacial cycles (Fujita et al., 2014).

## 4 Constructing the WAIS Divide WD2014 chronology

### 4.1 Annual layer count (0–31.2 ka)

A first layer-counted chronology for the upper 2800 m of the WD core based on electrical conductivity measurements (ECM), named WDC06A-7, was presented by WAIS Divide Project Members (2013). The WAIS chronology presented

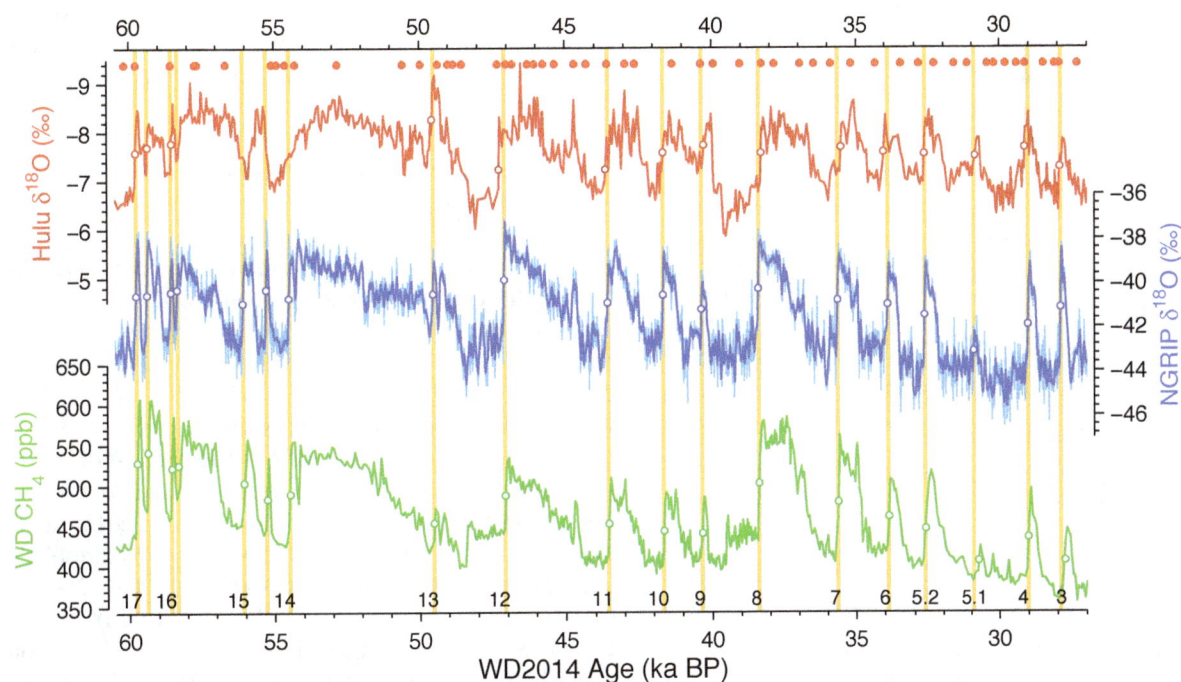

**Figure 5.** Records of abrupt DO climate variability, **(a)** revised Hulu Cave speleothem $\delta^{18}O$ record on Hulu chronology with U/Th ages above the time series (red dots), **(b)** NGRIP ice core $\delta^{18}O$ on $1.0063 \times$ GICC05 chronology, and **(c)** WD $CH_4$ on WD2014 (discrete data). DO numbering is given in the bottom of the figure following Rasmussen et al. (2014). White dots denote the midpoints of the stadial–interstadial transitions; the orange vertical lines show the timing of the NGRIP tie points (on $1.0063 \times$ GICC05). For DO 3, 4, and 5.1 the WD2014 chronology is based on annual-layer counting, and minor timing differences exist between WD and NGRIP.

in this work, WD2014, uses an updated layer count for the upper 2850 m, based on new data and analyses that have become available since publication of WDC06A-7. These updates are as follows:

1. a reassessment of the dating in the upper 577 m (2.4 ka) using high-resolution multi-parameter chemistry data in combination with automated layer detection algorithms (Winstrup et al., 2012);

2. a reassessment of the dating between 577 and 2300 m (2.4–15.3 ka) using high-resolution multi-parameter chemistry data in combination with ECM;

3. a reassessment of the dating between 2300 and 2800 m (15.3–29.5 ka) using ECM and dust particle measurements, with the ECM having increasing importance with depth;

4. an extension of the annual-layer dating between 2800 and 2850 m (29.5–31.2 ka) using ECM.

Details on the updated WD layer count and the layer counting methodology are presented in part 2 of the WD2014 papers.

### 4.2 Methane synchronization (31.2–68 ka)

For the deep part of the core where an annual-layer count is not available, we date WD by synchronization to well-

dated Northern Hemisphere (NH) climate records of abrupt DO variability using the WD record of globally well-mixed $CH_4$ (Fig. 5). This process consists of several steps:

1. Determine the midpoint of the abrupt DO transitions in WD $CH_4$, NGRIP $\delta^{18}O$, and Hulu speleothem $\delta^{18}O$.

2. Assign a gas age to the WD $CH_4$ tie points (i.e., the DO transitions).

3. Apply the WD $\Delta$age (Sect. 3) to find the corresponding ice age at the depth of the $CH_4$ tie points.

4. Interpolate between the ice age constraints to find the WD depth–age relationship.

5. Redo the $\Delta$age calculations on the new ice age scale.

6. Repeat steps 3–5 iteratively until the depth–age relationship is stable within 1 year. At WD this happened after three iterations.

These steps are described in more detail in the following sections.

### 4.3 Establishing the midpoint in abrupt DO transitions

The procedure for determining the midpoint of the abrupt DO warming transitions is depicted in Fig. 6. For each of

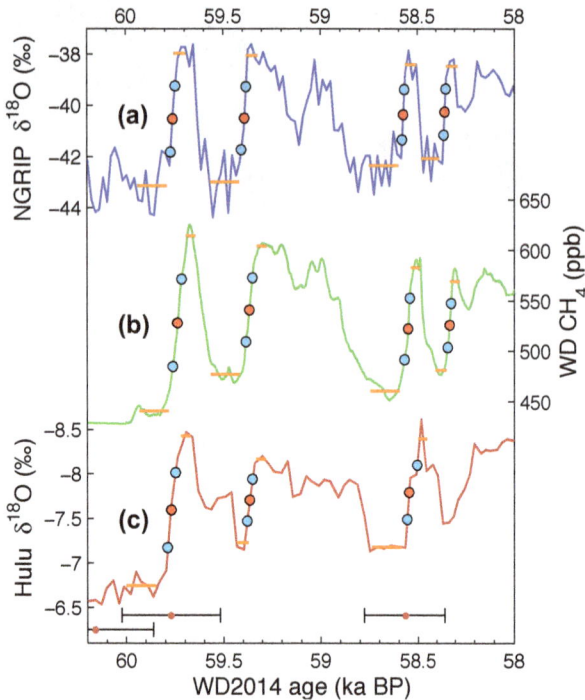

**Figure 6.** Determining the midpoint for the abrupt warming phases of (from oldest to youngest) DO 17.2, 17.1, 16.2, and 16.1 in (**a**) NGRIP $\delta^{18}$O (on 1.0063×GICC05), (**b**) WD CH$_4$ (continuous data, on WD2014), and (**c**) Hulu $\delta^{18}$O with U / Th ages beneath the time series (red dots with error bars). Red dots give the midpoint (50 %) of the DO transition; the blue dots give the 25 and 75 % marks in the DO transitions. The DO transition at 58.35 ka was not used in Hulu, where it is much more gradual than in the other records (possibly because calcite sampling was not perfectly perpendicular to the stalagmite isochrones, or because growth rates were variable in between the U / Th ages).

the transitions we manually determine pre-event and post-event averages, as indicated by the orange lines. The averaging time is set to 150 and 50 years for stadial and interstadial periods, respectively; this difference in duration is used because (i) several of the interstadials are of short duration and (ii) Greenland $\delta^{18}$O is more variable during stadial climates, requiring longer averaging. For DO 16.1, the duration of the pre-event stadial baseline climate was shorter than 150 years, and the averaging time was reduced to 100 years (Fig. 6).

After determining the pre- and post-event averages, we use linear interpolation of the time series to find the time at which the variable of interest had completed 25, 50, and 75 % of the total transition (Fig. 6). We use the 50 % marker (red) as the midpoint of the transition, which is used in the methane synchronization. The 25 and 75 % markers (blue) are used as the ±1$\sigma$ uncertainty estimate. In rare cases the time series contain inversions within the transitions that lead to ambiguity in the timing of the markers; for these events we find the markers using a monotonic spline fit to the data.

The midpoints of abrupt interstadial terminations were determined in the same fashion (WD CH$_4$ and NGRIP only). Tables 1 and 2 give the results for NH warming and NH cooling, respectively.

## 4.4 Synchronizing WD to a NGRIP–Hulu hybrid chronology

Abrupt DO variability is expressed clearly in a great number of NH climate records (Voelker, 2002). For the purpose of methane synchronization, our interest is in high-resolution records that express the abrupt DO events very clearly, and are furthermore exceptionally well dated. We here use a combination of two such NH records (Fig. 5), namely the Greenland NGRIP $\delta^{18}$O record (NGRIP community members, 2004), and a refined version of the Hulu Cave speleothem $\delta^{18}$O record (Edwards et al., 2015; Reimer et al., 2013; Southon et al., 2012) with improved resolution and additional dating constraints (see Wang et al., 2001, for the original, lower resolution Hulu $\delta^{18}$O record). The DO events are resolved most clearly in the NGRIP $\delta^{18}$O record, which is available at 20-year resolution. We use the GICC05-modelext chronology for this core, which is based on annual-layer counting back to 60 ka BP and ice-flow modeling for ice older than 60 ka (Rasmussen et al., 2006; Svensson et al., 2006; Wolff et al., 2010). While annual-layer counting provides accurate relative ages (e.g., the duration of DO interstadials), it provides relatively inaccurate absolute ages due to the cumulative nature of counting uncertainty (Table 1). The refined Hulu $\delta^{18}$O record also shows the abrupt DO events in high temporal resolution (Fig. 6). The speleothem chronology is based on U / Th radiometric dating, providing much smaller uncertainty in the absolute ages than GICC05 (Table 1). The reason for selecting this record over other speleothem records is the large number of U / Th dates, the low detrital Th at the site, and the high sampling resolution of the $\delta^{18}$O record (Wang et al., 2001). In the Hulu data, as in other records of DO variability, the interstadial onsets are more pronounced and abrupt than their terminations. We therefore only use the timing of the former as age constraints, as they can be established more reliably. The onset of NH interstadial periods as expressed in Hulu $\delta^{18}$O is given in Table 1.

In both the NGRIP and Hulu Cave $\delta^{18}$O records we have determined the ages of the midpoints of the DO transitions (Fig. 6; Table 1); a plot of their difference (Hulu age minus NGRIP age) is shown in Fig. 7, where the error bars denote the root sum square of the NGRIP and Hulu midpoint determination uncertainty (Sect. 4.3). The Hulu ages are systematically older than the NGRIP ages, and the age difference increases going further back in time. Note that the Hulu–NGRIP age difference is smaller than the stated GICC05 counting uncertainty (832 to 2573 years) but larger than the Hulu age uncertainty (92 to 366 years). A linear fit through these data, forced to intersect the origin, is given by

**Table 1.** Overview of $CH_4$ tie points for NH warming events. WD ages printed in boldface are assigned as part of the $CH_4$ synchronization; all other ages are on their independent chronologies.

| | NGRIP | | | | Hulu | | | WD | | | |
|---|---|---|---|---|---|---|---|---|---|---|---|
| | Depth (m) | Age (years BP) | Age uncert. (years) | Midpoint (years) | Hulu age (years BP) | Age uncert. (years) | Midpoint (years) | Depth (m) | Gas age (years BP) | Ice age (years BP) | Midpoint (years) |
| YD-PB | 1490.89 | 11 619 | 98 | 23 | | | | 1983.02 | 11 546 | 11 740 | 33 |
| OD-BA | 1604.05 | 14 628 | 185 | 15 | | | | 2259.40 | 14 576 | 14 804 | 29 |
| DO 3 | 1869.00 | 27 728 | 832 | 12 | 27 922 | 95 | 39 | 2755.74 | 27 755 | 28 144 | 19 |
| DO 4 | 1891.27 | 28 838 | 898 | 14 | 29 134 | 92 | 21 | 2797.92 | 29 011 | 29 397 | 22 |
| DO 5.1 | 1919.48 | 30 731 | 1023 | 11 | 30 876 | 255 | 37 | 2848.38 | 30 730 | 31 186 | 22 |
| DO 5.2 | 1951.66 | 32 452 | 1132 | 15 | 32 667 | 236 | 21 | 2885.44 | **32 631** | **33 051** | 17 |
| DO 6 | 1974.48 | 33 687 | 1213 | 19 | 34 034 | 337 | 36 | 2913.01 | **33 874** | **34 283** | 18 |
| DO 7 | 2009.62 | 35 437 | 1321 | 16 | 35 532 | 299 | 20 | 2958.64 | **35 636** | **35 982** | 20 |
| DO 8 | 2069.88 | 38 165 | 1449 | 13 | 38 307 | 155 | 19 | 3021.37 | **38 381** | **38 681** | 33 |
| DO 9 | 2099.50 | 40 104 | 1580 | 13 | 40 264 | 241 | 42 | 3066.52 | **40 332** | **40 690** | 19 |
| DO 10 | 2123.98 | 41 408 | 1633 | 14 | 41 664 | 310 | 27 | 3094.17 | **41 643** | **41 980** | 18 |
| DO 11 | 2157.58 | 43 297 | 1736 | 17 | 43 634 | 144 | 26 | 3130.44 | **43 544** | **43 866** | 15 |
| DO 12 | 2221.96 | 46 794 | 1912 | 21 | 47 264 | 153 | 20 | 3195.25 | **47 064** | **47 335** | 16 |
| DO 13 | 2256.73 | 49 221 | 2031 | 17 | 49 562 | 251 | 52 | 3237.65 | **49 506** | **49 836** | 19 |
| DO 14 | 2345.39 | 54 164 | 2301 | 11 | | | | 3311.09 | **54 480** | **54 747** | 13 |
| DO 15.1 | 2355.17 | 54 940 | 2349 | 16 | | | | 3322.24 | **55 261** | **55 564** | 11 |
| DO 15.2 | 2366.15 | 55 737 | 2392 | 26 | | | | 3329.72 | **56 063** | **56 381** | 14 |
| DO 16.1 | 2398.71 | 57 988 | 2498 | 11 | | | | 3350.44 | **58 328** | **58 610** | 9 |
| DO 16.2 | 2402.25 | 58 210 | 2510 | 12 | 58 545 | 226 | 22 | 3352.59 | **58 552** | **58 848** | 14 |
| DO 17.1 | 2414.82 | 59 018 | 2557 | 15 | 59 364 | 366 | 18 | 3360.02 | **59 364** | **59 627** | 17 |
| DO 17.2 | 2420.35 | 59 386 | 2573 | 15 | 59 772 | 254 | 23 | 3363.42 | **59 735** | **59 997** | 25 |
| DO 18 | 2465.84 | 64 049 | 2611 | 30 | | | | 3388.73 | **64 428** | **64 773** | 15 |

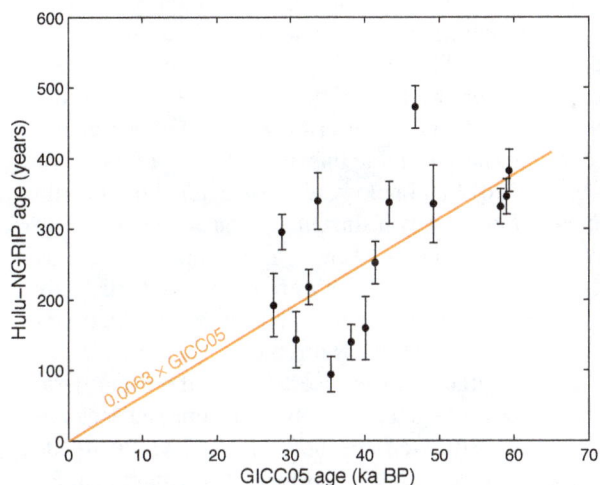

**Figure 7.** Hulu–NGRIP age offset at the midpoint of the DO $\delta^{18}O$ transitions. The error bars denote the root sum square of the midpoint determination uncertainty in NGRIP and Hulu $\delta^{18}O$ (Table 1). The GICC05 ages are placed on the BP 1950 scale rather than the b2k scale (years prior to 2000 CE).

$0.0063 \times$ GICC05 age, suggesting that the GICC05 annual-layer count on average misses 6.3 out of every 1000 layers. Because of this observation we use a linearly scaled version of the GICC05 chronology (GICC05 $\times$ 1.0063) as the target chronology for methane synchronization. This approach has several advantages. First, it respects both the superior relative ages (i.e., interval durations) of GICC05, as well as the superior absolute ages of the Hulu chronology. Second, it is very simple to convert between the WD2014 and GICC05 chronologies ($CH_4$-synchronized section of the chronology only); one simply needs to divide WD2014 ages by 1.0063 (and add 50 years to convert to the b2k reference date). Third, it still allows for direct synchronization of WD $CH_4$ to the NGRIP $\delta^{18}O$ record, providing more tie points than direct synchronization to the Hulu record would. Note that the GICC05 $\times$ 1.0063 target chronology respects the Hulu age constraints in an average sense only; the age of individual events differs between Hulu and our target chronology by up to 180 years. Our approach therefore represents only a first-order correction of a growing offset between GICC05 and Hulu; nonlinear temporal changes in the counted dating error may exist from one tie point to the next (Fleitmann et al., 2009).

The exercise of finding the transition midpoints and determining the GICC05–Hulu scaling factor was performed by two of the authors (J. P. Severinghaus and C. Buizert), independently of each other. The scaling factors obtained were 1.0063 and 1.0064, respectively, showing that, to first order, this result is insensitive to (subjective) judgment in identifying the transitions. The difference between the Hulu ages and 1.0063 $\times$ GICC05 ages are all within the stated Hulu $2\sigma$ dating error (Table 1). Consequently, our chronology is not in violation of any Hulu constraint as it respects the Hulu $2\sigma$ error at all of the tie points. In deriving the scaling we have assumed that the abrupt DO transitions observed in NGRIP and Hulu are simultaneous, which is not necessarily true. The variations in monsoon intensity represented by Hulu $\delta^{18}O$ are commonly explained by meridional movement of the Intertropical Convergence Zone (ITCZ) and tropical rainfall belts (Wang et al., 2001, 2006; Kanner et al., 2012); modeling work suggests such atmospheric readjustments occur on decadal timescales in response to NH high-latitude forcing (Chiang and Bitz, 2005; Broccoli et al., 2006; Cvijanovic and Chiang, 2013). Moreover, $CH_4$ emission changes are near-synchronous with Greenland $\delta^{18}O$ variations, which they lag by only a few decades on average (Huber et al., 2006; Baumgartner et al., 2014; Rosen et al., 2014). Since both $CH_4$ emissions and Hulu $\delta^{18}O$ are closely linked to tropical hydrology, timing lags between NGRIP and Hulu are also expected to be on decadal timescales. The uncertainty in the NGRIP–Hulu phasing is therefore probably small (decadal) relative to the correction we apply (up to 400 years).

Rather than synchronizing WD $CH_4$ to Greenland $CH_4$ records, we have chosen to synchronize directly to NGRIP $\delta^{18}O$, which varies in phase with $CH_4$ (but with a nearly constant time lag). We let the midpoint in the $CH_4$ transitions lag the midpoint in the NGRIP $\delta^{18}O$ transition by 25 years, as suggested by studies of Greenland $\delta^{15}N$-$CH_4$ phasing (Huber et al., 2006; Baumgartner et al., 2014; Rasmussen et al., 2013; Kindler et al., 2014; Rosen et al., 2014). The rationale behind this approach is threefold. First, throughout MIS 3 the NGRIP $\delta^{18}O$ record has both better precision and higher temporal resolution than any available Greenland $CH_4$ record (Baumgartner et al., 2014; Brook et al., 1996; Blunier et al., 2007). Second, the dating of Greenland gas records depends on the highly variable $\Delta$age function, which is not equally well constrained for all DO events (Schwander et al., 1997; Rasmussen et al., 2013). This reliance on Greenland $\Delta$age would introduce an additional source of uncertainty. The NGRIP $\delta^{18}O$ record, on the other hand, is accurately dated through the GICC05 layer count. Third, Greenland $CH_4$ records are more strongly impacted by firn smoothing than the WD $CH_4$ record, because glacial accumulation is lower in Greenland (Greenland glacial $\Delta$age is about 2–3 times as high as WD $\Delta$age during that time). In summary, our approach circumvents the uncertainties associated with using Greenland $CH_4$ as an intermediary, or, to state this another way, the uncertainty in the phasing between $CH_4$ and Greenland $\delta^{18}O$ is smaller than the uncertainty in the Greenland $\Delta$age.

### 4.5 Interpolation between age constraints

We can assign a gas age to each of the depths where an abrupt WD $CH_4$ transitions occurs; we do this for DO 4.1 through DO 18, i.e., the events prior to 31.2 ka BP (the onset of the WD layer count). The gas age we assign is equal to 1.0063 times the GICC05 age for the same event, with 25 years subtracted to account for the slight $CH_4$ lag behind Greenland $\delta^{18}O$. By adding $\Delta$age (Sect. 3) to this gas age we assign an

ice age. These assigned ages are printed in boldface in Tables 1 and 2.

To obtain a continuous depth–age relationship between these ice age constraints, we have to apply an interpolation strategy. This task amounts to estimating the annual-layer thickness $\lambda(z)$ along the deep part of the core. The simplest approach is to assume a constant accumulation rate in between the age constraints; this is shown in Fig. 8b for the case where we use the age constraints from NH warming events only (black) or the age constraints from both NH warming and cooling events (red). The disadvantage of this approach is that it results in discontinuities in $\lambda(z)$ (the first derivative of the depth–age relationship), which we consider highly unrealistic. A more realistic approach is therefore to assume that $\lambda(z)$ is continuous and smooth (Fudge et al., 2014); Fig. 8b shows two scenarios in which we use a spline function to estimate $\lambda(z)$, where again we have applied age constraints from NH warming events only (orange) or age constraints from both NH warming and cooling events (blue).

For comparison, past $A$ obtained from the firn densification model (Sect. 3) is plotted in green (Fig. 8b). While the $\delta^{15}N$-based $A$ follows the synchronization-based $A$ estimates broadly, the millennial-scale details do not agree. We want to point out that this is not unexpected, since both methods have their imperfections. In particular, any errors in the (stretched) GICC05 age model or in our modeled thinning function $f_\lambda(z)$ will strongly impact the synchronization-based $A$ estimates in Fig. 8b. The discrepancy is pronounced between 60 and 65 ka, where we have to strongly reduce $\lambda(z)$ in order to fit the age constraint(s) from DO 18, while $\delta^{15}N$ provides no evidence for low $A$ during this interval.

For the WD2014 chronology we have applied the smooth $\lambda(z)$ interpolation scheme using all age constraints (i.e., both NH warming and cooling events). The midpoint detection uncertainty is comparable for all events and systematically smaller at the start of interstadial periods than at the terminations (Tables 1 and 2). For short interstadials (e.g., DO 9) this leads to a large relative uncertainty in the event duration, and thereby a large uncertainty in the implied accumulation rates (Fig. 8b). We force the interpolation to fit all NH warming constraints perfectly, yet relax this requirement for NH cooling constraints to prevent large swings in $\lambda(z)$ for the short-duration events. The WD2014 chronology fits the NH warming and NH cooling age constraints with a 0- and 16-year rms offset, respectively. Because the duration of (inter)stadial periods is well constrained in the layer-counted GICC05 chronology, using both the NH warming and NH cooling tie points results in a more robust chronology. The duration of (inter)stadial periods is 0.63 % longer in WD2014 than in GICC05, which is well within the stated GICC05 counting error of 5.4 % (31.2–60 ka interval).

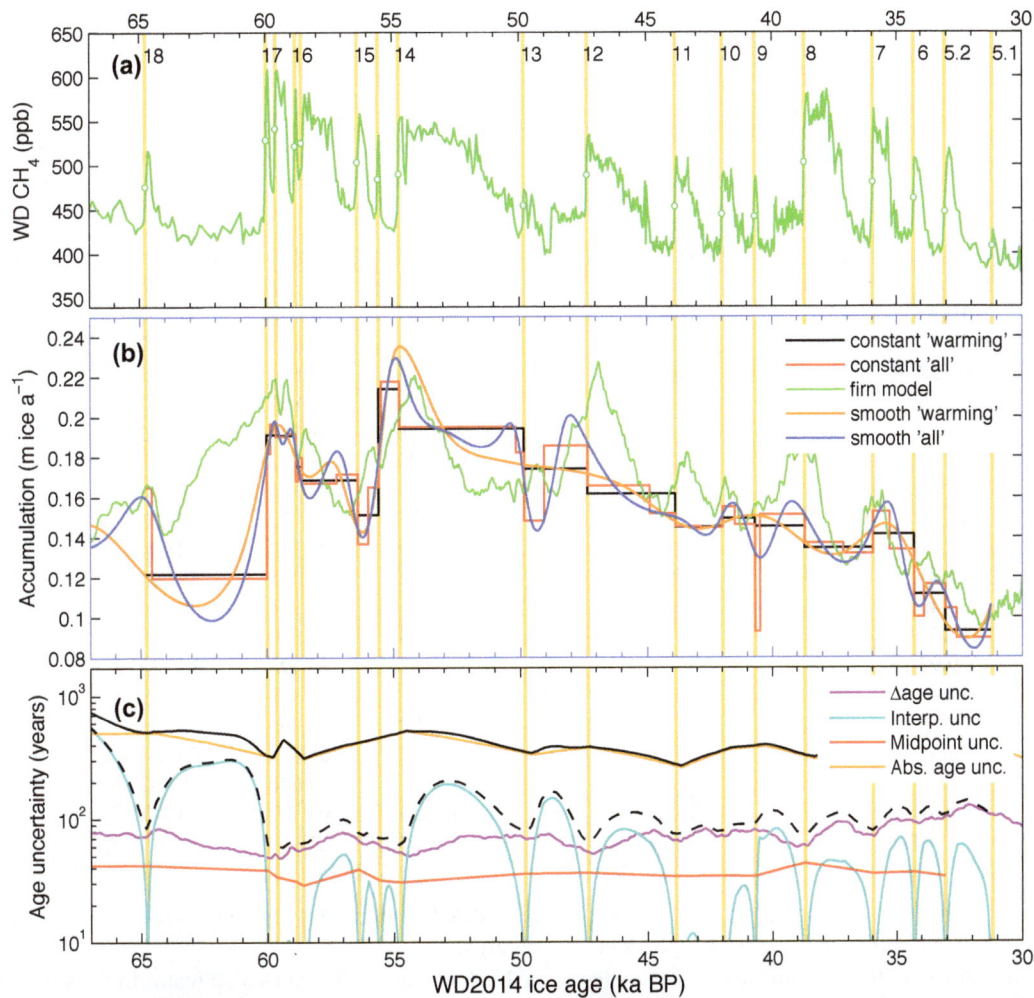

**Figure 8.** Interpolating between the $CH_4$ age constraints. **(a)** WD discrete $CH_4$ record with the abrupt stadial–interstadial transitions marked. DO numbering given at the top of the panel. **(b)** Different annual-layer thickness scenarios, converted to an accumulation rate for comparison to the $\delta^{15}N$-based firn model reconstructions. The interpolation strategy is to use either constant accumulation rates between tie points ("constant") or a smoothly varying $\lambda(z)$ ("smooth"); the age constraints used are either only the NH warming events ("warming"), or both the NH warming and cooling events ("all"). **(c)** Estimated $2\sigma$ uncertainties in the WD2014 chronology due to $\Delta$age, choice of interpolation scheme, midpoint detection, and the absolute age constraints used in the synchronization. Total absolute ice age uncertainty plotted in solid black; relative age uncertainty (i.e., with absolute age uncertainty in the Hulu–GICC05 master chronology withheld) plotted in dashed black.

## 4.6 Age uncertainty

The age uncertainty we assign to the deep part ($> 2850$ m) of the WD2014 chronology has four components.

The first source of uncertainty is the $\Delta$age calculation; we use the $2\sigma$ uncertainty obtained in the $\Delta$age sensitivity study (Sect. 3.2). The second source of uncertainty is the choice of interpolation scheme used to obtain a continuous chronology; here we use the standard deviation between the four different interpolation schemes of Fig. 8b as an uncertainty estimate. The third source of uncertainty is the difficulty in determining the timing of the abrupt events in the time series; we use the uncertainty in the midpoint evaluation (root sum square of WD $CH_4$ and NGRIP $\delta^{18}O$ estimates). The

last source of uncertainty is the age uncertainty in the hybrid NGRIP–Hulu chronology that we synchronize to. We use the stated Hulu age uncertainty plus 50 years to account for possible leads or lags in the NGRIP–Hulu $\delta^{18}O$ phasing, plus the absolute value of the offset between the Hulu ages and the $1.0063 \times$ GICC05 ages. For DO events where we do not have reliable Hulu age estimates (Table 1), we set the uncertainty to the Hulu age uncertainty of the nearest event, plus the uncertainty in the interval duration specified by the GICC05 layer count. For example, for DO 14 we do not have a reliable Hulu age estimate, and we use the Hulu age uncertainty of DO 16.2 (226 years) plus the uncertainty in the DO 14 to DO 16.2 interval duration on GICC05 (209 years), giving a total of $226 + 209 = 435$ years.

**Table 2.** Overview of CH$_4$ tie points for NH cooling events. WD ages printed in boldface are assigned as part of the CH$_4$ synchronization; all other ages are on their independent chronologies

| | NGRIP | | | | WD | | | |
|---|---|---|---|---|---|---|---|---|
| | Depth (m) | Age (years BP) | Age uncert. (years) | Midpoint (years) | Depth (m) | Gas age (years BP) | Ice age (years BP) | Midpoint (years) |
| BA-YD | 1524.21 | 12 775 | 136 | 81 | 2096.61 | 12 769 | 12 987 | 52 |
| DO 3 | 1861.91 | 27 498 | 822 | 52 | 2747.25 | 27 520 | 27 905 | 38 |
| DO 4 | 1882.59 | 28 548 | 887 | 17 | 2787.99 | 28 696 | 29 090 | 61 |
| DO 5.1 | 1916.45 | 30 571 | 1010 | 70 | 2845.37 | 30 618 | 31 067 | 50 |
| DO 5.2 | 1939.71 | 31 992 | 1108 | 13 | 2875.86 | **32 168** | **32 607** | 70 |
| DO 6 | 1964.52 | 33 323 | 1192 | 37 | 2905.55 | **33 508** | **33 905** | 60 |
| DO 7 | 1990.58 | 34 703 | 1286 | 13 | 2939.09 | **34 897** | **35 292** | 50 |
| DO 8 | 2027.43 | 36 571 | 1401 | 21 | 2986.58 | **36 776** | **37 172** | 32 |
| DO 9 | 2095.51 | 39 905 | 1572 | 42 | 3063.79 | **40 132** | **40 492** | 25 |
| DO 10 | 2112.53 | 40 917 | 1621 | 44 | 3083.89 | **41 150** | **41 508** | 44 |
| DO 11 | 2135.66 | 42 231 | 1685 | 27 | 3110.76 | **42 472** | **42 823** | 69 |
| DO 12 | 2171.17 | 44 308 | 1783 | 41 | 3149.89 | **44 562** | **44 904** | 47 |
| DO 13 | 2242.85 | 48 440 | 1996 | 27 | 3226.93 | **48 720** | **49 054** | 20 |
| DO 14 | 2261.49 | 49 552 | 2052 | 20 | 3243.03 | **49 839** | **50 165** | 65 |
| DO 15.1 | 2353.66 | 54 850 | 2339 | 18 | 3321.15 | **55 170** | **55 469** | 14 |
| DO 15.2 | 2359.92 | 55 369 | 2370 | 55 | 3326.47 | **55 693** | **55 983** | 45 |
| DO 16.1 | 2375.88 | 56 555 | 2435 | 49 | 3337.98 | **56 887** | **57 219** | 76 |
| DO 16.2 | 2400.56 | 58 123 | 2508 | 15 | 3351.80 | **58 465** | **58 756** | 9 |
| DO 17.1 | 2406.52 | 58 544 | 2530 | 35 | 3355.54 | **58 888** | **59 151** | 61 |
| DO 17.2 | 2417.77 | 59 257 | 2570 | 18 | 3362.26 | **59 606** | **59 862** | 24 |
| DO 18 | 2462.07 | 63 810 | 2611 | 14 | 3387.28 | **64 187** | **64 547** | 32 |

The uncertainties ($2\sigma$ values) are plotted in Fig. 8c (log scale). We assume these four uncertainties to be independent, and use their root sum square as the total uncertainty estimate on the WD2014 ice age scale (Fig. 8c, black curve). Note that the fourth source of uncertainty is only relevant when considering absolute ages; when evaluating relative ages (e.g., between WD ice and WD gas phase, or between WD and NGRIP), this last contribution does not need to be considered. For the deepest WD ice (3404 m depth) we thus find an age of $67.7 \pm 0.9$ ka BP.

## 5  Discussion

While the WAIS Divide ice core does not extend as far back in time as deep cores from the East Antarctic Plateau, its relatively high temporal resolution (due to the high snow accumulation rate) makes it an ice core of great scientific value. WD accumulation rate during the LGM ($\sim 10$ cm a$^{-1}$ ice equivalent) is still higher than the present-day accumulation rate at the EPICA (European Project for Ice Coring in Antarctica) Dronning Maud Land core (7 cm a$^{-1}$), which is generally considered a high-accumulation core (EPICA Community Members, 2006). With 68 ka in 3404 m of core, the core average $\lambda$ is 5 cm a$^{-1}$, at the onset of the last deglaciation (18 ka BP) $\lambda$ is around 4 cm a$^{-1}$, and near the bed $\lambda$ is around 0.4 cm a$^{-1}$. This high temporal resolution

provides the opportunity for obtaining very detailed climatic records.

High accumulation rates also result in a small $\Delta$age. Figure 9 compares $\Delta$age between several Antarctic cores (note the logarithmic scale). $\Delta$age at WD is approximately one-third of the $\Delta$age at EPICA DML (EDML) and Talos Dome (TALDICE), and one-tenth of the $\Delta$age at EPICA Dome C (EDC), Vostok, and Dome Fuji. Because the uncertainty in the $\Delta$age (or $\Delta$depth) calculation is typically on the order of 20 %, a smaller $\Delta$age allows for a more precise interhemispheric synchronization with Greenland ice core records using CH$_4$. The small WD $\Delta$age uncertainty during MIS 3 allows for investigation of the phasing of the bipolar seesaw (Stocker and Johnsen, 2003) at sub-centennial precision (WAIS Divide Project Members, 2015).

In comparing the shape of the $\Delta$age profiles, there are some interesting differences (Fig. 9). It is important to realize that not all the $\Delta$age histories shown were derived in the same way; WD and Dome Fuji $\Delta$age were derived using densification models, and the other four were derived using the $\Delta$depth approach (Parrenin et al., 2012) and a Bayesian inverse method that includes a wide range of age markers (Veres et al., 2013). We will therefore focus on comparing the WD and Dome Fuji results. $\Delta$age at WD shows more pronounced variability than at Dome Fuji, particularly during MIS 3. The reason is that the glacial firn pack at Dome

Fuji is about 4000 years old, and consequently the firn column integrates over 4000 years of climate variability, thereby dampening the $\Delta$age response to millennial-scale climatic variability. At WD the glacial firn layer is only about 350 years old, and therefore the firn is in near equilibrium with the millennial-scale climate variations. This difference in response time is also obvious during the deglaciation, where WD $\Delta$age transitions from glacial to interglacial values between 18 and 14.5 ka BP, while Dome Fuji takes more time (18–10 ka BP). Surprisingly, EDML $\Delta$age does not show a strong deglacial $\Delta$age response, unlike all the other cores.

The relatively small $\Delta$age at WAIS Divide also allows for precise investigation of the relative timing of atmospheric greenhouse gas variations and Antarctic climate (Barnola et al., 1991; Pedro et al., 2012; Caillon et al., 2003; Parrenin et al., 2013; Ahn et al., 2012). Recent works suggest that during the last deglaciation the rise in atmospheric $CO_2$ lagged the onset of pan-Antarctic warming by approximately 0 to 400 years (Pedro et al., 2012; Parrenin et al., 2013). This Antarctic warming around 18 ka BP is presumably driven by the bipolar seesaw, as it coincides with a reduction in Atlantic overturning circulation strength as seen in North Atlantic sediment records (McManus et al., 2004). The WD $\Delta$age at 18 ka (gas age) is $515 \pm 91$ years ($2\sigma$), much smaller than at central East Antarctic sites such as EPICA Dome C, where $\Delta$age is approximately $3850 \pm 900$ years (Veres et al., 2013, with the $\Delta$age uncertainty taken to be the difference between the gas age and ice age uncertainties). The precision with which one can determine the relative phasing of climatic (i.e., $\delta^{18}O$ of ice) and atmospheric signals is set by the uncertainty in $\Delta$age (or equivalently, the uncertainty in $\Delta$depth). High-resolution WD records of $CO_2$ and $CH_4$ (Marcott et al., 2014) place the onset of the deglacial rise in the atmospheric mixing ratio of these greenhouse gases on the WD2014 chronology at 18 010 and 17 820 years BP, respectively. However, evaluating the relative phasing of $CO_2$ and Antarctic climate is complicated by the observation of asynchronous deglacial warming across the Antarctic continent (WAIS Divide Project Members, 2013). Attempts to capture the climate–$CO_2$ relationship in a single lead-lag value may be an oversimplification of deglacial climate dynamics.

An important next step will be to synchronize the WD chronology with other Antarctic cores via volcanic matching and other age markers (e.g., Severi et al., 2007; Sigl et al., 2014). Because of the annual-layer count and possibility of tight synchronization to Greenland ice cores, WD could contribute to an improved absolute dating of Antarctic cores, as well as improved cross-dating between cores. Such cross-dating could help inform the WD chronology as well, particularly in the deepest part of the core, where the ice is potentially highly strained, as suggested by the interpolation difficulties in the 60–65 ka interval (Fig. 8b). With a synchronized chronology, WD could improve the representation of West Antarctic climate in Antarctic ice core stacks (Pedro

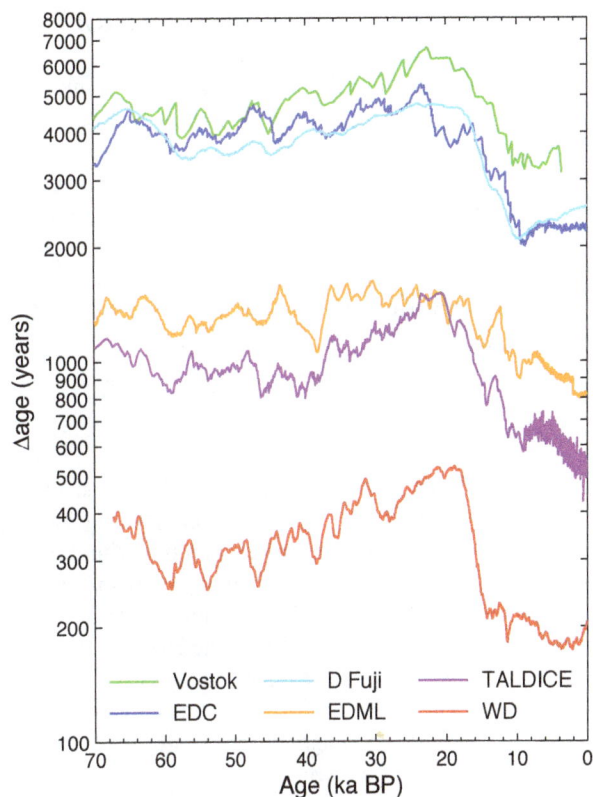

**Figure 9.** Comparison of $\Delta$age for different Antarctic cores, plotted on the gas age scale. Dome Fuji $\Delta$age from Kawamura et al. (2007); WD from Sect. 3; all others from Bazin et al. (2013); Veres et al. (2013).

et al., 2011; Parrenin et al., 2013), and provide a more refined pan-Antarctic picture of the climate–$CO_2$ relationship.

## 6 Summary and conclusions

We have presented a first chronology for the deep (> 2850 m) section of the WAIS Divide ice core, which is based on stratigraphic matching to Greenland ice cores using globally well-mixed methane. We use a dynamical firn densification model constrained by $\delta^{15}N$ data to calculate past $\Delta$age, and find that $\Delta$age was smaller than $525 \pm 120$ years for all of the core. Using high-resolution WD records of atmospheric $CH_4$, we synchronize WD directly to Greenland NGRIP $\delta^{18}O$ for the abrupt onset and termination of each of the DO interstadials. To each event we assign an age corresponding to 1.0063 times its GICC05 age, which brings the ages in agreement with the high-resolution U / Th-dated Hulu speleothem record. The uncertainty in the final chronology is based on the uncertainties in (i) the $\Delta$age calculations, as evaluated with a sensitivity study; (ii) the interpolation strategy, as evaluated by comparing four different interpolation methods; (iii) determining the timing of events in the different time series; and (iv) the ages of the hybrid NGRIP–Hulu chronology we are synchronizing to.

Due to the combination of a small $\Delta$age and a high-resolution methane record, the WAIS Divide ice core can be synchronized more precisely to Greenland records than any other Antarctic core to date. This is important when investigating interhemispheric climate relationships such as the bipolar seesaw. The small WD $\Delta$age furthermore provides valuable opportunities for precise investigation of the relative phasing of atmospheric greenhouse gas variations and Antarctic climate.

## Appendix A: Densification physics

The densification rates used in this work are based on the empirical steady-state model by Herron and Langway (1980) (the H-L model). We use the H-L model with minor modifications that allow it to be run dynamically (i.e., with time-variable $T$ and $A$) and to include the softening effect of impurities following Freitag et al. (2013a). The H-L model divides the firn column in two stages, separated at the critical density $\rho_c = 550 \, \mathrm{kg \, m^{-3}}$, occurring at the critical depth $z_c$.

For the upper firn ($\rho \leq \rho_c$, stage 1), the densification rates are given by

$$\frac{d\rho}{dt} = k_1 A (\rho_{\mathrm{ice}} - \rho), \tag{A1}$$

with

$$k_1 = 11 \exp\left(-\frac{E_1}{RT}\right), \tag{A2}$$

where $E_1 = 10.16 \, \mathrm{kJ \, mol^{-1}}$ is the activation energy for stage 1 and $R$ is the universal gas constant. Because both the sinking velocity of deposited layers ($w = dz/dt$) and the densification rate scale linearly with $A$, the resulting density–depth profile $\rho(z)$ in stage 1 becomes independent of $A$, and sensitive to $T$ variations only.

For the deeper firn ($\rho > \rho_c$, stage 2), we use Eq. (4c) from Herron and Langway (1980), which was first derived by Sigfús J. Johnsen. This equation gives the densification rate in terms of overburden load, which allows the model to be run dynamically. The stage 2 densification rates are given by

$$\frac{d\rho}{dt} = k_2^2 \frac{(\sigma_z - \sigma_{z_c})(\rho_{\mathrm{ice}} - \rho)}{\ln\left[(\rho_{\mathrm{ice}} - \rho_c)/(\rho_{\mathrm{ice}} - \rho)\right]}, \tag{A3}$$

with

$$k_2 = 575 \exp\left(-\frac{E_2}{RT}\right), \tag{A4}$$

where $E_2 = 21.4 \, \mathrm{kJ \, mol^{-1}}$ is the activation energy for stage 2 and $\sigma_z$ denotes the firn overburden load at a given depth in $\mathrm{Mg \, m^{-2}}$:

$$\sigma_z = \int_0^z \rho(z') dz' / 1000. \tag{A5}$$

Note that we divide by 1000 to convert from $\mathrm{kg \, m^{-3}}$ to $\mathrm{Mg \, m^{-3}}$, the units used by Herron and Langway (1980).

We use the mathematical description by Freitag et al. (2013a) to include the hypothesized firn softening effect of impurities. In this approach an increasing Ca concentration, as a proxy for mineral dust content, lowers the activation energy of firn, thereby enhancing densification rates. This is tantamount to stating that dusty firn behaves as if it were "warmer" than its climatological temperature. The H-L activation energies of Eqs. (A2) and (A4) are modified by [Ca] in the following way:

$$E^{\mathrm{Ca}} = E^{\mathrm{HL}} \times \alpha \left[1 - \beta \ln\left(\frac{[\mathrm{Ca}]}{[\mathrm{Ca}]_{\mathrm{crit}}}\right)\right], \tag{A6}$$

where $E^{\mathrm{Ca}}$ and $E^{\mathrm{HL}}$ are the Ca-modified and original H-L activation energies, respectively, $[\mathrm{Ca}]_{\mathrm{crit}} = 0.5 \, \mathrm{ng \, g^{-1}}$ is the minimum concentration at which impurities affect densification, and $\alpha$ and $\beta$ are calibration parameters. Whenever $[\mathrm{Ca}](z) < [\mathrm{Ca}]_{\mathrm{crit}}$, we set $[\mathrm{Ca}](z) = [\mathrm{Ca}]_{\mathrm{crit}}$.

The parameter $\beta$ sets the sensitivity to dust loading, and $\alpha$ is a normalization parameter that is included to account for the fact that the original H-L model was calibrated without the impurity effect. Consequently, if $\beta > 0$, one needs to compensate by setting $\alpha > 1$ to preserve the original H-L calibration. The work by Freitag et al. (2013a) recommends $\beta = 0.01$ and $\alpha = 1.025$ (which yields $E^{\mathrm{Ca}} = E^{\mathrm{HL}}$ at $[\mathrm{Ca}] = 5.73 \, \mathrm{ng \, g^{-1}}$).

Using the recommended value of $\alpha = 1.025$ at WD provides a poor fit to observations of present-day firn density and close-off depth. The optimal fit to present-day WD observations is obtained using an activation energy equal to $1.007 \times E^{\mathrm{HL}}$; this is in between the values suggested by Herron and Langway (1980) and Freitag et al. (2013a). In the experiment presented in Fig. 4 we changed the dust sensitivity $\beta$; it is clear that we need to simultaneously change $\alpha$ to keep the model well-calibrated to present-day conditions. Due to the fact that the mean late Holocene WD [Ca] is around $0.8 \, \mathrm{ng \, g^{-1}}$, we let $\alpha = 1.007/(1 - \beta \ln[0.8/0.5])$ in the experiment of Fig. 4. This approach ensures that the present-day $E^{\mathrm{Ca}}$ is invariant with $\beta$, and equals $E^{\mathrm{Ca}} = 1.007 \times E^{\mathrm{HL}}$. This means that whatever value we choose for $\beta$, we will obtain a good fit to the present-day $\Delta$age, $\delta^{15}\mathrm{N}$, and $A$ values that are well known from direct observations (Battle et al., 2011).

To validate the H-L model $\Delta$age simulations, we repeated the firn modeling using the densification physics of Arnaud et al. (2000), which is also the basis of the model by Goujon et al. (2003). Our implementation of the Arnaud model is based on the description in the latter paper, with one modification at the critical density that we outline here.

In the Arnaud model, densification in the stage 1 follows the work of Alley (1987), and is given by

$$\frac{\mathrm{d}D}{\mathrm{d}t} = \gamma \left(\frac{P}{D^2}\right)\left(1 - \frac{5}{3}D\right), \tag{A7}$$

with $D$ the relative density $D = \rho/\rho_{\text{ice}}$, $P$ the overburden pressure, and $\gamma$ a scaling factor used to make the densification rates continuous across the critical density $D_c$. Stage 2 densification is given by

$$\frac{\mathrm{d}D}{\mathrm{d}t} = k_A \left(D^2 D_c\right)^{\frac{1}{3}} \left(\frac{a}{\pi}\right)^{\frac{1}{2}} \left(\frac{4\pi P}{3aZD}\right)^3, \tag{A8}$$

with

$$k_A = 4.182 \times 10^4 \exp\left(-\frac{E_A}{RT}\right), \tag{A9}$$

where $a$ is the average contact area between the grains, $Z$ is the coordination number, and $E_A$ is the activation energy ($60\,\text{kJ}\,\text{mol}^{-1}$). Arnaud densification rates for stage 3 ($D \geq 0.9$) are describe elsewhere (Goujon et al., 2003; Arnaud et al., 2000).

The difficulty in implementing this model is the following. The densification rates of Eqs. (A7–A8) exhibit a discontinuity at the critical density $D = D_c = 0.6$ that cannot be remedied with the scaling factor $\gamma$. On approaching $D_c$, densification rates given by Eq. (A7) go to zero (due to the inclusion of the term $(1 - \frac{5}{3}D)$), while densification rates given by Eq. (A8) go to infinity because the contact area $a$ equals zero at $D = D_c$. Clearly neither equation gives a realistic result at $D = D_c$. Therefore, in our implementation of the Arnaud model we use the H-L densification rates of Eq. (A1) instead of Eq. (A7) in stage 1. We take the onset of stage 2 to be the density at which Eqs. (A1) and (A8) intercept, thus avoiding the singularity in Eq. A8. This approach has the additional advantages of removing dependence on ad hoc scaling factor $\gamma$ and introducing realistic temperature dependence for stage 1 densification. Because stage 1 spans just the top 10–20 % of the firn column, the modification has only a minor influence on the overall behavior of the Arnaud model. The Goujon model code avoids the singularity in Eq. (A8) by extending stage 1 to $D_c + \varepsilon$ (Anaïs Orsi, personal communication, 2014), a procedure not described in Goujon et al. (2003).

*Acknowledgements.* We thank Mark Twickler and Joseph Souney Jr. of the WAIS Divide Science Coordination Office; the Ice Drilling Design and Operations group at the University of Wisconsin for recovering the ice core; O. Maselli, N. Chellman, M. Grieman, J. D'Andrilli, L. Layman, and R. Grinstead for help in making the continuous $CH_4$ and Ca measurements; A. J. Schauer, S. W. Schoenemann, P. D. Neff, and B. Vanden Heuvel for help in making the water-isotopologue measurements; Raytheon Polar Services for logistics support in Antarctica; the 109th New York Air National Guard for airlift in Antarctica; Anaïs Orsi for fruitful discussions; Dan Muhs for feedback on the manuscript; and the dozens of core handlers in the field and at the National Ice Core Laboratory for the core processing. This work is funded by the US National Science Foundation through grants 0539232, 0537661 (to K. M. Cuffey), ANT05-38657 (to J. P. Severinghaus.), NSFC 41230524 (to H. Cheng. and R. L. Edwards), 0839093, 1142166 (to J. R. McConnell.), and 1043092 (to E. J. Steig.) and through grants 0230396, 0440817, 0944348, and 0944266 to the Desert Research Institute of Reno Nevada and University of New Hampshire for the collection and distribution of the WAIS Divide ice core and related tasks. We further acknowledge support by the NOAA Climate and Global Change fellowship program, administered by the University Corporation for Atmospheric Research (to C. Buizert.) and the Villum Foundation (to M. Winstrup.).

Edited by: H. Fischer

## References

Aciego, S., Bourdon, B., Schwander, J., Baur, H., and Forieri, A.: Toward a radiometric ice clock: uranium ages of the Dome C ice core, Quaternary Sci. Rev., 30, 2389–2397, 2011.

Ahn, J. and Brook, E. J.: Atmospheric $CO_2$ and climate on millennial time scales during the last glacial period, Science, 322, 83–85, 2008.

Ahn, J., Brook, E. J., Schmittner, A., and Kreutz, K.: Abrupt change in atmospheric $CO_2$ during the last ice age, Geophys. Res. Lett., 39, L18711, doi:10.1029/2012gl053018, 2012.

Alley, R. B.: Firn densification by grain-boundary sliding – a 1st model, Journal De Physique, 48, 249–256, 1987.

Arnaud, L., Barnola, J. M., and Duval, P.: Physical modeling of the densification of snow/firn and ice in the upper part of polar ice sheets, in: Physics of Ice Core Records, edited by Hondoh, T., 285–305, 2000.

Barnola, J. M., Pimienta, P., Raynaud, D., and Korotkevich, Y. S.: $CO_2$-climate relationship as deduced from the Vostok ice core: a re-examination based on new measurements and on a re-evaluation of the air dating, Tellus B., 43, 83–90, 1991.

Battle, M. O., Severinghaus, J. P., Sofen, E. D., Plotkin, D., Orsi, A. J., Aydin, M., Montzka, S. A., Sowers, T., and Tans, P. P.: Controls on the movement and composition of firn air at the West Antarctic Ice Sheet Divide, Atmos. Chem. Phys., 11, 11007–11021, doi:10.5194/acp-11-11007-2011, 2011.

Baumgartner, M., Kindler, P., Eicher, O., Floch, G., Schilt, A., Schwander, J., Spahni, R., Capron, E., Chappellaz, J., Leuenberger, M., Fischer, H., and Stocker, T. F.: NGRIP $CH_4$ concentration from 120 to 10 kyr before present and its relation to a $\delta^{15}N$ temperature reconstruction from the same ice core, Clim. Past, 10, 903–920, doi:10.5194/cp-10-903-2014, 2014.

Bazin, L., Landais, A., Lemieux-Dudon, B., Toyé Mahamadou Kele, H., Veres, D., Parrenin, F., Martinerie, P., Ritz, C., Capron, E., Lipenkov, V., Loutre, M.-F., Raynaud, D., Vinther, B., Svensson, A., Rasmussen, S. O., Severi, M., Blunier, T., Leuenberger, M., Fischer, H., Masson-Delmotte, V., Chappellaz, J., and Wolff, E.: An optimized multi-proxy, multi-site Antarctic ice and gas orbital chronology (AICC2012): 120–800 ka, Clim. Past, 9, 1715–1731, doi:10.5194/cp-9-1715-2013, 2013.

Bendel, V., Ueltzhoffer, K. J., Freitag, J., Kipfstuhl, S., Kuhs, W. F., Garbe, C. S., and Faria, S. H.: High-resolution variations in size, number and arrangement of air bubbles in the EPICA DML (Antarctica) ice core, J. Glaciol., 59, 972–980, 2013.

Bender, M. L.: Orbital tuning chronology for the Vostok climate record supported by trapped gas composition, Earth Planet. Sc. Lett., 204, 275–289, 2002.

Bender, M. L., Barnett, B., Dreyfus, G., Jouzel, J., and Porcelli, D.: The contemporary degassing rate of Ar-40 from the Solid Earth, P. Natl. Acad. Sci. USA, 105, 8232–8237, 2008.

Blunier, T. and Brook, E. J.: Timing of millennial-scale climate change in Antarctica and Greenland during the last glacial period, Science, 291, 109–112, 2001.

Blunier, T. and Schwander, J.: Gas enclosure in ice: age difference and fractionation, in: Physics of Ice Core Records, edited by: Hondoh, T., Hokkaido University Press, Sapporo, 307–326, 2000.

Blunier, T., Chappellaz, J., Schwander, J., Dallenbach, A., Stauffer, B., Stocker, T. F., Raynaud, D., Jouzel, J., Clausen, H. B., Hammer, C. U., and Johnsen, S. J.: Asynchrony of Antarctic and Greenland climate change during the last glacial period, Nature, 394, 739–743, 1998.

Blunier, T., Spahni, R., Barnola, J.-M., Chappellaz, J., Loulergue, L., and Schwander, J.: Synchronization of ice core records via atmospheric gases, Clim. Past, 3, 325–330, doi:10.5194/cp-3-325-2007, 2007.

Broccoli, A. J., Dahl, K. A., and Stouffer, R. J.: Response of the ITCZ to Northern Hemisphere cooling, Geophys. Res. Lett., 33, L01702, doi:10.1029/2005gl024546, 2006.

Brook, E. J., Sowers, T., and Orchardo, J.: Rapid variations in atmospheric methane concentration during the past 110,000 years, Science, 273, 1087–1091, 1996.

Buizert, C., Martinerie, P., Petrenko, V. V., Severinghaus, J. P., Trudinger, C. M., Witrant, E., Rosen, J. L., Orsi, A. J., Rubino, M., Etheridge, D. M., Steele, L. P., Hogan, C., Laube, J. C., Sturges, W. T., Levchenko, V. A., Smith, A. M., Levin, I., Conway, T. J., Dlugokencky, E. J., Lang, P. M., Kawamura, K., Jenk, T. M., White, J. W. C., Sowers, T., Schwander, J., and Blunier, T.: Gas transport in firn: multiple-tracer characterisation and model intercomparison for NEEM, Northern Greenland, Atmos. Chem. Phys., 12, 4259–4277, doi:10.5194/acp-12-4259-2012, 2012.

Buizert, C., Sowers, T., and Blunier, T.: Assessment of diffusive isotopic fractionation in polar firn, and application to ice core trace gas records, Earth Planet. Sc. Lett., 361, 110–119, doi:10.1016/j.epsl.2012.11.039, 2013.

Buizert, C., Baggenstos, D., Jiang, W., Purtschert, R., Petrenko, V. V., Lu, Z.-T., Muller, P., Kuhl, T., Lee, J., Severinghaus, J. P., and Brook, E. J.: Radiometric $^{81}$Kr dating identifies 120,000-year-old ice at Taylor Glacier, Antarctica, P. Natl. Acad. Sci. USA, 111, 6876–6881, 2014a.

Buizert, C., Gkinis, V., Severinghaus, J. P., He, F., Lecavalier, B. S., Kindler, P., Leuenberger, M., Carlson, A. E., Vinther, B., Masson-Delmotte, V., White, J. W. C., Liu, Z., Otto-Bliesner, B., and Brook, E. J.: Greenland temperature response to climate forcing during the last deglaciation, Science, 345, 1177–1180, 2014b.

Caillon, N., Severinghaus, J. P., Jouzel, J., Barnola, J.-M., Kang, J., and Lipenkov, V. Y.: Timing of atmospheric $CO_2$ and antarctic temperature changes across termination III, Science, 299, 1728–1731, doi:10.1126/science.1078758, 2003.

Capron, E., Landais, A., Chappellaz, J., Schilt, A., Buiron, D., Dahl-Jensen, D., Johnsen, S. J., Jouzel, J., Lemieux-Dudon, B., Loulergue, L., Leuenberger, M., Masson-Delmotte, V., Meyer, H., Oerter, H., and Stenni, B.: Millennial and sub-millennial scale climatic variations recorded in polar ice cores over the last glacial period, Clim. Past, 6, 345–365, doi:10.5194/cp-6-345-2010, 2010.

Capron, E., Landais, A., Buiron, D., Cauquoin, A., Chappellaz, J., Debret, M., Jouzel, J., Leuenberger, M., Martinerie, P., Masson-Delmotte, V., Mulvaney, R., Parrenin, F., and Prié, F.: Glacial–interglacial dynamics of Antarctic firn columns: comparison between simulations and ice core air-$\delta^{15}$N measurements, Clim. Past, 9, 983–999, doi:10.5194/cp-9-983-2013, 2013.

Chappellaz, J., Stowasser, C., Blunier, T., Baslev-Clausen, D., Brook, E. J., Dallmayr, R., Faïn, X., Lee, J. E., Mitchell, L. E., Pascual, O., Romanini, D., Rosen, J., and Schüpbach, S.: High-resolution glacial and deglacial record of atmospheric methane by continuous-flow and laser spectrometer analysis along the NEEM ice core, Clim. Past, 9, 2579–2593, doi:10.5194/cp-9-2579-2013, 2013.

Chiang, J. C. H. and Bitz, C. M.: Influence of high latitude ice cover on the marine Intertropical Convergence Zone, Clim. Dynam., 25, 477–496, 2005.

Cuffey, K. M. and Clow, G. D.: Temperature, accumulation, and ice sheet elevation in central Greenland through the last deglacial transition, J. Geophys. Res., 102, 26383–26396, 1997.

Cuffey, K. M. and Paterson, W. S. B.: The Physics of Glaciers, 4th edn., Butterworth-Heinemann, Oxford, UK, 2010.

Cuffey, K. M., Clow, G. D., Alley, R. B., Stuiver, M., Waddington, E. D., and Saltus, R. W.: Large arctic temperature change at the Wisconsin-Holocene glacial transition, Science, 270, 455–458, 1995.

Cvijanovic, I. and Chiang, J. C. H.: Global energy budget changes to high latitude North Atlantic cooling and the tropical ITCZ response, Clim. Dynam., 40, 1435–1452, 2013.

Dahl-Jensen, D., Mosegaard, K., Gundestrup, N., Clow, G. D., Johnsen, S. J., Hansen, A. W., and Balling, N.: Past temperatures directly from the greenland ice sheet, Science, 282, 268–271, 1998.

Dansgaard, W. and Johnsen, S.: A flow model and a time scale for the ice core from Camp Century, Greenland, J. Glaciol., 8, 215–223, 1969.

Dlugokencky, E., Myers, R., Lang, P., Masarie, K., Crotwell, A., Thoning, K., Hall, B., Elkins, J., and Steele, L.: Conversion of NOAA atmospheric dry air $CH_4$ mole fractions to a gravimetrically prepared standard scale, J. Geophys. Res.-Atmos., 110, D18306, doi:10.1029/2005JD006035, 2005.

Dreyfus, G. B., Jouzel, J., Bender, M. L., Landais, A., Masson-Delmotte, V., and Leuenberger, M.: Firn processes and $\delta^{15}$N: potential for a gas-phase climate proxy, Quaternary Sci. Rev., 29, 28–42, 2010.

Edwards, R. L., Cheng, H., Wang, Y. J., Yuan, D. X., Kelly, M. J., Severinghaus, J. P., Burnett, A., Wang, X. F., Smith, E., and

Kong, X. G.: A Refined Hulu and Dongge Cave Climate Record and the Timing of Climate Change during the Last Glacial Cycle, Earth Planet. Sci. Lett., in review, 2015.

EPICA Community Members: One-to-one coupling of glacial climate variability in Greenland and Antarctica, Nature, 444, 195–198, 2006.

Etheridge, D. M., Pearman, G. I., and Fraser, P. J.: Changes in tropospheric methane between 1841 and 1978 from a high accumulation-rate Antarctic ice core, Tellus B, 44, 282–294, 1992.

Fleitmann, D., Cheng, H., Badertscher, S., Edwards, R. L., Mudelsee, M., Gokturk, O. M., Fankhauser, A., Pickering, R., Raible, C. C., Matter, A., Kramers, J., and Tuysuz, O.: Timing and climatic impact of Greenland interstadials recorded in stalagmites from northern Turkey, Geophys. Res. Lett., 36, L19707, doi:10.1029/2009gl040050, 2009.

Freitag, J., Kipfstuhl, J., Laepple, T., and Wilhelms, F.: Impurity-controlled densification: a new model for stratified polar firn, J. Glaciol., 59, 1163–1169, 2013a.

Freitag, J., Kipfstuhl, S., and Laepple, T.: Core-scale radioscopic imaging: a new method reveals density-calcium link in Antarctic firn, J. Glaciol., 59, 1009–1014, 2013b.

Fudge, T. J., Waddington, E. D., Conway, H., Lundin, J. M. D., and Taylor, K.: Interpolation methods for Antarctic ice-core timescales: application to Byrd, Siple Dome and Law Dome ice cores, Clim. Past, 10, 1195–1209, doi:10.5194/cp-10-1195-2014, 2014.

Fujita, S., Hirabayashi, M., Goto-Azuma, K., Dallmayr, R., Satow, K., Zheng, J., and Dahl-Jensen, D.: Densification of layered firn of the ice sheet at NEEM, Greenland, J. Glaciol., 60, 905–921, doi:10.3189/2014JoG14J006, 2014.

Goujon, C., Barnola, J. M., and Ritz, C.: Modeling the densification of polar firn including heat diffusion: Application to close-off characteristics and gas isotopic fractionation for Antarctica and Greenland sites, J. Geophys. Res.-Atmos., 108, 18 pp., doi:10.1029/2002jd003319, 2003.

Gow, A. J., Meese, D. A., Alley, R. B., Fitzpatrick, J. J., Anandakrishnan, S., Woods, G. A., and Elder, B. C.: Physical and structural properties of the Greenland Ice Sheet Project 2 ice core: a review, J. Geophys. Res.-Oceans, 102, 26559–26575, 1997.

Gundestrup, N. S., Dahl-Jensen, D., Hansen, B. L., and Kelty, J.: Bore-hole survey at Camp Century, 1989, Cold Reg. Sci. Technol., 21, 187–193, 1993.

Herron, M. M. and Langway, C. C.: Firn densification: an empirical model, J. Glaciol., 25, 373–385, 1980.

Horhold, M. W., Laepple, T., Freitag, J., Bigler, M., Fischer, H., and Kipfstuhl, S.: On the impact of impurities on the densification of polar firn, Earth Planet. Sc. Lett., 325–326, 93–99, 2012.

Huber, C., Leuenberger, M., Spahni, R., Fluckiger, J., Schwander, J., Stocker, T. F., Johnsen, S., Landals, A., and Jouzel, J.: Isotope calibrated Greenland temperature record over Marine Isotope Stage 3 and its relation to $CH_4$, Earth Planet. Sc. Lett., 243, 504–519, 2006.

Johnsen, S. J., Dahl-Jensen, D., Gundestrup, N., Steffensen, J. P., Clausen, H. B., Miller, H., Masson-Delmotte, V., Sveinbjörnsdottir, A. E., and White, J.: Oxygen isotope and palaeotemperature records from six Greenland ice-core stations: Camp Century, Dye-3, GRIP, GISP2, Renland and NorthGRIP, J. Quaternary Sci., 16, 299–307, 2001.

Kanner, L. C., Burns, S. J., Cheng, H., and Edwards, R. L.: High-latitude forcing of the south American summer monsoon during the last glacial, Science, 335, 570–573, 2012.

Kaspers, K. A., van de Wal, R. S. W., van den Broeke, M. R., Schwander, J., van Lipzig, N. P. M., and Brenninkmeijer, C. A. M.: Model calculations of the age of firn air across the Antarctic continent, Atmos. Chem. Phys., 4, 1365–1380, doi:10.5194/acp-4-1365-2004, 2004.

Kawamura, K., Severinghaus, J. P., Ishidoya, S., Sugawara, S., Hashida, G., Motoyama, H., Fujii, Y., Aoki, S., and Nakazawa, T.: Convective mixing of air in firn at four polar sites, Earth Planet. Sc. Lett., 244, 672–682, 2006.

Kawamura, K., Parrenin, F., Lisiecki, L., Uemura, R., Vimeux, F., Severinghaus, J. P., Hutterli, M. A., Nakazawa, T., Aoki, S., Jouzel, J., Raymo, M. E., Matsumoto, K., Nakata, H., Motoyama, H., Fujita, S., Goto-Azuma, K., Fujii, Y., and Watanabe, O.: Northern Hemisphere forcing of climatic cycles in Antarctica over the past 360 000 years, Nature, 448, 912–916, 2007.

Kindler, P., Guillevic, M., Baumgartner, M., Schwander, J., Landais, A., and Leuenberger, M.: Temperature reconstruction from 10 to 120 kyr b2k from the NGRIP ice core, Clim. Past, 10, 887–902, doi:10.5194/cp-10-887-2014, 2014.

Lal, D., Jull, A. J. T., Donahue, D. J., Burtner, D., and Nishiizumi, K.: Polar ice ablation rates measured using in situ cosmogenic C-14, Nature, 346, 350–352, 1990.

Landais, A., Barnola, J. M., Kawamura, K., Caillon, N., Delmotte, M., Van Ommen, T., Dreyfus, G., Jouzel, J., Masson-Delmotte, V., Minster, B., Freitag, J., Leuenberger, M., Schwander, J., Huber, C., Etheridge, D., and Morgan, V.: Firn-air delta N-15 in modern polar sites and glacial-interglacial ice: a model-data mismatch during glacial periods in Antarctica?, Quaternary Sci. Rev., 25, 49–62, 2006.

Marcott, S. A., Bauska, T. K., Buizert, C., Steig, E. J., Rosen, J. L., Cuffey, K. M., Fudge, T. J., Severinghaus, J. P., Ahn, J., Kalk, M. L., McConnell, J. R., Sowers, T., Taylor, K. C., White, J. W. C., and Brook, E. J.: Centennial-scale changes in the global carbon cycle during the last deglaciation, Nature, 514, 616–619, 2014.

Martinerie, P., Lipenkov, V. Y., Raynaud, D., Chappellaz, J., Barkov, N. I., and Lorius, C.: Air content paleo record in the Vostok ice core (Antarctica): A mixed record of climatic and glaciological parameters, J. Geophys. Res.-Atmos., 99, 10565–10576, 1994.

McConnell, J. R., Lamorey, G. W., Lambert, S. W., and Taylor, K. C.: Continuous ice-core chemical analyses using inductively coupled plasma mass spectrometry, Environ. Sci. Technol., 36, 7–11, 2002.

McConnell, J. R., Edwards, R., Kok, G. L., Flanner, M. G., Zender, C. S., Saltzman, E. S., Banta, J. R., Pasteris, D. R., Carter, M. M., and Kahl, J. D. W.: 20th-century industrial black carbon emissions altered arctic climate forcing, Science, 317, 1381–1384, 2007.

McManus, J. F., Francois, R., Gherardi, J. M., Keigwin, L. D., and Brown-Leger, S.: Collapse and rapid resumption of Atlantic meridional circulation linked to deglacial climate changes, Nature, 428, 834–837, 2004.

Mischler, J. A., Sowers, T. A., Alley, R. B., Battle, M., McConnell, J. R., Mitchell, L., Popp, T., Sofen, E., and

Spencer, M. K.: Carbon and hydrogen isotopic composition of methane over the last 1000 years, Global Biogeochem. Cy., 23, GB4024, doi:10.1029/2009gb003460, 2009.

Mitchell, L., Brook, E., Lee, J. E., Buizert, C., and Sowers, T.: Constraints on the late holocene anthropogenic contribution to the atmospheric methane budget, Science, 342, 964–966, 2013.

Mitchell, L. E., Brook, E. J., Sowers, T., McConnell, J. R., and Taylor, K.: Multidecadal variability of atmospheric methane, 1000–1800 C.E, J. Geophys. Res., 116, G02007, doi:10.1029/2010jg001441, 2011.

NEEM community members: Eemian interglacial reconstructed from a Greenland folded ice core, Nature, 493, 489–494, 2013.

NGRIP community members: High-resolution record of Northern Hemisphere climate extending into the last interglacial period, Nature, 431, 147–151, 2004.

Orsi, A. J.: Temperature reconstruction at the West Antarctic Ice Sheet Divide, for the last millennium, from the combination of borehole temperature and inert gas isotope measurements, Ph. D. thesis, University of California, San Diego, 2013.

Parrenin, F., Barker, S., Blunier, T., Chappellaz, J., Jouzel, J., Landais, A., Masson-Delmotte, V., Schwander, J., and Veres, D.: On the gas-ice depth difference (Δdepth) along the EPICA Dome C ice core, Clim. Past, 8, 1239–1255, doi:10.5194/cp-8-1239-2012, 2012.

Parrenin, F., Masson-Delmotte, V., Kohler, P., Raynaud, D., Paillard, D., Schwander, J., Barbante, C., Landais, A., Wegner, A., and Jouzel, J.: Synchronous change of atmospheric $CO_2$ and antarctic temperature during the last deglacial warming, Science, 339, 1060–1063, 2013.

Pedro, J. B., van Ommen, T. D., Rasmussen, S. O., Morgan, V. I., Chappellaz, J., Moy, A. D., Masson-Delmotte, V., and Delmotte, M.: The last deglaciation: timing the bipolar seesaw, Clim. Past, 7, 671–683, doi:10.5194/cp-7-671-2011, 2011.

Pedro, J. B., Rasmussen, S. O., and van Ommen, T. D.: Tightened constraints on the time-lag between Antarctic temperature and $CO_2$ during the last deglaciation, Clim. Past, 8, 1213–1221, doi:10.5194/cp-8-1213-2012, 2012.

Petrenko, V. V., Severinghaus, J. P., Brook, E. J., Reeh, N., and Schaefer, H.: Gas records from the West Greenland ice margin covering the Last Glacial Termination: a horizontal ice core, Quaternary Sci. Rev., 25, 865–875, 2006.

Rasmussen, S. O., Andersen, K. K., Svensson, A. M., Steffensen, J. P., Vinther, B. M., Clausen, H. B., Siggaard-Andersen, M. L., Johnsen, S. J., Larsen, L. B., Dahl-Jensen, D., Bigler, M., Röthlisberger, R., Fischer, H., Goto-Azuma, K., Hansson, M. E., and Ruth, U.: A new Greenland ice core chronology for the last glacial termination, J. Geophys. Res., 111, D06102, doi:10.1029/2005jd006079, 2006.

Rasmussen, S. O., Abbott, P. M., Blunier, T., Bourne, A. J., Brook, E., Buchardt, S. L., Buizert, C., Chappellaz, J., Clausen, H. B., Cook, E., Dahl-Jensen, D., Davies, S. M., Guillevic, M., Kipfstuhl, S., Laepple, T., Seierstad, I. K., Severinghaus, J. P., Steffensen, J. P., Stowasser, C., Svensson, A., Vallelonga, P., Vinther, B. M., Wilhelms, F., and Winstrup, M.: A first chronology for the North Greenland Eemian Ice Drilling (NEEM) ice core, Clim. Past, 9, 2713–2730, doi:10.5194/cp-9-2713-2013, 2013.

Rasmussen, S. O., Bigler, M., Blockley, S. P., Blunier, T., Buchardt, S. L., Clausen, H. B., Cvijanovic, I., Dahl-Jensen, D., Johnsen, S. J., Fischer, H., Gkinis, V., Guillevic, M., Hoek, W. Z., Lowe, J. J., Pedro, J. B., Popp, T., Seierstad, I. K., Steffensen, J. P., Svensson, A. M., Vallelonga, P., Vinther, B. M., Walker, M. J., Wheatley, J. J., and Winstrup, M.: A stratigraphic framework for abrupt climatic changes during the Last Glacial period based on three synchronized Greenland ice-core records: refining and extending the INTIMATE event stratigraphy, Quaternary Sci. Rev., 106, 14–28, 2014.

Reimer, P., Bard, E., Bayliss, A., Beck, J., Blackwell, P., Ramsey, C. B., Buck, C., Cheng, H., Edwards, R. L., Friedrich, M., Grootes, P., Guilderson, T., Haflidason, H., Hajdas, I., Hatté, C., Heaton, T., Hoffmann, D., Hogg, A., Hughen, K., Kaiser, K., Kromer, B., Manning, S., Niu, M., Reimer, R., Richards, D., Scott, E., Southon, J., Staff, R., Turney, C., and van der Plicht, J.: IntCal13 and Marine13 Radiocarbon Age Calibration Curves 0–50 000 Years cal BP, Radiocarbon, 55, 1869–1887, 2013.

Rhodes, R. H., Fain, X., Stowasser, C., Blunier, T., Chappellaz, J., McConnell, J. R., Romanini, D., Mitchell, L. E., and Brook, E. J.: Continuous methane measurements from a late Holocene Greenland ice core: atmospheric and in-situ signals, Earth Planet. Sc. Lett., 368, 9–19, 2013.

Rhodes, R. H. Brook, E. J., Chiang, J. C. H., Blunier, T., Maselli, O. J., McConnell, J. R., Romanini, D., Severinghaus, J. P.: Enhanced tropical methane production in response to iceberg discharge in the North Atlantic, in review, 2015.

Rosen, J. L., Brook, E. J., Severinghaus, J. P., Blunier, T., Mitchell, L. E., Lee, J. E., Edwards, J. S., and Gkinis, V.: An ice core record of near-synchronous global climate changes at the Bolling transition, Nat. Geosci., 7, 459–463, 2014.

Schwander, J. and Stauffer, B.: Age difference between polar ice and the air trapped in its bubbles, Nature, 311, 45–47, 1984.

Schwander, J., Sowers, T., Barnola, J. M., Blunier, T., Fuchs, A., and Malaize, B.: Age scale of the air in the summit ice: Implication for glacial-interglacial temperature change, J. Geophys. Res.-Atmos., 102, 19483–19493, 1997.

Seierstad, I., Abbott, P., Bigler, M., Blunier, T., Bourne, A., Brook, E., Buchardt, S. L., Buizert, C., Clausen, H. B., Cook, E., Dahl-Jensen, D., Davies, S., Guillevic, M., Johnsen, S., Pedersen, D., Popp, T., Rasmussen, S. O., Severinghaus, J., Svensson, A., and Vinther, B.: Consistently dated records from the Greenland GRIP, GISP2 and NGRIP ice cores for the past 104 ka reveal regional millennial-scale isotope gradients with possible Heinrich Event imprint, Quaternary Sci. Rev., 106, 29–46, 2014.

Severi, M., Becagli, S., Castellano, E., Morganti, A., Traversi, R., Udisti, R., Ruth, U., Fischer, H., Huybrechts, P., Wolff, E., Parrenin, F., Kaufmann, P., Lambert, F., and Steffensen, J. P.: Synchronisation of the EDML and EDC ice cores for the last 52 kyr by volcanic signature matching, Clim. Past, 3, 367–374, doi:10.5194/cp-3-367-2007, 2007.

Severinghaus, J. P., Beaudette, R., Headly, M. A., Taylor, K., and Brook, E. J.: Oxygen-18 of $O_2$ records the impact of abrupt climate change on the terrestrial biosphere, Science, 324, 1431–1434, 2009.

Sigl, M., McConnell, J. R., Layman, L., Maselli, O., McGwire, K., Pasteris, D., Dahl-Jensen, D., Steffensen, J. P., Vinther, B., Edwards, R., Mulvaney, R., and Kipfstuhl, S.: A new bipolar ice core record of volcanism from WAIS Divide and NEEM and implications for climate forcing of the last 2000 years, J. Geophys. Res.-Atmos., 118, 1151–1169, 2013.

Sigl, M., McConnell, J. R., Toohey, M., Curran, M., Das, S. B., Edwards, R., Isaksson, E., Kawamura, K., Kipfstuhl, S., Krüger, K., Layman, L., Maselli, O. J., Motizuki, Y., Motoyama, H., Pasteris, D. R., and Severi, M.: Insights from Antarctica on volcanic forcing during the Common Era, Nat. Clim. Change, 4, 693–697, 2014.

Southon, J., Noronha, A. L., Cheng, H., Edwards, R. L., and Wang, Y.: A high-resolution record of atmospheric $^{14}$C based on Hulu Cave speleothem H82, Quaternary Sci. Rev., 33, 32–41, 2012.

Sowers, T., Bender, M., and Raynaud, D.: Elemental and isotopic composition of occluded $O_2$ and $N_2$ in polar ice, J. Geophys. Res.-Atmos., 94, 5137–5150, 1989.

Sowers, T., Bender, M., Raynaud, D., and Korotkevich, Y. S.: $\delta^{15}$N of $N_2$ in air trapped in polar ice: A tracer of gas transport in the firn and a possible constraint on ice age-gas age differences, J. Geophys. Res.-Atmos., 97, 15683–15697, 1992.

Steig, E. J., Ding, Q., White, J. W. C., Kuttel, M., Rupper, S. B., Neumann, T. A., Neff, P. D., Gallant, A. J. E., Mayewski, P. A., Taylor, K. C., Hoffmann, G., Dixon, D. A., Schoenemann, S. W., Markle, B. R., Fudge, T. J., Schneider, D. P., Schauer, A. J., Teel, R. P., Vaughn, B. H., Burgener, L., Williams, J., and Korotkikh, E.: Recent climate and ice-sheet changes in West Antarctica compared with the past 2000 years, Nat. Geosci., 6, 372–375, 2013.

Stocker, T. F. and Johnsen, S. J.: A minimum thermodynamic model for the bipolar seesaw, Paleoceanography, 18, 1087, doi:10.1029/2003pa000920, 2003.

Stowasser, C., Buizert, C., Gkinis, V., Chappellaz, J., Schüpbach, S., Bigler, M., Faïn, X., Sperlich, P., Baumgartner, M., Schilt, A., and Blunier, T.: Continuous measurements of methane mixing ratios from ice cores, Atmos. Meas. Tech., 5, 999–1013, 2012, http://www.atmos-meas-tech.net/5/999/2012/.

Svensson, A., Andersen, K. K., Bigler, M., Clausen, H. B., Dahl-Jensen, D., Davies, S. M., Johnsen, S. J., Muscheler, R., Rasmussen, S. O., Rothlisberger, R., Steffensen, J. P., and Vinther, B. M.: The Greenland Ice Core Chronology 2005, 15–42 ka. Part 2: Comparison to other records, Quaternary Sci. Rev., 25, 3258–3267, 2006.

Veres, D., Bazin, L., Landais, A., Toyé Mahamadou Kele, H., Lemieux-Dudon, B., Parrenin, F., Martinerie, P., Blayo, E., Blunier, T., Capron, E., Chappellaz, J., Rasmussen, S. O., Severi, M., Svensson, A., Vinther, B., and Wolff, E. W.: The Antarctic ice core chronology (AICC2012): an optimized multi-parameter and multi-site dating approach for the last 120 thousand years, Clim. Past, 9, 1733–1748, doi:10.5194/cp-9-1733-2013, 2013.

Voelker, A. H. L.: Global distribution of centennial-scale records for Marine Isotope Stage (MIS) 3: A database, Quaternary Sci. Rev., 21, 1185–1212, 2002.

WAIS Divide Project Members: Onset of deglacial warming in West Antarctica driven by local orbital forcing, Nature, 500, 440–444, 2013.

WAIS Divide Project Members: Precise interpolar phasing of abrupt climate change during the last ice age, in press, 2015.

Wang, X. F., Auler, A. S., Edwards, R. L., Cheng, H., Ito, E., and Solheid, M.: Interhemispheric anti-phasing of rainfall during the last glacial period, Quaternary Sci. Rev., 25, 3391–3403, 2006.

Wang, Y. J., Cheng, H., Edwards, R. L., An, Z. S., Wu, J. Y., Shen, C. C., and Dorale, J. A.: A high-resolution absolute-dated late pleistocene monsoon record from Hulu Cave, China, Science, 294, 2345–2348, 2001.

Winstrup, M., Svensson, A. M., Rasmussen, S. O., Winther, O., Steig, E. J., and Axelrod, A. E.: An automated approach for annual layer counting in ice cores, Clim. Past, 8, 1881–1895, doi:10.5194/cp-8-1881-2012, 2012.

Wolff, E. W., Fischer, H., Fundel, F., Ruth, U., Twarloh, B., Littot, G. C., Mulvaney, R., Rothlisberger, R., de Angelis, M., Boutron, C. F., Hansson, M., Jonsell, U., Hutterli, M. A., Lambert, F., Kaufmann, P., Stauffer, B., Stocker, T. F., Steffensen, J. P., Bigler, M., Siggaard-Andersen, M. L., Udisti, R., Becagli, S., Castellano, E., Severi, M., Wagenbach, D., Barbante, C., Gabrielli, P., and Gaspari, V.: Southern Ocean sea-ice extent, productivity and iron flux over the past eight glacial cycles, Nature, 440, 491–496, 2006.

Wolff, E. W., Chappellaz, J., Blunier, T., Rasmussen, S. O., and Svensson, A.: Millennial-scale variability during the last glacial: the ice core record, Quaternary Sci. Rev., 29, 2828–2838, 2010.

# Permissions

# List of Contributors

**D. J. Hill**
School of Earth and Environment, University of Leeds, Leeds, UK
British Geological Survey, Keyworth, Nottingham, UK

**A. M. Haywood**
School of Earth and Environment, University of Leeds, Leeds, UK

**D. J. Lunt**
School of Geographical Sciences, University of Bristol, Bristol, UK

**S. J. Hunter**
School of Earth and Environment, University of Leeds, Leeds, UK

**F. J. Bragg**
School of Geographical Sciences, University of Bristol, Bristol, UK

**C. Contoux**
Laboratoire des Sciences du Climat et de l'Environnement, Saclay, France
Sisyphe, CNRS/UPMC Univ. Paris 06, Paris, France

**C. Stepanek**
Alfred Wegener Institute Helmholtz Centre for Polar and Marine Research, Bremerhaven, Germany

**L. Sohl**
Columbia University – NASA/GISS, New York, NY, USA

**N. A. Rosenbloom**
National Center for Atmospheric Research, Boulder, Colorado, USA

**W.-L. Chan**
Atmosphere and Ocean Research Institute, University of Tokyo, Kashiwa, Japan

**Y. Kamae**
Graduate School of Life and Environmental Sciences, University of Tsukuba, Tsukuba, Japan

**Z. Zhang**
UniResearch and Bjerknes Centre for Climate Research, Bergen, Norway
Nansen-zhu International Research Centre, Institute of Atmospheric Physics, Chinese Academy of Sciences, Beijing, China

**A. Abe-Ouchi**
Atmosphere and Ocean Research Institute, University of Tokyo, Kashiwa, Japan
Japan Agency for Marine-Earth Science and Technology, Yokohama, Japan

**M. A. Chandler**
Columbia University – NASA/GISS, New York, NY, USA

**A. Jost**
Sisyphe, CNRS/UPMC Univ. Paris 06, Paris, France

**G. Lohmann**
Alfred Wegener Institute Helmholtz Centre for Polar and Marine Research, Bremerhaven, Germany

**B. L. Otto-Bliesner**
National Center for Atmospheric Research, Boulder, Colorado, USA

**G. Ramstein**
Laboratoire des Sciences du Climat et de l'Environnement, Saclay, France

**H. Ueda**
Graduate School of Life and Environmental Sciences, University of Tsukuba, Tsukuba, Japan

**C. Meyer-Jacob**
Department of Ecology and Environmental Science, Umeå University, 901 87 Umeå, Sweden

**H. Vogel**
Institute of Geology and Mineralogy, University of Cologne, Zuelpicher Str. 49a, 50674 Cologne, Germany
Institute of Geological Sciences&Oeschger Centre for Climate Change Research, University of Bern, Baltzerstrasse 1+3,

**A. C. Gebhardt**
Alfred Wegener Institute Helmholtz Centre for Polar and Marine Research, Columbusstraße, 27515 Bremerhaven, Germany

**V. Wennrich**
Institute of Geology and Mineralogy, University of Cologne, Zuelpicher Str. 49a, 50674 Cologne, Germany

**M. Melles**
Institute of Geology and Mineralogy, University of Cologne, Zuelpicher Str. 49a, 50674 Cologne, Germany

**P. Rosén**
Department of Ecology and Environmental Science, Umeå University, 901 87 Umeå, Sweden
Climate Impacts Research Centre (CIRC), Abisko Scientific Research Station, 981 07 Abisko, Sweden

**S. Kasper**
NIOZ Royal Netherlands Institute for Sea Research, Department of Marine Organic Biogeochemistry, P.O. Box 59, 1790 AB Den Burg (Texel), the Netherlands

**M. T. J. van der Meer**
NIOZ Royal Netherlands Institute for Sea Research, Department of Marine Organic Biogeochemistry, P.O. Box 59, 1790 AB Den Burg (Texel), the Netherlands

**A. Mets**
NIOZ Royal Netherlands Institute for Sea Research, Department of Marine Organic Biogeochemistry, P.O. Box 59, 1790 AB Den Burg (Texel), the Netherlands

**R. Zahn**
Institució Catalana de Recerca i Estudis Avançats, ICREA, Barcelona, Spain
Universitat Autònoma de Barcelona, Institut de Ciència i Tecnologia Ambientals (ICTA) and Departament de Física, 08193 Bellaterra, Spain

**J. S. Sinninghe Damsté**
NIOZ Royal Netherlands Institute for Sea Research, Department of Marine Organic Biogeochemistry, P.O. Box 59, 1790 AB Den Burg (Texel), the Netherlands

**S. Schouten**
NIOZ Royal Netherlands Institute for Sea Research, Department of Marine Organic Biogeochemistry, P.O. Box 59, 1790 AB Den Burg (Texel), the Netherlands

**K. Rehfeld**
Potsdam Institute for Climate Impact Research, P.O. Box 601203, 14412 Potsdam, Germany
Department of Physics, Humboldt-Universität zu Berlin, Newtonstr. 15, 12489 Berlin, Germany

**J. Kurths**
Potsdam Institute for Climate Impact Research, P.O. Box 601203, 14412 Potsdam, Germany
Department of Physics, Humboldt-Universität zu Berlin, Newtonstr. 15, 12489 Berlin, Germany
Institute for Complex Systems and Mathematical Biology, University of Aberdeen, Aberdeen AB243UE, UK

**M. Heinze**
Max Planck Institute for Meteorology, Bundesstrasse 53, 20146 Hamburg, Germany
International Max Planck Research School on Earth System Modelling, Hamburg, Germany

**T. Ilyina**
Max Planck Institute for Meteorology, Bundesstrasse 53, 20146 Hamburg, Germany

**M. J. Herrero**
Departamento de Petrología y Geoquímica, Fac. Ciencias Geológicas, Universidad Complutense Madrid, C/Jose Antonio Novais 2, 28040 Madrid, Spain

**J. I. Escavy**
Departamento de Petrología y Geoquímica, Fac. Ciencias Geológicas, Universidad Complutense Madrid, C/Jose Antonio Novais 2, 28040 Madrid, Spain

**B. C. Schreiber**
Department of Earth and Space Sciences, University of Washington, Seattle, WA 98195, USA

**O. Cartapanis**
Aix-Marseille Université, CNRS, IRD, Collège de France, CEREGE UM34, 13545 Aix en Provence, France

**K. Tachikawa**
Aix-Marseille Université, CNRS, IRD, Collège de France, CEREGE UM34, 13545 Aix en Provence, France

**O. E. Romero**
Instituto Andaluz de Cs. de la Tierra (CSIC-UGR), Ave. de las Palmeras 4, 18100 Armilla-Granada, Spain
McGill University, Department of Earth and Planetary Sciences, 3450 University Street, Montreal, H3A 0E8, Quebec, Canada
MARUM, Center for Marine Environmental Sciences, University of Bremen, Leobener Str., 28359 Bremen, Germany

**E. Bard**
Aix-Marseille Université, CNRS, IRD, Collège de France, CEREGE UM34, 13545 Aix en Provence, France

**H. C. Steen-Larsen**
Laboratoire des Sciences du Climat et de l'Environnement, UMR8212, CEA-CNRS-UVSQ/IPSL, Gif-sur-Yvette, France
Cooperative Institute for Research in Environmental Sciences, University of Colorado, Boulder, USA
Centre for Ice and Climate, Niels Bohr Institute, University of Copenhagen, Copenhagen, Denmark

**V. Masson-Delmotte**
Laboratoire des Sciences du Climat et de l'Environnement, UMR8212, CEA-CNRS-UVSQ/IPSL, Gif-sur-Yvette, France

**M. Hirabayashi**
National Institute of Polar Research, Tokyo, Japan

**R. Winkler**
Laboratoire des Sciences du Climat et de l'Environnement, UMR8212, CEA-CNRS-UVSQ/IPSL, Gif-sur-Yvette, France

**K. Satow**
National Institute of Polar Research, Tokyo, Japan

**F. Prié**
Laboratoire des Sciences du Climat et de l'Environnement, UMR8212, CEA-CNRS-UVSQ/IPSL, Gif-sur-Yvette, France

**N. Bayou**
Cooperative Institute for Research in Environmental Sciences, University of Colorado, Boulder, USA

**E. Brun**
Meteo-France – CNRS, CNRM-GAME UMR 3589, GMGEC, Toulouse, France

**K. M. Cuffey**
Department of Geography, Center for Atmospheric Sciences, 507 McCone Hall, University of California, Berkeley, CA 94720-4740, USA

**D. Dahl-Jensen**
Centre for Ice and Climate, Niels Bohr Institute, University of Copenhagen, Copenhagen, Denmark

**M. Dumont**
Meteo-France – CNRS, CNRM-GAME UMR 3589, CEN, Grenoble, France

**M. Guillevic**
Laboratoire des Sciences du Climat et de l'Environnement, UMR8212, CEA-CNRS-UVSQ/IPSL, Gif-sur-Yvette, France
Centre for Ice and Climate, Niels Bohr Institute, University of Copenhagen, Copenhagen, Denmark

**S. Kipfstuhl**
Alfred Wegener Institute for Polar and Marine Research, Bremerhaven, Germany

**A. Landais**
Laboratoire des Sciences du Climat et de l'Environnement, UMR8212, CEA-CNRS-UVSQ/IPSL, Gif-sur-Yvette, France

**T. Popp**
Centre for Ice and Climate, Niels Bohr Institute, University of Copenhagen, Copenhagen, Denmark

**C. Risi**
Laboratoire de Météorologie Dynamique, Jussieu, Paris, France

**K. Steffen**
Cooperative Institute for Research in Environmental Sciences, University of Colorado, Boulder, USA
ETH, Swiss Federal Institute of Technology, Zurich, Switzerland

**B. Stenni**
Department of Geological, Environmental and Marine Sciences, University of Trieste, Trieste, Italy

**A. E. Sveinbjörnsdottír**
Institute of Earth Sciences, University of Iceland, Reykjavik, Iceland

**A. Matsikaris**
University of Birmingham, Edgbaston, Birmingham B15 2TT, UK
Max Planck Institute for Meteorology, Hamburg, Germany

**M. Widmann**
University of Birmingham, Edgbaston, Birmingham B15 2TT, UK

**J. Jungclaus**
Max Planck Institute for Meteorology, Hamburg, Germany

**G. E. A. Swann**
School of Geography, University of Nottingham, University Park, Nottingham, NG7 2RD, UK

**A. M. Snelling**
NERC Isotope Geosciences Facilities, British Geological Survey, Keyworth, Nottingham, NG12 5GG, UK

**G. A. Schmidt**
NASA Goddard Institute for Space Studies, 2880 Broadway, New York, NY 10025, USA

**J. D. Annan**
Research Institute for Global Change, JAMSTEC, Yokohama Institute for Earth Sciences, Yokohama, Japan

**P. J. Bartlein**
University of Oregon, Eugene, OR 97403, USA

**B. I. Cook**
NASA Goddard Institute for Space Studies, 2880 Broadway, New York, NY 10025, USA

**E. Guilyardi**
NCAS-Climate, University of Reading, Whiteknights, P.O. Box 217, Reading, Berkshire, RG6 6AH, UK

Laboratoire d'Océanographie et du Climat: Expérimentation et Approches Numériques/Institut Pierre Simon Laplace, CNRS-IRD-UPMC – UMR7617, 4 place Jussieu, 75252 Paris Cedex 05, France

**J. C. Hargreaves**
Research Institute for Global Change, JAMSTEC, Yokohama Institute for Earth Sciences, Yokohama, Japan

**S. P. Harrison**
Centre for Past Climate Change and School of Archaeology, Geography and Environmental Sciences (SAGES), University of Reading, Whiteknights, P.O. Box 217, Reading, Berkshire, RG6 6AH, UK
Macquarie University, Sydney, NSW 2109, Australia

**M. Kageyama**
Laboratoire des Sciences du Climat et de l'Environnement, Institut Pierre Simon Laplace, CEA-CNRS-UVSQ – UMR8212, CE Saclay l'Orme des Merisiers, 91191 Gif-sur-Yvette, France

**A. N. LeGrande**
NASA Goddard Institute for Space Studies, 2880 Broadway, New York, NY 10025, USA

**B. Konecky**
Brown University, 324 Brook St., P.O. Box 1846, Providence, RI 02912, USA

**S. Lovejoy**
McGill University, 805 Sherbrooke Street West Montreal, Quebec, H3A 0B9, Canada

**M. E. Mann**
Pennsylvania State University, University Park, PA 16802, USA

**V. Masson-Delmotte**
Laboratoire des Sciences du Climat et de l'Environnement, Institut Pierre Simon Laplace, CEA-CNRS-UVSQ – UMR8212, CE Saclay l'Orme des Merisiers, 91191 Gif-sur-Yvette, France

**C. Risi**
Laboratoire de Météorologie Dynamique/Institut Pierre Simon Laplace, 4, place Jussieu, 75252 Paris Cedex 05, France

**D. Thompson**
University of Arizona, Department of Geosciences, Gould-Simpson Building #77, 1040 E 4th St., Tucson, AZ 85721, USA

**A. Timmermann**
University of Hawaii, 2525 Correa Road, Honolulu, HI 96822, USA

**L.-B. Tremblay**
McGill University, 805 Sherbrooke Street West Montreal, Quebec, H3A 0B9, Canada

**P. Yiou**
Laboratoire des Sciences du Climat et de l'Environnement, Institut Pierre Simon Laplace, CEA-CNRS-UVSQ – UMR8212, CE Saclay l'Orme des Merisiers, 91191 Gif-sur-Yvette, France

**C. Ehlert**
GEOMAR Helmholtz Centre for Ocean Research Kiel, Kiel, Germany

**P. Grasse**
GEOMAR Helmholtz Centre for Ocean Research Kiel, Kiel, Germany

**D. Gutiérrez**
Instituto del Mar del Perú (IMARPE), Dirección de Investigaciones Oceanográficas, Callao, Peru

**R. Salvatteci**
Instituto del Mar del Perú (IMARPE), Dirección de Investigaciones Oceanográficas, Callao, Peru
Institute of Geoscience, Department of Geology, Kiel University, Ludewig-Meyn-Str. 10, 24118 Kiel, Germany
Max Planck Research Group for Marine Isotope Geochemistry, Institute for Chemistry and Biology of the Marine Environment (ICBM), University of Oldenburg, Oldenburg, Germany

**M. Frank**
GEOMAR Helmholtz Centre for Ocean Research Kiel, Kiel, Germany

**C. Buizert**
College of Earth, Ocean, and Atmospheric Sciences, Oregon State University, Corvallis, OR 97331, USA

**K. M. Cuffey**
Department of Geography, University of California, Berkeley, CA 94720, USA

**J. P. Severinghaus**
Scripps Institution of Oceanography, University of California, San Diego, La Jolla, CA 92093, USA

**D. Baggenstos**
Scripps Institution of Oceanography, University of California, San Diego, La Jolla, CA 92093, USA

**T. J. Fudge**
Quaternary Research Center and Department of Earth and Space Sciences, University of Washington, Seattle, WA 98195, USA

**E. J. Steig**
Quaternary Research Center and Department of Earth and Space Sciences, University of Washington, Seattle, WA 98195, USA

**B. R. Markle**
Quaternary Research Center and Department of Earth and Space Sciences, University of Washington, Seattle, WA 98195, USA

**M. Winstrup**
Quaternary Research Center and Department of Earth and Space Sciences, University of Washington, Seattle, WA 98195, USA

**R. H. Rhodes**
College of Earth, Ocean, and Atmospheric Sciences, Oregon State University, Corvallis, OR 97331, USA

**E. J. Brook**
College of Earth, Ocean, and Atmospheric Sciences, Oregon State University, Corvallis, OR 97331, USA

**T. A. Sowers**
Department of Geosciences and Earth and Environmental Systems Institute, Pennsylvania State University, University Park, PA 16802, USA

**G. D. Clow**
US Geological Survey, Boulder, CO 80309, USA

**H. Cheng**
Institute of Global Environmental Change, Xi'an Jiaotong University, Xi'an 710049, China
Department of Geology and Geophysics, University of Minnesota, Minneapolis, MN 55455, USA

**R. L. Edwards**
Department of Geology and Geophysics, University of Minnesota, Minneapolis, MN 55455, USA

**M. Sigl**
Desert Research Institute, Nevada System of Higher Education, Reno, NV 89512, USA

**J. R. McConnell**
Desert Research Institute, Nevada System of Higher Education, Reno, NV 89512, USA

**K. C. Taylor**
Desert Research Institute, Nevada System of Higher Education, Reno, NV 89512, USA

www.ingramcontent.com/pod-product-compliance
Lightning Source LLC
Chambersburg PA
CBHW080703200326
41458CB00013B/4947